11th EASN International Conference on Innovation in Aviation & Space to the Satisfaction of the European Citizens

11th EASN International Conference on Innovation in Aviation & Space to the Satisfaction of the European Citizens

Editors

Spiros Pantelakis
Andreas Strohmayer
Liberata Guadagno

MDPI • Basel • Beijing • Wuhan • Barcelona • Belgrade • Manchester • Tokyo • Cluj • Tianjin

MDPI

Editors

Spiros Pantelakis
University of Patras
Greece

Andreas Strohmayer
University of Stuttgart
Germany

Liberata Guadagno
University of Salerno
Italy

Editorial Office
MDPI
St. Alban-Anlage 66
4052 Basel, Switzerland

This is a reprint of articles from the Special Issue published online in the open access journal *Aerospace* (ISSN 2226-4310) (available at: https://www.mdpi.com/journal/aerospace/special_issues/11th_EASN).

For citation purposes, cite each article independently as indicated on the article page online and as indicated below:

LastName, A.A.; LastName, B.B.; LastName, C.C. Article Title. *Journal Name* **Year**, *Volume Number*, Page Range.

ISBN 978-3-0365-6346-6 (Hbk)
ISBN 978-3-0365-6347-3 (PDF)

Contents

About the Editors

Spiros Pantelakis

Spiros Pantelakis is Prof. Emeritus, University of Patras. He has been Director of the Laboratory of Technology and Strength of Materials (2007–2020), Chairman of the Department of Mechanical Engineering and Aeronautics (2009–2013), and Vice President of the Board of Executives of the Research Committee of the University of Patras (2010–2013). He is a founding member of the European Aeronautics Science Network (EASN) Association since its establishment in 2008 and served as its Chairman up to 2019. In 2019, he was awarded the title of Honorary Chairman of the EASN Association. He has been Representative of the European aeronautics' academia in the Plenary of ACARE and Chairman of ACARE's Working Group on Human Resources. He has been a member of the Clean Sky Academy since its establishment in 2015. He participated in the authors' team of the Clean Aviation Partnership Program and co-chaired the group, dealing with the exploration and maturation of technologies. He has been Member of the Board of Directors of Hellenic Aerospace Industry (2015-2020) and is a former Chairman of the Board of Directors of the Greek Metallurgical Society (2008-2011). He has about 45 years' experience in the field of materials and structures and has been involved in more than 100 international aeronautics-related research projects. He is the author of several scientific books, chapters in international textbooks, and more than 250 scientific publications in peer-reviewed international journals and conference proceedings, and he has been the supervisor of more than 20 PhD Theses. He is chairman of about 30 International Scientific Conferences and organizer of several decades of International Workshops. Currently, he is proactively participating in the preparation of the Clean Aviation Joint Undertaking as representative of academia at the decision-making level and as a member of the core writing team of the Clean Aviation work program.

Andreas Strohmayer

Andreas Strohmayer is a professor of aircraft design at the Institute of Aircraft Design, University of Stuttgart, with a research focus on manned (hybrid) electric flight ("Icaré 2" and "e-Genius") and scaled UAS flight testing. He studied Aeronautical Engineering at TU Munich and graduated in 2001 under Dr. Ing in the field of conceptual aircraft design. From 2002 to 2008, he was director of Grob Aerospace in Mindelheim, Germany, responsible for the design, production, and support of the Grob fleet of all-composite aircraft: specifically, the development of a four-seat aerobatic turboprop, a seven-seat turboprop, and the SPn business jet. From 2009 to 2013, he was program director for a 19-seater commuter aircraft project at Sky Aircraft in Metz, France, and then VP Programs at SST Flugtechnik in Memmingen, Germany, setting up an EASA-approved design organization holding the TC for a six-seater all-composite turbo-prop aircraft. He then left the industry to join the University of Stuttgart in 2015, teaching aircraft design and promoting electric flight. In 2016, he became a member of the Board of Directors of the European Aeronautics Science Network (EASN); since 2019, he has been an EASN Chairman. In this position, he also represents academia in the ACARE General Assembly.

Liberata Guadagno

Liberata Guadagno is Full Professor of "Foundations of Chemistry for Technologies" at the Department of Industrial Engineering of Salerno University. She has held her current role as an eminent scientist since 2015, according to the decision DM 276/2011. Her research activity has mainly been directed toward the design and development of smart polymeric composites (self-generating materials for autonomic damage control in thermosetting resins; multifunctional composites; self-responsive bulk and skin materials; energy-saving manufacturing processes, etc.). Her research activities are documented in 224 scientific papers in international journals, chapters of books, and 150 conference proceedings. She has extensive experience in publicly and industrially funded projects. Recently, Prof. Guadagno has been actively involved in many industrial research and EU-funded projects, of which she has played the role of European coordinator and WPs coordinator. With regard to recent EU projects led by Prof. Guadagno, the European Commission's Innovation Radar highlighted excellent innovations and recognized "UNISA" team as a European 'Key Innovator' in the following topics: (1) "smart thermosetting and thermoplastic composites based on electrical conductivity properties"; (2) new mesoscopic models to simulate the electroactive and sensing behavior of innovative materials; (3) thermoset materials with self-curing capabilities (CFRP and GFRP composites); and (4) self-deicing thermoset materials for aerospace and automotive applications. Prof. Guadagno, in collaboration with Leonardo SpA, has produced 27 international patents (with concessions in Europe and in the USA). On November 15, 2018, she was the winner of the Leonardo SpA Aircraft Division 2018 "Innovation Award for the best Patent". Prof. Guadagno currently works in the role of the "National Contact Point for Italy of the EASN (European Aeronautics Science Network)".

aerospace

MDPI

Editorial

Special Issue "11th EASN International Conference on Innovation in Aviation & Space to the Satisfaction of the European Citizens"

Liberata Guadagno [1,*], Spiros Pantelakis [2,*] and Andreas Strohmayer [3,*]

1 Department of Industrial Engineering, University of Salerno, Via Giovanni Paolo II, 132, 84084 Fisciano, Italy
2 Department of Mechanical Engineering and Aeronautics, University of Patras, Panepistimioupolis Rion, 26500 Patras, Greece
3 Department of Aircraft Design, Institute of Aircraft Design (IFB), University of Stuttgart, Pfaffenwaldring 31, 70569 Stuttgart, Germany
* Correspondence: lguadagno@unisa.it (L.G.); pantelak@upatras.gr (S.P.); strohmayer@ifb.uni-stuttgart.de (A.S.)

Citation: Guadagno, L.; Pantelakis, S.; Strohmayer, A. Special Issue "11th EASN International Conference on Innovation in Aviation & Space to the Satisfaction of the European Citizens". *Aerospace* **2022**, *9*, 808. https://doi.org/10.3390/aerospace9120808

Received: 8 December 2022
Accepted: 8 December 2022
Published: 9 December 2022

Publisher's Note: MDPI stays neutral with regard to jurisdictional claims in published maps and institutional affiliations.

This Special Issue contains selected papers from works presented at the 11th EASN International Conference on "Innovation in Aviation & Space to the Satisfaction of the European Citizens" (http://easnconference.eu/2021/home (accessed on 1 September 2022)), which was successfully held between the 1st and the 3rd of September 2021. Due to pandemic-related restrictions, this EASN Conference took place virtually. The event included 9 keynote lectures given by distinguished personalities of the European Aviation & Space Community and more than 370 technical presentations distributed in 69 virtual sessions. It is worth noting that in the frame of the conference, 85 Aviation and Space projects disseminated their latest research results as well as the future trends in their respective technological fields. In total, more than 420 remote participants from 31 countries worldwide joined the 11th EASN International Conference. These numbers make the 11th EASN Conference the second largest event in the series of the EASN Conferences.

From the numerous contributions to the 11th EASN Virtual Conference, a number of 19 papers, with more than 1000 views each up to now, were accepted for publication in the present Special Issue, following peer review.

In the work of Gelhausen et al. [1], the authors provided a first overview of the environmental benefits of Clean Sky 2 technologies, by employing DLR's forecast model. A substantial emissions reduction through the use of Clean Sky 2 innovations was highlighted. Under the umbrella of the Clean Sky 2 framework, De Martin et al. [2] developed a high-fidelity system model to support the design of an innovative iron bird developed in the frame of a Clean Sky 2 project and contribute to the definition of its control laws. The simulation results highlighted the stability of the proposed control scheme and provided a preliminary assessment of the system's performance.

On the engine side, De Giorgi et al. [3] presented an engine health monitoring system using artificial intelligence. The system was found to be capable of predicting the so-called performance parameters of some of the principal components constituting a turboshaft. In a second work focused on aircraft engines, Erario et al. [4], developed a numerical model of a micro gas turbine intended for the prediction and prognostics of engine performance. The results demonstrated the suitability of the model for predicting microturbine performance. Finally, in another engine-related work, Mehdi et al. [5] investigated the experimental and numerical characterization of flow dynamics and flame re-ignition in a rectangular burner of an aeroengine, particularly at high altitude conditions. To this end, a ring-needle type plasma actuator was considered and run by a high-voltage (HV) nanopulsed plasma generator.

There are two works on novel propulsion in this Issue. In the work of Seitz et al. [6], a fuel cell–gas turbine hybrid propulsion concept was introduced and assessed. The performed initial assessment demonstrated the potential of future solid oxide fuel cell technology synergistically combined with Gas Turbine engines for application in transport aircraft propulsion and power systems. Spinelli et al. [7] introduced a novel set-based design space exploration methodology in order to analyze the effects of different strategies on the fuel consumption, NOx, and take-off mass of a hybrid-electric aircraft. The proposed energy management strategies indicated reduction in fuel consumption and NOx emissions.

Two works focused on aircraft wings are included in the present Special Issue. In the first one, Castilla et al. [8] developed a methodology to estimate the structural mass breakdown of a wing while taking into account the effect of distributed electric propulsion. To achieve this, a semi-analytical methodology was adapted for use in the conceptual and preliminary design stages. In the second work, performed by Kafkas et al. [9], a multi-fidelity optimization framework for current state-of-the-art composite aircraft wings, based on the Mixed Integer Distributed Ant Colony Optimization (MIDACO), was presented. The proposed methodology demonstrated a successful blend of current state-of-the-art multi-fidelity structural, aerodynamic, and fluid–structure interaction analysis for structural optimization.

From the materials side, four works can be found in the present Special Issue. Valvez et al. [10] investigated the compression behavior of 3D-printed PETG composites reinforced with carbon or Kevlar fibers. The authors underlined that the addition of fibers to the polymers led to higher stress relaxations and compressive displacements. In another study focused on materials, Lahbacha et al. [11] investigated the electro-thermal properties of commercial films made by pressed graphene nano-platelets (GNPs), in view of their use as heating elements in innovative de-icing systems for aerospace applications. The first results showed that the targets requested for de-icing and anti-icing applications were met. The work of Norrefeldt et al. [12] investigated the effect of particle foam insulation and its installation procedure on the hygrothermal conditions behind the aircraft cabin sidewall. It was highlighted that careful installation of the particle foam frame insulation provided a similar level of moisture protection as the current state-of-the-art insulation. Finally, Kartsiropoulos et al. [13] presented a novel hybrid thermoplastic prepreg material enabling the fabrication of next-generation recyclable composite aerostructures produced by affordable, automated technologies. The results of the study showed that the developed hybrid thermoplastic material represents a viable manufacturing option from an industrial point of view and that its implementation in structural component manufacturing leads to clear cost and environmental advantages.

The work by Zajdel et al. [14] presents an airplane flight stabilization system, demonstrating its safety and efficiency during all ground phases of development. The presented stabilization system is expected to substantially increase the safety of general aviation airplanes performing the flights according to the free sky conditions. In [15], Welcer et al. assessed the impact of sensor errors on flight stability and underlined the importance of such information for the validation of the designed control system. In the work of Neumaier et al. [16], the authors proposed a method for an automatic pipe routing that automatically generates pipe designs with an integrated design optimization encoded in graph-based design languages and executable in a design compiler. The capabilities of their approach were demonstrated and compared to the series pipework in an Airbus A320 main landing gear bay. In [17], Todorov et al. propose the integration of fuzzy sets to model uncertainties during the evaluation of technological options by the experts at the conceptual design phase. The authors underlined that the suggested methodology can be defined as one of the integral stages necessary for the development of a much wider design method. Pérez-Castán et al. [18] developed a data-driven approach that predicts separation infringements between aircraft within airspace, by exploiting machine learning algorithms. The results showed that the said algorithms could perform conflict prediction with high-accuracy metrics. Finally, in the review paper of Kühnelt et al. [19], an overview of recent approaches for structural batteries was provided, assessing their multifunctional performance, and identifying gaps in technology development towards their introduction of commercial aeronautic applications.

The editors of this Special Issue would like to thank the authors for their high-quality contributions and for making this Special Issue manageable. Furthermore, the editors would like to express their gratitude to Ms. Linghua Ding and the *Aerospace* editorial team for their professional support.

Conflicts of Interest: The authors declare no conflict of interest.

References

1. Gelhausen, M.C.; Grimme, W.; Junior, A.; Lois, C.; Berster, P. Clean Sky 2 Technology Evaluator—Results of the First Air Transport System Level Assessments. *Aerospace* **2022**, *9*, 204. [CrossRef]
2. De Martin, A.; Jacazio, G.; Sorli, M. Simulation of Runway Irregularities in a Novel Test Rig for Fully Electrical Landing Gear Systems. *Aerospace* **2022**, *9*, 114. [CrossRef]
3. De Giorgi, M.G.; Strafella, L.; Menga, N.; Ficarella, A. Intelligent Combined Neural Network and Kernel Principal Component Analysis Tool for Engine Health Monitoring Purposes. *Aerospace* **2022**, *9*, 118. [CrossRef]
4. Erario, M.L.; De Giorgi, M.G.; Przysowa, R. Model-Based Dynamic Performance Simulation of a Microturbine Using Flight Test Data. *Aerospace* **2022**, *9*, 60. [CrossRef]
5. Mehdi, G.; Bonuso, S.; De Giorgi, M.G. Plasma Assisted Re-Ignition of Aeroengines under High Altitude Conditions. *Aerospace* **2022**, *9*, 66. [CrossRef]
6. Seitz, A.; Nickl, M.; Troeltsch, F.; Ebner, K. Initial Assessment of a Fuel Cell—Gas Turbine Hybrid Propulsion Concept. *Aerospace* **2022**, *9*, 68. [CrossRef]
7. Spinelli, A.; Enalou, H.B.; Zaghari, B.; Kipouros, T.; Laskaridis, P. Application of Probabilistic Set-Based Design Exploration on the Energy Management of a Hybrid-Electric Aircraft. *Aerospace* **2022**, *9*, 147. [CrossRef]
8. Alonso Castilla, R.; Lutz, F.; Jézégou, J.; Bénard, E. Wing Structural Model for Overall Aircraft Design of Distributed Electric Propulsion General Aviation and Regional Aircraft. *Aerospace* **2022**, *9*, 5. [CrossRef]
9. Kafkas, A.; Kilimtzidis, S.; Kotzakolios, A.; Kostopoulos, V.; Lampeas, G. Multi-Fidelity Optimization of a Composite Airliner Wing Subject to Structural and Aeroelastic Constraints. *Aerospace* **2021**, *8*, 398. [CrossRef]
10. Valvez, S.; Silva, A.P.; Reis, P.N.B. Compressive Behaviour of 3D-Printed PETG Composites. *Aerospace* **2022**, *9*, 124. [CrossRef]
11. Lahbacha, K.; Sibilia, S.; Trezza, G.; Giovinco, G.; Bertocchi, F.; Chiodini, S.; Cristiano, F.; Maffucci, A. Electro-Thermal Parameters of Graphene Nano-Platelets Films for De-Icing Applications. *Aerospace* **2022**, *9*, 107. [CrossRef]
12. Norrefeldt, V.; Riedl, G. Investigation of the Impact of a Particle Foam Insulation on Airflow, Temperature Distribution, Pressure Profile and Frost Buildup on the Aircraft Structure. *Aerospace* **2021**, *8*, 359. [CrossRef]
13. Katsiropoulos, C.V.; Pantelakis, S.G.; Barile, M.; Lecce, L. Development of a Novel Hybrid Thermoplastic Material and Holistic Assessment of Its Application Potential. *Aerospace* **2021**, *8*, 351. [CrossRef]
14. Zajdel, A.; Krawczyk, M.; Szczepański, C. Pre-Flight Test Verification of Automatic Stabilization System Using Aircraft Trimming Surfaces. *Aerospace* **2022**, *9*, 111. [CrossRef]
15. Welcer, M.; Szczepański, C.; Krawczyk, M. The Impact of Sensor Errors on Flight Stability. *Aerospace* **2022**, *9*, 169. [CrossRef]
16. Neumaier, M.; Kranemann, S.; Kazmeier, B.; Rudolph, S. Automated Piping in an Airbus A320 Landing Gear Bay Using Graph-Based Design Languages. *Aerospace* **2022**, *9*, 140. [CrossRef]
17. Todorov, V.T.; Rakov, D.; Bardenhagen, A. Enhancement Opportunities for Conceptual Design in Aerospace Based on the Advanced Morphological Approach. *Aerospace* **2022**, *9*, 78. [CrossRef]
18. Pérez-Castán, J.A.; Pérez-Sanz, L.; Serrano-Mira, L.; Saéz-Hernando, F.J.; Rodríguez Gauxachs, I.; Gómez-Comendador, V.F. Design of an ATC Tool for Conflict Detection Based on Machine Learning Techniques. *Aerospace* **2022**, *9*, 67. [CrossRef]
19. Kühnelt, H.; Beutl, A.; Mastropierro, F.; Laurin, F.; Willrodt, S.; Bismarck, A.; Guida, M.; Romano, F. Structural Batteries for Aeronautic Applications—State of the Art, Research Gaps and Technology Development Needs. *Aerospace* **2022**, *9*, 7. [CrossRef]

aerospace

MDPI

Article

Clean Sky 2 Technology Evaluator—Results of the First Air Transport System Level Assessments

Marc Christopher Gelhausen *, Wolfgang Grimme, Alf Junior, Christos Lois and Peter Berster

Institute for Air Transport and Airport Research, German Aerospace Center (DLR), 51147 Cologne, Germany; wolfgang.grimme@dlr.de (W.G.); alf.junior@dlr.de (A.J.); christos.lois@dlr.de (C.L.); peter.berster@dlr.de (P.B.)
* Correspondence: marc.gelhausen@dlr.de; Tel.: +49-2203-601-2463

Abstract: The authors have adjusted the DLR forecast model to evaluate the environmental benefits in terms of CO_2 and NO_x emissions of Clean Sky 2 technology innovations. The paper briefly describes the model employed: it consists of a passenger/flight volume forecast, a fleet model, and emission modelling. The novelty of the forecast approach compared to previous studies is that it is based on airport pairs instead of larger aggregates like countries or regions. Therefore, a separate breakdown on airports is unnecessary in the case of a more detailed analysis is needed, and it enables us to include airport capacity constraints which affect demand and flight volume, as well as the fleet development at constrained and unconstrained airports. We eventually present the forecast results in terms of passenger and flight volume, fleet development, and CO_2 and NO_x emissions. The results show that emissions can be reduced substantially by the use of Clean Sky 2 technology compared to a reference case which represents the status quo.

Keywords: air transport forecasting; Clean Sky 2; Technology Evaluator; emission modelling; fleet modelling

Citation: Gelhausen, M.C.; Grimme, W.; Junior, A.; Lois, C.; Berster, P. Clean Sky 2 Technology Evaluator—Results of the First Air Transport System Level Assessments. *Aerospace* **2022**, *9*, 204. https://doi.org/10.3390/aerospace9040204

Academic Editors: Spiros Pantelakis and Alexei Sharpanskykh

Received: 14 December 2021
Accepted: 6 April 2022
Published: 9 April 2022

1. Introduction

Clean Sky 2 (CS2) was established in order to reduce the environmental impact of air transport by the introduction of new advanced aeronautical technologies. CS2 aims to improve mobility and support a globally competitive European aeronautical industry in Europe by speeding up technological developments. CS2 demonstrates reduction potentials with regard to CO_2 and NO_x emissions of between 20% and 30% compared to state-of-the-art aircraft in 2014.

These general goals have been translated into specific emission reduction objectives for the CS2 concept aircraft models according to their expected entry into service, and ranging from 19-seat commuter to short- and long-range high-capacity commercial aircraft. On the level of the global air transport system, those new concept aircraft have been embedded into two fleet projections up until 2050, and were then compared to a projection without new CS2 aircraft, in order to understand the emission reduction potentials of a future global fleet in 2050, and the long-term aviation footprint. This is the topic of this paper.

A major improvement compared to previous studies, e.g., Clean Sky 1 or manufacturer forecasts like Airbus or Boeing, is the development of a new forecast method. This method is based upon airport pairs instead of more aggregated regions like countries or route groups, as in the case of the approach of the International Civil Aviation Organization (ICAO). In this way, it is possible to include airport capacity constraints, which influence demand volume, the number of flights, and the fleet development at constrained and unconstrained airports [1]. For this paper, all airports with scheduled flight services are included, which is in total about 4000 airports. Furthermore, this approach is more appropriate for airport or aircraft mission-level analysis, as there is no need for a separate breakdown of an aggregate forecast on the study airports. However, airport and aircraft mission-level analysis are not the topic of this paper. We have set out a global forecast until 2035, and two scenarios from 2035 until 2050 on the air transport system level. Nevertheless, the global results are based on the airport pair level, so that it is essentially a bottom-up approach.

In Section 2, we begin with a brief discussion of the methods and models employed for the global impact assessment, which follows in Section 3. Section 4 closes the paper with a discussion of the results, and a series of conclusions.

2. Model Overview

In this chapter we present a brief overview of the approach for the passenger and flight forecast, the fleet model, and the emission modelling. For the time period 2014–2035 we create a single forecast, and from 2035 to 2050 two scenarios are employed. These differ basically in terms of aviation technology and market development. The scenario assumptions have two effects: first, aviation technology development determines the technology of future aircraft. This in turn influences the development of future airfares, which have an effect on passenger demand and, finally, flight volume. For the time period up to 2035 we assume a decline in real airfares, i.e., inflation-adjusted airfares, of 1.5% per year based on data analysis of Sabre AirVision Market Intelligence and inflation data from World Development Indicators [2]. For the time period 2035–2050, we have therefore created two scenarios: in the High Scenario, we assume a more favorable technological and market development, which results in a decline of real airfares of 2.0% per year. On the other hand, in the Low Scenario, we assume a decline of 0.5% per year because of a less favorable development. At this stage, the technological development is rather generic, i.e., it is not related to specific aircraft types. As a result, the two scenarios differ in terms of their demand development, and we simply call them the "High Scenario" and "Low Scenario" in Section 2.1.

In Sections 2.2 and 2.3, i.e., the fleet and emissions modelling, we use the terms "Reference Scenario" and "CS2 Scenario". "Reference Scenario" means that only aircraft that were already in service or available for order in 2014 or earlier are employed ("state-of-the-art aircraft" of 2014 or earlier). "CS2 Scenario", on the other hand, allows the introduction of newer aircraft, especially CS2 concept aircraft. These scenarios are related to specific aircraft types, and they will be explained in more detail later in this paper. "CS2 Scenario" and "Reference Scenario" are therefore specific aircraft technology related scenarios.

These two types of scenarios can be combined, e.g., the CS2 Scenario with the High Scenario or the Reference Scenario with the Low Scenario. In total, we have four combinations, and they are important for the assessment of the emission reduction potential of CS2 concept aircraft. The emission reduction potential of CS2 aircraft is assessed relative to the state-of-the-art aircraft of 2014 in order to identify the contribution of CS2 technology. The results of the Reference Scenario are not reported on their own; they serve as a reference point for the results of the CS2 Scenario.

2.1. Passenger and Flight Forecasts

In this section we briefly cover the methodology of the passenger and flight forecasts. For a more detailed description, the reader is referred to [2]. Figure 1 illustrates the model approach:

- In the first step, unconstrained passenger demand and flight volume is forecast for each airport pair.
- In the next step, for each airport pair, the flight volume is compared with the current and expected airport capacity. The forecast passenger and flight volume, as well as the constraint situation at airports, will influence the average future aircraft size, which is forecast for each airport pair.
- In the final step, the expected passenger and flight volume is balanced with airport capacity and aircraft size development to yield the constrained passenger and flight volume. This might result in some unaccommodated passenger demand and flight volume, depending on the severity of airport capacity constraints and the potential of employing larger aircraft.

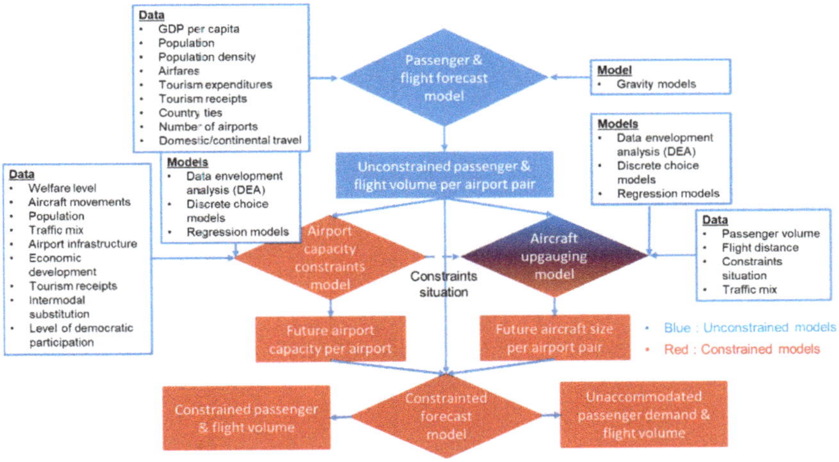

Figure 1. Overview of the passenger and flight forecast model.

The relationships between passenger demand, aircraft size and airport capacity are particularly important in a world in which future airport capacity tends to fall short of demand, at least at some airports. Regarding capacity constraints, we mainly focus on the runway system, as this is the most critical element for overall airport capacity in the long term. Especially in Western countries, the enlargement of the runway system requires a plannig approval procedure which involves the public, who are typically opposed to such plans because of environmental concerns, e.g. increased noise levels. On the other hand, other aspects of airport capacity, such as terminal capacity, can be enlarged as part of the airports' business decisions without the involvement of the public [3]. Employing bigger aircraft is an effective option to mitigate capacity shortages; however, in particular at hub airports, existing network carriers rather tend to increase flight frequency in order to restrict competitors' access to core markets [4,5]. On the other hand, there are certainly limits to ever increasing aircraft size and transporting more and more passengers per flight. In the case of a high share of feeder traffic, in particular to smaller airports, frequencies are typically higher [6].

Figure 2 displays the relationships between the passenger demand (annual passenger volume), airport capacity (annual flight volume) and average aircraft size (passengers per aircraft) on a very general level: in order to accommodate the forecast passenger demand, airport capacity in terms of flight movements is required, in combination with a pre-determined average aircraft size (which is endogenous to the model). Both the average aircraft size and the airport capacity limit the maximum number of passengers that can be handled. Aircraft size and airport capacity can be substituted for each other to some degree: if the airport capacity is insufficient to serve a given passenger demand, increasing aircraft size can compensate for a lack of airport capacity to serve that demand. A lack in terms of aircraft size can be compensated for by more airport capacity, so that more flights—but with, on average, fewer passengers per flight—can be handled. However, as already explained earlier, typically airport capacity is the bottleneck.

The model accounts for these interrelations between passenger demand, airport capacity and aircraft size by adjusting all three elements in a constrained forecast. In contrast, an unconstrained forecast always assumes the best case regarding the future development of airport capacity. Moreover, unconstrained forecasts underestimate the long-term increase of average aircraft size, i.e., the average number of passengers transported per flight, as a measure to compensate for the shortfall in airport capacity.

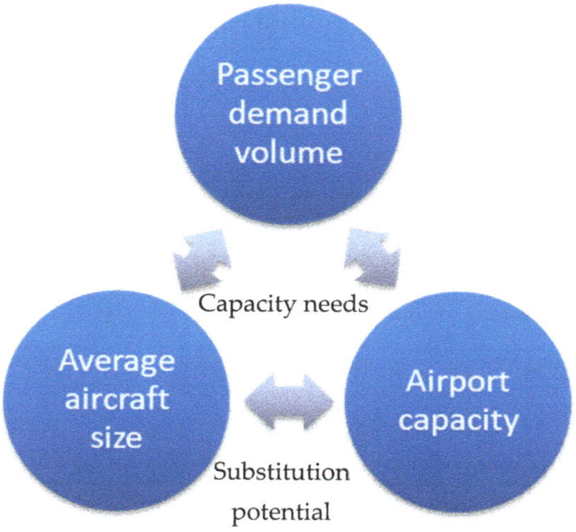

Figure 2. Relationship between passenger demand, airport capacity and aircraft size. Reprinted from Ref. [7].

Shifting traffic to neighbour airports can be another option to mitigate capacity constraints. Gelhausen [8] shows that, from a passenger's perspective, such a shift can theoretically happen in a decentralised or multi-airport environment. Gudmundsson et al. [9] identify spillover effects between London Heathrow and Gatwick, and to a lesser degree between Heathrow and Stansted (and even to the more remote airports Birmingham and Manchester). However, if they control for low-cost carrier effects, they only find spillover effects between Gatwick and London City airports in European traffic. Redondi and Gudmundsson [10] find significant spillover effects of constraints at London Heathrow and Frankfurt airports in stop-over traffic in European and Gulf hub airports. Dennis [11], on the other hand, argues that shifting flights to less-congested airports is simply not compatible with the demands of passengers and hub airlines. Currently, this is a very difficult topic to implement in a sound way, especially on a global level, and therefore certainly needs more research. The main questions are: What share of excess traffic is shifted to which airports? Is excess traffic only partially or fully shifted to neighbour airports? Currently, it seems that the potential of shifting traffic to neighbour airports is rather limited, and the model currently does not contain a (full) shift of excess traffic to airports in the vicinity.

In order to illustrate this point in more detail, we have chosen the London area, which serves as a prime example of a region with a heavily constrained hub airport (London Heathrow), surrounded by another major airport (Gatwick) and two secondary airports (Stansted and Luton) with ample capacity reserves. We have excluded London City airport from the analysis, as it has only limited means of absorbing excess traffic from London Heathrow because of its restricted location in the London Royal Docks, and it can be served only by small aircraft.

London Heathrow has already been constrained for about 20 years, and there has been a long debate about its expansion [12]. Figure 3 illustrates the development of passenger volume and aircraft movements at these four airports, as well as for the London area and the UK between 2006 and 2019 relative to the volumes of 2006. If we look at the UK or the London area as a whole, aircraft movements have remained more or less constant in the UK, and increased by 10% in the London area. Passenger volumes, on the other hand, increased by 33% between 2006 and 2019, both in the UK and in the London area. Of the smaller airports, only Luton increased its flight volume significantly, by around 50%, and in the case of Stansted there was even a decline in the number of flights. The flight volume remained constant at Heathrow, and increased at Gatwick by about 30% since 2006; however, this was much less than passenger volume, which increased by almost 60% since 2006. As we can see, a large share of the passenger volume growth has been accommodated by increasing the number of passengers per flight.

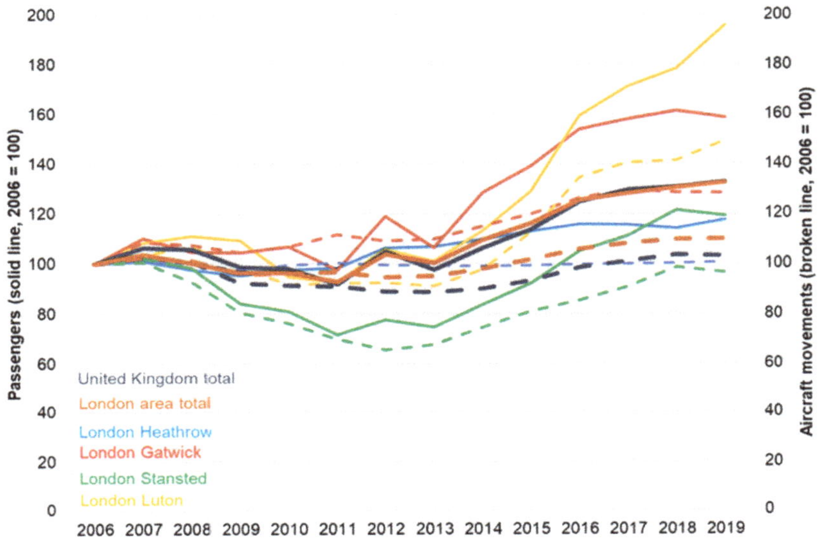

Figure 3. Relative development of passenger volume and aircraft movements at airports in the London area between 2006 and 2019 (2006 = 100). Reprinted from Ref. [7].

Our forecast model consists of four distinct sub-modules:

- Air passenger demand, which is origin–destination (OD) passenger flows, and the total passenger flows including transfer passengers, between countries as well as airports.
- Airport capacity and capacity utilisation.
- Airport capacity enlargements and limits.
- Average aircraft size: the average number of passengers per flight.

2.1.1. Air Passenger Demand

Well-known long-term forecasts of global air traffic are typically conducted by aircraft manufacturers such as Airbus [13], Boeing [14], and supranational organisations like the ICAO [15]. In the academic literature, the gravity model is still the workhorse of spatial demand modelling. One of the first gravity models employed in air transport research was developed by Harvey [16] to analyse airline traffic patterns in the USA. Grosche et al. [17] employed two gravity models to estimate the air passenger volume of city pairs without any air service. Tusi and Fung [18] analysed passenger flows at Hong Kong International Airport, and focused on a single airport. Matsumoto [19] and Shen [20] based their gravity models on network analysis: Matsumoto's model was used to estimate passenger and cargo flows between large cities such as Tokyo, London, Paris and New York, while Shen's was used to analyse inter-city airline passenger flows in a 25-node US network. Bhadra and Kee [21] employed a gravity model to analyse demand characteristics, such as the fare and income elasticities of the US origin–destination market over time. Endo [22] developed a gravity model to analyse the impact of a bilateral aviation policy between the USA and Japan on passenger air transport, while Hazledine [23] used a gravity model to analyse border effects in international air travel. The gravity model has also been applied in air freight modelling: Alexander and Merkert [24] developed a gravity model to evaluate the air freight market in Australia. Later, Alexander and Merkert [25] analysed US international air freight markets in light of the 2008/09 financial crisis. Baier et al. [26] developed a gravity model with airport fixed effects to model global air freight between airports.

The air passenger demand model that is employed in this paper is also based on the gravity model. Whilst variables such as distance, population and gross domestic product (GDP) are common in gravity models, we expect further insights from the inclusion of an airfare variable. The model was estimated on Sabre AirVison Market Intelligence data [27], which includes information on actual airfares paid by air passengers. From the model estimation, we found airfare elasticity to be −1.11%, i.e., if airfare decreases by 1% then OD air travel demand rises by 1.11%; it is thus slightly elastic. For better discrimination between different types of origin and destination, we included variables such as tourism receipts and expenditure, and population density. Tourism receipts and expenditures are measures for the tourism affinity of the destination and origin country. They both have a positive influence on OD demand; however, tourism receipts have a higher impact. Population density, which is only relevant for domestic air travel, is a measure for mode substitutability: the higher the population density is, the better developed and more relevant the road and rail network for domestic travel is typically in terms of travel time. Take for example the US, which has a population density of about 36 people per km^2, and Germany, which has a population density of about 232 people per km^2. The rail and road networks are relatively better developed in Germany compared to the US. For example, travelling from Hamburg to Munich by car or train is more relevant than for the journey from New York City to Los Angeles: for the trip from Hamburg to Munich, you need about one and a half hours by plane (excluding access and egress times), around eight hours by car, and five and a half hours by train (again, excluding access and egress times). For the trip from New York City to Los Angeles, you need nearly six hours by plane, 41 h by car, and more than two and a half days by train. Of course, the distance is much longer between New York City and Los Angeles than between Hamburg and Munich (about 600 vs. 4000 km great circle distance), but cars and trains are relatively faster for Hamburg–Munich, and the planes are relatively faster in the case of New York City–Los Angeles, even though the relative advantage is substantially larger for planes than for cars and trains. As a result, population density has a slightly negative impact on OD air travel demand. The model contains further variables of lesser importance, like distance and country ties [2].

For model estimation, we employed a Poisson pseudo-maximum likelihood (PPML) estimator to produce better and more reliable forecasts [28]. As a result, a series of relevant demand elasticities (see Figure 4) were estimated, e.g., if the GDP per capita increases in the origin country by 1%, then OD passenger demand rises by 0.45%, and if GDP per capita rises in the destination region by 1%, the OD demand grows by 0.23%.

Due to a lack of consistent global data, there is no differentiation between business and leisure passengers. There are only individual surveys covering particular markets like Germany and the UK. We do have information about the share of ticket categories for each airport pair, e.g., from Sabre AirVision Market Intelligence, but it is not possible to infer the share of business travellers from the share of business tickets, as many business travellers choose the economy class. For example, in Germany, 69% of domestic air travel in 2003 was business, and 28% of all European and intercontinental trips from Germany were for business purposes [29]. However, based on the analyses of such surveys, Mason [30] estimates the share of business passengers to be about 30%, but it is not possible to break down this number to the country or airport pair level, which is needed for consistent model estimation.

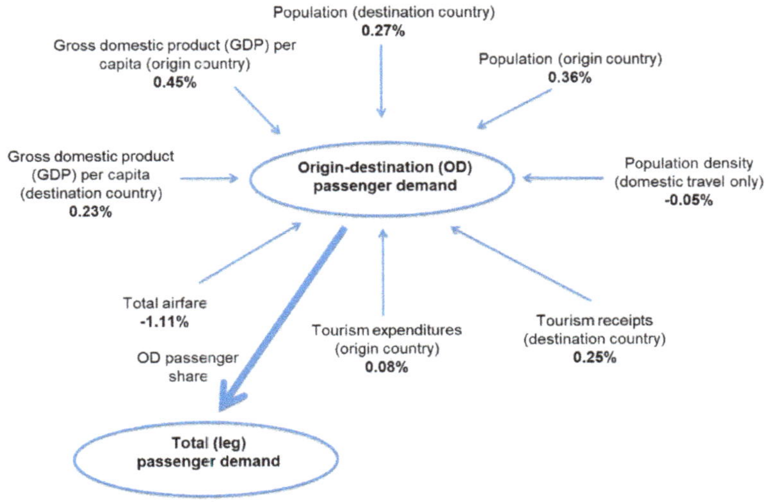

Figure 4. Elasticities of air passenger demand. Reprinted from Ref. [7].

2.1.2. Airport Capacity and Capacity Utilisation

The global airport capacity model is based upon data envelopment analysis (DEA) and regression analysis. DEA is a non-parametric empirical method used in operations research to estimate production frontiers employing linear programming techniques. DEA is a standard tool for efficiency analyses, and for the benchmarking of so-called decision-making units (DMUs) [31,32]. DEA allows us to compare DMUs which differ in their input and output structures. Examples of such DMUs include hospitals, energy production, or the cost-/profit centers of large organisations and, in our case, airports. The model analyses airport capacity utilisation with one main question in mind: Which airport achieves what output given a particular input structure? In our model approach, this enables us to compute the current and future annual service volumes of airports worldwide. The model allows the forecasting of the 5% peak hour volume and the average hourly volume of an airport in a situation of the highest possible capacity utilisation.

The model produces robust results using input information which is generally available, e.g., the runway system, as this typically limits overall airport capacity in the long term. For our generic airport capacity model, we aimed for an as-high-as-possible degree of accuracy on average on the global level, but we do not pretend that it is as accurate as detailed airport capacity analyses for specific airports. Nevertheless, the results are surprisingly accurate on the airport level. For a comparison with the US and some European and Asian airports, the reader is referred to [2].

Figure 5 provides an overview of the generic airport capacity model process:

- The first step is the use of the aforementioned DEA to estimate the current airport capacity for airports of interest.
- In a second step, the average number of aircraft movements per runway and per operating hour at the highest possible level of capacity utilisation for each airport is calculated.
- The last step is to perform a regression analysis based on the results of the DEA.

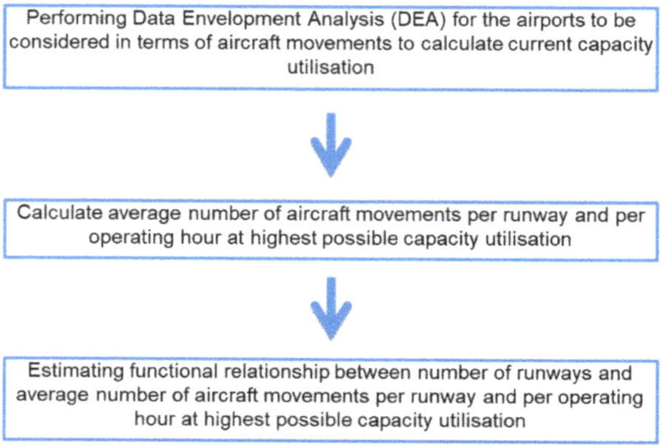

Figure 5. Estimation of the generic airport capacity model.

2.1.3. Airport Capacity Enlargements and Limits

With the model of forecasting the realisation probabilities of airport capacity enlargements and limits, we have introduced an approach which incorporates the enlargement of limited airport capacity in air transport forecasts. If the forecast number of flights exceeds the current airport capacity, the model analyses whether adding new runways is possible, and if this is the case, how long this process is expected to take. This analysis is conducted for each airport and each new runway at that airport. The model is based on the idea that there is a particular degree of opposition to airport expansion from the population living in proximity to the airport. This depends on factors such as noise annoyance, pollution, welfare level, economic opportunities, participation level, and intermodal substitution. The degree of opposition may range from almost none to such an intense opposition that airport expansion is virtually impossible. As a result, the model enables us to estimate the probability of the realisation of a new runway, which can be transformed into an expected value in terms of delay. The approach used is a probabilistic one based on Markov chains [33] and discrete choice theory [34]. The Markov chain comprises of two situations that an airport can face (see Figure 6):

- Situation one: forecast demand < airport capacity.
- Situation two: forecast demand $\geq$ airport capacity.

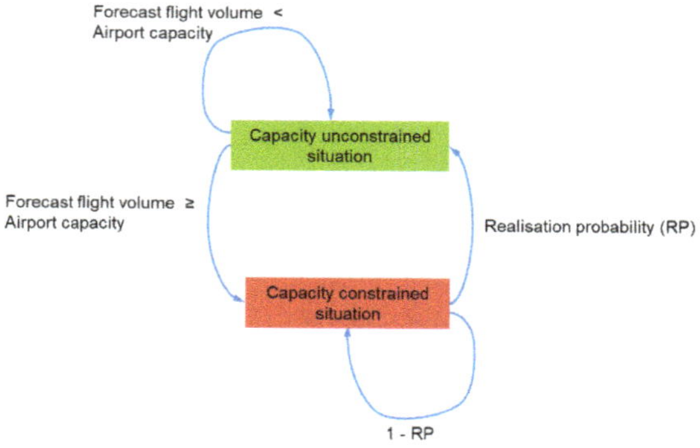

Figure 6. Markov chain of runway expansion.

Entry into situation two is triggered by the underlying demand forecast and the current capacity of the airport. If the airport is in situation two, the realisation probability (RP) of an additional runway corresponds to the transition probability from situation two to situation one. Eventually, based on the probability of delay, we can calculate the expected delay of runway capacity expansion, as it is the inverse of the transition probability minus one, because 100% RP is defined as no delay in the model with, theoretically, the new runway being instantly available. The realisation probability of runway expansion is modelled by means of a binary logit model [35]. Of course, it is difficult to define an exact threshold above which an airport is capacity constrained. It is a more or less a smooth transition from an unconstrained into a constrained situation, where participation in the general traffic growth becomes increasingly difficult. We can assume that an (artificial) threshold that separates constrained from unconstrained airports lies in a range of 75% capacity utilisation [36]. In the light of this definition, there were ten of about 4000 airports which were capacity constrained in 2008, such as Shanghai (SHA), London Heathrow (LHR) and New York LaGuardia (LGA) [36]. Until 2050, we expect the number of airports to increase to 24 in the Low Scenario and 36 in the High Scenario. These calculations already include new runways that have been built since 2008 (e.g., the fourth runway at Frankfurt airport in 2011) or are expected to be realised by 2050 according to the model's forecast, shifting traffic to off-peak hours and employing larger aircraft.

2.1.4. Average Aircraft Size: Average Number of Passengers per Flight

The fourth sub-model is on the forecast of passengers per flight by airport pair. As in the case of airport capacities, the approach is highly problem-specific, and cannot be a substitute for a detailed flight route analysis in terms of aircraft fleet characteristics and their future development. The method is very similar to the modelling of airport capacities. The basic analysis is a DEA, which is further refined by regression analyses to generalise the results, so that they can be used for forecasting purposes.

The average number of passengers per flight between an airport pair is determined by the passenger volume of that airport pair, the flight distance, and the constraint situation at the origin and destination airport [2]. The results serve as an input for the more elaborate fleet modelling presented in the next section.

Figure 7 summarises the various steps with regard to the aircraft size model. First, a DEA is performed for all of the airport pairs under consideration to obtain the current values of passenger capacity potential, i.e., the maximum number of passengers that can theoretically be transported per year, and its utilisation. In the next step, these values are updated on the basis of the passenger demand forecast by means of the passenger capacity potential utilisation model and the passenger capacity potential model. Finally, we can calculate the future number of passengers per flight, which is translated into the average aircraft size by applying a seat load factor.

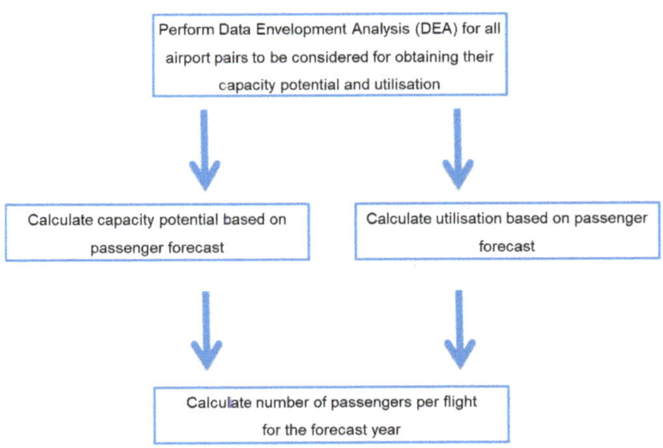

Figure 7. Forecasting the passengers per aircraft for each airport pair.

2.2. Aircraft Fleet Forecast

Aircraft fleet modelling is a complex task that has received a substantial amount of research in the past. A fundamental question is the trade-off between aircraft size and flight frequency on a route with particular characteristics, such as distance, passenger volume and the competitive situation: airlines can basically choose between smaller aircraft with a corresponding higher flight frequency and larger aircraft with a lower flight frequency to serve a given passenger volume. This has already been discussed in Section 2.1 (see Figure 2). For example, according to Wei and Hansen [37], airlines prefer to increase flight frequency instead of aircraft size to attract more passengers. On the other hand, Presto et al. [38] analysed four different frequency regulation strategies: they performed differently with regard to air traffic flow management delay, cash operating costs of airlines, the net travel time balance of passengers, and the fuel consumption of aircraft. Route characteristics that influence the trade-off between flight frequency and aircraft size are typically passenger volume and flight distance [39], in addition to airline competition [40], whether the origin or destination airport is a hub, what kind of airports are connected (e.g., hub-to-hub or hub-to-spoke) and the season of travel [41]. Pai [42] conducted, for the US market, a much more detailed analysis regarding the choice of aircraft. He included market demographics (e.g., population and income), airport characteristics (e.g., runway length), airline characteristics (e.g., low-cost carriers) and route characteristics (e.g., distance). In order to actually implement an aircraft choice model, Bhadra [41,43,44] employs a multinomial logit model. Kölker et al. [39] developed an approach based on categories: they identified a distribution of aircraft size according to passenger volume and distance.

Typically, an external passenger demand volume forecast is needed for aircraft choice models. As the case may be, the passenger volume forecast has to be broken down to the airport or route level to apply the aircraft choice model. Therefore, applying these models for a global forecast can be challenging. A particular strength of our approach is the integrated modelling of passenger volume, number of flights and fleet mix on the airport pair level irrespective of the number of airports included. Compared to the approaches discussed, we include the variables of the passenger volume, flight distance and airport capacity utilisation (see Section 2.1), as well as the aircraft distribution of the base year of the forecast. The number of passengers per flight for each airport pair serves as the major input for the fleet modelling. Applying a seat load factor transforms the passengers per flight into seats per flight. The fleet modelling concerns the distribution of aircraft types on each airport pair based on this input. In detail, the passenger aircraft fleet forecast is based on the following inputs and assumptions:

- The passenger traffic forecast, including the future number of passengers and flights per airport pair (Section 2.1).
- The seat load factor forecast for the conversion of the passengers per flight to the seats offered per flight.
- The base year, i.e., 2014 flight schedules as a list of flight operations by airport pair and aircraft type.
- The base-year fleet data.
- Aircraft retirement curves.
- Aircraft utilisation assumptions.
- A list of available aircraft (production window = the time between entry into service and the out-of-production date of an aircraft type) in each seat category.

The base-year fleet data originate from Cirium's Fleets Analyzer [45], and the schedules are taken from OAG [46] and Innovata [47]. The aircraft that make up the base-year fleet are limited to Cirium's definition of primary usage "Passenger", "Combi/Mixed (Passenger/Cargo)", "Quick-Change/Convertible (Passenger/Cargo)", and a limited number of aircraft used as "Business-Air Taxi/Air Charter", where operators also carry out scheduled flights. This limitation is in line with the objective of conducting a forecast with regard to the mainliner, regional and commuter airline fleet which is operating scheduled passenger services. With regard to this limitation, we exclude aircraft used as business/VIP aircraft, aircraft in private use, cargo aircraft, and other uses where aircraft are operating unscheduled services. In terms of the number of such aircraft, about 33,000 business/air taxi aircraft are excluded from this analysis, which only account for a fraction of the emissions of scheduled passenger flights. Moreover, as only passenger traffic is the scope of the analysis, approximately 3200 cargo aircraft in service in 2014 are not considered.

Table 1 provides an overview of the aircraft included in the analysis for the base year 2014. The total number of aircraft in the base-year fleet is 24,017 (as of mid-year 2014).

Table 1. The 2014 base year aircraft fleet by seat class.

Aircraft Seat Class	Number of Aircraft in 2014
1–19 Seats	1885
20–50 Seats	2424
51–70 Seats	1111
71–85 Seats	1240
86–100 Seats	148
101–125 Seats	1359
126–150 Seats	3346
151–175 Seats	3496
176–210 Seats	5273
211–300 Seats	2144
301–400 Seats	1435
401–500 Seats	156
Total	24,017
Total >19 Seats	22,132

For subsequent years, the connection between the fleet and the schedule is established as follows: schedule data provide information on the aircraft types and variants being operated on a particular flight. This, however, does not include individual aircraft. In the forecast years, with regard to each aircraft, the survival probability, taken from the respective retirement curve model, is applied, resulting in the average percentage of surviving aircraft for each aircraft type. This average survival percentage is applied to the frequencies of the base-year flight schedule, such that the number of flights that can be operated with the surviving fleet of the base and the previous years' aircraft can be operated, and the number of flights for which new aircraft will be required can be estimated.

The seat load factor forecast is employed for the conversion of the number of passengers per flight to the number of seats offered. Here, a global model using an s-curve is fitted to empirical data from Sabre AirVision Market Intelligence [27] and ICAO [48,49] using empirical data from 1996 onwards (see Figure 8). From this model, we forecast a maximum seat load factor of 88% in 2050 for all routes.

For aircraft retirement modelling, the approach of International Civil Aviation Organization Committee on Aviation Environmental Protection (ICAO CAEP)/12 was employed (see Figure 9). Here, aircraft are subdivided into the following categories: turboprop, regional jet, narrowbody jet and widebody jet. Each type has different coefficients which influence the probability of retirement for a particular aircraft age.

The aircraft utilisation model is needed for the calculation of the number of aircraft that is required to serve the forecast schedule. It is specified as:

$$Utilisation = \frac{a \times Average\ stage\ length}{(Average\ stage\ length + b)}$$

The total annual aircraft utilisation for each aircraft is estimated by the annual distance flown as a function of the average mission distance with a and b being parameters that are estimated with empirical data. Utilisation typically depends on the average flight length. The longer the flights are, on average, the higher the aircraft utilisation typically is, because the total number of annual turn-arounds with the associated ground times is smaller. The data used to calibrate the model, as shown in Figure 10, originate from ADS-B data collected by Flightradar24 [50] in 2019. ADS-B coverage can be considered to be very high globally, especially because equipage has become mandatory for most operation types In Europe, about 91% of flights are covered by ADS-B as of early 2022 [51].

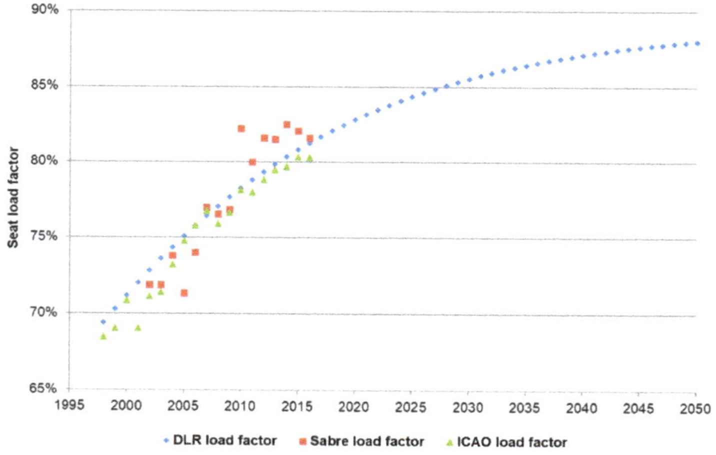

Figure 8. Seat load factor data and forecast [27,48,49].

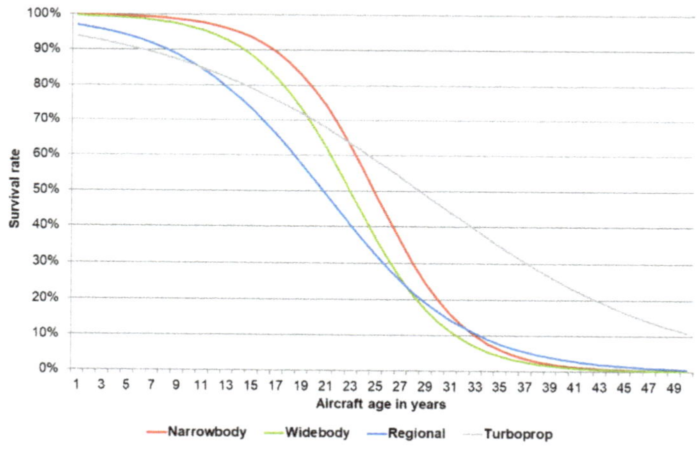

Figure 9. Survival probabilities of aircraft in relation to the aircraft category and age.

The accuracy of the model is shown in Table 2, in which we applied the model to the flight schedule for the year 2019. The model is able to estimate the number of aircraft in the global fleet with an accuracy of 96%. However, forecast results should be taken with a pinch of salt in the case of aircraft-specific deliveries, because there are significant deviations in some cases, e.g., Boeing 757 and 767. A negative value for the deviation means that more aircraft than forecast are needed for a given flight schedule. Nevertheless, in total, i.e., for all aircraft types, this more or less levels out. Furthermore, this only affects deliveries of particular aircraft types but not the total CO_2 and NO_x emissions produced by aircraft in service in this paper. Emissions are calculated by the number of flights per airport pair and aircraft type.

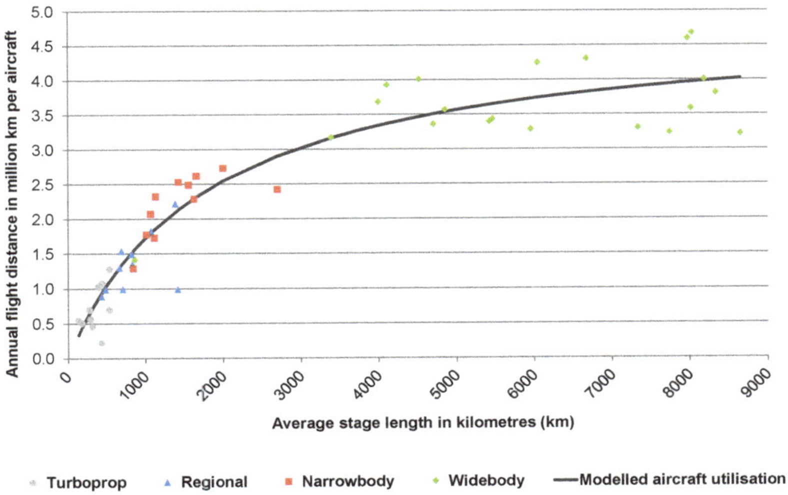

Figure 10. Aircraft utilisation model.

Table 2. Aircraft utilisation model accuracy (data from [45,47]).

Aircraft Type	Active Fleet—Cirium Fleets Analyzer	Estimated Number of Aircraft Based on Innovata Schedule	% Deviation
Airbus A320 Family	7893	7997	1.3%
Airbus A330	1,209	1108	−8.4%
Airbus A340	135	106	−21.3%
Airbus A350	281	315	12.0%
Airbus A380	231	220	−4.6%
ATR 42/72	877	720	−17.9%
Boeing 737	6827	6623	−3.0%
Boeing 747	128	130	1.8%
Boeing 757	352	261	−25.7%
Boeing 767	418	290	−30.6%
Boeing 777	1255	1257	0.2%
Boeing 787	803	836	4.1%
Bombardier CRJ	1227	1078	−12.2%
Embraer E-Series	1397	1378	−1.4%
deHavilland Dash 8	830	639	−23.0%
Total	23,863	22,957	−3.8%

Table 3 shows the aircraft types used for the Reference Scenario. Only aircraft that were available for order or already in service no later than 2014 are used in the Reference Scenario. They represent state-of-the-art aircraft technology for 2014. Originally, in CS2, only six reference aircraft were included. However, the modelling approach is based upon the 14 ICAO seat classes. The ICAO forecast [15] is one of the leading forecasts in the field, and thus is a point of reference for our study. Furthermore, 14 seat classes are a more appropriate resolution for fleet modelling from our point of view. In order to make the reference aircraft compatible with the modelling approach, additional aircraft were added. In the Reference Scenario, these aircraft are employed during the whole forecasting period from 2014 to 2050. A drawback of the model approach is that aircraft types are assigned permanently to a particular seat category for the full forecast period. Over time, aircraft of an already existing type enter the market, or existing aircraft are upgraded with more seats. The Airbus A320neo is such an example, which was assigned to the 151–175 seats category in 2014, but the aircraft in service hit the upper limit of the class in 2022 [48].

Table 3. Reference aircraft.

Original CS2 Reference Aircaft	ICAO Seat Category	CS2 Reference Aircraft
X	1–19	Do228
	20–50	ATR42-500
X	51–70	CASA C295 Civil (2014 Multi-Mission)
	71–85	Bombardier Dash-8-400
X	86–100	ATR72 Scaled to 90 Seats
	101–125	Embraer E195
X	126–150	Airbus A220-300
	151–175	Airbus A320neo
X	176–210	Airbus A321neo (SMR 2014 ref)
	211–300	Boeing 787-8
X	301–400	Airbus A350-900 (LR 2014 ref)
	401–500	Airbus A380-800 (up to 2021)/Boeing 777-9 (from 2022)

The list of available aircraft in the CS2 Scenario is determined by the environmental goals breakdown table (Table 4). Aircraft with advanced and ultra-advanced CS2 technologies gradually enter the market according to the entry into service schedule (Table 5).

Table 4. Clean Sky 2 aircraft type and entry into service.

Conceptual Aircraft/ Air Transport Type	Reference Aircraft	Window [1]	ΔCO_2	ΔNO_x	Δ Noise	Target [2] TRL @ CS2 Close
Advanced Long-range (A-LR)	LR 2014 ref	2030	20%	20%	20%	4
Ultra advanced Long-range (UA-LR)	LR 2014 ref	2035+	30%	30%	30%	3
Advanced Short/Medium-range (A-SMR)	SMR 2014 ref	2030	20%	20%	20%	5
Ultra-advanced Short/Medium-range (UA-SMR)	SMR 2014 ref	2035+	30%	30%	30%	4
Innovative Turboprop (TP), 130 Pax	2014 130 Pax ref	2035+	19 to 25%	19 to 25%	20 to 30%	4
Advanced Turboprop (A-TP), 90 Pax	2014 TP ref	2025+	35 to 40%	>50%	60 to 70%	5
Regional Multi-Mission TP, 70 Pax	2014 Multi-mission	2025+	20 to 30%	20 to 30%	20 to 30%	6
19-Pax Commuter	2014 19 Pax a/c	2025	20%	20%	20%	4-5

[1] All of the key enabling technologies at TRL 6 with a potential entry into service five years later. [2] Key enabling technologies at the major system level.

As with the reference aircraft, there are only eight original CS2 aircraft classes (see Table 5). In order to fill all of the ICAO seat classes, additional aircraft with similar characteristics to the original CS2 aircraft were introduced. An overview of the aircraft and in-production windows is shown in Table 5.

Table 5. Overview of the CS2 Scenario aircraft and the in-production window.

Original CS2 Aircraft	Aircraft Scenario Category	ICAO Seat Category	Reference Aircraft	CS2 Aircraft Type	Entry into Service in Forecast Model	Out of Production
x	Reference	1–19	2014 Pax a/c	19-Pax Reference Aircraft	2014	2029
x	CS2	1–19		19-Pax Commuter	2030	2050
	Reference	20–50		ATR42-500	2014	2029
	CS2	20–50		ATR42-500 Advanced	2030	2050
x	Reference	51–70	2014 Multi-Mission	CASA C295 Civil	2014	2029
x	CS2	51–70		Regional Multi-Mission TP 70 seats	2030	2050
	Reference	71–85		Bombardier Dash-8-400	2014	2029
	CS2	71–85		Bombardier Dash-8-400 Advanced	2030	2050
x	Reference	86–100	2014 TP ref	ATR72 Scaled to 90 seats	2014	2029
x	CS2	86–100		Advanced TP90	2030	2050
	Reference	101–125		Embraer E195	2014	2020
	CS2	101–125		Embraer E195–E2	2021	2034
	CS2	101–125		A-SMR-Embraer E195	2035	2039
	CS2	101–125		UA-SMR-Embraer E195	2040	2050
x	Reference	126–150	2014 130 Pax ref	Airbus A220 300	2014	2039
x	CS2	126–150		Innovative Turboprop	2040	2050
	Reference	151–175		Airbus A320neo	2014	2034
	CS2	151–175		A-SMR-Airbus A320neo	2035	2039
	CS2	151–175		UA-SMR-Airbus A320neo	2040	2050
x	Reference	176–210	SMR 2014 ref	Airbus A321neo	2014	2034
x	CS2	176–210		A-SMR	2035	2039
x	CS2	176–210		UA-SMR	2040	2050
	Reference	211–300		Boeing 787-8	2014	2034
	CS2	211–300		A-LR-Boeing 787-8	2035	2039
	CS2	211–300		UA-LR-Boeing 787-8	2040	2050
x	Reference	301–400	LR 2014 ref	Airbus A350-900	2014	2034
x	CS2	301–400		Airbus A350-900neo	2035	2039
x	CS2	301–400		UA-LR	2040	2050
	Reference	401–500		Airbus A380-800	2014	2021
	Reference	401–500		Boeing 777-9	2022	2034
	CS2	401–500		Airbus A350-2000neo	2035	2039
	CS2	401–500		UA-LR	2040	2050

2.3. Emission Modelling

For the calculation of air transport emissions, various modelling techniques can be employed. In this paper we use the commercial flight performance software Piano-X (Project Interactive Analysis and Optimization) [52] for all reference aircraft for which no emission profiles were provided by the System Platform Developers (SPDs). It is based on the aircraft analysis tool Piano 5, and features flight performance indicators such as fuel burn, CO_2 and NO_x emissions for a set of more than 500 aircraft types. It is widely used in the aviation community, and various studies have validated the accuracy of the tool [53,54]. Generally, the results of calculations using Piano-X have shown a very good alignment with actual fuel consumption and emissions values, as a comparison with the flight data recorder data shows. For CS2 concept aircraft, wherever available, emissions profiles provided by the SPDs were used. In all other cases, reference aircraft with specific reduction factors were employed.

Figure 11 provides an overview of the emissions modelling at the air transport system level. The main inputs are the flight volume and fleet forecast by airport, and the emission profiles from SPD models and Piano-X. The emission profiles are provided by the SPD models and Piano-X. Piano-X is employed for already existing aircraft and the SPD models for the reference and CS2 aircraft. The aircraft models contain encrypted databases from the manufacturers concerning engine thermodynamic and aerodynamic data as well as engine NO_x and fuel flow values. For the calculation of NO_x, the Boeing Fuel Flow Method2 [55] is employed. The aircraft performance module first calculates a trajectory and then the emissions.

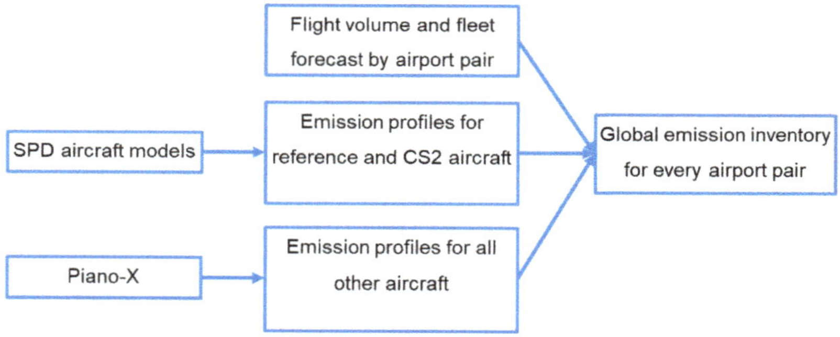

Figure 11. Overview of air transport system level emissions calculation.

The final step is to calculate the global emission inventory using the inputs from the aircraft models for every airport pair.

The main objective of the TE within CS2 is to assess the contribution of aircraft and engine technology to the environmental goals; therefore, we focused on these effects and omitted other contributing factors, which nevertheless play a role in the total fuel consumption and aircraft emissions. For example, we did not consider exact flight routes, but based all of our calculations on great circle distances between origin and destination airports. Any additional distance covered in the terminal maneuvering area around airports, as well as additional cruise distance flown due to detours because of weather, wind or air traffic management restrictions, as well as voluntary detours by airlines in order to save air navigation fees (a typical behavior observed in the EUROCONTROL area, where unit rates can differ substantially in different neighboring countries) are not considered in our analysis. Furthermore, the impact of sub-optimal flight levels or the degradation of aircraft due to age or contamination are not considered. This was to avoid any confounding effects which might negatively influence the identification of aircraft/engine technology effects on emissions.

3. Results

In this section we present the forecast results of:

- the passenger demand and flight volume (Section 3.1).
- the aircraft fleet (Section 3.2).
- the aircraft emissions (Section 3.3).

3.1. Passenger Demand and Flight Volume

Figure 12 displays the results of the passenger demand and flight forecasts. The passenger volume increases on average by 3.9% per year in the High Scenario and 3.3% per year in the Low Scenario. The flight volume rises, on average, by 2.0% per year in the High Scenario and 1.7% per year in the Low Scenario. The difference between the two growth rates, i.e., between passenger and flight volume growth rates, illustrates the increase in the number of passengers per flight. As a result, there is an increase from 109 passengers per flight in 2014 to 214 (1.9% per year) in the High Scenario, and 196 (1.6% per year) in the Low Scenario in 2050. Thus, almost 80% more passengers will be transported per flight in 2050 in the Low Scenario, and 96% more in the High Scenario. In order to manage the passenger volume growth, we have three options:

- we can increase the number of passengers per flight (and hold the numbers of flights constant),
- we can increase the number of flights (and hold the passengers per flight transported constant),
- we can mix both options.

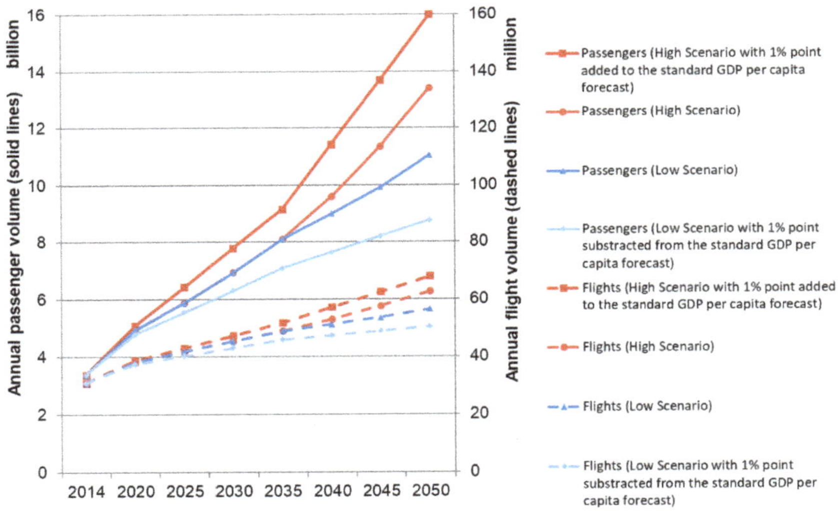

Figure 12. Results of the passenger demand and flight volume forecast.

The forecast result is that about 50% of the passenger volume growth is managed by transporting more passengers per flight, and the other 50% is managed by increasing the flight volume. In order to handle more flights, the airport capacity needs to be utilised better and, in some cases, even enlarged. Still, aircraft and airport capacity in 2050 will not be sufficient to avoid any unaccommodated demand: about 7% of passenger demand and 9% of flights will be lost in the High Scenario in 2050 due to capacity constraints. In the Low Scenario, around 4% of the passenger demand and 5% of the flights will be lost in 2050 because of airport capacity shortages.

Furthermore, Figure 12 presents a forecast with a variation of ±1 percentage point of GDP per capita. This illustrates that the forecast is quite sensitive to the GDP per capita assumptions. Because of airport capacity constraints and their impact on flight volume, passenger volume is more sensitive to variations in GDP per capita than flight volume.

Figure 13 shows a comparison of our forecast with the Airbus [56], Boeing [57] and ICAO forecasts [15], which include both domestic and international markets. Those are the leading forecasts in the field, which is the reason why we have chosen them as points of reference. The High Scenario is about 10% lower than the ICAO CAEP/11 forecast. The Airbus and Boeing forecasts are significantly higher: the High Scenario is about 20% lower than the Airbus and Boeing values in 2037 and 2038 as a result of limited airport capacity, which is an aspect which the other three forecasts do not include.

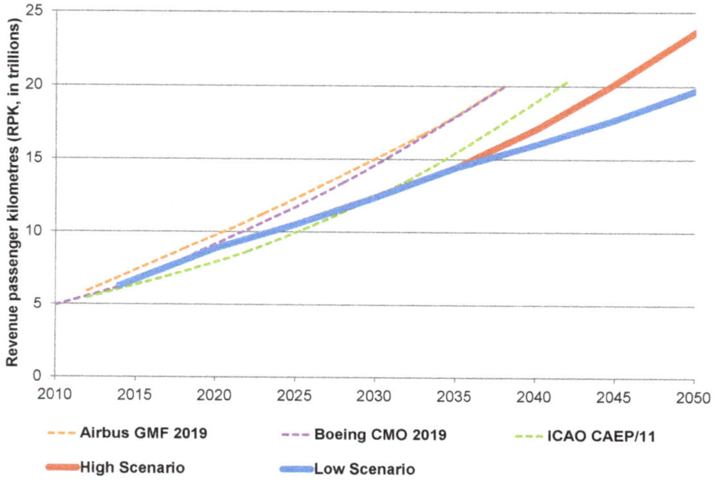

Figure 13. Comparison with the Airbus, Boeing and ICAO forecasts.

3.2. *Aircraft Fleet*

Figure 14 displays the fleet forecast by seat class for the Low Scenario, and Figure 15 shows the results for the High Scenario. In both cases we expect a stagnating or even declining regional aircraft market, except for the segment of 86 to 100 seats. For mainliner aircraft, we forecast a substantial shift towards aircraft with more seats, and this tendency is particularly strong in the High Scenario. As a result, we expect the average seats per flight to increase by 1.6% per year in the High Scenario and 1.4% per year in the Low Scenario between 2014 and 2050. By comparison, the seats per flight increased by 1.7% per year between 2000 and 2014 [46].

In the Low Scenario, we forecast a declining importance in terms of aircraft with 101 to 175 seats, and a shift to aircraft with 176 to 500 seats for the mainliner segment. Here, the number of flights with aircraft of 211 to 300 seats, e.g., a Boeing 787–8, grows particularly strong.

In the High Scenario, flight volume with mainliner aircraft of 101 to 175 seats is expected to decline until 2050, while the number of flights with more seats grow in importance. Here, as in the Low Scenario, volume of flights with 211 to 300 seats grows particularly strong. Compared to the Low Scenario, there is a much stronger growth of the number of flights with 301 to 500 seats.

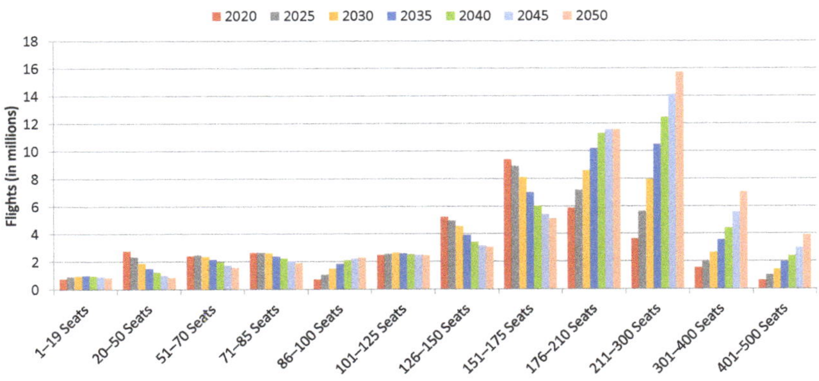

Figure 14. Fleet forecast results in the Low Scenario.

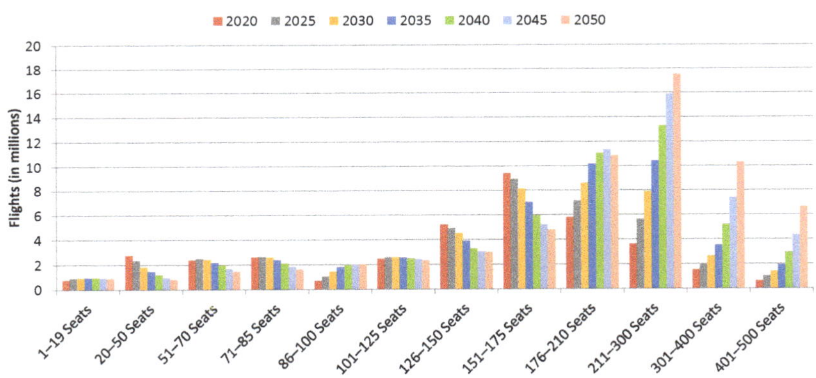

Figure 15. Fleet forecast results in the High Scenario.

In both scenarios, the shift towards larger aircraft is a result of airlines' economic considerations and limited airport capacity, especially at large airports which handle a substantial portion of the global flight volume: the largest 120 airports, which correspond to 3% of all airports worldwide with scheduled flights, handle about 50% of the global flight volume. This has not changed significantly since 2000 [2]. However, the shift towards larger aircraft is more pronounced in the High Scenario due to the stronger increase in passenger demand.

Figure 16 compares the forecast of aircraft deliveries until 2038 for the Airbus Global Market Forecast (GMF), which only considers aircraft of over 100 seats, with the results of the High Scenario. According to the Airbus GMF, aircraft are subdivided in three classes: small aircraft have 101 to 210 seats, medium aircraft have 211 to 300 seats, and large aircraft have over 300 seats [56]. Aircraft deliveries are expected to be lower in the High Scenario compared to the Airbus GMF because of the lower number of flights that are forecast (about 30% lower). This is a result of the airport capacity shortage and the related shift to larger aircraft. We expect only about 10,000 new deliveries in the small aircraft class, instead of around 30,000 as predicted by Airbus, but 2400 more in the medium aircraft class and 2900 more in the large aircraft class.

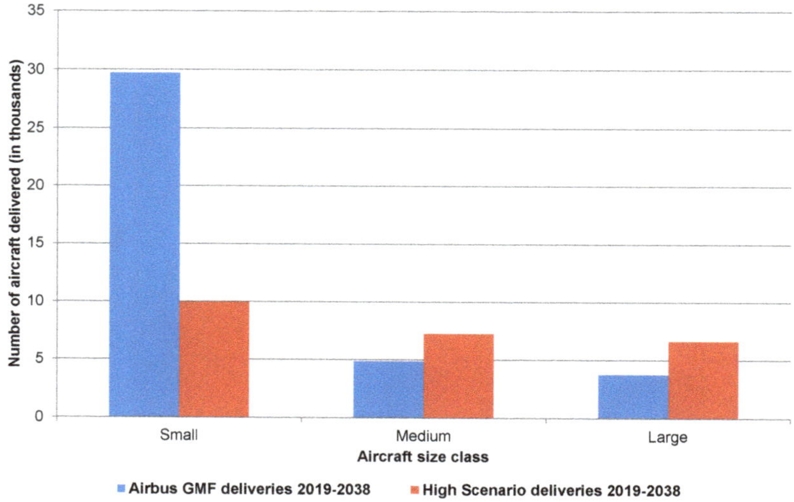

Figure 16. Comparison of the aircraft deliveries forecast of the Airbus GMF and the High Scenario forecast for the time period 2019–2038.

Figure 17 displays the available seat kilometres (ASK) in terms of the technology level in the Low Scenario. As a result of the traffic growth and the replacement of retired aircraft, 52% of the total global ASK will be provided by aircraft with ultra-advanced CS2, and 19% will be provided by aircraft with advanced CS2 technologies in 2050. In total, 29% of the global ASK will be offered by 2014 reference aircraft. Aircraft from the base year and older have almost vanished.

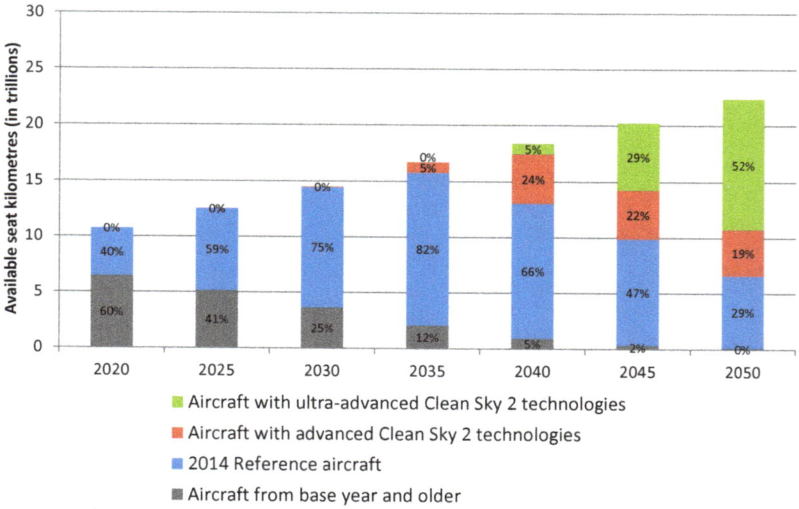

Figure 17. Available seat kilometres (ASK) by technology level in the Low Scenario. The grey, blue, red and green bars add up to the total ASK volume for the respective year.

Figure 18 displays the ASK by technology level in the High Scenario. Due to traffic growth and the replacement of retired aircraft, in the year 2050, 56% of the total global ASK will be provided by aircraft with ultra-advanced CS2, and 19% will be provided by aircraft with advanced CS2 technologies. In total, 24% of the global ASKs will be offered by 2014 reference aircraft. Aircraft from the base year and older have virtually vanished. Therefore, in the High Scenario, more CS2 technology aircraft—especially those with ultra-advanced technologies—will be employed in 2050, as there will be more passenger demand, and thus more need for flights.

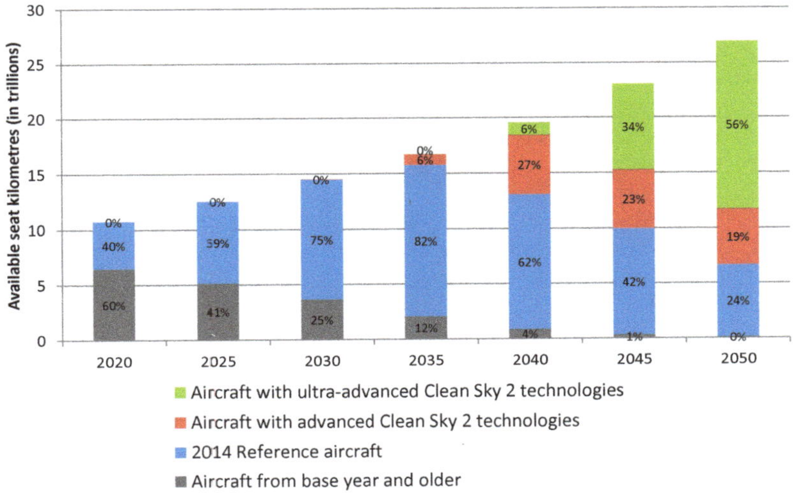

Figure 18. Available seat kilometres (ASK) by technology level in the High Scenario. The grey, blue, red and green bars add up to the total ASK volume for the respective year.

Figures 19 and 20 display the breakdown of the passenger in-service fleet for the base-year fleet (including reference aircraft already in service in 2014), with replacement and growth for the Low Scenario and High Scenario, respectively.

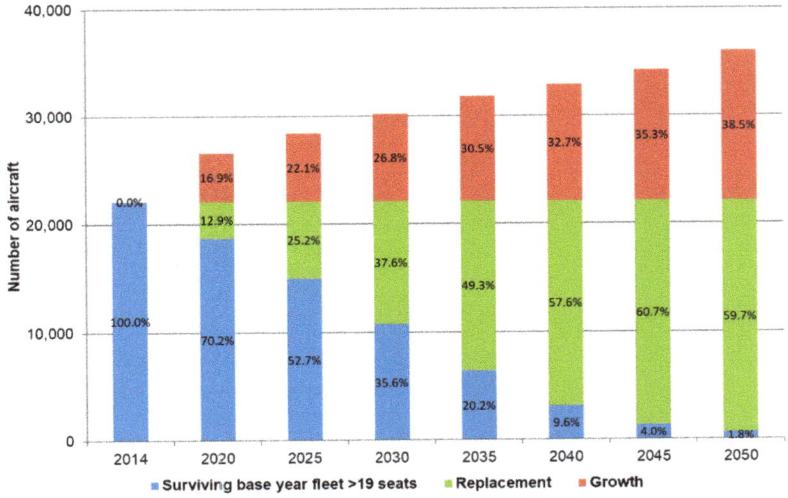

Figure 19. Passenger-in service fleet breakdown in the Low Scenario.

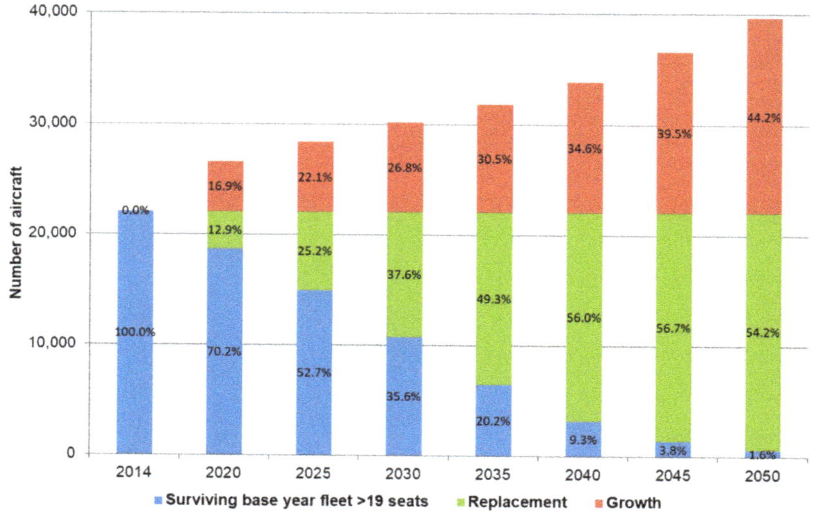

Figure 20. Passenger-in service fleet breakdown in the High Scenario.

The surviving base-year fleet and the replacement volume are exactly the same in both scenarios. Naturally, the difference lies in the growth volume, which is substantially larger in the High Scenario due to greater passenger demand leading to a need for more flights. In both scenarios, the base-year fleet has almost vanished in 2050.

3.3. Aircraft Emissions

Table 6 displays the aircraft emission results for the Low and High Scenario for the time period 2035–2050. Here, emission reductions are assessed using a scenario in which CS2 aircraft are introduced over time (Table 5), with a scenario in which only reference aircraft are introduced over time (Table 4). In both the Low and High Scenario there is a substantial reduction of CO_2 and NO_x emissions compared to the corresponding Reference Scenario. The relative emission reduction increases over time, as more and more CS2 aircraft enter service due to demand increasing and older aircraft being retired.

Table 6. Emission results of the Low and High Scenarios for 2035 to 2050.

Year	High Scenario		Low Scenario	
	ΔCO_2	ΔNO_x	ΔCO_2	ΔNO_x
2035 CS2 vs. Reference	−0.8%	−1.9%	−0.8%	−1.9%
2040 CS2 vs. Reference	−4.6%	−12.0%	−4.1%	−10.2%
2045 CS2 vs. Reference	−10.1%	−23.0%	−9.2%	−20.1%
2050 CS2 vs. Reference	−14.6%	−31.0%	−13.8%	−29.0%

Although there are about 22% more passengers transported and 11% more flights in the High compared to the Low Scenario, the emission reductions compared to the Reference Scenario are very similar in both scenarios. The reason for this is that there is greater traffic growth in the High Scenario, and consequently more CS2 aircraft will be employed until 2050. For example, in the High Scenario, CS2 has a share of around 75% in terms of ASK in 2050, while it is only about 70% in the Low Scenario. As a result, in the High Scenario, there is more traffic because of greater demand, but this is offset by more-advanced and ultra-advanced CS2 aircraft. Overall, CO_2 emissions will be reduced by about 15% and NO_x emissions by around 30% until 2050, compared to the Reference Scenario.

Figures 21 and 22 illustrate the distribution of CO_2 and NO_x emissions by the Airbus seat classes (see also Figure 16) and flight distance, as well as the share of flights and revenue passenger kilometres (RPK) for each distance band and seat class, respectively. The results are virtually the same for both the Low and High Scenario. In 2050, about 55% of CO_2 and NO_x emissions will be from the flights of medium (211–300 seats) and large (over 300 seats) aircraft, on flights of less than 4000 km. In 2020, this share was only about 20%, and in 2014 it was around 10%. If we look at flights of up to 2000 km, medium and large aircraft account for 35% of CO_2 and 36% of NO_x emissions in 2050. The reason for this development is the substantial increase in passenger demand, and the related growth of aircraft size, i.e., passengers and seats per flight, especially on rather short routes.

Flight distance (km)

Seats	0-1000	1000-2000	2000-3000	3000-4000	4000-5000	5000-6000	6000-7000	7000-8000	8000-9000	9000-10,000	>10,000	CO_2	Flights	RPK
0-19												0.04%	1.5%	0.01%
20-100												1.1%	9.6%	0.8%
101-210												11.7%	33.7%	12.3%
211-300												26.8%	28.1%	22.0%
>300												60.3%	27.2%	64.8%
CO_2	19%	26%	14%	8%	4%	5%	5%	4%	4%	4%	6%			
Flights	54%	28%	8%	3%	1%	1%	1%	1%	1%	1%	1%			
RPK	15%	25%	14%	9%	5%	5%	6%	5%	5%	5%	8%			

Share of total CO_2 emissions in 2050: 0%-0.001% | 0.001%-0.1% | 0.1%-2% | 2%-5% | 5%-10% | 10%-15%

Figure 21. Distribution of CO_2 emissions by seat class and flight distance in 2050.

Flight distance (km)

Seats	0-1000	1000-2000	2000-3000	3000-4000	4000-5000	5000-6000	6000-7000	7000-8000	8000-9000	9000-10000	>10000	No_x	Flights	RPK
0-19												0.02%	1.5%	0.01%
20-100												1.1%	9.6%	0.8%
101-210												12.8%	33.7%	12.3%
211-300												21.2%	28.1%	22.0%
>300												65.0%	27.2%	64.8%
NO_x	21%	26%	14%	8%	4%	4%	5%	4%	4%	4%	6%			
Flights	54%	28%	8%	3%	1%	1%	1%	1%	1%	1%	1%			
RPK	15%	25%	14%	9%	5%	5%	6%	5%	5%	5%	8%			

Share of total NO_x emissions in 2050: 0%-0.001% | 0.001%-0.1% | 0.1%-2% | 2%-5% | 5%-10% | 10%-15%

Figure 22. Distribution of NO_x emissions by seat class and flight distance in 2050.

4. Discussion and Conclusions

When considering the two scenarios, the technological (and thus the demand) development is much more favorable in the High Scenario, in which passenger demand increases by 297% and flight volume increases by 102% between 2014 and 2050, while in the Low Scenario, passenger demand rises by 227% and flight volume rises by 82%. The difference between the passenger and flight volume development illustrates that there is a substantial increase in the number of passengers per flight until 2050 in both scenarios, and therefore a strong tendency to employ larger aircraft in the future.

Over time, there will be a constant shift towards bigger aircraft, with the class of 211 to 300 seats being the most important in terms of flights in 2050. Larger aircraft with up to 500 seats are becoming more important as well, especially in the High Scenario, because of the stronger passenger demand development. The number of seats per flight is expected to increase by 1.6% per year in the High Scenario and 1.4% per year in the Low Scenario.

However, due to the CS2 aircraft which enter into service over time as older aircraft are replaced and additional new aircraft are needed because of the increasing passenger demand, there will be a considerable drop in CO_2 and NO_x emissions compared to the Reference Scenario, in which only aircraft from 2014 or older are employed: in both scenarios, CO_2 emissions can be reduced by about 15% and NO_x emissions can be reduced by around 30% until 2050 compared to the Reference Scenario. Despite the substantial difference in passenger and flight volumes between the two scenarios, the emission results of the High Scenario are even better than those of the Low Scenario, as more-advanced and ultra-advanced CS2 aircraft will have entered into service due to the higher level of passenger demand and flight volume. This is a very important result; it shows the environmental potential of CS2 aircraft. Nevertheless, in 2050, medium (211–300 seats) and large (over 300 seats) aircraft will produce a substantial share of their CO_2 and NOx emissions on flights of up to 2000 km. This is essentially a result of the large increase of passenger demand volume on rather short routes, which is not optimal for medium and large aircraft. This naturally raises a question about the development of a large-capacity aircraft for short- and medium-distance flights, say of up to 3000 or 4000 km.

Nevertheless, there are of course some serious limitations of the study, such that further research is needed. First, the effects of the COVID-19 pandemic are not considered. The pandemic might have an effect on the long-term travel behaviour of leisure as well as business passengers, such that future demand development will be different compared to this study. Furthermore, there is uncertainty about when air traffic volume will have been recovered to pre-pandemic levels. First analyses have already been conducted [7] and will be part of an update of this study.

Another limitation of the study is that there is no differentiation between business and leisure travellers. This is because of a lack of consistent global data on the share of business and leisure travellers. Recent trends, like virtual meetings, which have been more and more established during the COVID-19 pandemic, might continue after the pandemic to some degree, and thus might influence the need for business travel.

In our model, aircraft types are assigned to particular seat classes for the whole forecast period and cannot switch to another class. However, we have observed that aircraft of an already existing type can enter the market with more seats, or aircraft which are already in service can be upgraded. The Airbus A320neo is such an example: it is assigned to the 151–175 seats category in 2014, but aircraft in service already hits the upper limit of the class in 2022 [48]. While this seems to be more of a relatively minor limitation of the study, it is an area where the model can be improved. However, it is a challenge to anticipate seat capacity upgrades of existing aircraft, and to differentiate between whether such an upgraded aircraft or an already existing aircraft with larger seat capacity is employed. This is an issue in which complexity quickly increases.

Lastly, the approach to model multi-airport regions like London Heathrow and their potential to mitigate capacity constraints needs to be improved. This has an effect on the air passenger demand level, and in the course of this on aircraft fleet composition. However, based on our research so far, we do not expect this to have a major influence on demand level, as airlines seem to prefer to increase the number of passengers per flight instead of shifting flights to neighbour airports. This can be achieved by fitting more seats in existing aircraft or employing larger aircraft. Employing larger aircraft is not only about very large aircraft like the Airbus A380 (which is out of production now) or the Boeing 777–9 but also about employing larger aircraft step-by-step across the whole range from small to large aircraft. Here, aircraft in the range of about 211 to 300 seats play a particularly important role, like the Airbus A321neo or the Boeing 787–8. On the other hand, we see some potential for more nonstop long-haul connections with the introduction of smaller, efficient, long-haul aircraft like the Airbus A321XLR [58,59].

These limitations should be taken seriously, and naturally raise the question about the benefits of this study. Despite the need for improvements, we have shown that the model developed can assess the environmental benefits of aircraft technology end-to-end, i.e., from the generation of air passenger demand to aircraft fleet modelling, and finally emissions modelling. A particular strength is the bottom-up approach, i.e., the approach is based on airport pairs instead of larger aggregates, which enables the consideration of airport capacity constraints.

Furthermore, the study provides a first overview of the environmental benefit of CS2 technology. Additional technical improvements will be considered in the second (and final) assessment in 2024. The COVID-19 pandemic-related limitations are expected to lower future passenger and flight volume, and thus the overall environmental impact of aviation, but this does not change the benefit of CS2 technology substantially.

Author Contributions: Conceptualization, M.C.G., W.G. and A.J.; methodology, M.C.G., W.G. and A.J.; software, M.C.G., W.G., A.J. and C.L.; validation, M.C.G., W.G. and A.J.; formal analysis, M.C.G., P.B., W.G. and A.J.; investigation, M.C.G., W.G. and A.J.; resources, M.C.G., W.G. and A.J.; data curation, M.C.G., W.G., A.J. and P.B.; writing—original draft preparation, M.C.G., W.G. and A.J.; writing—review and editing, M.C.G., P.B., W.G. and A.J.; visualization, M.C.G., W.G. and A.J.; supervision, M.C.G., W.G. and A.J.; project administration, M.C.G. All authors have read and agreed to the published version of the manuscript.

Funding: This research received funding from the Clean Sky 2 project.

Institutional Review Board Statement: Not applicable.

Informed Consent Statement: Not applicable.

Data Availability Statement: Not applicable.

Conflicts of Interest: The authors declare no conflict of interest.

References

1. Berster, P.; Gelhausen, M.C.; Wilken, D. Is increasing aircraft size common practice of airlines at congested airports? *J. Air Transp. Manag.* **2015**, *46*, 40–48. [CrossRef]
2. Gelhausen, M.C.; Berster, P.; Wilken, D. *Airport Capacity Constraints and Strategies for Mitigation: A Global Perspective*; Academic Press: New York, NY, USA, 2019; pp. 1–327.
3. Wilken, D.; Berster, P.; Gelhausen, M.C. New empirical evidence on airport capacity utilisation: Relationships between hourly and annual air traffic volumes. *Res. Transp. Bus. Manag.* **2011**, *1*, 118–127. [CrossRef]
4. Givoni, M.; Rietveld, P. The environmental implications of airlines' choice of aircraft size. *J. Air Transp. Manag.* **2010**, *16*, 159–167. [CrossRef]
5. Pitfield, D.E.; Caves, R.E.; Quddus, M.A. Airline strategies for aircraft size and airline frequency with changing demand and competition: A simultaneous-equations approach for traffic on the north Atlantic. *J. Air Transp. Manag.* **2010**, *16*, 151–158. [CrossRef]
6. Brueckner, J.K.; Zhang, Y. A model of scheduling in airline networks: How a hub-and-spoke system affects flight frequency, fares and welfare. *J. Transp. Econ. Policy* **2001**, *35*, 195–222.
7. Gelhausen, M.C.; Berster, P.; Wilken, D. Post-COVID-19 Scenarios of Global Airline Traffic until 2040 That Reflect Airport Capacity Constraints and Mitigation Strategies. *Aerospace* **2021**, *8*, 300. [CrossRef]
8. Gelhausen, M.C. Modelling the effects of capacity constraints on air travellers' airport choice. *J. Air Transp. Manag.* **2011**, *17*, 116–119. [CrossRef]

9. Gudmundsson, S.V.; Paleari, S.; Redondi, S. Spillover effects of the development constraints in London Heathrow Airport *J. Transp. Geogr.* **2014**, *35*, 64–74. [CrossRef]
10. Redondi, R.; Gudmundsson, S.V. Congestion spill effects of Heathrow and Frankfurt airports on connection traffic in European and Gulf hub airports. *Transp. Res. Part A Policy Pract.* **2016**, *92*, 287–297. [CrossRef]
11. Dennis, N. Airline hub operations in Europe. *J. Transp. Geogr.* **1994**, *2*, 219–233. [CrossRef]
12. Airports Commission. *Airports Commission: Final Report*; Airports Commission: London, UK, 2015.
13. Airbus. *Global Market Forecast 2021–2040*; Airbus: Blagnac, France, 2021.
14. Boeing. *Commercial Market Outlook 2021–2040*; Boeing: Seattle, WA, USA, 2021.
15. International Civil Aviation Organization (ICAO). *ICAO Long-Term Traffic Forecast—Passenger and Cargo*; ICAO: Montreal, QC, Canada, 2016.
16. Harvey, D. Airline passenger traffic patterns within the United States. *J. Air Law Commer.* **1951**, *18*, 157–165.
17. Grosche, T.; Rothlauf, F.; Heinzl, A. Gravity models for airline passenger volume estimation. *J. Air Transp. Manag.* **2007**, *13*, 175–183. [CrossRef]
18. Tusi, W.H.K.; Fung, M.K.Y. Analysing passenger network changes: The case of Hong Kong. *J. Air Transp. Manag.* **2016**, *50*, 1–11.
19. Matsumoto, H. International urban systems and air passenger and cargo flows: Some calculations. *J. Air Transp. Manag.* **2004**, *10*, 241–249. [CrossRef]
20. Shen, G. Reverse-fitting the gravity model to inter-city airline passenger flows by an algebraic simplification. *J. Transp. Geogr.* **2004**, *12*, 219–234. [CrossRef]
21. Bhadra, D.; Kee, J. Structure and dynamics of the core US air travel markets: A basic empirical analysis of domestic passenger demand. *J. Air Transp. Manag.* **2008**, *14*, 27–39. [CrossRef]
22. Endo, N. International trade in air transport services: Penetration of foreign airlines into Japan under the bilateral aviation policies of the US and Japan. *J. Air Transp. Manag.* **2007**, *13*, 285–292. [CrossRef]
23. Hazledine, T. Border effects for domestic and international Canadian passenger air travel. *J. Air Transp. Manag.* **2009**, *15*, 7–13. [CrossRef]
24. Alexander, D.W.; Merkert, R. Challenges to domestic air freight in Australia: Evaluating air traffic markets with gravity modelling. *J. Air Transp. Manag.* **2017**, *61*, 41–52. [CrossRef]
25. Alexander, D.W.; Merkert, R. Applications of gravity models to evaluate and forecast US international air freight markets post-GFC. *Transp. Policy* **2021**, *104*, 52–62. [CrossRef]
26. Baier, F.; Berster, P.; Gelhausen, M.C. Global cargo gravitation model: Airports matter for forecasts. *Int. Econ. Econ. Policy* **2022**, *19*, 219–238. [CrossRef]
27. Sabre AirVision Market Intelligence. Data Based on Market Information Data Tapes (MIDT). Available online: https://www.sabre.com/products/market-intelligence/ (accessed on 8 November 2021).
28. Silva, J.M.C.S.; Tenreyro, S. The log of gravity. *Rev. Econ. Stat.* **2006**, *88*, 641–658. [CrossRef]
29. Wilken, D.; Berster, P.; Gelhausen, M.C. Airport choice in Germany: New empirical evidence of the 2003 German air traveller survey. *Airpt. Manag.* **2007**, *1*, 165–179.
30. Mason, K.J. Corona Markiert Eine Zäsur Bei Geschäftsreisen. Available online: https://www.airliners.de/airline-geschaeftsmodelle-88-corona-markiert-zaesur-geschaeftsreisen/61658 (accessed on 23 August 2021).
31. Charnes, A.; Cooper, W.W.; Rhodes, E. Measuring the efficiency of decision-making units. *Eur. J. Oper. Res.* **1978**, *2*, 429–444. [CrossRef]
32. Cooper, W.W.; Seiford, L.M.; Tone, K. *Data Envelopment Analysis—A Comprehensive Text with Models, Applications, References and DEA-Solver Software*; Springer: New York, NY, USA, 2007.
33. Markov, A.A. An Example of Statistical Investigation of the Text Eugene Onegin Concerning the Connection of Samples in Chains. *Sci. Context* **2006**, *19*, 591–600. [CrossRef]
34. McFadden, D. Conditional Logit Analysis of Qualitative Choice Behavior. In *Frontiers in Econometrics*; Zarembka, P., Ed.; Academic Press: New York, NY, USA, 1974.
35. Ben-Akiva, M.; Lerman, S.R. *Discrete Choice Analysis*; MIT Press: Cambridge, MA, USA, 1985.
36. Gelhausen, M.C.; Berster, P.; Wilken, D. Do airport capacity constraints have a serious impact on the future development of air traffic? *J. Air Transp. Manag.* **2013**, *28*, 3–13. [CrossRef]
37. Wei, W.; Hansen, M. Impact of aircraft size and seat availability on airlines' demand and market share in duopoly markets. *Transp. Res. Part E Logist. Transp. Rev.* **2005**, *41*, 315–327. [CrossRef]
38. Presto, F.; Gollnick, V.; Lütjens, K. Fleet upgauging and reducing flight frequency. *J. Air Transp.* **2022**, *30*, 23–36. [CrossRef]
39. Kölker, K.; Bießlich, P.; Lütjens, K. From passenger growth to aircraft movements. *J. Air Transp. Manag.* **2016**, *56*, 99–106. [CrossRef]
40. Givoni, M.; Rietveld, P. Airline's choice of aircraft size—Explanations and implications. *Transp. Res. Part A* **2009**, *43*, 500–510. [CrossRef]
41. Bhadra, D. Choice of aircraft fleets in the U.S. domestic scheduled air transportation system: Findings from a multinomial logit analysis. *J. Transp. Res. Forum* **2005**, *44*, 143–162. [CrossRef]
42. Pai, V. On the factors that affect airline flight frequency and aircraft size. *J. Air Transp. Manag.* **2010**, *16*, 169–177. [CrossRef]
43. Bhadra, D. Choice of aircraft fleets in the US NAS: Findings from a multinomial logit analysis. In Proceedings of the 3rd AIAA Aviation Technology, Integration, and Operations Conference, Denver, CO, USA, 17–19 November 2003.

44. Bhadra, D. Demand for Air Travel in the United States: Bottom-Up Econometric Estimation and Implications for Forecasts by Origin-Destination Pairs. In Proceedings of the 2nd AIAA Aviation Technology, Integration, and Operations Conference, Los Angeles, CA, USA, 1–3 October 2002.
45. Cirium. Fleets Analyzer. Available online: https://www.cirium.com/products/views/fleets-analyzer/ (accessed on 8 November 2021).
46. Official Airline Guide (OAG). Flight Schedules. Available online: https://www.oag.com/ (accessed on 8 November 2021).
47. Flightglobal. Innovata Flight Schedules. Available online: https://www.flightglobal.com/ (accessed on 8 November 2021).
48. International Civil Aviation Organization (ICAO). Annual Report of the Council (Doc 9916). 2009. Available online: https://www.icao.int/publications/Documents/9916_en.pdf (accessed on 22 February 2022).
49. International Civil Aviation Organization (ICAO). Annual Report of the Council, Presentation of 2016 Air Transport Statistical Results. 2017. Available online: https://www.icao.int/annual-report-2016/Pages/the-world-of-air-transport-in-2016-statistical-results.aspx (accessed on 22 February 2022).
50. Flightradar24. Live Air Traffic. Available online: https://www.flightradar24.com/50.83,6.91/6 (accessed on 8 November 2021).
51. Eurocontrol. Automatic Dependent Surveillance–Broadcast Airborne Equipage Monitoring. Available online: https://www.eurocontrol.int/service/adsb-equipage (accessed on 22 February 2022).
52. Lissys. Piano-X. Available online: https://www.lissys.uk/PianoX.html (accessed on 8 November 2021).
53. Dons, J.; Mariens, J.; O'Callaghan, G.D. *Use of Third-Party Aircraft Performance Tools in the Development of the Aviation Environmental Design Tool (AEDT)*; Report of the U.S. Department of Transportation, Research and Innovative Technology Administration; John A. Volpe National Transportation Systems Center, Environmental Measurement and Modeling Division: Washington, NJ, USA, 2011.
54. Senzig, D.A.; Dons, J.; Mariens, J.; O'Callaghan, G.D.; Iovinelli, R.J. *Extending Aircraft Performance Modeling Capabilities in the Aviation Environmental Design Tool (AEDT)*; Report of the John A. Volpe National Transportation Systems Center; Federal Aviation Administration—Office of Environment and Energy: Washington, NJ, USA, 2011.
55. DuBois, D.; Paynter, G.C. Fuel Flow Method2 for estimating aircraft emissions. *SAE Int. J. Aerosp.* **2006**, *115*, 1–14.
56. Airbus. *Global Market Forecast 2019–2038*; Airbus: Blagnac, France, 2019.
57. Boeing. *Commercial Market Outlook 2019–2038*; Boeing: Seattle, WA, USA, 2019.
58. Grimme, W.; Maertens, S.; Bingemer, S.; Gelhausen, M.C. Estimating the market potential for long-haul narrowbody aircraft using origin-destination demand and flight schedules data. *Transp. Res. Procedia* **2021**, *52*, 412–419. [CrossRef]
59. Wilken, D.; Berster, P.; Gelhausen, M.C. Analysis of demand structures on intercontinental routes to and from Europe with a view to identifying potential for new low-cost services. *J. Air Transp. Manag.* **2016**, *56B*, 79–90. [CrossRef]

aerospace

MDPI

Article

Simulation of Runway Irregularities in a Novel Test Rig for Fully Electrical Landing Gear Systems

Andrea De Martin *, Giovanni Jacazio and Massimo Sorli

Department of Mechanical and Aerospace Engineering, Politecnico di Torino, 10129 Turin, Italy; giovanni.jacazio@polito.it (G.J.); massimo.sorli@polito.it (M.S.)
* Correspondence: andrea.demartin@polito.it

Abstract: The E-LISA research project, under way within the Clean Sky 2 framework, has the objective of developing an innovative iron bird dedicated to executing tests on the landing gear of a small aircraft transport equipped with an electro-mechanical landing gear and electrical brake. Such tests include the simulation of complete landing procedures under different operating conditions such as runway friction, presence of periodical defects along the runway, variable aircraft weight, and approach speed. To this end, the iron bird requires novel solutions in both its architecture and its control scheme. This paper details an innovative solution that is being implemented in the E-LISA iron bird to enable the execution of tests on a landing gear, reproducing the effects of any type of runway irregularity. First, the rig architecture is presented in detail, with particular care toward the hybrid position/force control system that manages its operations. Then, a simulation model is introduced with the objective of verifying the control system stability and the test rig capability to reproduce on the test articles the effects produced on the landing gear leg of periodical runway irregularities. Simulations results are presented, highlighting the stability of the proposed control scheme and providing a preliminary assessment of the system performances.

Keywords: test bench; landing gear; electrical brake; force control; runway irregularities

Citation: De Martin, A.; Jacazio, G.; Sorli, M. Simulation of Runway Irregularities in a Novel Test Rig for Fully Electrical Landing Gear Systems. *Aerospace* **2022**, *9*, 114. https://doi.org/10.3390/aerospace9020114

Academic Editors: Spiros Pantelakis, Andreas Strohmayer and Liberata Guadagno

Received: 14 January 2022
Accepted: 17 February 2022
Published: 21 February 2022

1. Introduction

The E-LISA research project, under way within the Clean Sky 2 framework, has the objective of developing an innovative iron bird dedicated to executing tests on the landing gear of a small aircraft transport equipped with an electro-mechanical landing gear and electrical brake. The E-LISA iron bird consists of a multi-functional intelligent test facility integrating hardware and software, allowing all the tests and analyses perceived as fundamental to be performed to demonstrate the maturity of an electro-mechanical landing gear, hence paving the way for its implementation in a small passenger aircraft and will include prognostics and health management (PHM) functionalities for the electrical brake system. Such tests include the simulation of complete landing procedures under different operating conditions such as runway friction (wet/dry), presence of waving and irregularities along the runway, variable aircraft weight, and approach speed.

The development and certification process of landing gear systems involves the characterization and analysis of several components, often tested separately on different test benches. A test rig dedicated to the study of the fatigue behavior of a landing gear is reported in [1], while a hydraulic solution to execute the drop-test is reported in [2]. In [3], the authors present the implementation of a high-dynamics force control loop to test the actuation system of a main landing gear within the M-346 iron bird. In [4], the authors performed the experimental validation of a simulation model of a landing gear leg through a comparison with the results obtained with a dedicated experimental setup. In this work, the landing gear leg, complete with braking system, was mounted horizontally and pressed with a pre-defined force set against a rotating cylinder to evaluate the performance of the anti-skid system. Literature on the development of PHM systems for landing gear legs is thus far limited to preliminary simulation studies for the actuation system [5], or to certain components of the electrical brakes [6], while a preliminary comparative analysis of data-driven methods was reported in [7]. The proposed iron bird is set to provide meaningful experimental data by collecting signals difficult to extract from the currently in-service aircraft. To this end, the E-LISA iron bird is able to reproduce fully comprehensive test scenarios and conditions, following the workflow described in [8]. Such scenarios include the dynamic representation of the loads on the landing gear actuator and of aircraft velocity during the entire landing phase as well as taxi and take-off, aircraft inertia and weight, lift and drag forces acting on the aircraft during landing, taxiing, and take-off, and real contact between the landing gear wheel and runway. This last subject is of particular importance to determine the expected aircraft behavior during take-off and landing procedures [9]. In [10], the authors provide an evaluation of the effects that the most common runway defects have on the contact between thee wheel and pavement, while an analysis on the effects of pavement anti-skid properties is provided in [11]. With specific reference to the last topic, an important condition to be explored is the effect of runway irregularities on the landing gear while the aircraft is moving on the runway at different speeds. Such functionalities are not usually present in classic test-bench architecture, which are usually not concerned with the simulation of the aircraft dynamics during the landing procedure. The aim of reproducing such dynamic behavior has a cascade effect on the test bench architecture, fostering the introduction of a software-in-the-loop approach typical of iron birds where the aircraft behavior is simulated in real time to compute the load pattern applied to the landing gear during the test. The runway is reproduced in the iron bird by a rotating cylinder with a large diameter, whose tip velocity is equal to that of the aircraft, and the contact force between simulated runway and landing gear wheel is generated by a hydraulic force control system, with the contact force continuously modulated as a function of aircraft conditions (weight, speed, attitude, etc.). The effect of runway irregularities is reproduced by superimposing fluctuations of the controlled force, which are a function of the irregularities and of the aircraft operating conditions to the progressively varying contact force resulting from the variation in aircraft speed and attitude during landing. A dedicated algorithm computes the instantaneous force command, and a highly responsive force control system generates the required contact force. The paper first details the test bench architecture and its control system, highlighting the innovative solutions employed to meet the program objective. The simulation model prepared to evaluate the system behavior is then presented in detail and used to study the control system stability and its expected performances for different combinations of command frequency and amplitude. The dynamic model was also used to obtain the dynamic requirements that the test-bench must meet to be capable of reproducing the effects of periodical runway irregularities on the landing gear. The expected performances were then compared against these requirements with nominal and increased aircraft weight, finally leading to the preliminary assessment of the test-bench capabilities.

2. Materials and Methods

The E-LISA iron bird is currently under development following a model-based design approach, where a high-fidelity dynamic model of the rig was prepared in MATLAB/Simulink to support the choice of its components, define its control system, and allow the preliminary study of its performances. This paragraph details the test-bench architecture and presents the employed control system, with particular care toward the novel solutions employed to meet the program ambitions. The dynamic model employed to represent the expected behavior of the test rig is then presented and its most significant features discussed.

2.1. Test-Bench Architecture

The iron bird, whose schematic is depicted in Figure 1, is based upon a fixed mechanical structure and a moving platform integral with the landing gear leg, complete with a wheel and electrical brake. The platform slides along low friction vertical guides according to the force provided by an electro-hydraulic servoactuator. A 160 L/min three-stage servovalve is used to control the flowrates according to the commands provided by the test rig control system. A calibrated by-pass orifice connects the two hydraulic lines serving the actuator to improve the dynamic response of the force-controlled system. The sensor suite includes one linear variable differential transformer (LVDT) sensor monitoring the hydraulic actuator travel, and a load cell measuring the force exchanged between the actuator and the moving platform. The hydraulic power necessary for the actuator operation is provided by a dedicated power generation unit operating at 207 bar.

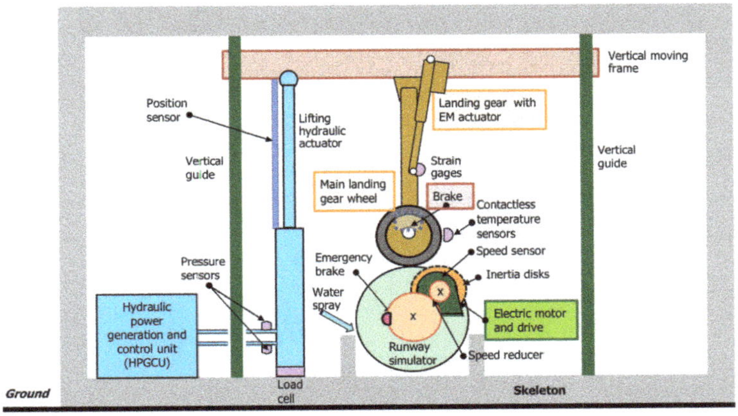

Figure 1. Iron bird schematics.

The contact between the landing gear wheel and the runway is represented through a runway simulator, a rotating disk, connected to a selected number of inertia disks, representative of the aircraft inertia, through a gearbox. The diameter of the rotating cylinder is to be representative of the expected linear speed of the aircraft along the runway during the landing procedure and must be higher than the wheel diameter. The gearbox is interposed to significantly reduce the mass and the encumbrance of the flywheels, the number of which can be increased or decreased to scale-up or -down the weight of the simulated aircraft. The runway simulator was designed with the possibility to change the external coating to allow for a change in the friction coefficient between the wheel and rotating disk, while a sprinkler can be activated to reproduce the wet-runway conditions. An electric motor is used to accelerate the runway simulator up to the angular frequency corresponding to the aircraft horizontal speed given the diameter of the rotating disk.

The iron bird operation is managed by an engineering test station (ETS), which accepts the inputs from a central control unit (CCU) that in turn receives the commands from an operator via a user interface. As depicted in Figure 2, both the input data and the rig signals are sent to a dedicated computer running the real-time (RT) simulation of the aircraft conditions. The control system was designed to combine a position control loop to descend the moving platform until the wheel makes contact with the runway simulator and a force control loop to simulate the effects of the landing procedure on the landing gear leg. The real-time simulation was based upon three modules: the aircraft dynamics module receives as input the linear speed of the runway simulator and computes the expected position of the vehicle along the runway, along with the forces exchanged between the airframe and the landing gears. The runway module is used to generate the contact surface at the interface between the wheel and the runway, allowing for the introduction of waving or pavement irregularities. Finally, real-time landing gear models are used to link the aircraft with the runway and compute the force that must be imposed by the hydraulic load actuator (HLA) to replicate the aircraft behavior foreseen by the aircraft real-time model in the article.

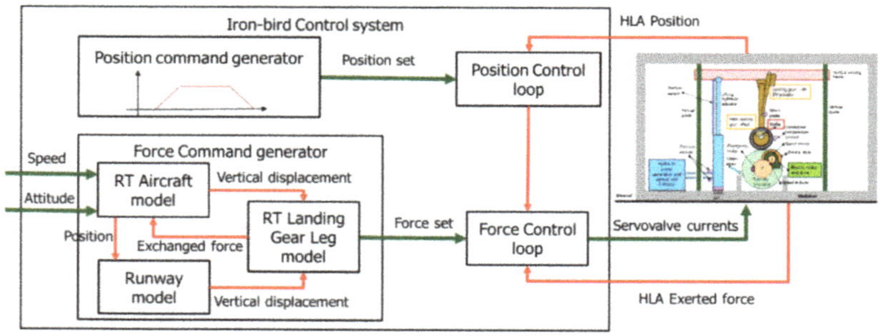

Figure 2. Iron bird control structure.

2.2. Control System

Control of the moving platform is obtained through a combination of a basic control loop, regulating the force exerted by the hydraulic actuator, and a modulating control loop which is used to manage the moving platform approach and retrieval at the beginning and at the end of each test. The general architecture of the two control loops and their interactions are described in Figure 3. The basic control loop implements a closed loop force control that compares the vertical force command with the actual force exerted by the HLA measured by the load cell installed on the test rig. A force error is then obtained and processed by a suitable control law, thus generating a control signal for the three-stage servovalve. Before reaching the servovalve, this signal is limited by a dynamic saturation block, where the upper and lower bounds are defined by the modulating control loop.

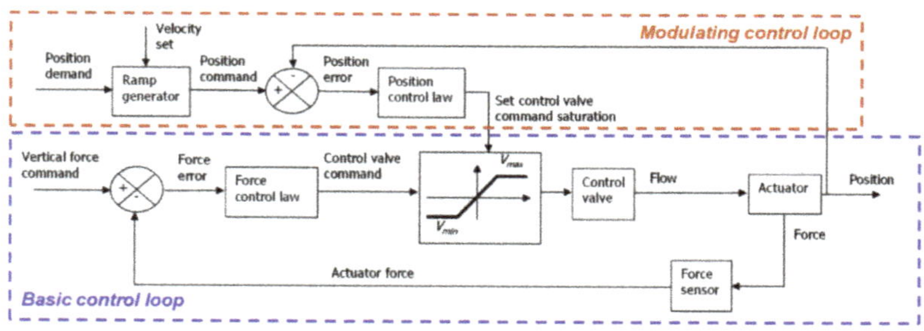

Figure 3. Schematics of the test rig control system.

To better understand the reasoning behind the control scheme, let us consider the test rig starting conditions. At the beginning of each test, the hydraulic actuator is fully extended, and a command is given to bring the landing gear wheel in contact with the runway simulator. In these conditions, three inputs are given to the hydraulic actuator control system. A force command equal to the force required at the initial contact between the landing gear wheel and ground is computed by the aircraft simulator computer. Then, a position demand equal to an actuator position lower than the one corresponding to the contact between the landing gear wheel and runway simulator is obtained; the actuator is hence commanded to move beyond what can be physically attainable. Finally, a velocity set that establishes the actuator lowering velocity is required. As long as there is no contact between the landing gear wheel and runway simulator, the error of the force control loop is large, and a maximum control valve command is thus generated. This would cause a fast acceleration of the hydraulic actuator, which is, however, prevented because the modulating control loop limits the maximum value of the servovalve control current in a way that maintains the actuator velocity at the value established by the ramp generator. This process continues until the landing gear wheel touches the runway simulator. When the landing gear wheel makes contact with the runway simulator, the actuator movement stops, and the position error increases. The control valve saturation limit consequently reaches the maximum and the force developed by the actuator increases to the value established by the force command. For the remainder of the test, the control valve saturation limit remained at the maximum and the basic control loop performed the continuous force control according to the force commands received from the aircraft simulation computer. An opposite process takes place at the end of the test to raise the landing gear assembly.

The control law for the modulating control loop is shown in Figure 4. Since this loop is active only during the ascent/descent portion of the test, there are no strict requirements on its dynamic performance, and a simple first-order transfer function with saturation is used.

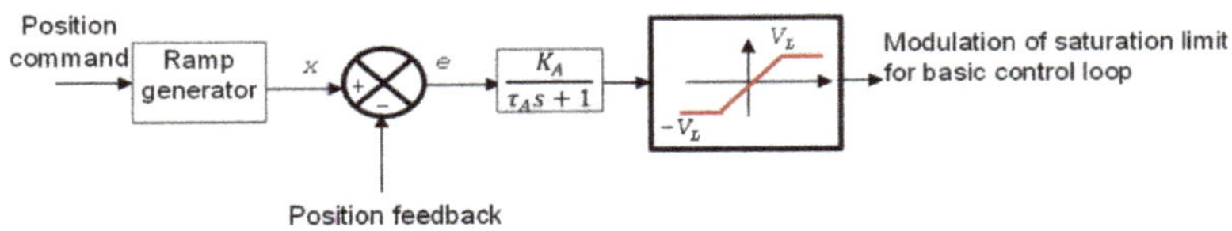

Figure 4. Control law of the modulating control loop.

The control law for the basic control loop is instead depicted in Figure 5. The employed architecture implements a force control system similar to those provided in other demanding applications such as iron birds for flight control systems [12,13] and test rigs for actuators under PHM experimentation [14]. The control law is based upon a modified PID controller, where the integral contribute K_I/s is computed on the error between the force set F_c and force feedback F_z, while both the proportional (K_p) and the derivative ($K_D s$) signals act directly on the feedback, while a feedforward (K_F) of the force command is added to improve the system dynamics. As described in Equation (1), a feedforward signal of the time derivative of the force command, saturated between zero and a maximum value, is then used in conjunction with the fixed gain K_0 to increase the system response to quick variations of the set by increasing the voltage V_x applied to the servovalve coils.

$$\overline{V_x} = \left[\left(\overline{F_c} - \overline{F_z}\right) \frac{K_I}{s} + K_F \overline{F_c} - \left(K_P + K_D s\right) \overline{F_z} \right] \left(K_x s \left|\overline{F_c}\right| + K_0\right) \tag{1}$$

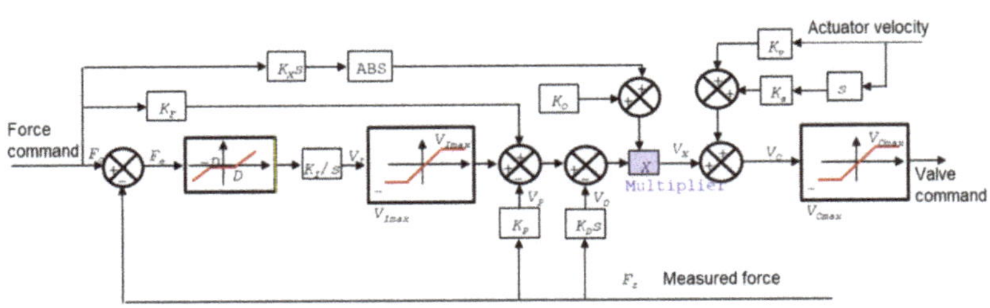

Figure 5. Control law of the force control loop.

For this specific application, the actuator velocity and acceleration compensations may not be necessary and are provisionally not proposed, although present in the scheme; as the project development continues, more refined data on the parameters of the test rig and of the equipment under test will become available.

2.3. Force Command Generator

The force command generator links the test rig with the control system, allowing the definition of the hardware-in-the-loop architecture of the test rig. As stated in Section 2.1, the force command generator is comprised of three modules: a simplified aerodynamic model of the aircraft; the dynamic representation of the landing gears; and a runway model The aircraft dynamics can be described according to the two-dimensional representation depicted in Figure 6. The model is prepared to solve the vertical dynamics of the system, along with the rotation around the pitch axis as:

$$\begin{cases} -\sum_i F_{MLG,i} - F_{NLG} - W + L\,cos\vartheta_y + L_t cos\vartheta_y + Tsin\vartheta_y - Dsin\vartheta_y = m_{air}\ddot{z} \\ -W\,Y_G\,cos\vartheta_y - W\,X_G\,sin\vartheta_y - T\,H_t + D\,H_D - L\,H_L - L_T\,H_{Lt} - F_{NLG}H_{NLG} = I_{y,air}\ddot{\vartheta}_y \end{cases} \tag{2}$$

where $F_{MLG,i}$ is the force exerted by each leg of the main landing gear (MLG) and the analogous F_{NLG} of the nose landing gear (NLG). L and L_t are the lift components due to the wing and the tail; D is the drag force; and T the engine thrust. The aircraft mass is m_{air}, while I_y is its moment of inertia. The parameters $H_{T,L,Lt}$ are the distances of the application points of each considered force from the MLG leg. L_{air} is the aircraft pitch, while (X_G, Y_G) are the center of mass coordinates with respect to the MLG. The model assumes symmetric loading conditions on the two legs of the main landing gear and a null roll angle. Being a two-dimensional model, the NLG steering is not represented. The aerodynamic actions are computed through the aerodynamic coefficients of lift (C_L,C_{Lt}) and drag (C_D), the data for which were provided by the industrial partners of the project as a function of the aircraft attitude:

$$\begin{cases} L = C_L(t)A_{wing}v^2 \\ L_T = C_{Lt}(t)A_{tail}v^2 \\ D = C_D(t)A_{drag}(t)v^2 \end{cases} \tag{3}$$

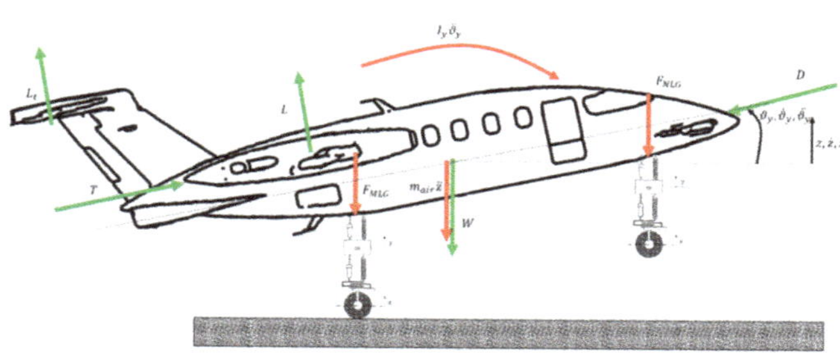

Figure 6. Two-dimensional model of the aircraft dynamics.

Each force is proportional to the area of the aerodynamic surface of interest, function of the aircraft attitude, and to the square of the horizontal speed of the aircraft along the runway v. On the test rig, this signal is derived from the angular frequency of the runway simulator and used as input to the aerodynamic mode. The presence of surface irregularities on the runway is modeled as a vertical movement of the pavement function of the position of the simulated aircraft along the runway. At first, only periodical runway irregularities were considered and thus have been introduced according to the sinusoidal law hereby provided, where L_0 is the irregularities pitch and X_0 their amplitude.

$$x_{rw} = X_0 \sin\left(2\,\pi \frac{v}{L_0}\,t\right) \tag{4}$$

2.4. Test Rig Simulation Model

To simulate the test bench behavior under a variety of operating conditions, a non-linear mathematical model representative of the test-bench dynamics was prepared.

The test bench was first divided into a selected number of functional subsystems. The test bench software comprised the real-time model used to simulate the aircraft landing, which was responsible for the definition of the command issued to the test bench and for the control of the moving platform, and the control system, which implemented the position and force control law. On the hardware side, it is possible to identify the electro-hydraulic servo actuator, the moving platform and test article assembly, the runway simulator, and the braking system of the landing gear. The model of both the command generator and the control system followed the equations and block schemes provided in Section 2.2. A simple transport delay model was added to represent the effects of the computational time over the test bench dynamics. The model of the electro-hydraulic actuator was prepared in accordance with the high-fidelity representation provided in [15], where the dynamics of the servovalve, the cylinder, and the supports are represented without resorting to black box modeling. The internal parameters of the servovalve model were tuned to mimic the expected characteristics of the considered model as provided by the manufacturer's catalogue. The displacement of the rod x_a was obtained as a function of the pressure drop across the chambers $(p_1 - p_2)$, itself dependent on the balance between the flowrates Q_1 and Q_2 exchanged between the actuator and the servovalve.

$$\begin{cases} Q_1 - A_c\dot{x}_a - Q_{l,i} - Q_{bp} = \frac{\beta}{V_0 + A_{c,1}x_a}\frac{dp_1}{dt} \\ Q_2 + A_c\dot{x} + Q_{l,i} + Q_{bp} = \frac{\beta}{V_0 - A_{c,2}x_a}\frac{dp_2}{dt} \\ p_1A_{c,1} - p_2A_{c,2} - F_{fr} - \gamma\dot{x}_a - k_{mp}(x_a - z) - c_{mp}(\dot{x}_a - \dot{z}) - m_a g = m_a\ddot{x}_a \\ Q_{bp} = sign(p_1 - p_2)C_d(Re)A_{bp}\sqrt{\frac{2(p_1 - p_2)}{\rho}} \end{cases} \tag{5}$$

where $Q_{l,i}$ is the flow rate lost due to internal leakage. Q_{bp} is instead the flow through the by-pass orifice, computed as a function of the pressure drop across the actuator chambers, the area of metering section A_{bp} and the discharge coefficient, dependent on the local Reynolds number, C_d. The trust areas of the two chambers are assumed equal to $A_{c,1}$ and $A_{c,2}$, respectively, while the bulk modulus and the chamber volume at mid stroke are addressed by β and V_0. The bulk modulus is computed as a function of the fluid temperature and air fraction, assumed to be respectively equal to 40 °C and 0.5% in the preliminary analysis. Friction forces due to contact between the sealing elements and the cylinder F_{fr} can be computed according to [4], while γ is the viscous friction coefficient. Finally, we address with z the vertical displacement of the moving platform and with k_{mp}, c_{mp} the stiffness and the damping coefficient of the attachments, respectively. Considering the two attachments equal, the dynamic equilibrium of the cylinder can be obtained as

$$p_2A_{c,2} - p_1A_{c,1} - F_{fr} - \gamma\dot{x}_c - k_{mp}(x_c) - c_{mp}(\dot{x}_c) - m_c g = m\ddot{x}_c \tag{6}$$

where x_c is the cylinder movement. The attachments on both sides of the actuator were modeled as viscoelastic components to account for their deformation. The vertical dynamics of the assembly including the moving platform and the test article were modeled according to the scheme provided in Figure 7. In the absence of a detailed mass breakdown, the moving platform is connected to the mass m, representative of the single landing gear (LG) leg, by a non-linear spring-damper element representative of the shock-absorber. On the other side, the eventual viscoelastic contact between the LG tires and the runway simulator is represented. The dynamic equilibrium of the moving platform can then be represented considering the following equation, where y_{leg} is the vertical displacement of the leg, while k_{leg} and c_{leg} are the shock absorber stiffness and damping. F^*_{leg} is the preload of the shock absorber.

$$k_{mp}(x_a - z) + c_{mp}(\dot{x}_a - \dot{z}) + k_{mp}\left(y_{leg} - z\right) + c_{mp}\left(\dot{y}_{leg} - \dot{z}\right) - m_{mp}g = m_{mp}\ddot{z} \tag{7}$$

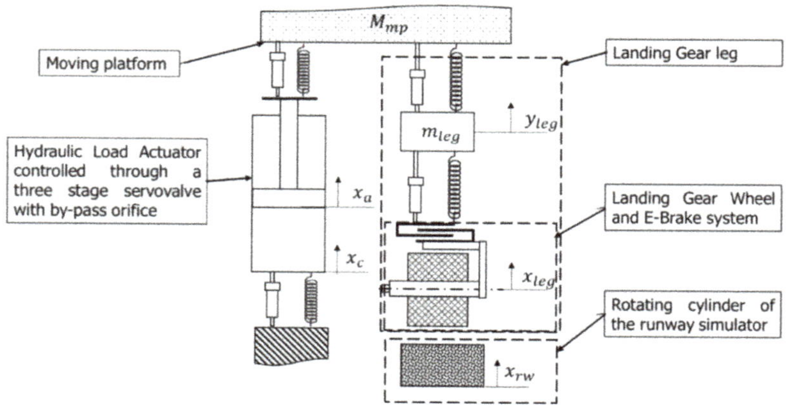

Figure 7. Schematics of the test rig assembly model.

The shock absorber characteristics are known and provided by the landing gear supplier. Similarly, the vertical dynamics of the leg can be represented as follows:

$$\begin{cases} k_{leg}\left(x_{leg} - y_{leg}\right) + c_t\left(\dot{x}_{leg} - \dot{y}_{leg}\right) - k_{mp}\left(y_{leg} - z\right) + F^*_{leg} + c_{mp}\left(\dot{y}_{leg} - \dot{z}\right) = m_{leg}\ddot{y}_{leg} \\ k_t\left(x_{leg} - x_{rw}\right) + c_t\left(\dot{x}_{leg} - \dot{x}_{rw}\right) - k_{leg}\left(x_{leg} - z\right) - F^*_{leg} - c\left(\dot{x}_{leg} - \dot{z}\right) = m_w\ddot{x}_{leg} \end{cases} \tag{8}$$

where k_t and c_t are the stiffness and damping coefficient representative of the tire-runway simulator contact, respectively, while x_{leg} and x_{rw} are respectively the vertical displacement of the wheel and that of the runway simulator, by default equal to zero. The wheel mass is m_w. Please note that the force exerted by the contact between the tire and the rotating cylinder can only apply a compressive load on the leg. The characteristics of the tire are derived from data provided by the industrial partners. The rotational dynamics of the wheel can be described according to the free-body diagram of Figure 8. With $F_n = k_t\left(x_{leg} - x_{rw}\right) + c_t\left(\dot{x}_{leg} - \dot{x}_{rw}\right)$, the vertical force exchanged between the wheel and the runway simulator, with $F_t = F_n\mu$ the friction force and with u_{rw} the rolling friction parameter, function of the wheel angular frequency, and of the tire pressure [16], we can obtain:

$$F_n\mu\left[\frac{D_w}{2} - \left(x_{leg} - x_{rw}\right)\right]\mathrm{sign}(\lambda) - F_n u \tanh\dot{\vartheta}_w - c_w\dot{\vartheta}_w - T_{brk} = I_w\ddot{\vartheta}_w \tag{9}$$

where ϑ_w is the wheel rotation in [rad]; I_w is the moment of inertia of the wheel assembly; D_w its diameter; and c_w is the viscous friction coefficient roughly representative of the dissipation in the wheel supports. The friction coefficient μ is evaluated according to the Burckhardt model [17] as a function of the slip factor between the wheel and runway simulator.

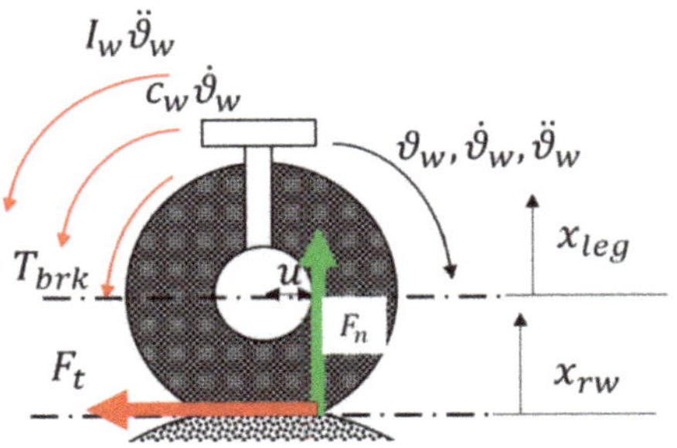

Figure 8. Free-body diagram of the wheel–runway simulator interaction.

The runway simulator is modeled as a cylinder, where the moment of inertia I_{rw} is representative of the mass of the decelerating aircraft. Although the chosen configuration is different, the effects of the adoption of a geared solution over the system behavior are negligible since they only interest the acceleration phase of the cylinder and the eventual activation of the emergency brake. Both the process of accelerating the cylinder up to the required speed and that of decelerating the runway simulator down until a complete stop due to the activation of the emergency brake are expected to be fairly long processes and thus not suitable for simulation through a complex model of the test bench. As such, by defining the diameter of the runway simulator with D_{rw}, it is possible to express its rotational dynamics as follows:

$$-F_n\mu\frac{D_{rw}}{2}\text{sign}(\lambda) - F_n u \tanh\dot{\vartheta}_{rw} - c_{rw}\dot{\vartheta}_{rw} = I_{rw}\ddot{\vartheta}_{rw} \tag{10}$$

Within the simulation model, the runway simulator is brought to the required speed by means of a simplified speed control loop, which is then disconnected as a non-zero position command is issued to the moving platform.

2.5. Test-Article Simulation Model

The E-Brake system is made of four electro-mechanical actuators (EMAs) controlled in force and acting in parallel on a disk brake. The functional scheme of the system is reported in Figure 9. Each actuator is driven by a brushless-DC motor connected to the braking pad though a gearbox and a ballscrew each. Each actuator was equipped with a dedicated load cell used to measure the exerted force and close the force control loop. The model of the actuators control system was similar to that of the testbench, where a sequence of proportional-integrative controllers operated on the force control loop and the current control loop of each brushless motor. The sensor suit was modeled through second order transfer functions replicating the expected dynamics of the load cell and of the Hall-effect sensors employed to monitor the angular position of the brushless-DC rotor. The simulation of the measure chain is completed with the model of the employed A/D converters. The dynamic model of each EMA is derived from the simulation models previously employed for prognostic activities on flight control systems [18]. The electronic power converter was modeled as a three-phase inverter with pulse width modulation (PWM) following the functional modeling provided in [19]. An initial value of 10 kHz was chosen for the carrier frequency of the PWM modulator. The electrical dynamics of the motor can be described according to a streamlined three-phase model of the system.

$$[\mathrm{V_{a,b,c}}] = [\mathbf{R_{a,b,c}}(T_w)][i_{a,b,c}] + [\mathbf{L}(T_w)]\frac{d}{dt}[i_{a,b,c}] + \frac{d}{dt}[\phi_{a,b,c}(\vartheta_{el})] \tag{11}$$

where $[\mathbf{R_{a,b,c}}]$ is the electric resistance matrix, the elements of which depend on the windings' temperature (T_w). $[\mathbf{L}]$ is the inductance matrix, accounting for self-induction and mutual induction phenomena, along with the effect of magnetic flux dispersion. Finally $[\phi_{a,b,c}]$ is the concatenated magnetic flux provided by the permanent magnets, function of the electrical angle (ϑ_{el}). The torque at the motor shaft, and thus the dynamic equilibrium of the rotor, can be computed as:

$$\sum_{a,b,c} \frac{d\phi}{dt} i_{a,b,c} - c\dot{\vartheta}_m - k_m\left(\vartheta_m - \vartheta_{gb}\right) - c_m\left(\dot{\vartheta}_m - \dot{\vartheta}_{gb}\right) = I_m\ddot{\vartheta}_m \tag{12}$$

where ϑ_m is the motor rotation, while ϑ_{gb} is the angular position of the gear shaft and I_m is the moment of inertia of the rotor. The parameters k_m and c_m represent the torsional stiffness of the motor shaft and its associated damping. The gear pair is described within the test bench model as a rotational mass–spring–damper system, thus leading to the following equation

$$\begin{aligned} k_m\left(\vartheta_m - \vartheta_{gb}\right) + c_m\left(\dot{\vartheta}_m - \dot{\vartheta}_{gb}\right) - \tfrac{1}{\tau}\left[k_{gb}\left(\vartheta_{gb} - \vartheta_{rs}\right) + c_{gb}\left(\dot{\vartheta}_{gb} - \dot{\vartheta}_{rs}\right)\right] - T_{fr,gb} \\ = I_{gb}\ddot{\vartheta}_{gb} \end{aligned} \tag{13}$$

where τ is the transmission ratio; $T_{fr,gb}$ is the friction torque, while ϑ_{rs} is the angular position of the rotating part of the screw. The friction torque is computed as the sum of three components: one dependent on the acting load, one related to the viscous friction, and a drag torque component. The power-screw is modeled as two-degrees of freedom elements, where the rotating part is connected to the translating element through a viscoelastic element. Defining the position of the translating portion of the screw pertaining to the *i-th* actuator with $x_{rs,i}$, it becomes possible to describe the brake dynamics. The force provided by each actuator $F_{n,i}$ to the brake disks is modeled as a viscoelastic backlash, where no force is exerted until contact is made. Addressing the stiffness and the preload of the return spring with k_{eb} and F^*, respectively, it is possible to evaluate the braking torque acting on the landing gear wheel as a function of the translating mass of the brake disks m_{eb}, its translation x_{eb}, and the angular speed of the wheel $\dot{\vartheta}_w$ as:

$$\begin{cases} T_{brk} = f_{eb}F_n \leftrightarrow \dot{\vartheta}_w \neq 0 \\ T_{brk} = \min\left(f_{eb}^*F_n, F_n\mu\left[\frac{D_w}{2} - \left(x_{leg} - x_{rw}\right)\right]\text{sign}(\lambda) - F_n u \tanh\dot{\vartheta}_w\right) \end{cases} \tag{14}$$

where f_{eb} and f_{eb}^* are the friction coefficients under sliding and adhering conditions. λ is the slip parameter, computed as the ratio between the horizontal speed of the aircraft along the runway (the linear speed of the runway simulator in the test rig) and linear velocity of the landing gear tires. The brake system features an anti-skid logic that modulates the braking force command depending on the required slip factor, chosen as a function of the runway conditions.

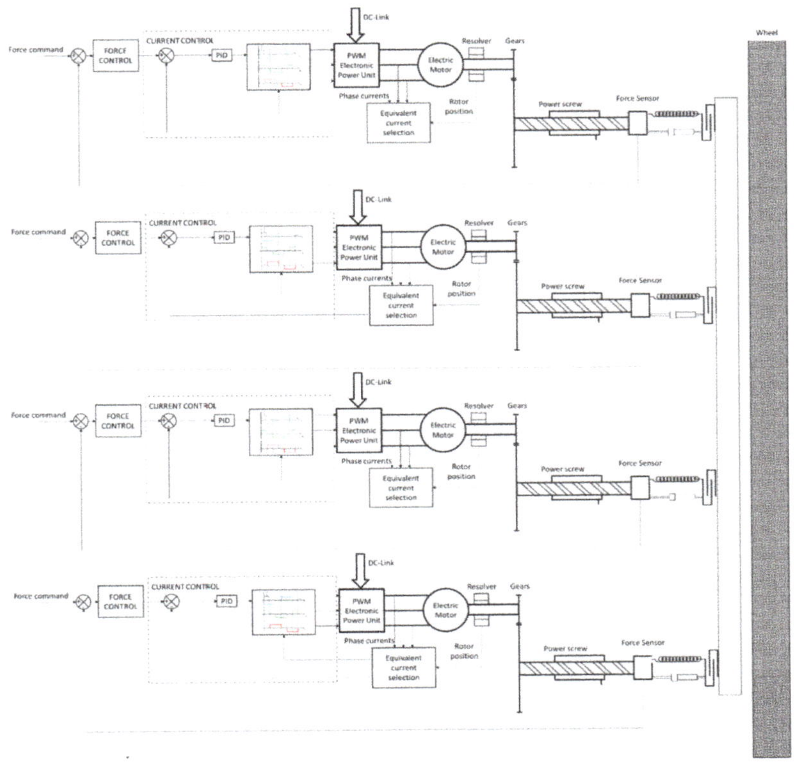

Figure 9. Schematics of the E-Brake system.

3. Results

The non-linear model of the test rig was implemented in MATLAB/Simulink and used to evaluate the performances of the control system and to study its capability to replicate the effects of the presence of different kinds of runway irregularities under different operating conditions.

3.1. Control Stability and Performances

The only purpose of the position control loop is to manage the moving platform at the start and at the end of each test, conceding priority to the force control loop during the landing test procedure. During these phases, the system is commanded through slow position ramps, with the sole purpose of limiting the descent speed of the moving platform, thus removing the need for strict requirements over its dynamic performance. The force control loop is critical to the system behavior and is thus studied in detail. The selected gains of the force control loop provide sufficient stability margins. As depicted in Figure 10, the gain margin G_m can be estimated as 25.6 dB, while the phase margin Φ_m is equal to 82.7°, which are well above the usual requirements of 6 dB and 60°, respectively.

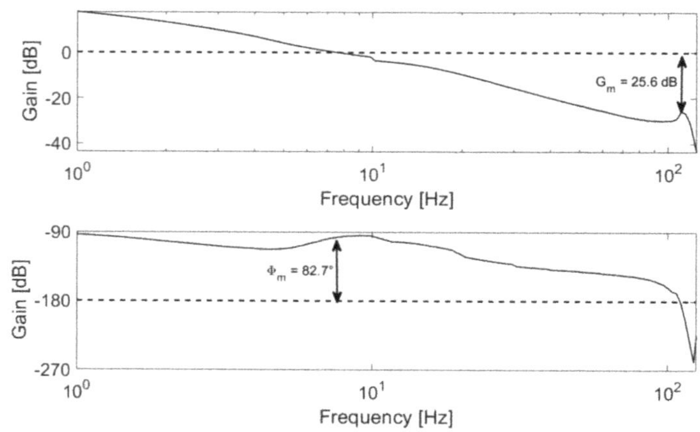

Figure 10. Open loop frequency response of the force control system.

The closed loop response of the force control system is depicted in Figure 11 as a function of the command amplitude. The bandwidth, defined as the frequency associated with a closed loop gain equal to −3 dB, is heavily dependent on the amplitude of the force command imposed on the system, reaching 8.9 Hz for a ±5 kN sinusoidal set while falling below 1 Hz for command amplitudes higher than 40 kN. The system exhibits the tendency to slightly overshoot within frequency ranges dependent on the command itself. However, these peaks are deemed acceptable since they are modest in value, lower than 1 dB, and mostly affect the low-amplitude commands, tending to disappear as the command amplitude rises.

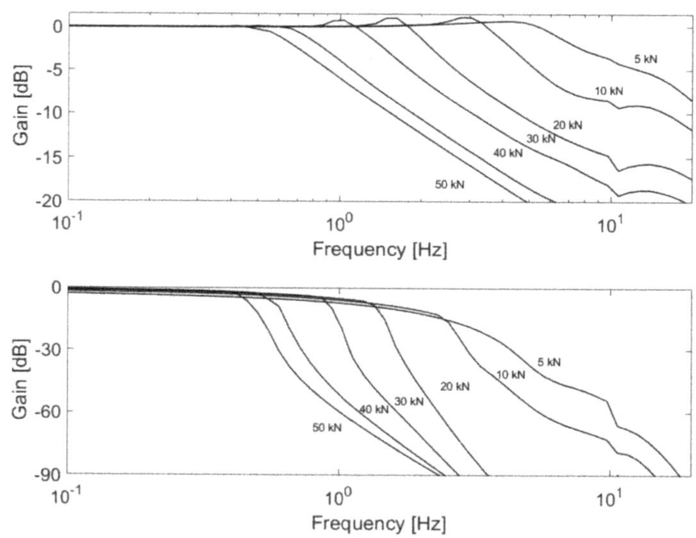

Figure 11. Close loop frequency response for the force control system—mean value of 0 kN.

To better highlight the test bench capabilities, the maximum amplitude of the force exerted by the hydraulic actuator was compared with the amplitude of the sinusoidal load acting on the landing gear leg due to runway irregularities with varying frequency of occurrence and different amplitudes X0 in Figure 12. Since the frequency of the dynamic load due to the sinusoidal runway irregularities depends on the aircraft speed along the runway, the test bench must be able to provide a force at least equal to the required one for the entire frequency range representative of the aircraft operating conditions. For example, to reproduce the landing of an aircraft approaching the runway with an expected horizontal speed of 60 m/s in the presence of periodical irregularities with 10 m pitch and given amplitude, the test rig must be able to generate, on the landing gear, a sinusoidal force at least equal to the required one for the whole frequency range up to 6 Hz. The most critical conditions occur in correspondence of the first resonant frequency of the landing gear leg, which occurs in the proximity of 2.25 Hz. As depicted in Figure 12, the test bench is expected to reproduce a maximum amplitude of periodical runway irregularities equal or lower than 6 mm. Higher amplitude irregularities can still be reproduced, provided that the ratio between the aircraft speed and the waving pitch causes force oscillations at lower frequencies. For example, irregularities with 20 mm amplitudes can be successfully replicated if their frequency drops below 1.67 Hz, which corresponds to a waving pitch of 35.9 m for an aircraft approaching the runway with a horizontal speed of 60 m/s. Figure 13 details the operating envelope of the test rig, depicting for several values of runway waving amplitude X_0 and the minimum pitch L that allows its representation on the iron bird.

Such an envelope means that the test bench is capable of describing runways with an International Roughness Index (IRI) higher than 2 mm/m, which is usually considered the threshold value for pavements in good conditions. Moreover, the test bench is also able to reproduce, in most instances, the effects of runways with IRI higher than 3.5 mm/m, which is conventionally assumed as the threshold value for failed pavements [20].

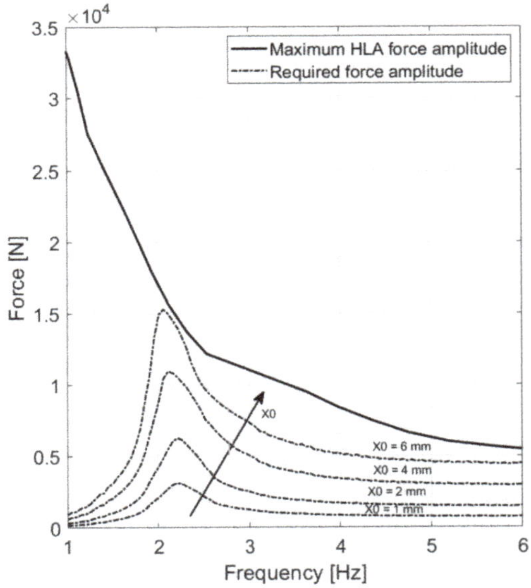

Figure 12. Verification of the test bench capability to reproduce periodical runway irregularities.

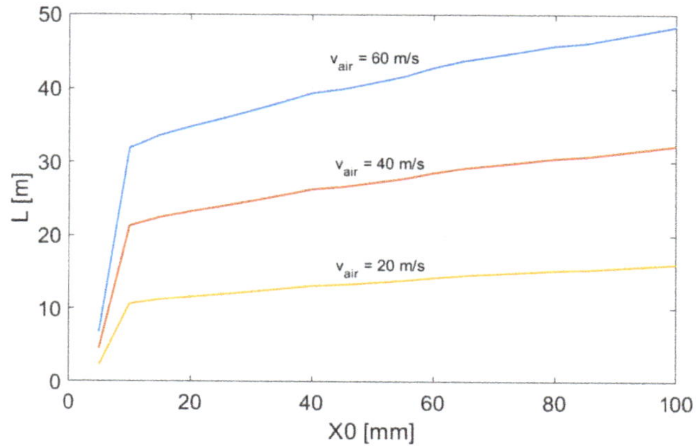

Figure 13. Test bench capability to reproduce periodical runway irregularities: minimum pitch (L) vs waving amplitude (X0).

3.2. Landing Test Simulation

To best demonstrate the expected test bench behavior and to study its projected performances in the time domain, several simulations mirroring the test procedures defined for the physical test rig were performed, starting with the representation of landing conditions on a dry and perfectly plane runway. The simulation considered an aircraft of total mass at landing around 6 tons, approaching the runway with a relative horizontal speed of 55 m/s, which are the most probable landing conditions for the considered vehicle.

As depicted in Figure 14, the test begins by supplying a positive force command to the force control loop and a set signal equal to zero to the position control loop, both of which are needed to balance the combined weight of the landing gear leg, the hydraulic actuator rod, and the moving platform itself. At the same time, the runway simulator is accelerated through a dedicated electric motor, until its tip speed reaches the value coherent with the desired horizontal speed at landing. Please be mindful that this particular phase is expected to require a few minutes due to the high resistance opposed by the inertia disks and was artificially accelerated in the simulations. Once the descent command is provided, the real-time aircraft model is activated and begins to provide the force set as defined in Section 2, causing the slow descent of the moving platform. After contact is made between the wheel and the runway simulator, the position set is quickly brought to a value to negate the effects of the position control loop over the force regulation, thus allowing the system test rig to apply the load calculated by the real-time aircraft model. In response to a dedicated command provided by the iron bird control unit, the electromechanical brakes are engaged and start decelerating the runway simulator. The dynamics of this process is depicted in Figure 15, where the behavior of the quadrature currents $(i_{q1}, i_{q2}, i_{q3}, i_{q4})$ of the four electric motors acting on the E-Brake system can be observed. The delay between the braking command and the effective brake engagement is due to the dynamics of the electro-mechanical actuators and the need to cover the initial gap separating the brake's pads from the brake's disks. The resultant braking force is then modulated by the anti-skid system to avoid the wheel blockage until the runway simulator is brought to a standstill.

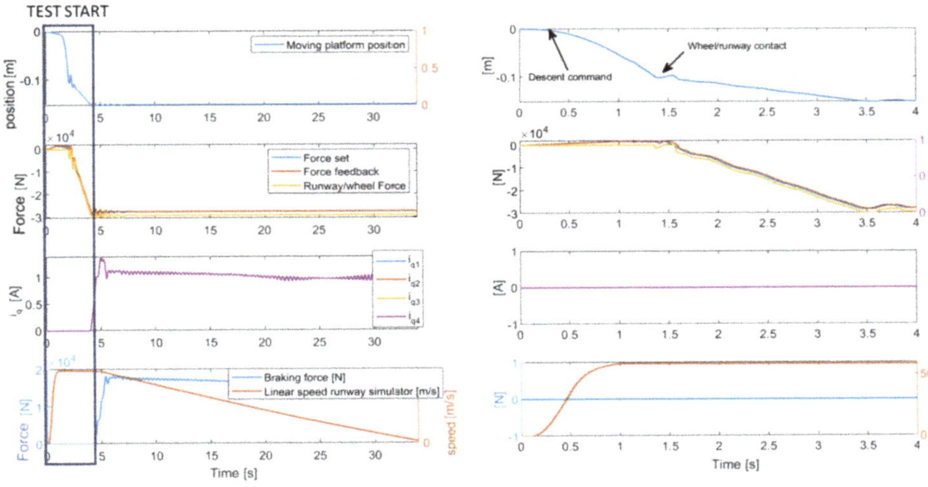

Figure 14. Landing test simulation: moving platform descent and wheel/runway simulator touchdown.

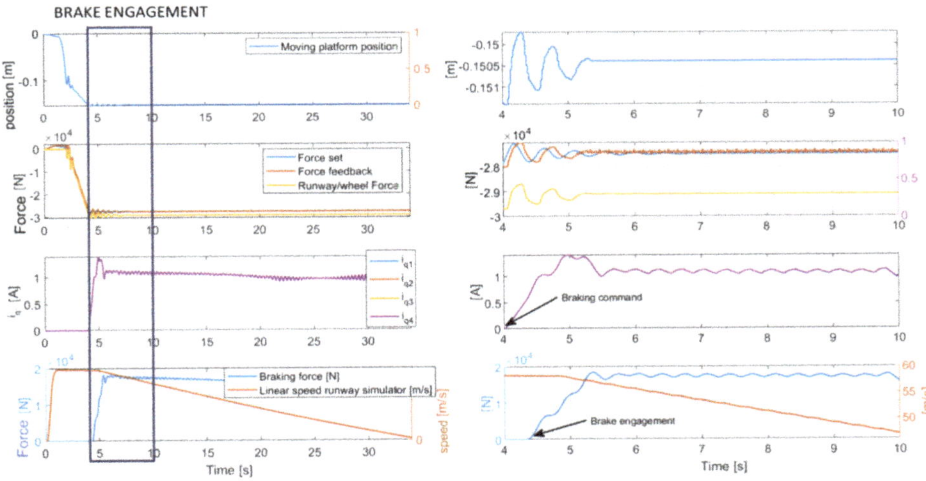

Figure 15. Landing test simulation: brake engagement and aircraft deceleration.

3.3. Simulation of a Landing Test in Presence of Runway Irregularities

To address the test rig capability of representing the effects of periodical runway irregularities on the landing gear system within the time domain, the same landing test simulations performed for ideal runway conditions were repeated in the presence of sinusoidal runway irregularities. The same initial conditions such as the aircraft mass at landing and the horizontal speed at touchdown were applied. Results from a simulation performed considering periodical irregularities of amplitude equal to ±5 mm and 10 m pitch are presented in Figures 16 and 17. In these simulations, the aircraft approached the runway with a 60 m/s horizontal speed, and the simulation runway irregularities were added at a pre-defined time instant to highlight their effects on the test rig behavior. As shown in Figure 16, the force exerted by the HLA closely followed the force command provided by the real-time model of the aircraft from the onset of the runway irregularities onward. As the test progresses, the braking action decelerates the runway simulator, where tip speed is continuously monitored by the real-time model of the aircraft. As the simulated aircraft slows down, the frequency of the dynamic force representative of the effects of the runway irregularities decreases, falling in the proximity of the first resonant frequency of the landing gear leg; for the considered simulation, this situation is expected to occur when the tip speed of the runway simulator approaches the 22 m/s mark. Depicted in Figure 17 the test rig behavior confirms the expectations, being able to replicate the force command with minimal delay and no overshoots.

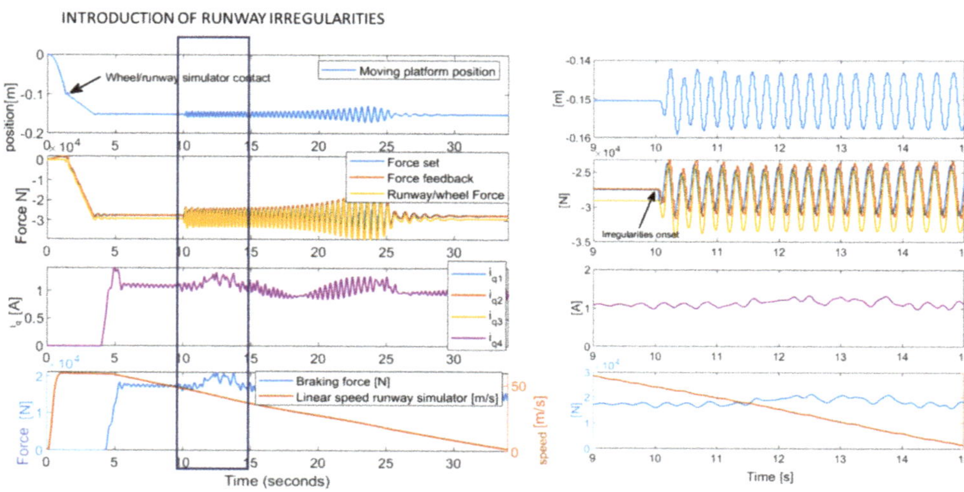

Figure 16. Landing test simulation: introduction of periodical runway irregularities (amplitude 5 mm, pitch 10 m).

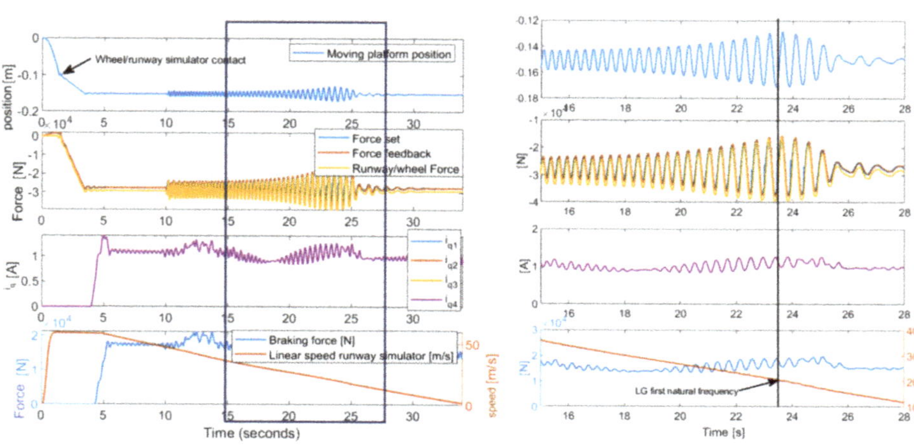

Figure 17. Landing test simulation: dynamic load close to the first LG resonance due to periodical runway irregularities (amplitude 5 mm, pitch 10 m).

3.4. Scalability

Although the test bench was designed for a well-determined platform, it is desirable to verify its scalability to different aircraft configurations or operating conditions. This objective serves multiple purposes. For once, the aircraft mass at landing is significantly affected by the amount of remaining fuel, embarked passengers, or unused military equipment, leading to more representative test campaigns, thus more effectively supporting the definition of prognostic routines for the employed components.

Moreover, the development of a scalable test rig can reduce the cost of testing and certification of future platforms. Finally, although not foreseen within the boundaries of this research program, a scalable test-bench could represent the aircraft behavior and thus the landing gear and E-Brake system operations during challenging conditions such as crosswind landing. Additional inertia disks required to represent the increased weight of the aircraft can be easily mounted on the runway simulator, provided that the mechanical components of the structure are properly sized. The hydraulic system is, on the other hand, capable of sustaining a modest increase in the aircraft mass, and thus the load, without significant loss in performance or in functionality. For the executive transport aircraft under consideration, an additional mass up to 15% of the previously considered value is considered compatible with an emergency landing executed close to take-off. As depicted in Figure 18, the test rig is able to reproduce the effects of runway irregularities, even in the presence of this additional mass.

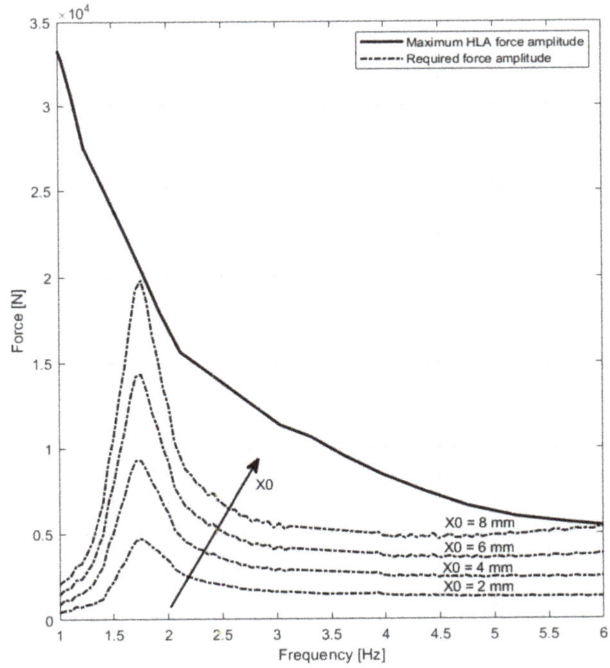

Figure 18. Verification of the test bench capability to reproduce periodical runway irregularities with increased aircraft weight (15% increase).

4. Discussion and Conclusions

The E-LISA iron bird has been designed with the double purpose of fulfilling the certification procedures for a new, fully electrical landing gear system and act as a technological demonstrator for prognostic techniques developed for a few key components.

As such, the need to closely reproduce a wide array of possible operating conditions including the presence of runway irregularities was assessed as a necessary feature. These requirements had a direct effect on the test bench architecture and on the definition of its control system, which involves a real-time simulation model of the aircraft to continuously compute the force command used by the test rig to reproduce a realistic loading pattern on the landing gear under analysis. The definition of a novel control scheme, based on the combination of a simple position control loop with an advanced force control loop was also required. To support the design of the iron bird and contribute to the definition of its control laws, a high-fidelity model of the system was prepared according to well established equations. Simulation results showed that the iron bird control system was stable, and that its expected performances allowed for the representation of the effects of runway irregularities on the test-article, although with some limitation. The test bench, in particular, is expected to be able to successfully replicate the effects of periodical runway irregularities with an amplitude equal or lower to 6 mm independently from the aircraft horizontal speed during landing and taxing. Moreover, the system has proven a certain degree of scalability; simulation results suggests that the system can successfully replicate the loading conditions on the landing gear leg considering a 30% increase in aircraft weight, even in the presence of runway irregularities. Further work is required to properly characterize the test rig capability to simulate localized runway defects of different severity and irregular geometry such as cracks and lobes. Once the test bench is built, further work will be directed at the proper identification of the dynamic parameters of the simulation model, the validation of the proposed control scheme, and the experimental verification of the iron bird performance.

Author Contributions: Conceptualization, A.D.M., G.J., and M.S.; methodology, A.D.M., G.J., and M.S.; software, A.D.M.; validation, A.D.M.; formal analysis, A.D.M. and G.J.; investigation, A.D.M. and G.J.; resources, A.D.M., G.J., and M.S.; data curation, A.D.M.; writing—original draft preparation, A.D.M.; writing—review and editing, G.J. and M.S.; visualization, A.D.M.; supervision, M.S.; project administration, G.J. and M.S.; funding acquisition, G.J. and M.S. All authors have read and agreed to the published version of the manuscript.

Funding: The research work presented in this paper was performed within the E-LISA project, which has received funding from the Clean Sky 2 Joint Undertaking under the European Union's Horizon 2020 Research and Innovation Program under grant agreement number 887222.

Conflicts of Interest: The authors declare no conflict of interest. The funders had no role in the design of the study; in the collection, analyses, or interpretation of data; in the writing of the manuscript, or in the decision to publish the results.

References

1. Schmidt, R.K.; Eng, P. An Integrated Modular Test Rig for Landing Gear Fatigue and Strength Testing. In Proceedings of the ICAS2002 CONGRESS, Toronto, ON, Canada, 8–13 September 2002; pp. 1–6.
2. Han, J.; Lee, Y.-S.; Ahn, O.-S. Design & Fabrication for Small Aircraft Landing Gear Drop Test Rig System. *J. Korean Soc. Aeronaut. Sp. Sci.* **2008**, *36*, 1121–1125. [CrossRef]
3. Jacazio, G.; Balossini, G. A Mechatronic Active Force Control System. In Proceedings of the ASME 2009 International Design Engineering Technical Conferences & Computers and Information in Engineering Conference, San Diego, CA, USA, 30 August 30–2 September 2009; pp. 1–8.
4. Dravico, L.; Tanelli, M.; Savaresia, S.M. Experimental validation of landing-gear dynamics for anti-skid control design. In Proceedings of the 2018 European Control Conference ECC, Limassol, Cyprus, 12–15 June 2018; pp. 2751–2756. [CrossRef]
5. Byington, C.S.; Watson, M.; Edwards, D. Data-driven neural network methodology to remaining life predictions for aircraft actuator components. In Proceedings of the IEEE Aerospace Conference Proceedings, Big Sky, MT, USA, 6–13 March 2004.
6. Ramesh, G.; Garza, P.; Perinpanayagam, S. Digital simulation and identification of faults with neural network reasoners in brushed actuators employed in an e-brake system. *Appl. Sci.* **2021**, *11*, 9171. [CrossRef]
7. Oikonomou, A.; Eleftheroglou, N.; Freeman, F.; Loutas, T.; Zarouchas, D. Remaining Useful Life Prognosis of Aircraft Brakes. *Int. J. Progn. Health Manag.* **2022**, *13*, 1–11.
8. Autin, S.; De Martin, A.; Jacazio, G.; Socheleau, J.; Vachtsevanos, G.J. Results of a feasibility study of a Prognostic System for Electro-Hydraulic Flight Control Actuators. *Int. J. Progn. Health Manag.* **2021**, *12*, 1–18. [CrossRef]

9. Wesołowski, M.; Blacha, K. Evaluation of airfield pavement micro and macrotexture in the light of skid resistance (friction coefficient) measurements. *MATEC Web Conf.* **2019**, *262*, 05017. [CrossRef]
10. Wesołowski, M.; Blacha, K.; Pietruszewski, P.; Iwanowski, P. Analysis of the Actual Contact Surface of Selected Aircraft Tires with the Airport Pavement as a Function of Pressure and Vertical Load. *Coatings* **2020**, *10*, 591. [CrossRef]
11. Zieja, M.; Wesołowski, M.; Blacha, K.; Iwanowski, P. Analysis of the anti-skid properties of new airfield pavements in aspect of applicable requirements. *Coatings* **2021**, *11*, 778. [CrossRef]
12. Jacazio, G.; Balossini, G. A high performance force control system for dynamic loading and fast moving actuators. In Proceedings of the Power Transmission and Motion Control, University of Bath, Bath, UK, 7–9 September 2005.
13. Jacazio, G.; Balossini, G. Real-time loading actuator control for an advanced aerospace test rig. *Proc. Inst. Mech. Eng. Part I J. Syst. Control Eng.* **2007**, *221*, 199–210. [CrossRef]
14. Chiavaroli, P.; De Martin, A.; Evangelista, G.; Jacazio, G.; Sorli, M. Real Time Loading Test Rig for Flight Control Actuators Under PHM Experimentation. In Proceedings of the ASME International Mechanical Engineering Congress and Exposition, Proceedings (IMECE), Pittsburgh, PA, USA, 9–15 November 2018; Volume 1, p. V001T03A032.
15. De Martin, A.; Dellacasa, A.; Jacazio, G.; Sorli, M. High-Fidelity Model of Electro-Hydraulic Actuators for Primary Flight Control Systems. In Proceedings of the 2018 Bath/ASME Symposium on Fluid Power and Motion Control FPMC2018, Bath, UK, 12–14 September 2018; p. V001T01A058.
16. Carbone, G.; Putignano, C. A novel methodology to predict sliding and rolling friction of viscoelastic materials: Theory and experiments. *J. Mech. Phys. Solids* **2013**, *61*, 1822–1834. [CrossRef]
17. Burckhardt, M. *Fahrwerktechnik: Radschlupf-Regelsysteme*; Vogel Verlag: Würzbg, Germany, 1993.
18. De Martin, A.; Jacazio, G.; Vachtsevanos, G. Windings Fault Detection and Prognosis in Electro-Mechanical Flight Control Actuators Operating in Active-Active Configuration. *Int. J. Progn. Health Manag.* **2017**, *8*. [CrossRef]
19. Mohan, N.; Undeland, T.M.; Robbins, W.P. *Power Electronics*, 3rd ed.; John Wiley and Sons, Inc.: Hoboken, NJ, USA, 2005.
20. Di Mascio, P.; Ragnoli, A.; Portas, S.; Santoni, M.; Sever, D.; Doler, D.; Kovačič, B. Monitor activity for the implementation of a pavement—Management system at cagliari airport. *Appl. Sci.* **2021**, *11*, 8697. [CrossRef]

aerospace

MDPI

Article

Intelligent Combined Neural Network and Kernel Principal Component Analysis Tool for Engine Health Monitoring Purposes

Maria Grazia De Giorgi *, Luciano Strafella, Nicola Menga and Antonio Ficarella

Department of Engineering for Innovation, The University of Salento, 73100 Lecce, Italy; luciano.strafella@unisalento.it (L.S.); nicola.menga@unisalento.it (N.M.); antonio.ficarella@unisalento.it (A.F.)
* Correspondence: mariagrazia.degiorgi@unisalento.it

Abstract: An efficient maintenance plan is an important aspect for aeronautical companies to increase flight safety and decrease costs. Modern technologies are widely used in the Engine Health Monitoring (EHM) discipline to develop intelligent tools capable of monitoring the health status of engines. In this work, Artificial Neural Networks (ANNs) and in-detail Feed-Forward Neural Networks (FFNNs) were exploited in addition to a Kernel Principal Component Analysis (KPCA) to design an intelligent diagnostic tool capable of predicting the Performance Parameters (PPs) of the main components used as their health index. For this purpose, appropriate datasets containing information about degraded engines were generated using the Gas Turbine Simulation Program (GSP). Finally, the original datasets and the reduced datasets obtained after the application of KPCA to the original datasets were both used in the training and testing process of neural networks, and results were compared. The goal was to obtain a reliable intelligent tool useful for diagnostic purposes. The study showed that the degraded component detection and estimation of its performance achieved by using the hybrid KPCA–FFNNs were predicted with accurate and reliable performance, as demonstrated through detailed quantitative confusion matrix analysis.

Keywords: engine health monitoring; diagnostics; artificial neural network; principal component analysis; artificial intelligence

Citation: De Giorgi, M.G.; Strafella, L.; Menga, N.; Ficarella, A. Intelligent Combined Neural Network and Kernel Principal Component Analysis Tool for Engine Health Monitoring Purposes. *Aerospace* **2022**, *9*, 118. https://doi.org/10.3390/aerospace9030118

Academic Editors: Spiros Pantelakis, Andreas Strohmayer and Liberata Guadagno

Received: 15 January 2022
Accepted: 21 February 2022
Published: 24 February 2022

1. Introduction

Diagnostics is nowadays a key concept used to ensure flight safety and cost reduction. Engine Health Monitoring is a discipline that takes care of keeping the state of health of the engine under control. Such a strategy leads to a health-based maintenance plan that stops the engine only when necessary, unlike a maintenance plan based on flight hours, as used in the past [1]. Degradation is a phenomenon caused by various factors that have a more or less serious impact on engine performance. Degrading phenomena that impact gas turbines (industrial and aeronautical) are numerous, such as fouling, erosion, abrasion, corrosion, increase in blade tip clearance, object ingestion (foreign object damage or domestic object damage), and exhaustive explanations are available in [2,3]. The effect on engine performance varies according to the component affected by the degradation and the phenomenon that has occurred. Compressors and turbines are the most impacted components by fouling and erosion, and their degradation is most investigated in the literature. Fouling is predominant in compressors, and it is the most prevalent degradation problem [4]. Fouling results in a decrease in the flow of both capacity and efficiency if it occurs on compressors and turbines. Erosion has the same effect of fouling on the compressor, while in the case of turbine erosion, efficiency continues to decrease but flow capacity increases [5]. For example, as reported in [6], a compressor fouling on an industrial gas turbine leads to a drop of 5% in mass flow and of 1.8% in compressor efficiency, respectively, in turn resulting in a 7% reduction in output power and 2.5% increase in heat rate. Meanwhile, [5] reported an increase in flow capacity of 4% and a decrease in efficiency of 2% due to turbine erosion. Erosion damage on compressors and turbines blades can be reduced by applying a coating. The results reported in [7], where a study has been conducted on the protective capabilities of various coatings against erosion and corrosion on compressor blades, show how coatings obtained by physical vapor deposition, especially the (Ti, Al)N coatings, provide the best results both in erosion and corrosion protection. In works [8,9], an Ingested Debris Monitoring System (IDMS) and an Engine Distress Monitoring System (EDMS) are used to monitor the debris present in the airflow ingested by a jet engine. The environmental impact on engine degradation is studied in [10,11], where the effect of sand and dust and the effect of volcanic ash on engines are illustrated, respectively. In the former, two experiments on the effect of sand ingestion are illustrated, one considering an "accelerate" approach, in which a big amount of sand is injected into the engine inlet in a short time, and the other considering an amount of sand approaching the one typical of a real case. In each case, experiments result in a decrease in thrust and an increase in specific fuel consumption. In the second, fan blade erosion due to volcanic ash is investigated using computational fluid dynamics tools considering two different approaches, one in which the erosion rate is considered constant and the other in which it was updated over time. In both cases, the results show an increase in material removed when moving from the midspan toward the tip. Furthermore, on the suction side, for regions close to the hub the highest material removal is located around the leading edge, this then moves toward the midchord for the outer midspan sections while the pressure side experiments with a lower amount of erosion, which is focused near the midchord of the tip. EHM techniques widely exploit computer algorithms for their purposes, which are efficient diagnostic tools based on information from sensors installed in the engine itself, such as temperatures or pressures. In [12], Remaining Useful Life (RUL) is predicted by means of genetic algorithms, decision trees, and fuzzy logic to create a hybrid model. Findings show good results in RUL prediction of a few time series. The authors underline how the performances start to be good around the 150th cycle and state that the result is expected because, as stated in [13], RUL prediction becomes accurate after some cycles have already been performed by the units, allowing the degradation to accumulate. In [14], MultiGene Genetic Programming (MGGP) is used to obtain a relationship between Exhaust Gas Temperature (EGT) and other variables characterizing an engine, and in the same work, Nonlinear AutoRegressive with eXogenous inputs neural network are exploited for a one-step-ahead prediction tool. The authors also investigated the effect of the number and the nature of the parameters used in the input for MGGP and ANN algorithms, showing the network's ability to deliver accurate results even with fewer input parameters. Two Nonlinear AutoRegressive neural networks are used in [15] for specific fuel consumption estimation. In particular, the first network was used to predict fuel flow rate, compressor rotational speed, turbine inlet temperature, and compressor pressure ratio, providing in input ambient conditions, Mach number, and Power Lever Angle (PLA), while the second network was used to predict the target performance parameter, i.e., specific fuel consumption, providing in input the parameters predicted by the first network plus ambient condition and Mach number. Predictions were performed with the engine in healthy condition and simulating a compressor degradation. In each case, results showed good behavior of the developed system, with an adaption process in the case of compressor degradation. In [16], Auto-Associative Neural Networks (AANNs) are used to implement a nonlinear Principal Component Analysis (PCA) for early warning of impending gas turbine failure, and results are compared with the one obtained by applying a standard linear PCA. Results show that although the training of an AANN model takes more time than it takes for a linear PCA, the former is more suitable to be applied in nonlinear systems. PCA was exploited for feature extraction purposes in [17] to develop a PCA–ANN fault classifier for diagnostic purposes. Results show very good performance compared with other algorithms, at the cost of higher computation times. Finally, neural networks are used in [18,19] to distinguish between different degradation conditions, i.e., the compressor fouling and the turbine erosion, with good results. In particular, in [18], results show component degradation and health condition identification EGT seems more accurate than fuel flow rate to be used as the input of the neural network, and as also evident in [19], results are better when the networks are trained with more information related to the engine (at least for the cases analyzed in these works).

Hence, Artificial Neural Networks are among the data-driven methods most widely used and very promising in the field of component degradation and fault diagnosis thanks to the solid ability in learning non-linear changes between a set of inputs and a set of outputs, as in the case of an aeronautical engine.

In this work, an FFNN is used to predict the PPs of some selected components constituting a turboshaft engine, i.e., polytropic efficiency and flow capacity for the compressor and high-pressure turbine, polytropic efficiency for the low-pressure turbine, pneumatic and combustion efficiencies for the burner, and mechanical efficiencies for shafts (as called "spools" in the present article). In addition to compressor and turbines, in our work, we also include burner and shafts transmissions, for which the degradation diagnostic is less investigated in the literature. GSP software was used to generate datasets containing information about degraded engines to be used to train neural networks. The just-mentioned datasets are noise-free. The datasets were obtained performing a series of steady-state simulations, each characterized by a different degradation condition (fault type) and severity. The degradation was implemented by changing the value of the PPs in the GSP model used for simulations. The degraded values of the PPs were obtained by an adequate equation between a maximum and a minimum value. Finally, the KPCA technique was used to reduce the amount of data to be supplied to the neural networks. In the past, several studies have used Principal Component Analysis (PCA). However, this technique has a reduced performance when applied to strongly non-linear problems, such as in the case of health monitoring of an aircraft engine. For this reason, Kernel–PCA (KPCA) has been used in this paper to overcome the limitations of conventional PCA to linear problems. ANNs were trained both with the original dataset (as obtained from simulations) and with the reduced datasets (after application of KPCA) and prediction results were compared. Compressor degradation was stimulated by a decrease in polytropic efficiency and flow capacity. Turbine degradation was considered a decrease in polytropic efficiency with both an increase and a decrease in flow capacity. For burner and shafts, degradation was simulated by a decrease in the efficiencies that characterize them.

2. Engine Block Model

The software exploited to model the engine used in this work was GSP, which has a drag and drop interface capable of simulating engine performance in various conditions, both in steady-state and transient conditions. More details about GSP are available in [20].

Modeled Engine: Features and Design Point

The engine used in this work was a double-spool, lightweight turboshaft designed for helicopter application, derived from the PW200 Pratt & Whitney Canada family. A high-pressure spool connects a single-stage centrifugal compressor to a single-stage high-pressure turbine (HPT), while the low-pressure spool is used to power the rotor (with a nominal speed of 6000 rpm) through a low-pressure turbine (LPT). The model validation was performed in standard sea-level static condition, i.e., temperature = 288.25 K, pressure = 1.013 bar, altitude = 0, and Mach = 0. The airflow rate was changed until power indicated on the engine reference database was obtained. The engine block model is shown in Figure 1, while Table 1 reports some engine parameters in design conditions. The indicated stations in Figure 1 are the following: 1—engine inlet; 2—compressor inlet; 3—compressor outlet; 4—HPT inlet; 45—LPT inlet; 46—LPT outlet; 9—exhaust gases.

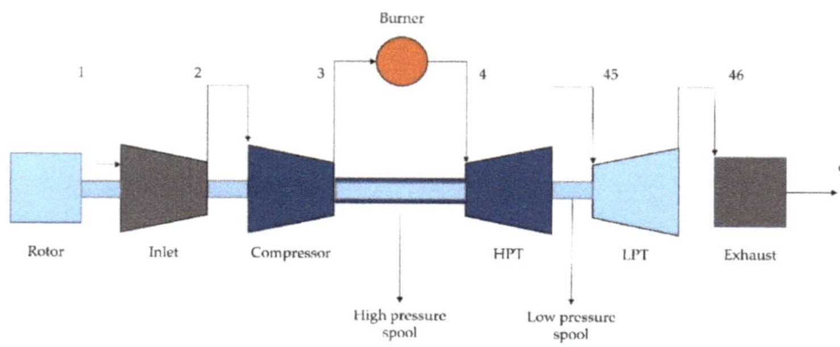

Figure 1. Block model of the used engine.

Table 1. Engine parameters in design point.

Parameter	Value
Rotor shaft power (P_r)	295 KW
Intake air mass flow rate (w_1)	2.05 Kg/s
Compressor pressure ratio (β_c)	7.17
Compressor polytropic efficiency (η_c)	0.82
HPT polytropic efficiency (η_{HPT})	0.88
LPT polytropic efficiency (η_{LPT})	0.88
Burner pneumatic efficiency ($\eta_{b,p}$)	0.96
Burner combustion efficiency ($\eta_{b,c}$)	0.985
Fuel mass flow rate (w_f)	0.0314 Kg/s
High-pressure shaft speed (ω_{HPS})	40,891 rpm
Low-pressure shaft speed (ω_{LPS})	6000 rpm

3. Dataset Generation: Simulation of Degraded Engines

The degradation scenarios simulated in this work were compressor degradation, HPT degradation, LPT degradation, burner degradation, shafts degradation, and multi-component degradation, i.e., a simultaneous degradation of the components previously analyzed individually. Table 2 describes the variables used as PPs of each component together with the related values in healthy condition at the design point and the range of percentage variation from the healthy condition. The dataset used to train ANNs was generated with GSP through a series of steady-state simulations performed on the working points described in Table 3, which were taken from the mission profiles present in [21]. For each degradation condition, 4500 steady-state simulations were performed (750 times each working point). In detail, the results of 1500 of 4500 simulations were used as testing cases, while the remaining 3000 were used as training cases. However, some simulations were discharged because they represent extreme conditions in which the software fails to converge. Each simulation is characterized by a different level of degradation, i.e., a different percentage variation of the PPs with respect to a healthy condition. The degraded PPs were obtained using Equation (1), which generates a random value of the PPs between a maximum and a minimum:

$$pp_d = \left[1 - \left(1 - \frac{MIN}{MAX}\right) RAN^3\right] MAX \quad (1)$$

where pp_d is the degraded value of the PP, MIN and MAX are the lower and the upper limits, respectively, and RAN is a randomly generated number between 0 and 1. Equation (1) was used for each PP. Input and output variables for neural network training were chosen between the ones present in the datasets obtained from GSP simulations. In detail, as the goal of the work was to predict the PP of the degraded component (i.e., for each degradation condition), the neural networks must be trained on the correlation between the PPs and some characteristics variables of the engine operating status.

Table 2. Performance Parameters of each degraded component. η_c: compressor polytropic efficiency; f_c: compressor flow capacity; η_{HPT}: HPT polytropic efficiency; f_{HPT}: HPT flow capacity; η_{LPT}: LPT polytropic efficiency; $\eta_{b,p}$: burner pneumatic efficiency; $\eta_{b,c}$: burner combustion efficiency; η_{HPS}: high-pressure shaft mechanical efficiency; η_{LPS}: low-pressure shaft mechanical efficiency.

Component	Degraded Performance Parameters	Healthy Condition	Range of Percentage Variation
Compressor	η_c f_c	0.82 1	−20–0% −20–0%
HPT	η_{HPT} f_{HPT}	0.88 1	−20–0% −10–+10%
LPT	η_{LPT}	0.88	−20–0%
Burner	$\eta_{b,p}$ $\eta_{b,c}$	0.96 0.985	−20–0% −20–0%
Shaft transmissions	η_{HPS} η_{LPS}	0.99 0.99	−20–0% −20–0%

Table 3. Working point simulated in steady-state simulations.

Working Point Name	Speed (m/s)	Altitude (m)	Power (KW)
START-A	30.6	0	48
START-B	0	1150	173
MAXPOW-A	30.6	0	291
MAXPOW-B	3.96	1154	273
MAXDUR-A	30.6	492	169
MAXDUR-B	0	2550	176

The variables that were chosen as input for ANNs are summarized in Table 4. These variables represent the input matrix for each neural network used while the output variables change from case to case, representing the PPs to be predicted for each degradation case, i.e., the PPs associated with the degraded component. The KPCA technique was used in this paper to reduce the input matrix dimension. Such a strategy could lead to a more efficient EHM system because the amount of data to be processed is lower compared to the case in which a data reduction technique is not used. In detail, for each of the six degradation cases, two neural networks were trained, one with the dataset consisting of the simulation results (called original dataset) and the other with the dataset obtained by applying KPCA to the original dataset (called reduced dataset).

Table 4. Variables used as input for the developed EHM system. Tot. T: Total Temperature; Tot. P.: Total Pressure.

Inputs	
Flight Mach (M)	Tot. T. at HPT inlet (TT_4)
Ambient Tot. P. (TP_a)	Tot. P. at HPT inlet (TP_4)
Ambient Tot. T. (TT_a)	Tot. T. at HPT outlet (TT_{45})
Power lever angle (PLA)	Tot. P. at HPT outlet (TP_{45})
Tot. T. at comp. inlet (TT_2)	Tot. T. at LPT outlet (TT_{46})
Tot. P. at comp. inlet (TP_2)	Tot. P. at LPT outlet (TP_{46})
Tot. T. at comp. outlet (TT_3)	LPT torque (T_{LPT})
Tot. P. at comp. outlet (TP_3)	Overall Pressure Ratio (OPR)
ω_{HPS}	Fuel mass flow rate/power lever angle (ρ)

4. EHM System: Combined KPCA + FFNN Algorithm

In this section, there is a brief description of how KPCA and FFNN work. Subsequently, the setup that was used is described.

4.1. Feed_Forward Neural Network: Principle of Operation and Setup

ANNs are artificial intelligence techniques designed to mimic the capacity of the human nervous system to learn from experience. The FFNN is the network type used in this work. FFNNs are constituted by layers (input layer, output layer, and hidden layers) and neurons, connected by links each having their own "weight". The input layer is the one in which information given to the network enters and the number of its neurons is equal to the number of variables given in the input. The output layer is the one from which the network predictions exit, and the number of its neurons is equal to the number of variables to be predicted. There are hidden layers between input and output layers, each constituted by a certain number of hidden neurons. The network performance depends on the number of hidden layers and hidden neurons. Equation (2) represents the output of hidden and output neurons, intended as the sum of the information (each characterized by its connection weight) deriving from the previous neurons and a bias.

$$y = \sum_{i=1}^{n} W_i X_i + b \tag{2}$$

where y denotes the neuron output, W_i is the weight of the connection between the neuron and the i-th previous neuron sending it information, X_i is the information obtained from the i-th previous neuron, n is the number of previous neurons, and b is the bias. Finally, an activation function is applied to normalize the results of Equation (2). To train a neural network means providing the network with a dataset constituted by input variables and related output variables and letting the network compute the values of weights and biases in order to predict the output variables starting from the input ones. In Table 5, the neural networks trained with the original datasets are described, while in Table 6, the neural networks trained with the reduced datasets are described. Each neural network used has only one hidden layer. In the training process, a maximum of 5000 epochs was set. Figure 2a shows a typical structure of an FFNN. Figure 2b instead reports the neural networks implemented with MATLAB for each case analyzed, i.e., for each degradation condition both with original and reduced datasets and having a structure like the one shown in Figure 2a.

Table 5. Characteristics of neural networks trained with original datasets.

Degradation Condition	Input Neurons	Hidden Neurons	Output Neurons
Compressor degradation	18	26	2
HPT degradation	18	26	2
LPT degradation	18	26	1
Burner degradation	18	26	2
Shaft transmissions degradation	18	26	2
Multi-component degradation	18	34	9

Table 6. Characteristics of neural networks with reduced datasets.

Degradation Condition	Input Neurons	Hidden Neurons	Output Neurons
Compressor degradation	9	26	2
HPT degradation	9	26	2
LPT degradation	9	26	1
Burner degradation	9	26	2
Shaft transmissions degradation	9	26	2
Multi-component degradation	13	34	9

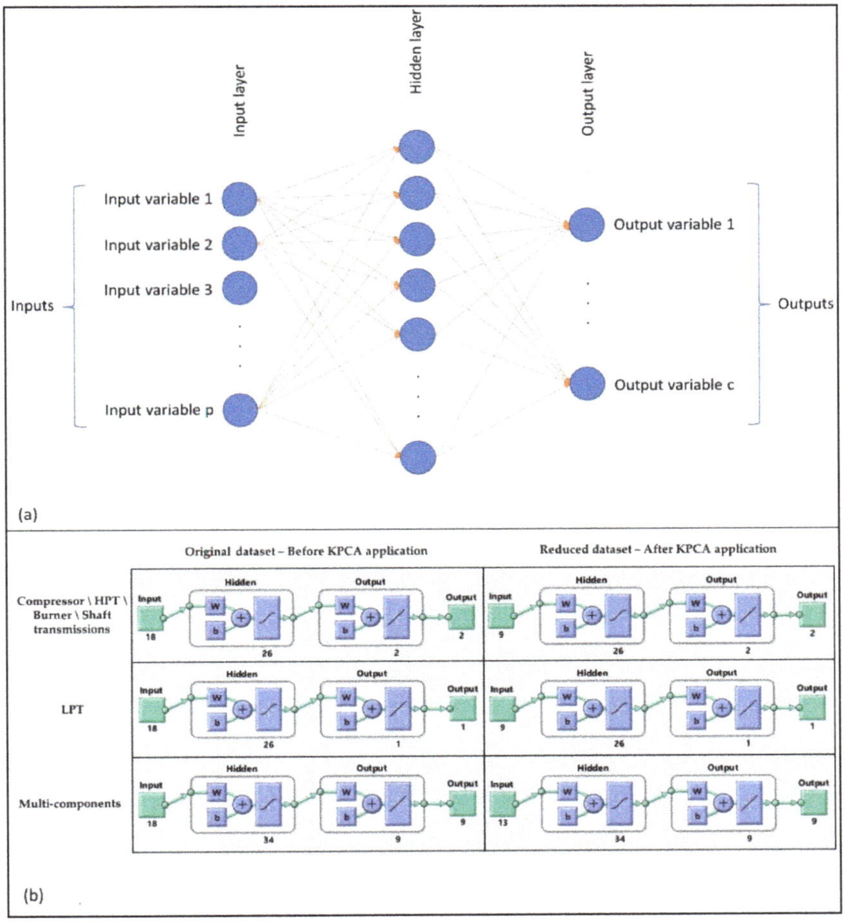

Figure 2. A typical structure of an FFNN (**a**) and neural networks used in this work (**b**), whose structure is based on the one represented in the top.

4.2. KPCA: A Data Reduction Technique

Principal Component Analysis is a data reduction technique used to reduce the dimension of datasets to be supplied to appropriate algorithms. However, PCA is more suitable for application in linear systems. For nonlinear systems, such as an aircraft engine, better performance is obtainable by KPCA, which combines the linear PCA with the Kernel method. The linear PCA creates a new dataset containing the same number of variables (called principal components) and the same amount of information [22] of the dataset of which is applied (original dataset). The new variables are numerically different from the ones located in the original dataset but depend on them. The principal components are ordered in the new dataset from the one containing the most information of the original dataset to the one containing the least. The strategy of the PCA is to discard the principal components having the least part of the information (the latest); in this manner, the new dataset is smaller than the original one, but carries most of the information contained in it.

The first step of PCA is the standardization of the original dataset. The new dataset X_{new} (formed by the principal components) is obtained from Equation (3):

$$X_{new} = X_{std}Q \quad (3)$$

where X_{std} is the original dataset after standardization and Q is the matrix formed by the sorted eigenvectors of the covariance matrix of X_{std}. The KPCA approach is to perform a mapping of the input data into a feature space using nonlinear mapping and subsequently perform a linear PCA in the feature space [23].

5. Results and Discussion

Here, the results obtained from the network predictions, both with and without the use of KPCA, are exposed. Figure 3 shows the comparison between the network predictions and the target values of the PPs obtained from GSP simulations for the first five degradation cases, i.e., for the single-component degradation cases. In more detail, the just-mentioned results are related to the predictions obtained from ANNs trained with the original datasets. Figure 4 reports the related percentage errors. From these first two figures, it is evident how the EHM system developed using only FFNNs shows very good results as R^2 (coefficient of determination) values are higher than 0.95; in particular, the lowest R^2 is equal to 0.9582 related to f_c. The high quality of the prediction is confirmed by the low values of the percentage errors reported in Figure 4. For the compressor case, for which the related PPs are η_c and f_c, a peak in error for a well-defined set of points is present at the working conditions MAXPOW-B, i.e., flight in altitude at high power. Percentage errors remain, however, below 6% for these two PPs. The remaining PPs are predicted with percentage errors all below 0.6%. In Figures 5 and 6 instead, the predictions of the same variables by the KPCA + FFNN EHM system are reported. In this case, the results are slightly worse than the previous ones but are still acceptable. The lowest R^2 is again related to f_c and is equal to 0.9869; therefore, the prediction of f_c is even better with the application of KPCA. Peak errors related to the MAXPOW-B working condition are still evident and remain below 6%. The remaining predictions have shown a percentage error below 0.5%. Figures 7–10 are related to the prediction of the PPs for the multi-component degradation case and report in more detail the comparison between predictions of FFNN without KPCA reduction and targets, the percentage errors for the predictions of FFNN without KPCA reduction, the comparison between the prediction of FFNN with KPCA reduction and targets, and the percentage errors for the predictions of FFNN with KPCA reduction, respectively. In this case, results are worse than those obtained in the prediction of the PPs in the single-component scenario. This is an unsurprising result because, in this case, the neural networks must predict the nine variables instead of two or only one, as in the case of LPT degradation. However, results for the prediction of FFNN without KPCA remain above 0.92 with an R^2 value related to f_c equal to 0.9211, which remains the lowest value. The percentage errors related to this case all remain above 5%, except for f_c, for which errors remain below 30%. However, it is important to note from Figure 8b that only some points are characterized by a high percentage error. Figures 9 and 10 show the performance for the prediction of the PPs from the KPCA + FFNN system in the multi-component degradation case. Additionally in this case, the lower performance in f_c prediction is confirmed with an R^2 value equal to 0.9211. Percentage errors of the case do not exceed 4%, except for the f_c prediction, whose percentage errors remain below 20%. However, also in this case, only some points are characterized by a higher percentage error.

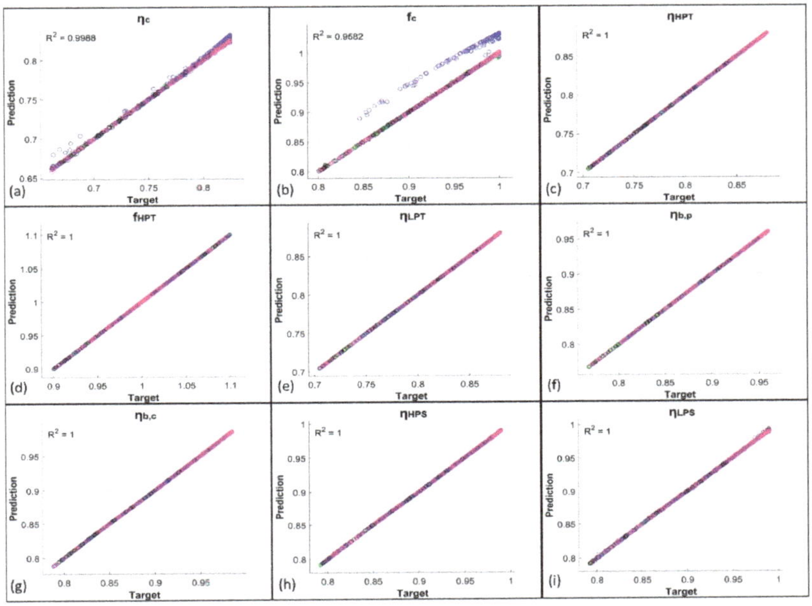

Figure 3. (**a–i**) Comparison between FFNN predictions and targets about the PPs related to the first five degradation conditions. Grey: START-A; brown: START-B; green: MAXPOW-A; blue: MAXPOW-B; black: MAXDUR-A; magenta: MAXDUR-B.

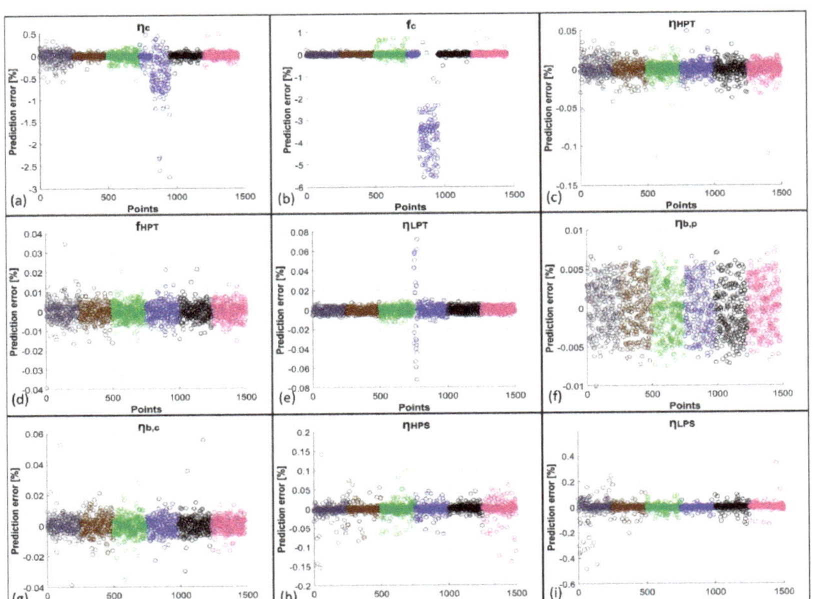

Figure 4. (**a–i**) Percentage errors between FFNN predictions and targets about the PPs related to the first five degradation conditions. Grey: START-A; brown: START-B; green: MAXPOW-A; blue: MAXPOW-B; black: MAXDUR-A; magenta: MAXDUR-B.

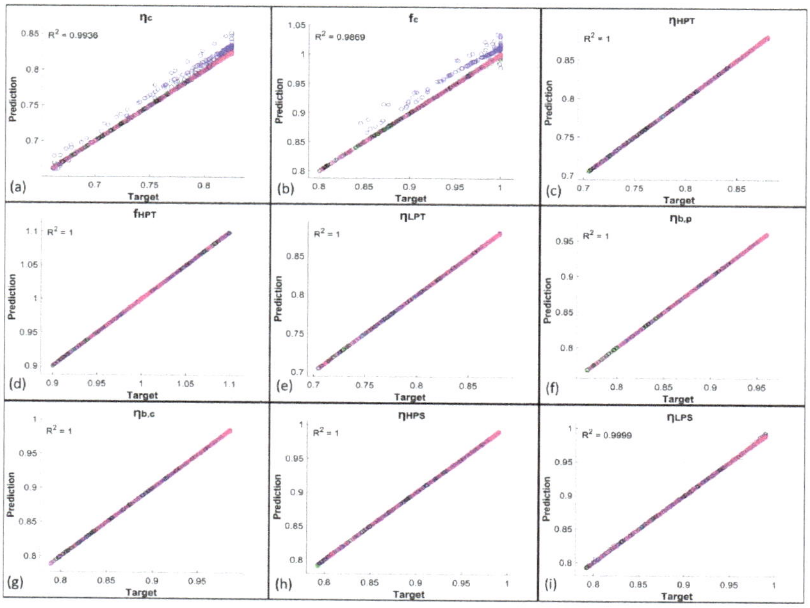

Figure 5. (**a–i**) Comparison between KPCA + FFNN predictions and targets about the PPs related to the first five degradation conditions. Grey: START-A; brown: START-B; green: MAXPOW-A; blue: MAXPOW-B; black: MAXDUR-A; magenta: MAXDUR-B.

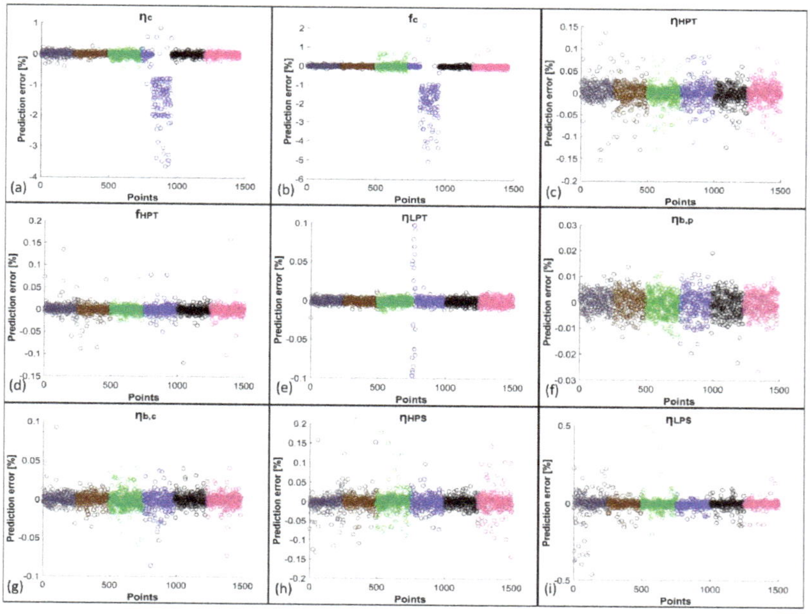

Figure 6. (**a–i**) Percentage errors between KPCA + FFNN predictions and targets about the PPs related to the first five degradation conditions. Grey: START-A; brown: START-B; green: MAXPOW-A; blue: MAXPOW-B; black: MAXDUR-A; magenta: MAXDUR-B.

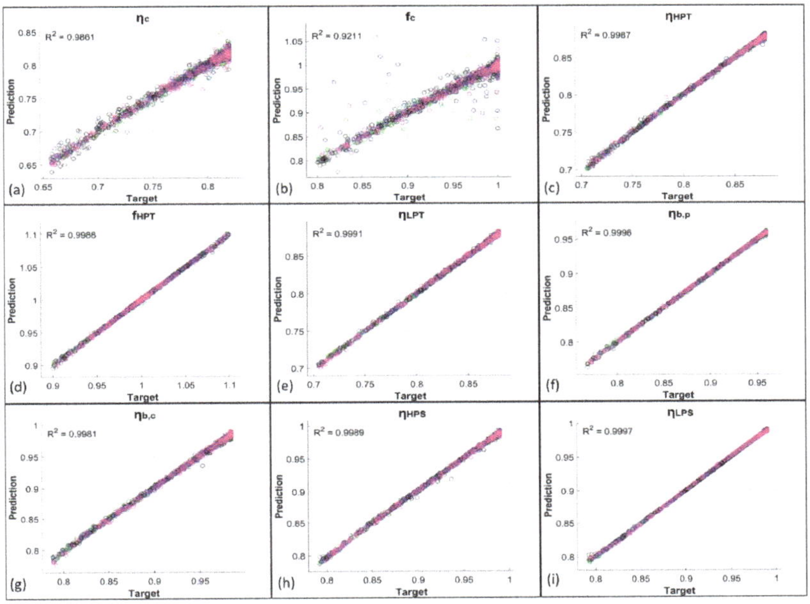

Figure 7. (**a–i**) Comparison between FFNN predictions and targets about the PPs related to the multi-component degradation condition. Grey: START-A; brown: START-B; green: MAXPOW-A; blue: MAXPOW-B; black: MAXDUR-A; magenta: MAXDUR-B.

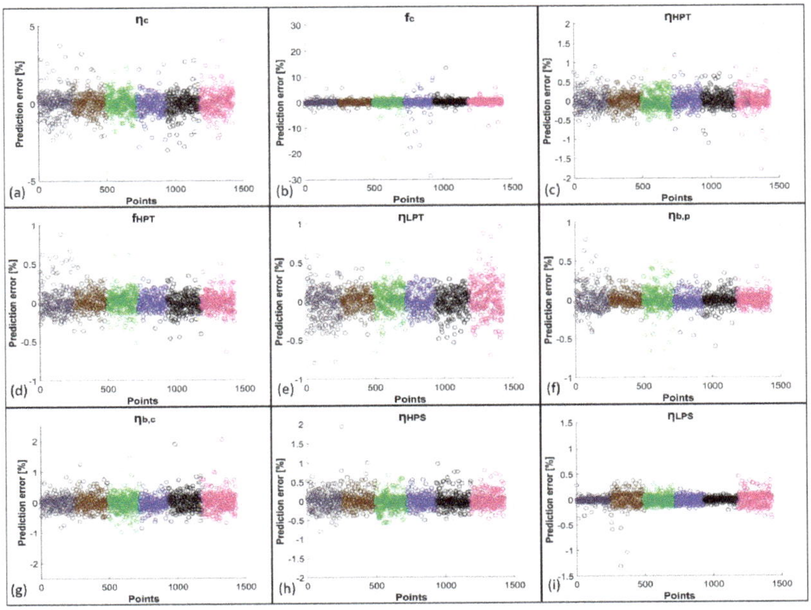

Figure 8. (**a–i**) Percentage errors between FFNN predictions and targets about the PPs related to the multi-component degradation condition. Grey: START-A; brown: START-B; green: MAXPOW-A; blue: MAXPOW-B; black: MAXDUR-A; magenta: MAXDUR-B.

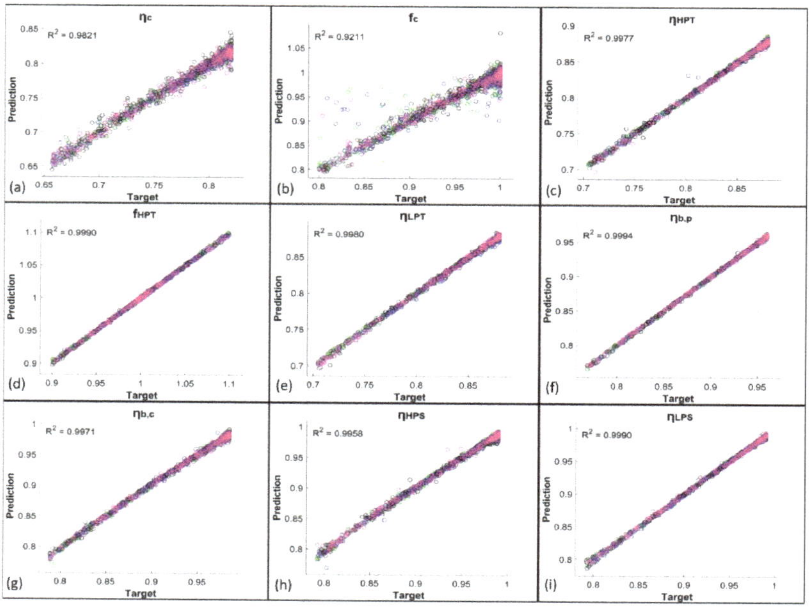

Figure 9. (**a–i**) Comparison between KPCA + FFNN predictions and targets about the PPs related to the multi-component degradation condition. Grey: START-A; brown: START-B; green: MAXPOW-A; blue: MAXPOW-B; black: MAXDUR-A; magenta: MAXDUR-B.

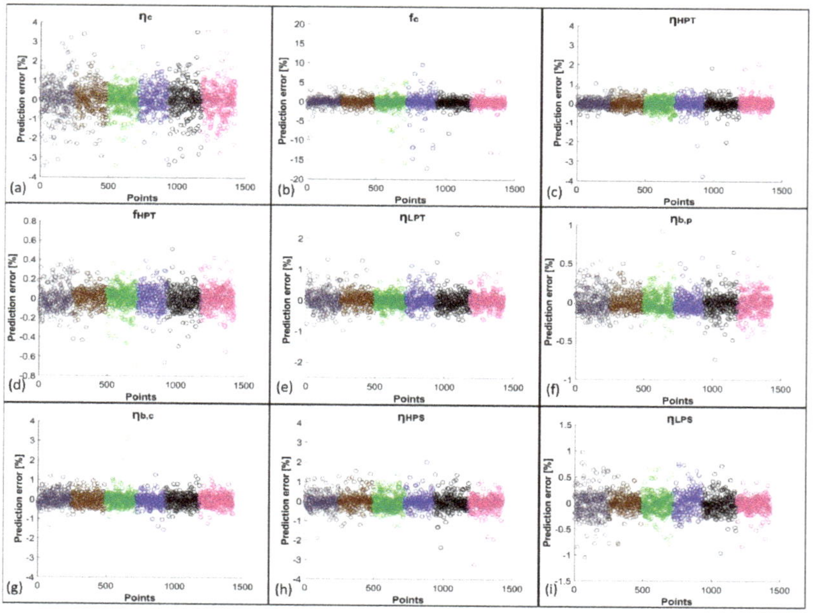

Figure 10. (**a–i**) Percentage errors between KPCA + FFNN predictions and targets about the PPs related to the multi-component degradation condition. Grey: START-A; brown: START-B; green: MAXPOW-A; blue: MAXPOW-B; black: MAXDUR-A; magenta: MAXDUR-B.

The performance of the implemented system in the classification of the degraded component was evaluated. For both single-component and multi-component degradation cases, a damage class was associated with each observation based on the percentage variation of PPs with respect to the healthy condition. The classification was carried out both for the actual PPs, obtained using Equation (1) (and used to carry out the simulations), and for the PPs predicted by the machine learning system. Confusion matrices were used to establish the accuracy of the prediction method to identify the degradation status of the different engine components. Table 7 shows the damage classes for the different components based on the two PPs that were used for the compressor (η_c, f_c), HPT (η_{HPT}, f_{HPT}), burner ($\eta_{b,p}$, $\eta_{b,c}$), and shaft transmissions (η_{HPS}, η_{LPS}), while Table 8 shows the classes for LPT, for which only one PP (η_{LPT}) was considered.

Table 7. Degradation classes used for compressor, HPT, burner, and shaft transmissions.

		PP1			
		PP1 < 5%	**5% < PP1 < 10%**	**10% < PP1 < 15%**	**15% < PP1**
PP2	PP2 < 5%	1	2	3	4
	5% < PP2 < 10%	2	3	4	5
	10% < PP2 < 15%	3	4	5	6
	15% < PP2	4	5	6	7

Table 8. Damage classes used for LPT.

PP1			
PP1 < 5%	**5% < PP1 < 10%**	**10% < PP1 < 15%**	**15% < PP1**
1	2	3	4

The percentages refer to the variation with respect to a healthy condition. For example, considering the compressor, if one of the two associated PPs (η_c and f_c) has a percentage variation with respect to the healthy case of 6% (column 2 in Table 7) and the other has one of 3% (row 1 in Table 7), the associated damage class is 2. The confusion matrices for the different components are shown in Figures 11–15. Regarding the compressor, the application of KPCA both in single-component and multi-component degradation cases leads to a slight drop in performance that is still acceptable. The difference between single-component and multi-component cases instead is more marked, passing from 96.7% to 88.7% and from 96.4% to 87.6% of well-classified results for FFNN and KPCA + FFNN methods, respectively. For HPT, LPT, burner, and shaft transmissions, results have a slight difference both in single and multi-component degradation and both using FFNN and KPCA + FFNN methods. Lower classification performance results are reported from the prediction of the burner's PPs in multi-component degradation using KPCA + FFNN, equal to 96.3%. Best results are obtained for LPT, for which there is a perfect classification in the single-component degradation case, both with FFNN and KPCA + FFNN. Based on these findings, further development of the present work would be achieved by the use of multiple single-output ANNs; in this way, it will be possible to better optimize the ANNs' settings to reduce the prediction errors in the case of multi-component degradation. Hence, it will be useful to combine the ANNs with KPCA to reduce the computational time, which increases switching from the multi-outputs single ANN approach to the multiple single-output ANNs.

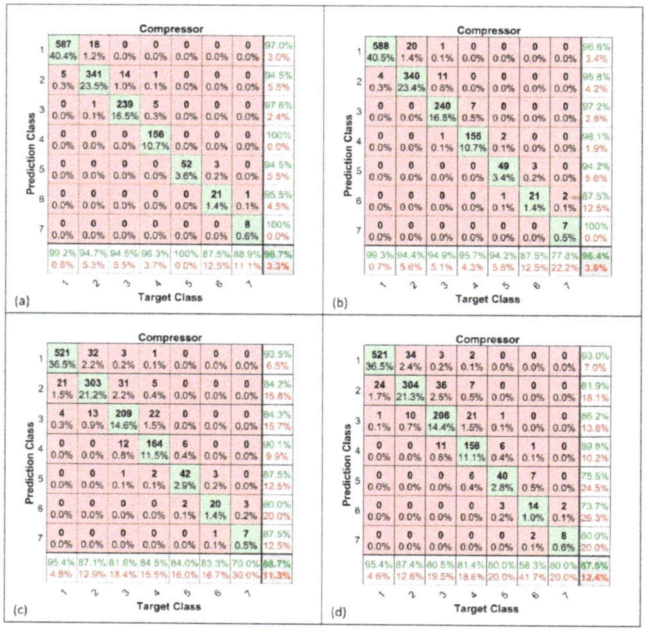

Figure 11. Confusion matrix for compressor. (**a**): Single-component FFNN; (**b**): single-component KPCA + FFNN; (**c**): multi-component FFNN; (**d**): multi-component KPCA + FFNN.

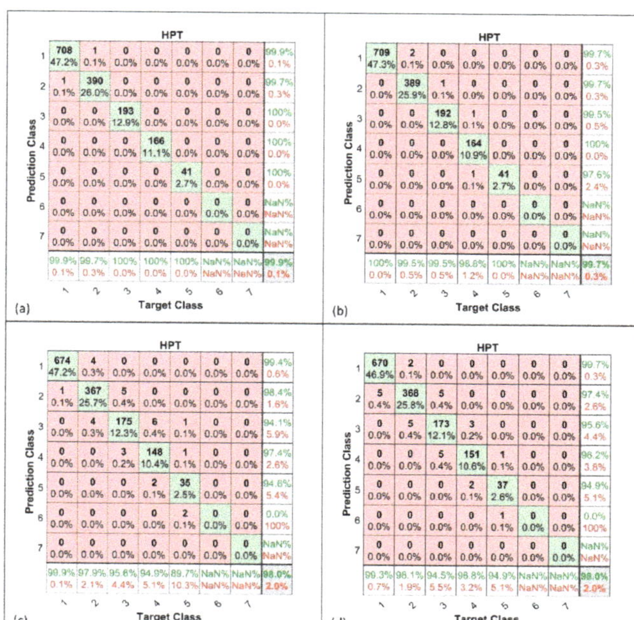

Figure 12. Confusion matrix for HPT. (**a**): Single-component FFNN; (**b**): single-component KPCA + FFNN; (**c**): multi-component FFNN; (**d**): multi-component KPCA + FFNN.

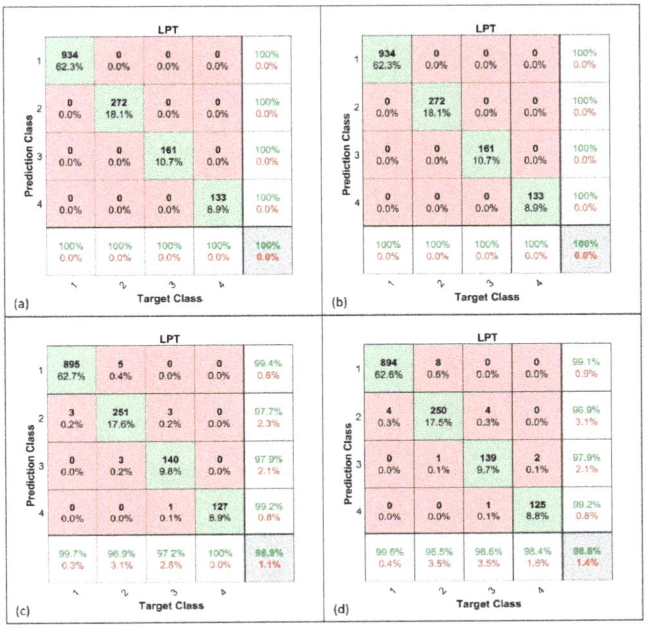

Figure 13. Confusion matrix for LPT. (**a**): Single-component FFNN; (**b**): single-component KPCA + FFNN; (**c**): multi-component FFNN; (**d**): multi-component KPCA + FFNN.

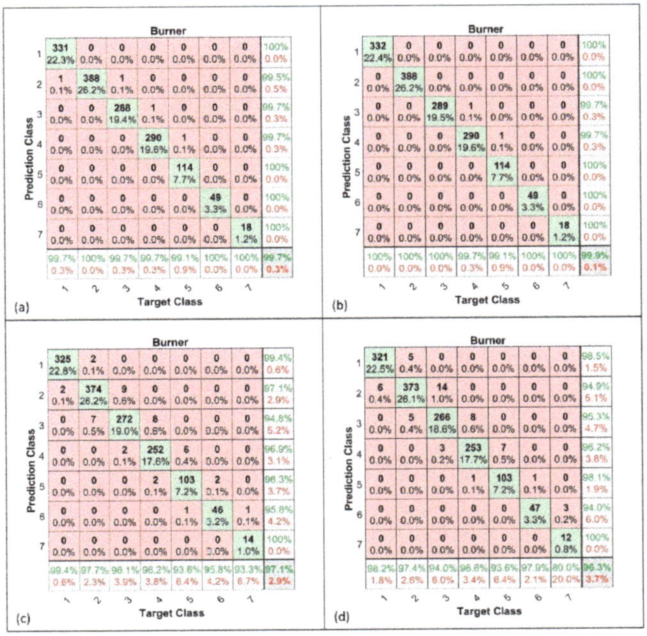

Figure 14. Confusion matrix for burner. (**a**): Single-component FFNN; (**b**): single-component KPCA + FFNN; (**c**): multi-component FFNN; (**d**): multi-component KPCA + FFNN.

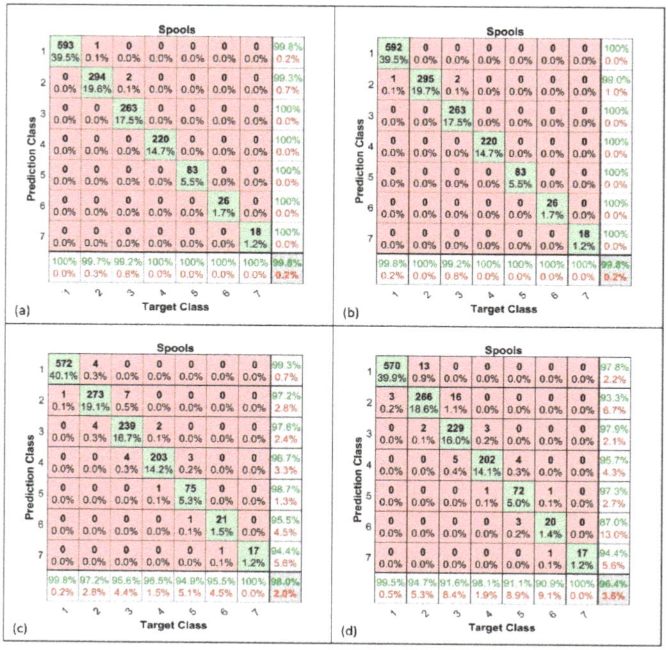

Figure 15. Confusion matrix for shaft transmissions. (**a**): Single-component FFNN; (**b**): single-component KPCA + FFNN; (**c**): multi-component FFNN; (**d**): multi-component KPCA + FFNN.

6. Conclusions

The goal of the study presented here was to obtain a valid EHM system capable of predicting the so-called Performance Parameters of some of the principal components constituting a turboshaft, i.e., compressor, high-pressure turbine, low-pressure turbine, burner, and shafts. Artificial intelligence has been used to reach the target. FFNNs were exploited and trained to predict the values of the PPs starting from the values of flight mission data and some state engine variables, which could be measured in a real engine by installed sensors. The input parameters included the flight Mach number, ambient total pressure and temperature, total pressures and total temperatures in various engine stations, high-pressure spool speed, low-pressure spool torque, overall pressure ratio and fuel flow-PLA ratio. In the present work, the datasets used to train neural networks were obtained from a series of steady-state GSP simulations, implementing the six different analyzed degradation cases. Each simulation was characterized by a different degradation level and condition. The degradation was implemented by changing the values of the PPs in the GSP block model of the simulated engine. The degraded values of the PPs were obtained using an adequate equation useful for obtaining a random value between a maximum and a minimum. The maximum and minimum acceptable values of the PPs were dictated by the considered range of percentage variation with respect to the healthy condition. The prediction of PPs by means of neural networks was performed in two ways; first by training the networks with the original datasets (as obtained from the simulations) and next by training the networks with the reduced datasets (obtained applying KPCA to the original datasets). The results were compared. The prediction of PPs in the case without the use of KPCA produced very good results, both for single-component degradation cases and for multi-component degradation cases. Application of the KPCA technique does not lead to a serious lowering of the reliability, both in the case of single-component and multi-component degradation. Finally, based on values of the PPs, a degraded component class was associated with each observation point in the datasets, both for the datasets obtained from the GSP simulations and for the one obtained from networks prediction. The results were shown using the confusion matrices. A data reduction technique is a valid method to reduce the number of computational efforts required, but if the use of these techniques compromises the efficiency of the system, it may be preferable not to use it. However, multiple single-output ANNs would be implemented to improve the prediction of multi-component degradation predictions; in this case, the use of KPCA will be useful to reduce the time of computations and should not impact the prediction performance, as shown in the single-component degradation case that presents few outputs (one or two) for each condition to be predicted. In the future, we will evaluate the robustness of the method in the presence of measurement uncertainty via input data corrupted by Gaussian noise.

Author Contributions: Conceptualization, M.G.D.G.; methodology, M.G.D.G. and L.S.; software, N.M. and L.S.; validation, N.M.; formal analysis, N.M.; investigation, N.M. and L.S.; resources, A.F.; data curation, L.S.; writing—original draft preparation, N.M. and M.G.D.G.; writing—review and editing, M.G.D.G.; supervision, M.G.D.G. and A.F.; project administration, A.F.; funding acquisition, A.F. All authors have read and agreed to the published version of the manuscript.

Funding: This research was funded by the Italian Ministry of University and Research, project PON "SMEA", code PON03PE_00067_5.

Institutional Review Board Statement: Not applicable.

Informed Consent Statement: Not applicable.

Conflicts of Interest: The authors declare no conflict of interest.

References

1. Vujić, D. Diagnostic systems in aircraft engines maintenance. *Sci. Technol. Rev.* **2005**, *55*, 30–34.
2. Kurz, R.; Brun, K. Gas Turbine Tutorial-Maintenance and Operating Practices Effects on Degradation and Life. In Proceedings of the 36th Turbomachinery Symposium, Texas A&M University, Turbomachinery Laboratories, Houston, TX, USA, 11–13 September 2007.
3. Fentaye, A.D.; Baheta, A.T.; Gilani, S.I.; Kyprianidis, K.G. A Review on Gas Turbine Gas-Path Diagnostics: State-of-the-Art Methods, Challenges and Opportunities. *Aerospace* **2019**, *6*, 83. [CrossRef]
4. Meher-Homji, C.B.; Chaker, M.; Bromley, A.F. The Fouling of Axial Flow Compressors: Causes, Effects, Susceptibility, and Sensitivity. *Turbo Expo Power Land Sea Air* **2009**, *48852*, 571–590.
5. Qingcai, Y.; Li, S.; Cao, Y.; Zhao, N. Full and Part-Load Performance Deterioration Analysis of Industrial Three-Shaft Gas Turbine Based on Genetic Algorithm. *Turbo Expo Power Land Sea Air* **2016**, *49828*, V006T05A016.
6. Diakunchak, I.S. Performance Deterioration in Industrial Gas Turbines. *J. Eng. Gas Turbines Power* **1992**, *114*, 161–168. [CrossRef]
7. Swadźba, L.; Formanek, B.; Gabriel, H.M.; Liberski, P.; Podolski, P. Erosion- and corrosion-resistant coatings for aircraft compressor blades. *Surf. Coat. Technol.* **1993**, *62*, 486–492. [CrossRef]
8. Powrie, H.; Fisher, C.E. Monitoring of Foreign Objects Ingested into the Intake of a Jet Engine. In *International Conference on Condition Monitoring Proceedings*; University of Wales: Swansea, UK, 1999; pp. 175–190.
9. Powrie, H.; Novis, A. Gas path debris monitoring for F-35 Joint Strike Fighter propulsion system PHM. In Proceedings of the IEEE Aerospace Conference, Big Sky, MT, USA, 4–11 March 2006; p. 8.
10. DeRemer, J. Sand and Dust Erosion in Aircraft Gas Turbines. *J. Am. Soc. Nav. Eng.* **1950**, *62*, 505–511.
11. De Giorgi, M.; Campilongo, S.; Ficarella, A. Predictions of Operational Degradation of the Fan Stage of an Aircraft Engine Due to Particulate Ingestion. *J. Eng. Gas Turbines Power* **2015**, *137*, 052603. [CrossRef]
12. Ishibashi, R.; Nascimento, C., Jr. GFRBS-PHM: A genetic fuzzy rule-based system for phm with improved interpretability. In Proceedings of the 2013 IEEE Conference on Prognostics and Health Management (PHM), Gaithersburg, MD, USA, 24–27 June 2013; pp. 1–7.
13. Heimes, F.O. Recurrent neural networks for remaining useful life estimation. In Proceedings of the 2008 International Conference on Prognostics and Health Management, Denver, CO, USA, 6–9 October 2008; pp. 1–6.
14. De Giorgi, M.G.; Quarta, M. Hybrid MultiGene Genetic Programming-Artificial neural networks approach for dynamic performance prediction of an aeroengine. *Aerosp. Sci. Technol.* **2020**, *103*, 105902. [CrossRef]

15. De Giorgi, M.G.; Strafella, L.; Ficarella, A. Neural Nonlinear Autoregressive Model with Exogenous Input (NARX) for Turboshaft Aeroengine Fuel Control Unit Model. *Aerospace* **2021**, *8*, 206. [CrossRef]
16. Palme, T.; Breuhaus, P.; Assadi, M.; Klein, A.; Kim, M. Early Warning of Gas Turbine Failure by Nonlinear Feature Extraction Using an Auto-Associative Neural Network Approach. *Turbo Expo Power Land Sea Air* **2011**, *54631*, 293–304.
17. Alozie, O.; Li, Y.G.; Pilidis, P.; Korakianitis, T.; Ren, W.; Shong, X.; Wu, X. An Integrated Principal Component Analysis, Artificial Neural Network and Gas Path Analysis Approach for Multi-Component Fault Diagnostics of Gas Turbine Engines. *Turbo Expo Power Land Sea Air* **2020**, *84140*, V005T05A023.
18. De Giorgi, M.G.; Ficarella, A.; De Carlo, L. Jet engine degradation prognostic using artificial neural networks. *Aircr. Eng. Aerosp. Technol.* **2019**, *92*, 296–303. [CrossRef]
19. De Giorgi, M.G.; Campilongo, S.; Ficarella, A. Development of a real time intelligent health monitoring platform for aero-engine. *MATEC Web Conf.* **2018**, *233*, 00007. [CrossRef]
20. GSP 11 User Manual. Available online: https://www.gspteam.com/GSPsupport/OnlineHelp/index.html?documentation.htm (accessed on 12 January 2022).
21. Donateo, T.; De Pascalis, C.L.; Strafella, L.; Ficarella, A. Optimal Energy Management of a Hybrid Electric Helicopter for Urban Air-Mobility. *IOP Conf. Ser. Mater. Sci. Eng.* **2021**, *1024*, 012074. [CrossRef]
22. Jolliffe, I.T. *Principal Component Analysis*; Springer: New York, NY, USA, 1986.
23. Kim, K.I.; Jung, K.; Kim, H.J. Face recognition using kernel principal component analysis. *IEEE Signal Process. Lett.* **2002**, *9*, 40–42.

aerospace

MDPI

Article

Model-Based Dynamic Performance Simulation of a Microturbine Using Flight Test Data †

Mario Leonardo Erario [1], Maria Grazia De Giorgi [1] and Radoslaw Przysowa [2,*]

1 Department of Engineering for Innovation, University of Salento, 73100 Lecce, Italy; erariomarioleonardo@gmail.com (M.L.E.); mariagrazia.degiorgi@unisalento.it (M.G.D.G.)
2 Instytut Techniczny Wojsk Lotniczych (ITWL), ul. Ksiecia Boleslawa 6, 01-494 Warsaw, Poland
* Correspondence: radoslaw.przysowa@itwl.pl

† This paper is an extended version of our paper published in 11th EASN Virtual Conference on "Innovation in Aviation & Space to the Satisfaction of the European Citizen".

Abstract: Microturbines can be used not only in models and education but also to propel UAVs. However, their wider adoption is limited by their relatively low efficiency and durability. Validated simulation models are required to monitor their performance, improve their lifetime, and to design engine control systems. This study aims at developing a numerical model of a micro gas turbine intended for prediction and prognostics of engine performance. To build a reliable zero-dimensional model, the available compressor and turbine maps were scaled to the available test bench data with the least squares method, to meet the performance of the engine achieved during bench and flight tests. A steady-state aeroengine model was implemented in the Gas turbine Simulation Program (GSP) and was compared with experimental operating points. The selected flight data were then used as input for the transient engine model. The exhaust gas temperature (EGT) and fuel flow were chosen as the two key parameters to validate the model, comparing the numerical predicted values with the experimental ones. The observed difference between the model and the flight data was lower than 3% for both EGT and fuel flow.

Keywords: microturbine; turbojet; component maps; map scaling; off-design; transient simulation; aerial target; unmanned aerial vehicle; flight data; digital twin

Citation: Erario, M.L.; De Giorgi, M.G.; Przysowa, R. Model-Based Dynamic Performance Simulation of a Microturbine Using Flight Test Data. *Aerospace* **2022**, *9*, 60. https://doi.org/10.3390/aerospace9020060

Academic Editor: Spiros Pantelakis

Received: 14 December 2021
Accepted: 20 January 2022
Published: 24 January 2022

1. Introduction

Microturbines are scaled-down turboshafts or turbojets with rotating components similar to those used in piston engine turbochargers [1,2]. They have a single- or double-stage radial compressor and a radial or axial turbine. The rotational speed is usually greater than 70,000 revolutions per minute, and for some applications it exceeds 200,000 rpm.

Microturbines have many promising applications. One major use is energy generation [3,4], primarily as standby or backup power, where power availability is critical. Testing alternative fuels is another area where microturbines are widely used: in fact, they make it possible to evaluate jet fuels synthesised in small quantities [5,6]. The use of propelling Unmanned Aerial Vehicles (UAV) or Light Personal Aircraft has been increasing over the last years [7–9], as more producers offer many types of micro and small turbojets or turboprops in a wide range of classes [10,11]. However, engine downsizing is accompanied by a significant increase in Reynolds number, resulting in a decrease in overall engine performance [12]. Because of this, research to understand the behaviour and performance of microturbines [13,14] is critical, specifically by implementing simulation models. They can be used, among others, to improve engine performance by converting micro turbojets to turbofans [15,16] or optimizing their exhaust nozzle [17,18].

Modelling the dynamics of a complex non-linear system, such as a gas turbine, makes it possible to control it effectively and monitor its performance [19]. Traditional engine models rely on the thermodynamic description of the engine, so they are called white-box or model-based approaches. Transient simulation requires representation of the changes of gas density in plenum volumes using the inter-component volume method [20–22]. Alternatively, the constant mass flow (CMF) iterative method can be used in dynamic simulations [23–25]. Chachurski et al. [26] developed a steady-state mathematical model of the JETPOL GTM 120 micro turbojet and validated it with test rig data. Several advanced models of JetCat engines were presented [27–29]. However, many published models of microturbines, for example, [30–32] do not use component maps. Such models are valid only for design-point or steady-state operation due to considerable simplifications in their structure. Engine control systems are also usually based on linearized models due to the required low response times [33–35].

A reliable transient model is necessary to simulate a gas-turbine engine propelling a highly maneuverable aircraft. Such a model needs component maps [19,36] to describe how the engine performs at nearly any design or off-design conditions. In practice, component maps for turbomachinery are generated in three different ways: by component rig testing, aerodynamic analysis, or scaling existing maps. For microturbines, component maps are primarily generated with CFD tools, for example, for radial compressors [28,37,38], turbines [39] and combustors [40,41]. What is more, a complete CFD model of the gas path was developed recently, which simulates engine performance without component maps [42].

Accurate maps are rarely available, so different map scaling and adaptation methods are commonly used [43–45]. The first-choice method is single point scaling [46] because it is built into GSP and other tools. However, it works properly only for components with similar geometry that is proportionally scaled up or down. Applying factors and deltas [19] is a linear method for three-dimensional map transformation. Such map operations should preserve the physical sense of the flow, in particular its Mach number [36]. Otherwise, scaled maps are valid only in the vicinity of experimental points they are based on.

During their operational life engine components are subject to several physical problems such as fouling, erosion, corrosion, blade damage, tip clearance increase, worn seals, combustor damage, and many others. The performance degradation of the engine depends on the type and severity of the deterioration and the components that are deteriorated. Generally, a deteriorated gas turbine provides less power for a certain amount of fuel mass flow or needs more fuel to produce a certain amount of power [19,47]. The severity of component deterioration can be characterized by the difference between the actual component condition parameters and their baseline. From a thermodynamic perspective the condition of gas path components is quantified in terms of isentropic efficiency η, mass flow W and pressure ratio PR.

Model-based gas path analysis systems make it possible to monitor engine performance parameters for fault diagnosis and manage component deterioration [48,49]. However, data-driven approaches are more and more common recently, not only in full-scale engines but also in microturbines [50,51]. Often, engine parameters are estimated or predicted by machine learning techniques which are trained with data obtained from simplified aeroengine gas-path performance prognostic models. For example, a transient model of the Viper 632-43 turbojet was implemented in our earlier project [52]. An integrated health monitoring platform for performance analysis and degradation diagnostics of gas turbine engines was demonstrated [47,53]. These studies underlined the suitability of Gas Path Analysis tools to predict aeroengine performance with a high accuracy. However, few works deal with the implementation of such models for small scale aeroengines.

To sum up, most publications on microturbine simulation are based on simplified models which hardly describe transient operation. Some micro turbojets were tested in a wind tunnel to study their performance in low-altitude flight [31,54]. Although several transient models of microturbines [20,55,56] were developed, they usually lacked validation with real flight data.

The purpose of this work was to create a numerical model of a microturbine under transient conditions, validated with experimental flight data. The model is intended for predicting emissions and generating training datasets for an artificial neural network (ANN), for the purpose of a planned engine health management system. There were three crucial tasks to develop the model:

- Use of test-rig data from two micro-turbines with the same thrust and size to define their operating points and create compressor and turbine maps. A least-squares scaling method was employed to transform the maps using parameters from four different engine operating points;
- Development and fine-tuning of the numerical model in GSP (Gas turbine Simulation Program) [57,58]: The model was used for simulation of the design point, steady states at various engine speeds, and also flight missions to model the behaviour of the engine under transient conditions;
- Validation of the model with the experimental data: The results of the steady-states simulation were compared with the test rig data of the reference microturbines. The comparison was made for different operating points referring to the in-flight engine operating range, even far from the design point, and with respect to various performance parameters (thrust, fuel flow, EGT, inlet air flow). The transient model was validated using data recorded by flight telemetry for four different missions, by comparing the EGT and fuel flow parameters.

2. Materials and Methods

2.1. Engine Specification

This research is based on extensive data gathered from test bench experiments and flight missions of a twin-engine target drone. Two types of micro turbojets were tested and modelled in this work: the Polish JETPOL GTM 140 and the German JetCat P140 Rxi-B. The experimental data belong to two different microturbines of the same thrust class, the Polish JETPOL GTM 140 and the German JetCat P140 Rxi-B. Specifically, complete steady-state data were available for the former engine, while flight data for the latter. Both engines were controlled by their original Electronic Control Unit (ECU), which was not modelled here.

Table 1 shows engine specifications. Looking at the values in the table, it is evident that the engine parameters are almost identical, except for the Engine Compression Ratio and the Design Maximum speed, which depend on design choices. The performance at the design point is very similar as well.

Table 1. Engine specifications.

Parameter	Unit	JETPOL GTM 140	Jetcat P140 Rxi-B
Overall Pressure Ratio		2.8	3.4
Air flow rate	kg/s	0.35	0.34
Maximum EGT	°C	700	720
Mass Flow	kg/s	0.35	0.34
Maximum Thrust	N	140	142
Design Speed	kRPM	120	125
Fuel consumption	g/s	7.0	7.33

It is noticeable that the performance of both engines is very similar, and the data available complement each other, providing all the necessary information to create a micro-turbine transient engine model that can be reliably validated. From the above observations, the following assumptions for implementing the engine model were taken:

- Maximum shaft speed (at Design Point) was set to 125,000 rpm, like for the JetCat P140 Rxi-B;
- Bench test measurements and the map scaling factor was derived from the JETPOL GTM 140;
- Transient performance was equal to those of JetCat, measured during flight missions.

2.2. Test Rig Data

For JETPOL GTM 140, test rig measurements [59] of temperatures and pressures at various engine stations, Exhaust Gas Temperature, air and fuel mass flow rate, and thrust were made at several shaft speeds (Figure 1). For JetCat P140 Rxi-B, a bench test measurement [60] of thrust and fuel flow was provided at different shaft speeds. When comparing these two last parameters (Figures 2 and 3), it is noticeable that the performance of the two engines is in good agreement. These two engines are not identical but their key components have the same geometry. There are significant differences in engine accessories, controls and instrumentation.

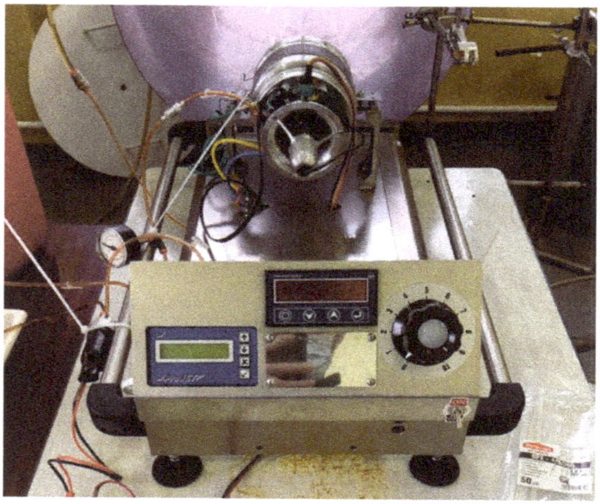

Figure 1. JETPOL GTM 140 at the test rig.

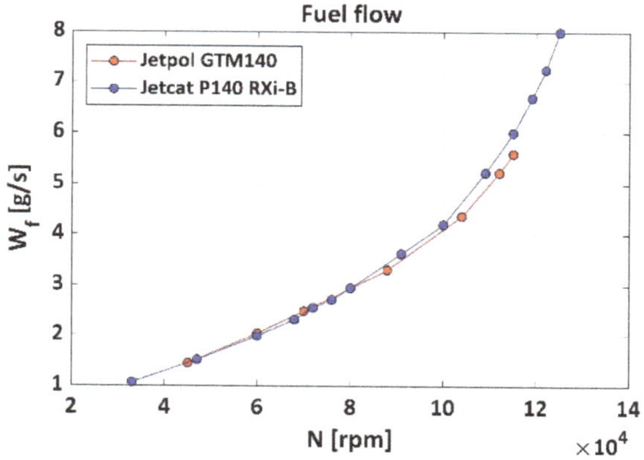

Figure 2. Fuel flow vs. rotational speed.

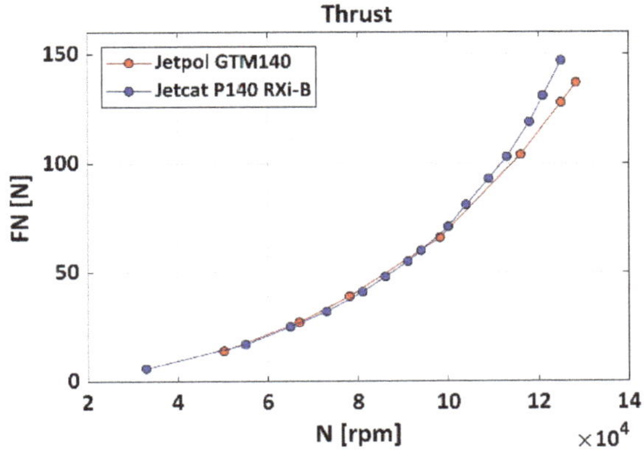

Figure 3. Thrust vs. rotational speed.

2.3. Flight Data

A prototype aerial target (Table 2, Figure 4) propelled by two JetCat P140 Rxi-B engines was flight-tested by the Air Force Institute of Technology (ITWL) in Poland [60,61]. In this work, flight data with a diverse mission profile were selected. The datasets represent a wide range of engine operations, so are well-suited for the effective validation of the engine transient model.

Figure 4. Aerial target developed by ITWL.

Table 2. Aircraft specifications.

Parameter	
Max take-off weight	85 kg
Wing span	2.85 m
Lenght	3.55 m
Operating speed	65–150 m/s
Climb speed	6 m/s
Altitude	1000–5000 m
Operating range	35 km
Endurance	60 min

The following parameters were sampled with the rate of 50 Hz and were acquired via a telemetry system [62]:

- Air speed
- Ambient temperature

- Altitude
- Shaft speed
- EGT
- Fuel flow

For a 30-min flight, the dataset is a table of about 90,000 rows, each of them representing a time step for which solving the system of equations, leads to prolonged computation in GSP. Furthermore, the recorded data included long intervals before the start and after the end of each flight. Those portions of the dataset, if included in the input, could completely invalidate a simulation process. Two approaches were used to reduce this amount of data (1) bringing the sample rate down to 1 Hz; (2) considering only the data with engine speed greater than 80,000 rpm because it never went below this threshold during flight operations. Consequently, the irrelevant records were removed from the dataset that corresponded to the moments:

- before take-off, when the engine idled but the aircraft had not been launched yet;
- after the end of the mission, when the parachute opened, the engine idled and was shut down after a while. The recording ended either when the telemetry was turned off or when the target drone was recovered.

The data reduction was considerable (Table 3).

Table 3. Length of original and reduced datasets.

Flight No	Original Data (Rows)	Reduced Data (Rows)
1	100,742	1325
2	221,372	3827
3	121,958	1811
4	135,389	2215

2.4. Compressor and Turbine Maps

It was necessary to build new compressor and turbine maps in this work since the generic maps included in the libraries of GSP and GasTurb were not appropriate. This process is presented in Figure 5. Our maps were generated using performance plots of a compressor and turbine provided by Rzeszow University of Technology, produced with CFD for a microturbine of similar geometry. For each component, there are two maps in the form of a bitmap: (1) Pressure ratio versus Mass Flow and (2) Efficiency vs Mass Flow. Then Smooth-C and Smooth-T tools from the GasTurb package were used, to digitize the plots, generate new interpolated speed lines, and create β-lines. The source maps and MATLAB scripts for their transformation are documented in thesis [63].

There are four parameters that are represented on a component map: (1) rotational speed; (2) mass flow; (3) pressure ratio (compression ratio for a compressor or expansion ratio for a turbine, respectively); and (4) efficiency. In addition, maps include β-lines, which are a family of curves that do not have a physical meaning but they make it possible to uniquely define an operating point when speed lines in the graph are parallel to either the pressure ratio or the mass flow rate axis. The compressor map also includes the surge line which limits the area of its stable operation.

For the compressor, the Smooth-C tool was used, to digitize the plots, generate new interpolated speed lines, and create β-lines. In this way, for each operating point of the component, values of rotational speed, mass flow, pressure ratio and efficiency were found, defining them univocally. Some numbers regarding the results of the point acquisition can be found in Table 4.

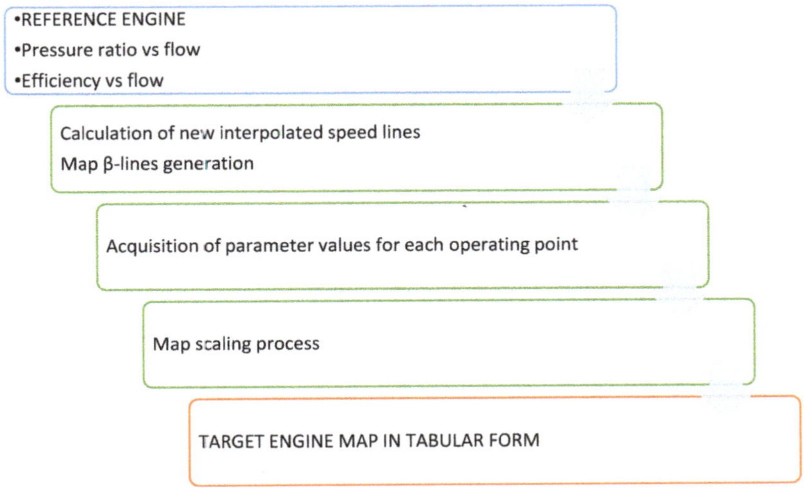

Figure 5. Map creation process.

Table 4. Elements of generated maps.

Map Element	Compressor Map	Turbine Map
Operating points	99	54
β-lines	11	9
Corrected speed lines	9	6

Then, single-point map scaling was attempted using the ratio of the design point value of the target engine and the design point value of the reference map. This was aimed at matching the values at the design point of the target engine without considering the differences on the other operating points. Unfortunately, this scaling failed since the model errors were above 40%, even when using radial compressor maps of similar sizes. Therefore, the alternative method known as applying factors and deltas was implemented. The least squares regression was employed to calculate scaling factors F_i and deltas Δ_i [64], fed by a number of operating points obtained experimentally. The factors and deltas were then used to find the values of the scaled map:

$$W_{target} = F_1 W_{ref} + \Delta_1 \tag{1}$$

$$PR_{target} = F_2 PR_{ref} + \Delta_2 \tag{2}$$

$$\eta_{target} = F_3 \eta_{ref} + \Delta_3. \tag{3}$$

For each map parameter, the values of more than two (four) operating points were considered, so that each of Equations (1)–(3) was an overdetermined system of equations with the scaling factor and delta as unknowns. Hence, as the name of the method suggests, these factors and deltas were obtained in a least square error sense. Although the scaled map does not exactly match the reference map, even at the design point, this method has the advantage of minimising the error between the different operating point data in the least square sense. The final target map can be found in Figure 6.

For the turbine, the Smooth-T tool was used to digitize the plots and generate new interpolated speed lines. However, in this case, instead of creating new β-lines, PR-constant lines were used as β-lines, because these were perpendicular to speed lines in the operating region of interest. Furthermore, no scaling process was required because the reference plots already had dimensionless parameters. The result can be seen in Figure 7, and some numbers regarding the results of the point acquisition can be found in the third column of Table 4. Finally, both maps written in the GasTurb tabular format [63] were loaded into GSP.

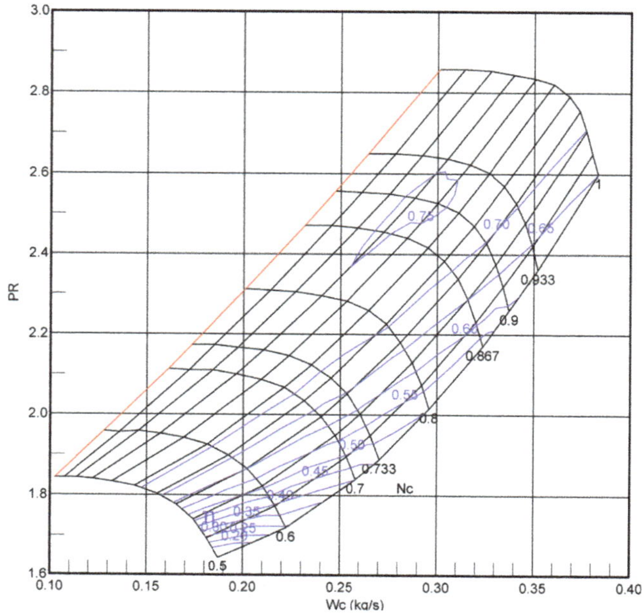

Figure 6. Compressor map.

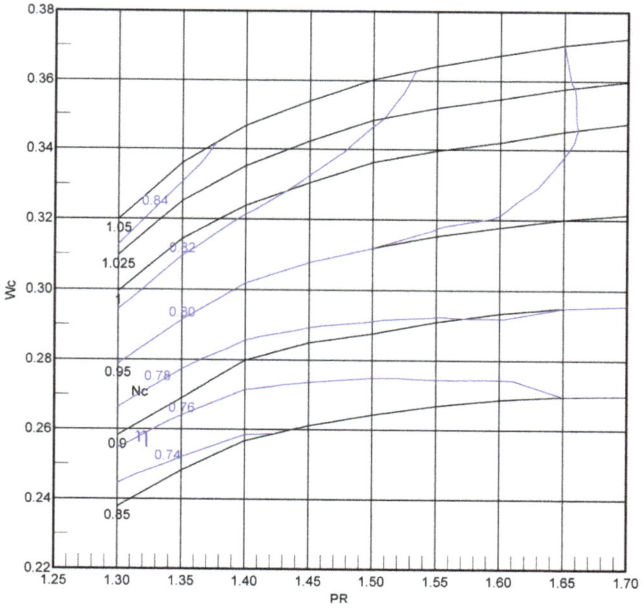

Figure 7. Turbine map.

2.5. GSP Model

The engine model was implemented in GSP, which is an object-oriented 0-D simulation environment [57] in which averaged flow properties are calculated only at the inlet and exit of components while the flow field inside them is not analysed. GSP implements and integrates a set of non-linear differential equations describing the thermodynamic cycle and rotor dynamics [65]. In particular, the acceleration or deceleration of the engine is the effect of power imbalance between the turbine and compressor:

$$I\omega \frac{d\omega}{dt} = P_t - P_c, \tag{4}$$

where P_t and P_c are turbine and compressor power, I is rotor's moment of inertia and ω is rotor angular speed.

The adopted structure of the engine model (Figure 8) follows the standard turbojet template with two Duct components added after the compressor and turbine, which was necessary for transient simulation. All the components were configured to accurately model the engine behaviour by setting their design parameters, such as the maximum speed, pressure ratio, mass flow rate, fuel flow rate, efficiencies and so forth.

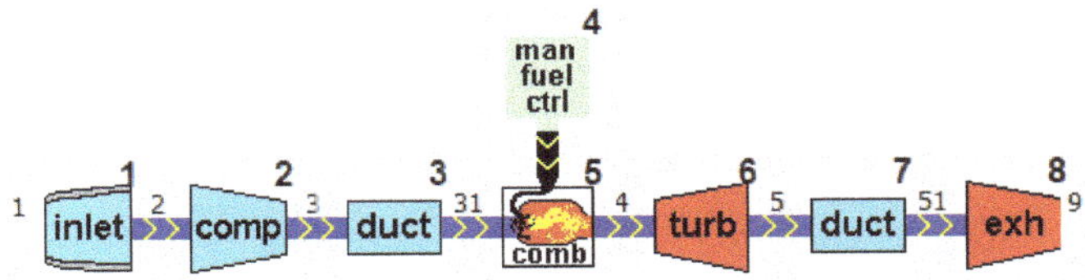

Figure 8. Engine model in GSP.

3. Results

3.1. Design Point Simulation

With Design Point simulation, the performance of the gas turbine design point is fixed to represent a particular gas turbine configuration. The components are 'sized' to the design point using data from the Design tab sheet of Component settings, and the component maps are not used. This kind of simulation is always necessary before Off-Design, Steady-State or Transient calculations since the design point is used as the reference point for off-design operating points. The results of the Design Point simulation are shown in Table 5.

Table 5. Design point simulation.

Rotor Speed rpm	TT_1 °C	PT_1 bar	PT_3 bar	EGT °C	Thrust N	W_f kg/s	W kg/s
125,000	15	1.0133	2.8371	628	142	0.0087	0.35

3.2. Steady-State Simulations

The model was then tuned to simulate a number of steady-state off-design points. They were chosen in accordance with the operating points captured at the rig [59]: 70,000 rpm, 88,000 rpm, 104,000 rpm, 112,000 rpm. It is noteworthy that speeds below 70,000 rpm were not analysed. Simulating those operating points has no practical use because, even if idle speed for both engines is about 35,000 rpm, the flight data show that the engine does not operate in the air under 80,000 rpm.

Figures 9 and 10 show the accuracy of the steady-state simulations: in the two plots, representing the fuel flow and the thrust, respectively, the green line, corresponding to the numerical model, is in agreement with the test bench measurements, especially in the regimes with shaft speed higher than 80,000 rpm, not exceeding a 5% difference in values. This means that the behaviour of the steady-state model is very similar to the real engines.

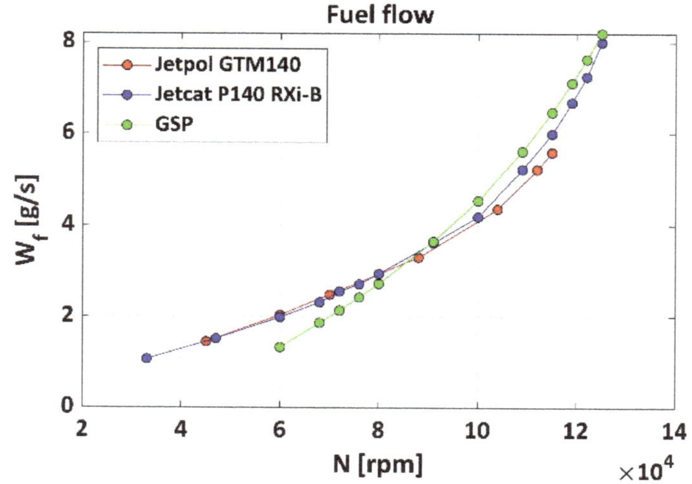

Figure 9. Fuel flow compared with GSP model.

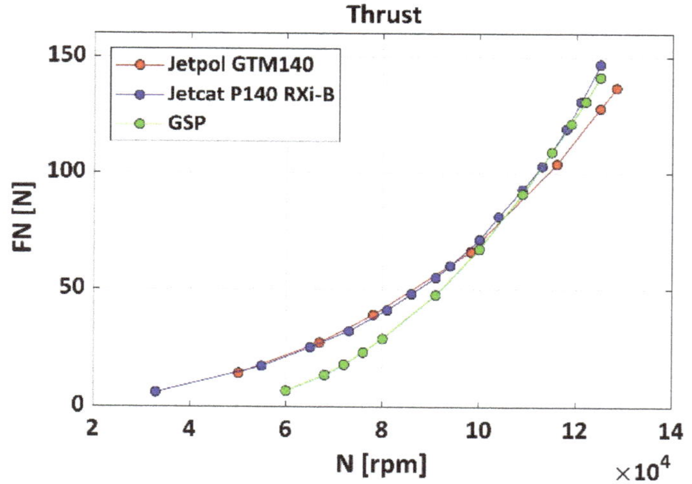

Figure 10. Thrust compared with GSP model.

3.3. Flight Mission Simulation

To assess the accuracy of a transient model, the time evolution of the state variables was observed in response to provided input data. Real engine data from a selected flight (Figures 11 and 12) with a climb and dive maneuver was used for this purpose.

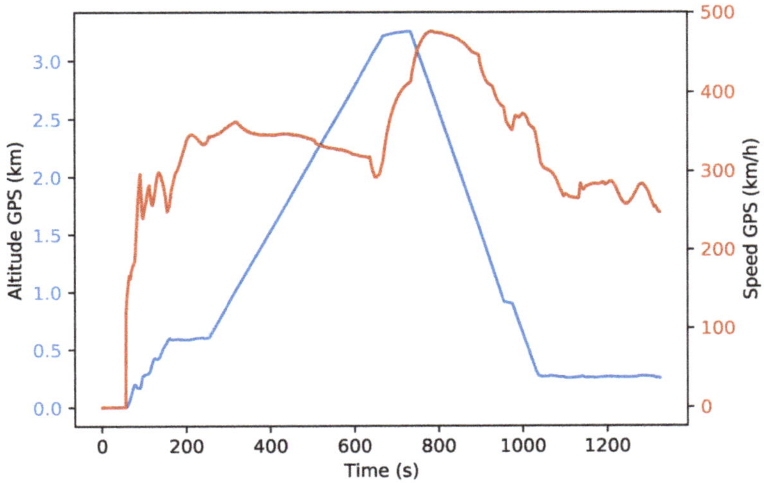

Figure 11. Aircraft speed and flight altitude.

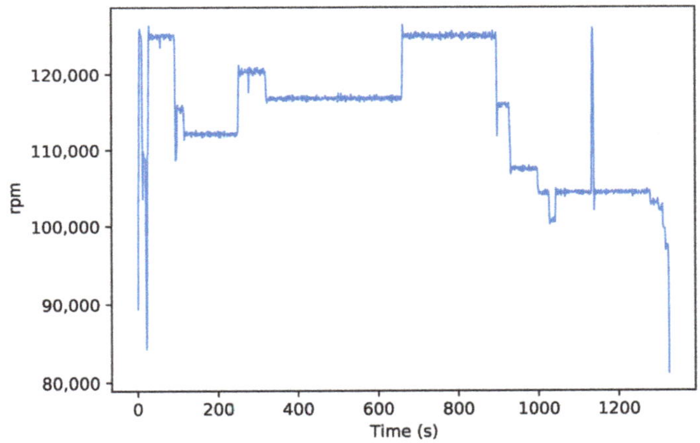

Figure 12. Rotational speed.

An input dataset for the model validation was created by choosing parameters which properly characterise the missions. The selected input variable was rotational speed accompanied by flight parameters such as altitude, Mach number and dTs, which is a temperature correction factor for the ISA model. The input dataset was loaded and processed by GSP using the Manual Case Control Component which was configured to run a transient simulation. After data reduction described in Section 2.3, the time required for the software to simulate a mission was in the order of an hour, which was similar to the real flight duration. Two validation parameters were selected:

1. Exhaust Gas Temperature (EGT)
2. Fuel flow

EGT is probably the most important variable for monitoring the engine health, and usually it is used alone as a validation parameter. The comparison of fuel flow values, in addition, helps to evaluate how the fuel control system reacts to sudden variations of input parameters during real-flight scenarios.

The plots in Figures 13 and 14 show the trend of the parameters chosen for validation as a function of mission time: the comparison demonstrates that the numerical model accurately predicts the behavior of the engine even in situations in which there is a sudden variation in ambient and engine parameters, such as non-stationary portions of a flight mission. Specifically, the graphs shown here are from mission 1 and its simulation. The average error for EGT and fuel flow is within 3%.

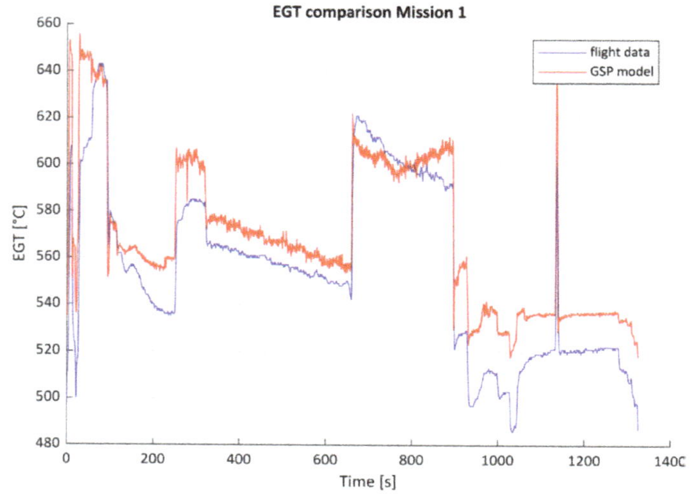

Figure 13. EGT comparison for Mission 1.

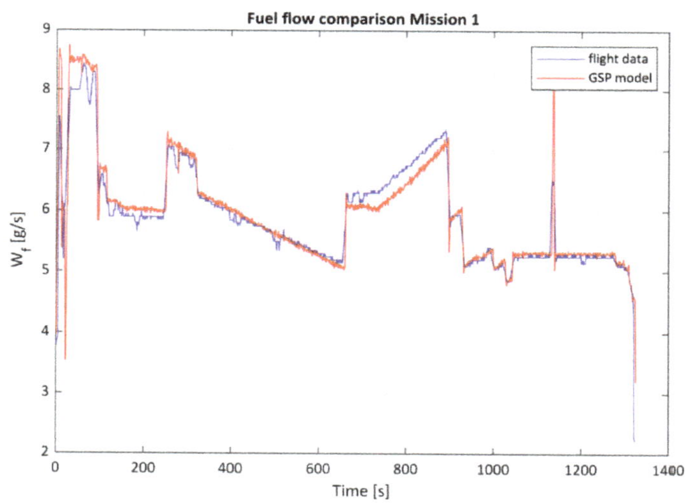

Figure 14. Fuel flow comparison for Mission 1.

4. Discussion

The performance modelling of full-scale engines has well-established methodologies and tools. However, they are not perfectly suited for micro turbojets, so their modelling may pose a challenge despite the basic structure of these engines. A full-size turbojet model cannot be simply scaled-down to a microturbine due to significant differences in efficiency and completely different dynamic response. Here, it was confirmed that library maps scaled at design point led to large model errors. Other minor issues were encountered e.g., the rotor's moment of inertia was close to the minimum value allowed by GSP (10^{-4} kg m^2). Due to very low inertia of the microturbine, the dynamics of transient responses observed in the experimental data were highly affected by sensors and the engine control system, which were not modelled here. The acceleration visible in Figure 12 at t = 1135 s took only a second, which is a negligible time when compared to a full-scale turbojet.

The correct simulation of off-design and transient operation needs compressor and turbine maps that accurately describe the performance of the component over a wide operating range. To describe the component without significant errors, the reference map must refer to a component that is quite similar to the one under analysis, especially in terms of geometry; furthermore, the number of operating points and the scaling method must be sufficient. The maps and model implemented here better describe the engine in the upper speed range used in flight but they are less precise in the low speed region. The reason for this was the low detail of source map plots there i.e., low number of speed lines, related to the general difficulty of modelling compressors and turbines at low speeds. The problem of map extension to zero speed was recently addressed by Kurzke [66] and Ferrer-Vidal et al. [67].

Scaling the maps based on the least squares regression was sufficient and much simpler and faster than the adaptation of maps using the nonlinear optimization method [68,69]. Adaptive modelling methods [70–72] will, however, be necessary for monitoring a fleet of engines with significant manufacturing tolerances or varying degrees of wear.

Considering that:

- Measurement uncertainty at a microturbine is not as low as the uncertainty at a full-scale engine due to the small size of sensors and lack of space for them;
- Performance can vary significantly from one engine to another due to its low cost that implies higher manufacturing tolerances of components;
- Model is zero-dimensional (0D) and does not take into account the geometry of engine components;
- Component maps were produced by CFD analysis for another micro turbojet because the actual geometry or experimental compressor and turbine maps were not available for both analysed engines;

the created GSP engine model showed satisfying results when compared to real flight data, with an average error of each output parameter under 5%. Therefore, it can be concluded that the model was validated.

Although it is not generally recommended, developing one model for two different designs was successful here because both compressors and turbines were almost identical. Thanks to this similarity, the difference in performance between JETPOL GTM 140 and JetCat P140 RXi-B was negligible, similar to the one related to manufacturing tolerances. For this reason, the developed model satisfactorily simulated the steady-state and transient operation of both engines. Its validation was a first step to implement a Digital Twin of the microturbine.

The digital twin model includes three parts: engine entity in physical space, digital engine in digital space, and data/information exchange between physical space and digital space (Figure 15). This virtual engine is a high-fidelity integrated model that dynamically modifies its digital model in real time with the help of real-time data to ensure that its prediction accuracy is close to the reality. Model-based and data-driven fusion methods are essential for the core technology of digital twin model.

The digital engine model will accurately predict the engine performance such as thrust, fuel consumption and emissions, as well as will have the ability of degradation and fault diagnostic and prognostic based on operational data. Engine degradation will be implemented by setting a variation of the isentropic efficiency, the mass flow capacity and the pressure ratio for each level of degradation severity [47]. Then the corresponding change of measurable variable (pressure and temperature) will be estimated. In this way a machine learning algorithm can be applied that based on operating conditions and measurable parameters such as pressure, temperature and rotational speeds can predict the health state conditions of each engine component. The algorithm will predict the component deterioration parameters and the level of degradation by a comparison with the healthy (no degradation) engine values, as well as the degraded performance in terms of thrust and specific fuel consumption.

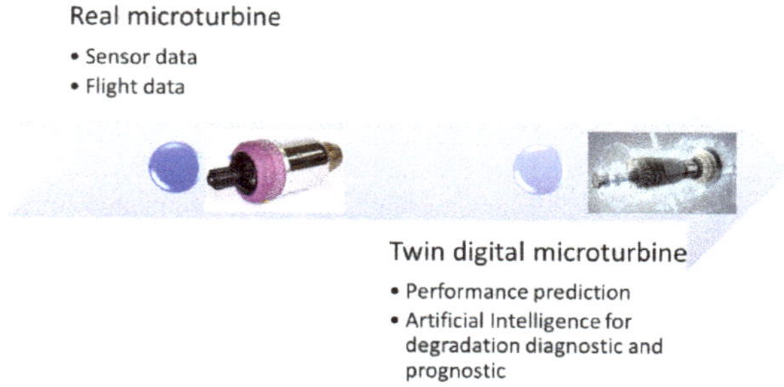

Figure 15. Digital twin of a micro turbine.

The physics-based approach based on numerical simulation methods, such as 3D CFD calculations, can be expensive in terms of computing resources and time. Hence the 0D model, as the one implemented in this work, has advantages in computational time and can be used to simulate the engine in real time. The model-based approach has the advantage of being able to reflect physical processes. However, the implementation of physics-based engine model is complex and it is not perfectly suited for the prediction of slow degradation In this case the data-driven method is more suitable and make it possible to estimate the engine performance according to the sensor data, even if the physical constraints and interpretability are weak. By using an hybrid approach employing physics-based and data-driven models [52], the digital twin model of the microturbine based on 0D performance model and flight data will lead to accurate performance monitoring and life prediction Furthermore, the model will permit a self-optimization ability to reduce fuel consumption when cruise and increase thrust when climb.

5. Conclusions

This paper presents the implementation of the numerical model of a microturbine in steady-state and transient conditions, and its validation using experimental data from a real flight mission. These are the main achievements of the work:

- Generation of suitable compressor and turbine maps, which match the engine performance values from experimental testing;
- Implementation and fine tuning of the model in GSP software, for the Design Point and Off-Design Steady State simulations, and the reproduction of a real flight mission which involved transient engine operation;
- Validation of the model with rig and flight test data.

The results show the suitability of the model for predicting the microturbine performance with a maximum error lower than 5% for a steady-state operation and 3% for a flight mission, thus making the model accurate enough to perform off-design and transient simulation.

To summarize, the approach demonstrated in this study based on using rig and flight data has proven to be effective in creating a transient 0-dimensional model of a microturbine, even without having exact compressor and turbine maps available. The model will be used for predicting emissions and generating training datasets for the developed engine health management system.

Author Contributions: Conceptualization, M.G.D.G. and R.P.; methodology, M.L.E., M.G.D.G. and R.P.; software, M.L.E.; validation, M.L.E., R.P. and M.G.D.G.; investigation, M.L.E.; resources, R.P.; data curation, M.L.E.; writing—original draft preparation, M.L.E. and R.P.; writing—review and editing, R.P. and M.G.D.G.; visualization, M.L.E.; supervision, M.G.D.G. and R.P. All authors have read and agreed to the published version of the manuscript.

Funding: This research received no external funding.

Institutional Review Board Statement: Not applicable.

Informed Consent Statement: Not applicable.

Data Availability Statement: Component maps and MATLAB scripts are documented in thesis [63]. Other data presented in this study are available on request from the corresponding author.

Acknowledgments: The authors would like to thank Michał Czarnecki, Bartosz Gawron, Jarosław Hajduk, Teresa Buczkowska-Murawska and Mariusz Żokowski for their support and providing the data necessary to accomplish this work. We are also grateful to Oscar Kogenhop and the GSP Team for providing the educational licence of GSP.

Conflicts of Interest: The authors declare no conflict of interest.

Abbreviations

The following symbols and abbreviations are used in this manuscript:

ANN	artificial neural network
APU	Auxiliary power unit
CFD	Computational fluid dynamics
CMF	Constant Mass Flow
dTs	temperature correction factor
EASN	European Aeronautics Science Network
ECU	Electronic Control Unit
EHM	engine health management
FN	Net Thrust
EGT	exhaust gas temperature
η	efficiency
GPA	gas path analysis
GPS	Global Positioning System
GSP	Gas turbine Simulation Program
I	moment of inertia
ITWL	The Air Force Institute of Technology in Warsaw
LS	Least Squares
N_c	corrected rotational speed
rpm	revolutions per minute
ω	angular speed
OP	operating point
P	power
PT	total pressure
PR	Pressure ratio
TT	total temperature
UAV	Unmanned Aerial Vehicle
W	mass flow rate
W_c	corrected mass flow rate
W_f	fuel flow

References

1. Schreckling, K. *Gas Turbine Engines for Model Aircraft*; Traplet: Upton upon Severn, UK, 1994.
2. Kamps, B.Y.T. *Model Jet Engines*; Traplet: Malvern, UK, 2005.
3. Gaonkar, D.N.; Patel, R.N. Modeling and simulation of microturbine based distributed generation system. In Proceedings of the 2006 IEEE Power India Conference, New Delhi, India, 10–12 April 2005; Volume 2005, pp. 256–260.

4. Badami, M.; Giovanni Ferrero, M.; Portoraro, A. Dynamic parsimonious model and experimental validation of a gas microturbine at part-load conditions. *Appl. Therm. Eng.* **2015**, *75*, 14–23. [CrossRef]
5. Przysowa, R.; Gawron, B.; Białecki, T.; Łęgowik, A.; Merkisz, J.; Jasiński, R. Performance and emissions of a microturbine and turbofan powered by alternative fuels. *Aerospace* **2021**, *8*, 25. [CrossRef]
6. Gawron, B.; Białecki, T.; Janicka, A.; Suchocki, T. Combustion and Emissions Characteristics of the Turbine Engine Fueled with HEFA Blends from Different Feedstocks. *Energies* **2020**, *13*, 1277. [CrossRef]
7. Alulema, V.; Valencia, E.; Cando, E.; Hidalgo, V.; Rodriguez, D. Propulsion sizing correlations for electrical and fuel powered unmanned aerial vehicles. *Aerospace* **2021**, *8*, 171. [CrossRef]
8. Adamski, M. Analysis of propulsion systems of unmanned aerial vehicles. *J. Mar. Eng. Technol.* **2018**, *16*, 291–297. [CrossRef]
9. Tang, W.; Wang, L.; Gu, J.; Gu, Y. Single neural adaptive PID control for small UAV micro-turbojet engine. *Sensors* **2020**, *20*, 345 [CrossRef]
10. Minijets. The Website for Fans of Light Jet Aircraft. Available online: https://minijets.org (accessed on 25 November 2021).
11. Pavlenko, D.; Dvirnyk, Y.; Przysowa, R. Advanced materials and technologies for compressor blades of small turbofan engines *Aerospace* **2021**, *8*, 1. [CrossRef]
12. Rodgers, C. *Some Effects of Size on the Performances of Small Gas Turbines*; Volume 3: Turbo Expo 2003; ASMEDC: New York, NY, USA, 2003; Volume 3, pp. 17–26. [CrossRef]
13. Oppong, F.; Spuy, S.J.V.D.; Diaby, A.L. An overview on the performance investigation and improvement of micro gas turbine engine. *R&D J. S. Afr. Inst. Mech. Eng.* **2015**, *31*, 35–41. [CrossRef]
14. Capata, R.; Saracchini, M. Experimental campaign tests on ultra micro gas turbines, fuel supply comparison and optimization *Energies* **2018**, *11*, 799. [CrossRef]
15. Large, J.; Pesyridis, A. Investigation of micro gas turbine systems for high speed long loiter tactical unmanned air systems *Aerospace* **2019**, *6*, 55. [CrossRef]
16. Kadosh, K.; Cukurel, B. Micro-Turbojet to Turbofan Conversion Via Continuously Variable Transmission: Thermodynamic Performance Study. *J. Eng. Gas Turbines Power* **2017**, *139*, 022603. [CrossRef]
17. Fulara, S.; Chmielewski, M.; Gieras, M. Variable geometry in miniature gas turbine for improved performance and reduced environmental impact. *Energies* **2020**, *13*, 5230. [CrossRef]
18. Villarreal-Valderrama, F.; Zambrano-Robledo, P.; Hernandez-Alcantara, D.; Amezquita-Brooks, L. Turbojet thrust augmentation through a variable exhaust nozzle with active disturbance rejection control. *Aerospace* **2021**, *8*, 293. [CrossRef]
19. Walsh, P.P.; Fletcher, P. *Gas Turbine Performance*, 2nd ed.; Blackwell Science Ltd.: Oxford, UK, 2004.
20. Davison, C.R.; Birk, A.M. *Comparison of Transient Modeling Techniques for a Micro Turbine Engine*; Volume 5: Marine; Microturbines and Small Turbomachinery; Oil and Gas Applications; Structures and Dynamics, Parts A and B; ASMEDC: New York, NY, USA, 2006; pp. 449–458. [CrossRef]
21. Wang, C.; Li, Y.; Yang, B. Transient performance simulation of aircraft engine integrated with fuel and control systems. *Appl. Therm. Eng.* **2017**, *114*, 1029–1037. [CrossRef]
22. Yepifanov, S.; Zelenskyi, R.; Sirenko, F.; Loboda, I. *Simulation of Pneumatic Volumes for a Gas Turbine Transient State Analysis*; Volume 6: Ceramics; Controls, Diagnostics and Instrumentation; Education; Manufacturing Materials and Metallurgy; American Society of Mechanical Engineers: Charlotte, NC, USA, 2017. [CrossRef]
23. Kong, C.; Ki, J.; Chung, S. Performance simulation of a turboprop engine for basic trainer. *KSME Int. J.* **2002**, *16*, 839–850. [CrossRef]
24. Tsoutsanis, E.; Meskin, N.; Benammar, M.; Khorasani, K. Dynamic performance simulation of an aeroderivative gas turbine using the matlab simulink environment. In *ASME International Mechanical Engineering Congress and Exposition, Proceedings (IMECE)*; ASME: New York, NY, USA, 2013. [CrossRef]
25. Wang, B.; Wang, X.; Liu, Y. A high by-pass ratio turbofan model based on CMF method. *Appl. Mech. Mater.* **2013**, *302*, 578–582. [CrossRef]
26. Chachurski, R.; Trzeciak, A.; Jedrowiak, B. Comparison of the results of mathematical modeling of a GTM 120 miniature turbine jet engine with the research results. *Combust. Engines* **2018**, *173*, 30–33. [CrossRef]
27. Ahmadian, N.; Khosravi, A.; Sarhadi, P. Adaptive control of a jet turboshaft engine driving a variable pitch propeller using multiple models. *Mech. Syst. Signal Process.* **2017**, *92*, 1–12. [CrossRef]
28. Fathy, T.; Elzahaby, A.; Khalil, M. Micro TJE centrifugal compressor performance prediction Tamer S. *J. Eng. Sci. Mil. Technol.* **2018**, *2*, 185–203. [CrossRef]
29. Grannan, N.D.; Hoke, J.; McClearn, M.J.; Litke, P.; Schauer, F. Trends in jetCAT microturbojet-compressor efficiency. In Proceedings of the AIAA SciTech Forum—55th AIAA Aerospace Sciences Meeting, Grapevine, TX, USA, 9–13 January 2017. [CrossRef]
30. Jensen, C.U.; Lebreton, H.; Nielsen, S.S.; Rasmussen, K.M. *Modeling and Validation of the SR-30 Turbojet Engine*; Aalborg University: Aalborg, Denmark, 2012; p. 107.
31. Villarreal-Valderrama, F. *Analysis and Modeling of Micro Turbojets a Comprehensive Model Based on Multiphysics Principles*; Autonomous University of Nuevo León: San Nicolás de los Garza, Mexico, 2019.
32. Kho, S.; Park, H. Design of the Electronic Engine Control Unit Performance Test System of Aircraft. *Aerospace* **2021**, *8*, 158. [CrossRef]

33. Polat, C. An Electronic Control Unit Design for a Miniature Jet Engine. Ph.D. Thesis, Middle East Technical University, Ankara, Turkey, 2009.
34. Pakmehr, M.; Fitzgerald, N.; Feron, E.; Paduano, J.; Behbahani, A. Physics-based dynamic modeling of a turboshaft engine driving a variable pitch propeller. *J. Propuls. Power* **2016**, *32*, 646–658. [CrossRef]
35. Andoga, R.; Fozo, L.; Kovács, R.; Beneda, K.; Moravec, T.; Schreiner, M. Robust control of small turbojet engines. *Machines* **2019**, *7*, 3. [CrossRef]
36. Kurzke, J.; Halliwell, I. Component Performance. In *Propulsion and Power: An Exploration of Gas Turbine Performance Modeling*; Springer International Publishing: Cham, Switzerland, 2018; pp. 439–575._12. [CrossRef]
37. Czarnecki, M.; Olsen, J. Combined methods in preliminary micro scale gas turbine diffuser design—A practical approach. *J. Appl. Fluid Mech.* **2018**, *11*, 567–575. [CrossRef]
38. Sebelev, A.A.; Tikhonov, A.S.; Aleksenskiy, V.A.; Shengals, A.A.; Klyavin, O.I. Aerodynamic analysis of the small-scaled centrifugal compressor for micro-turbojet engine applications. *J. Phys. Conf. Ser.* **2021**, *1891*, 012017. [CrossRef]
39. Bar, W.; Czarnecki, M. Design-point, off-design meanline performance analysis and cfd computations of the axial turbine to micro gas-turbine engine. *J. KONES Powertrain Transp.* **2009**, *16*, 9–16.
40. Suchocki, T.; Lampart, P.; Klonowicz, P. Numerical investigation of a GTM-140 turbojet engine. *Open Eng.* **2015**, *5*, 115–116. [CrossRef]
41. Briones, A.M.; Sykes, J.P.; Rankin, B.A.; Caswell, A.W. Steady-state cfd simulations of a small-scale turbojet engine from idle to cruise conditions. In Proceedings of the AIAA Scitech 2020 Forum, Orlando, FL, USA, 6–10 January 2020; pp. 1–18. [CrossRef]
42. Briones, A.M.; Caswell, A.W.; Rankin, B.A. Fully Coupled Turbojet Engine Computational Fluid Dynamics Simulations and Cycle Analyses Along the Equilibrium Running Line. *J. Eng. Gas Turbines Power* **2021**, *143*, 061019. [CrossRef]
43. Kong, C.; Ki, J.; Kang, M. A New Scaling Method for Component Maps of Gas Turbine Using System Identification. *J. Eng. Gas Turbines Power* **2003**, *125*, 979–985. [CrossRef]
44. Rademaker, E.R. *Scaling of Compressor and Turbine Maps on Basis of Equal Flow Mach Numbers and Static Flow Parameters*; Technical Report; National Aerospace Laboratory NLR: Amsterdam, The Netherlands, 2012.
45. Tsoutsanis, E.; Meskin, N.; Benammar, M.; Khorasani, K. *An Efficient Component Map Generation Method for Prediction of Gas Turbine Performance*; Volume 6: Ceramics; Controls, Diagnostics and Instrumentation; Education; Manufacturing Materials and Metallurgy; American Society of Mechanical Engineers: New York, NY, USA, 2014; p. V006T06A006. [CrossRef]
46. Coban, K.; Ekici, S.; Colpan, C.O.; Karakoç, T.H. Performance of a microjet using component map scaling. *Aircr. Eng. Aerosp. Technol.* 2021, ahead-of-p. [CrossRef]
47. De Giorgi, M.G.; Campilongo, S.; Ficarella, A. A diagnostics tool for aero-engines health monitoring using machine learning technique. *Energy Procedia* **2018**, *148*, 860–867. [CrossRef]
48. Fentaye, A.D.; Baheta, A.T.; Gilani, S.I.; Kyprianidis, K.G. A Review on Gas Turbine Gas-Path Diagnostics: State-of-the-Art Methods, Challenges and Opportunities. *Aerospace* **2019**, *6*, 83. [CrossRef]
49. Rahman, M.; Zaccaria, V.; Zhao, X.; Kyprianidis, K. Diagnostics-Oriented Modelling of Micro Gas Turbines for Fleet Monitoring and Maintenance Optimization. *Processes* **2018**, *6*, 216. [CrossRef]
50. Kumarin, A.; Kuznetsov, A.; Makaryants, G. Hardware-in-the-loop neuro-based simulation for testing gas turbine engine control system. In Proceedings of the 2018 Global Fluid Power Society PhD Symposium, GFPS 2018, Samara, Russia, 18–20 July 2018. [CrossRef]
51. L'Erario, G.; Fiorio, L.; Nava, G.; Bergonti, F.; Mohamed, H.A.O.; Benenati, E.; Traversaro, S.; Pucci, D.; L'Erario, G.; Fiorio, L.; et al. Modeling, Identification and Control of Model Jet Engines for Jet Powered Robotics. *IEEE Robot. Autom. Lett.* **2020**, *5*, 2070–2077. [CrossRef]
52. De Giorgi, M.G.; Quarta, M. Hybrid MultiGene Genetic Programming—Artificial neural networks approach for dynamic performance prediction of an aeroengine. *Aerosp. Sci. Technol.* **2020**, *103*, 105902. [CrossRef]
53. De Giorgi, M.G.; Campilongo, S.; Ficarella, A. Development of a real time intelligent health monitoring platform for aero-engine. *MATEC Web Conf.* **2018**, *233*, 00007. [CrossRef]
54. Derbel, K.; Beneda, K. Development of Airborne Test Environment for Micro Turbojet Engine—Part II: Remote Measurement System. In Proceedings of the NTinAD 2020—New Trends in Aviation Development 2020—15th International Scientific Conference, Proceedings, The High Tatras, Slovakia, 17–18 September 2020; pp. 43–48. [CrossRef]
55. Kong, C.; Park, J.; Kang, M. *A Study on Transient Performance Characteristics of the CRW Type UAV Propulsion System during Flight Mode Transition*; Volume 5: Turbo Expo 2005; ASMEDC: New York, NY, USA, 2005; Volume 5, pp. 163–170. [CrossRef]
56. Henke, M.; Monz, T.; Aigner, M. Introduction of a New Numerical Micro Gas Turbine Cycle Dynamics. *J. Eng. Gas Turbines Power* **2017**, *139*, 042601. [CrossRef]
57. Visser, W.P.J.; Broomhead, M.J. *GSP, A Generic Object-Oriented Gas Turbine Simulation Environment*; Volume 1: Aircraft Engine; Marine; Turbomachinery; Microturbines and Small Turbomachinery; American Society of Mechanical Engineers: New York, NY, USA, 2000. [CrossRef]
58. *GSP 11 User Manual*; NLR—Royal Netherlands Aerospace Centre: Amsterdam, The Netherlands, 2021.
59. Gawron, B.; Białecki, T. Measurement of exhaust gas emissions from miniature turbojet engine. *Combust. Engines* **2016**, *167*, 58–63. [CrossRef]

60. Hajduk, J.; Rykaczewski, D. Possibilities of Developing aerial target system JET-2 (Możliwości rozwoju zestawu odrzutowych celów powietrznych Zocp-Jet2). In *Mechanika w Lotnictwie ML-XVIII tom 2 (Mechanics in Aviation ML-XVIII Volume 2)*; Krzysztof, S., Ed.; PTMTS: Warszawa, Poland, 2018; pp. 139–154.
61. Rykaczewski, D. Implementation of national tests on the example of the aerial JET-2 target system (Realizacja badań państwowych na przykładzie zestawu odrzutowych celów powietrznych Zocp-Jet2). In *Mechanika w Lotnictwie ML-XVIII tom 2 (Mechanics in Aviation ML-XVIII Volume 2)*; Krzysztof, S., Ed.; PTMTS: Warszawa, Poland, 2018; pp. 221–230.
62. Buczkowska-Murawska, T.; Zokowski, M. Using the telemetry system as an element of the engine operation monitoring system of UAS. *J. KONBiN* **2021**, *52*, 25–33. [CrossRef]
63. Erario, M.L. *Model-Based Dynamic Simulation of a Microturbine and Performance Prognostics Using Artificial Neural Networks*; University of Salento: Lecce, Italy, 2021.
64. Sankar, B.; Shah, B.; Thennavarajan, S.; Vanam, V. On Gas Turbine Simulation Model Development. In Proceedings of the National Conference on Condition Monitoring (NCCM), Bangalore, India, 4–5 October 2013; pp. 1–17.
65. Visser, W.P.J.; Broomhead, M.J.; Kogenhop, O.; Rademaker, E.R. *Technical Manual of Gas Turbine Simulation Program*; Technical Report NLR-TR-2010-343-Issue-2; National Aerospace Laboratory NLR: Amsterdam, The Nederlands, 2010.
66. Kurzke, J. Turbine Map Extension—Theoretical Considerations and Practical Advice. *J. Glob. Power Propuls. Soc.* **2020**, *4*, 176–189. [CrossRef]
67. Ferrer-Vidal, L.E.; Pachidis, V.; Tunstall, R.J. Generating axial compressor maps to zero speed. *Proc. Inst. Mech. Eng. Part A J. Power Energy* **2021**, *235*, 956–973. [CrossRef]
68. Misté, G.A.; Benini, E. Turbojet Engine Performance Tuning with a New Map Adaptation Concept. In Proceedings of the ASME 2013 Gas Turbine India Conference, Bangalore, India, 5–6 December 2013; pp. 1–10. [CrossRef]
69. Khustochka, O.; Chernysh, S.; Yepifanov, S.; Ugryumov, M.; Przysowa, R. Estimation of performance parameters of turbine engine components using experimental data in parametric uncertainty conditions. *Aerospace* **2020**, *7*, 6. [CrossRef]
70. Visser, W.P.J.; Kogenhop, O.; Oostveen, M. A Generic Approach for Gas Turbine Adaptive Modeling. In *Turbo Expo: Power for Land, Sea, and Air*; Volume 2: Turbo Expo 2004; AMSE: New York, NY, USA, 2004; pp. 201–208. [CrossRef]
71. Bauwens, P. Gas Path Analysis for the MTT Micro Turbine. Ph.D. Thesis, TU Delft, Delft, The Netherlands, 2015.
72. Yan, B.; Hu, M.; Feng, K.; Jiang, Z. Enhanced component analytical solution for performance adaptation and diagnostics of gas turbines. *Energies* **2021**, *14*, 4356. [CrossRef]

aerospace

MDPI

Article

Plasma Assisted Re-Ignition of Aeroengines under High Altitude Conditions

Ghazanfar Mehdi, Sara Bonuso and Maria Grazia De Giorgi *

Department of Engineering for Innovation, University of Salento, Via per Monteroni, 73100 Lecce, Italy; ghazanfar.mehdi@unisalento.it (G.M.); sara.bonuso@unisalento.it (S.B.)
* Correspondence: mariagrazia.degiorgi@unisalento.it

Abstract: Re-ignition of aeroengines under high altitude conditions is of great importance to the safety and use of lean-burn flame. This study is focused on the experimental and numerical characterization of flow dynamics and flame re-ignition in a rectangular burner. A ring-needle type plasma actuator was considered and run by high-voltage (HV) nanopulsed plasma generator. The electrical power delivered to the fluid and an optimal value of reduced electric field (E_N) was calculated considering non-reactive flow. Smoke flow visualizations using a high-speed camera and proper orthogonal decomposition (POD) were performed to recognize the most dominant flow structures. Experimental results revealed the transport effects due to plasma discharge, such as the induced flow, that could have a strong impact on the recirculation zone near the corners of combustor, improving the mixing performance and reducing the ignition delay time. Two different numerical tools (ZDPlasKin and Chemkin) were used to investigate the ignition characteristics. ZDPlasKin calculated the thermal effect and the plasma kinetic of nanopulsed plasma discharge at the experimentally measured E_N. Finally, based on the output of ZDPlasKin, Chemkin estimated the flame ignition at low pressure and low temperature conditions. It was noticed that time required to achieve the maximum flame temperature with plasma actuation is significantly less than the auto-ignition time ('clean case', simulation result of the model without considering the plasma effect). Maximum reduction in ignition time was observed at inlet pressure 1 bar (3.5×10^{-5} s) with respect to the clean case (1.1×10^{-3} s). However, as the inlet pressure is reduced, the ignition delay time was increased. At 0.6 bar flame ignition occurred in clean case at 0.0048 s and at 0.0022 s in presence of the plasma actuation, a further decrease of the pressure up to 0.4 bar leads the ignition at 0.0027 s and 0.0063 s in clean and plasma actuation, respectively.

Keywords: plasma assisted ignition; flow structures; nanopulsed plasma; ignition delay time; low temperature; low pressure

Citation: Mehdi, G.; Bonuso, S.; De Giorgi, M.G. Plasma Assisted Re-Ignition of Aeroengines under High Altitude Conditions. *Aerospace* **2022**, *9*, 66. https://doi.org/10.3390/aerospace9020066

Academic Editors: Spiros Pantelakis, Andreas Strohmayer and Liberata Guadagno

Received: 29 December 2021
Accepted: 20 January 2022
Published: 26 January 2022

1. Introduction

Lean and premixing combustion techniques are one of the most prominent methods of lowering the flame temperature and subsequently NO_x emissions, however, this approach produces severe flame instabilities which have to be avoided. Moreover, at present, a major challenge of aerospace combustors is ignition at high altitude conditions characterized by low pressure and low temperature. In this context, several passive control strategies have been used to investigate flame stabilization by modifying the combustor geometry in a permanent way. However, plasma-assisted combustion technology is the most encouraging technique to actively control the combustion characteristics under real-time operating conditions [1].

These days, two different types of plasmas, one is the thermal plasma (equilibrium plasma) in which rotational and vibrational temperatures and particularly the electron temperature of active species are in thermal equilibrium and electron number density and neutral gas temperature are very high, for example plasma torch and spark plug. The other type of plasma is the non-thermal plasma (NTP). It is also known as non-equilibrium plasma or cold plasma. In NTP, the electrical power is primarily used to produce energetic electrons useful for activating the reagents, while the gas temperature remains constant Recently, the use of NTP is considered one of the most promising techniques to improve the flame ignition by shortening the ignition delay time, increasing lean flame stabilization, accelerating low-temperature oxidation, and extending lean blowout limits [1,2]. NTP has great capability to improve the flame ignition and combustion thanks to thermal effects (via temperature rise), kinetic effects (via plasma generated vibrationally and electronically excited molecules and neutral radicals), and transport effects. These last effects are given by diffusion transport enhancement effect via fuel reforming and low temperature oxidation and convective transport improvement due to plasma generated ionic wind, hydrodynamic instability, and flow motion via Coulomb and Lorentz forces [3].

Successful demonstrations of plasma-assisted ignition (PAI) has been performed by using different NTPs—such as atmospheric pressure discharge, microwave discharge, radio frequency discharge, corona discharge, surface discharge, direct current glow discharge (DCGD), laser ignition, and dielectric barrier discharge (DBD) [4]. Among NTP technologies, the dielectric barrier discharge (DBD) is widely considered and credit goes to the simplicity of its design and the wide range of operating modes. DBD plasma actuators showed great interest in aeronautics due to reattaching of separated flows and controlling of the boundary layer from transition to turbulence [5,6] due to the induced flow field, hence they could be used also to modify the flow dynamics close to the recirculation area of the combustors. The arrangement of DBD is based on two electrodes; one is connected with high voltage while the other one is grounded. The time varying voltage is required to drive the DBD in occurrence of capacitive coupling [4]. Basically, DBD is working into two different operations modes: the sinusoidal wave form and nanosecond repetitively pulsed discharge (NRPD). The NRPD has garnered much attention due to its wide capacity of generating excited states and active radicals by ionization, dissociation, and electron impact excitation reactions, which could be due to of high reduced electric field EN values which improved the flame ignition [7]. Moreover, NRPD also generate the fast gas heating which helps to improve the ignition via improvement in thermal pathways [8]. However, during the NRPD electron impact reactions (plasma chemistry) have different time scales than the gas phase reactions (combustion chemistry) [9].

Bernard [10,11] experimentally investigated a turbulent air jet under the effects of DBD driven by sinusoidal generator. The turbulent spectrum, turbulent kinetic energy, and velocity profiles have been analyzed; it was also noticed that DBD plasma actuation could lead to improve the mixing performance. Xu [12] studied the impacts of AC-DBD in coaxial methane/air burner. During the tests, the high-voltage electrode was around the outer burner tube whereas an inner burner tube acted as a grounded electrode. The authors noticed that flame structure was strongly affected by the electric field EN generated by means of the AC-DBD plasma discharge.

Many experimental and numerical studies have been performed to develop the kinetic model including plasma actuated reactions to investigate the ignition enhancement by NRPD. Aleksandrov [9] performed the plasma kinetic modeling of CH4/air mixture and revealed that the neutral radicals and active species produced through NRDP accelerated the ignition process. However, only atoms and free radicals were considered in their simulation work and the effects of excited species were neglected. Ju [7] studied the kinetic mechanism by only including electronical species of oxygen and hydrogen by neglecting the role of vibrational species. Kosarev [13] focused on the numerical and experimental approaches to characterize the ignition kinetics of CH4:O2:Ar mixture at temperature ranges of 1230 K to 1719 K and pressure of 0.3 bar to 1.1 bar by subjecting high-voltage NRDP in the shock tube. It has been noticed that ignition times were considerably reduced with the use of NRDP. However, this study neglected the evolution of vibrational and electronical excitation species. Starik [14] investigated the combustion mechanism of methane/air mixture, including kinetics of singlet oxygen but without the modeling of plasma discharge and vibrational species. Uddi [15] performed the experimental and comparative measurements of oxygen atom in air and methane–air mixture subjecting NRDP by using laser induced fluorescence, but they neglected the treatment of vibrational species.

Although many studies have been performed during the last decade, the accurate mechanism and fundamental knowledge of ignition enhancement by plasma discharge is still at exploratory stages. Furthermore, few studies focused on the re-ignition of aeroengine under high altitude conditions is of great importance to the safety and use of lean-burn.

Hence, this study performed a numerical characterization of the ignition delay time in a rectangular burner, particularly under high altitude conditions. A ring-needle type plasma actuator powered by a high-voltage nanosecond pulsed generator was used. Firstly, an experimental analysis was performed. The electrical power delivered to the fluid and an optimal value of reduced electric field (EN) was measured through dedicated experiments under non-reactive flow, which could be appropriate for the numerical analysis of flame ignition. Smoke flow visualizations were also carried out by using a high-resolution camera to investigate the fluid dynamics effects of plasma actuation on the flow behavior. The preliminary POD analysis was also performed to recognize the major dominant flow structures generated at different plasma actuation conditions. Finally, two different numerical tools (ZDPlasKin and Chemkin) were used to investigate the flame ignition characteristics under high altitude conditions.

2. Experimental Setup and Electrical Characterization

The experimental setup was implemented to study the effects on the flow of NRPD such as the induced flow in the recirculation zone. A ring-needle plasma actuator was installed in a rectangular burner and operated by nanosecond pulsed generator. The plasma actuator consisted of two copper electrodes, one is a ring type electrode with the inner diameter = 32 mm and the outer diameter = 52 mm, connected to the HV. The second one is the grounded electrode that is a needle with a length of 330 mm and the diameter equal to 1 mm, placed at the center of the ring electrode at the standoff distance s = 0 mm, and acted as grounded. The standoff distance referred to the position of the needle electrode from the upper edge of the ring electrode.

The effects of the output voltage of the HV power supply and of the standoff distance were analyzed, as reported in Table 1. At first, two different percentage values of the HV voltage were considered, in particular 20% and 40% of the maximum output voltage of the power supply V_{max}, by keeping constant the standoff distance equal to 0 mm. Secondly, the standoff distance varied from 0 mm to 20 mm and 40 mm at 40%V_{max} applied voltage and frequency.

Table 1. Plasma actuated experimental conditions.

Test Case No.	Actuator Configuration		Actuation Conditions			Standoff Distance (mm)
	High-Voltage	Ground	Generator	Amplitude (% V_{max})	Frequency (% F_{max})	
1.	Disk	Needle	Nanopulsed Generator	20%	30%	00
2.				40%	30%	00
3.				40%	30%	20
4.				40%	30%	40

Figure 1 shows the sketch of the rectangular burner with the position of grounded electrode (GND) and the high-voltage (HV) electrode, location of plasma, and expected instantaneous streamlines. Plasma was generated at the tip of needle electrode and at the corners of disk electrode as in Figure 1b. Figure 1c,d depicts the instantaneous velocity streamlines for the test case at high-voltage amplitude equal to 40%V_{max}, frequency equal to 30% and standoff distance equal to 0 mm and 20 mm, respectively. They clearly define the recirculation zone at the rectangular corners of the burner that could help to promote the combustion process. All the tests were conducted considering non-reactive flow at ambient conditions. A faraday cage was used to protect the system from electromagnetic field caused by high E_N. Figure 2 depicts a detailed view of experimental setup for combustion and electrical characterization. The electrical setup consisted of HV nanosecond pulsed generator, current transformer, HV probe, oscilloscope, and a personal computer. The current transformer and HV probe were connected to the oscilloscope to acquire the time dependent voltage and current signals.

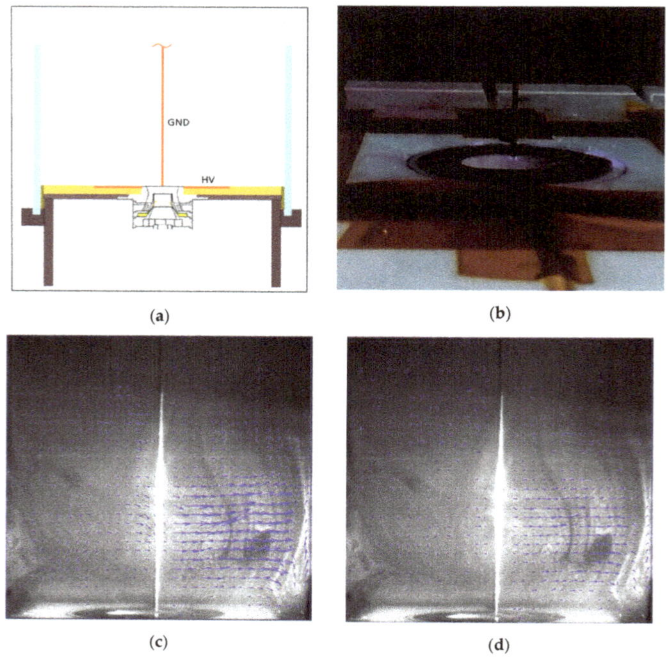

Figure 1. (**a**) A sketch of the burner indicating the electrodes, GHD: grounded electrode and HV: annular high-voltage electrode, (**b**) location of plasma, (**c**) instantaneous streamlines for the test case 2 (standoff distance 0 mm), and (**d**) instantaneous streamlines for the test case 3 (standoff distance 20 mm).

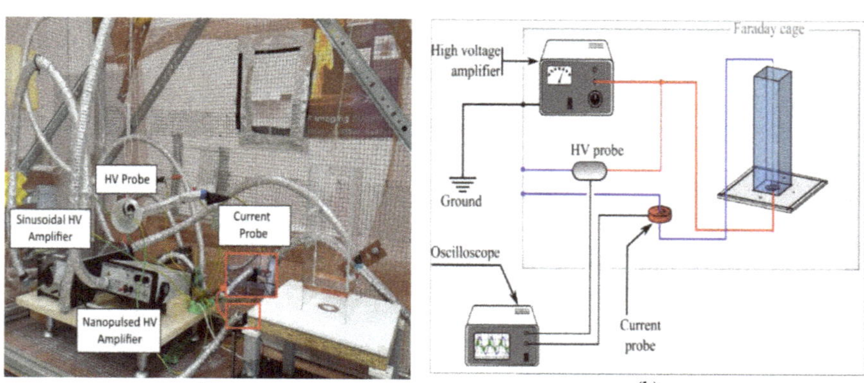

Figure 2. (**a**) Experimental setup; (**b**) Electrical characterization.

3. Numerical Procedure

Numerical modeling of PAI has been performed by using a zero-dimensional plasma kinetic model (ZDPlasKin) [16] and the chemical kinetic model (CHEMKIN) [17]. Figure 3 depicts the detailed explanation of numerical methodology. The extended version of chemical kinetics mechanisms for methane/air plasma assisted combustion was implemented from the literature by considering both electron impact reactions and gas phase reactions [18–21]. The mechanism consisted of 161 species and 1382 reactions—including charged transfer reactions, vibrationally excited reactions, electronically excited reactions, three body recombination reactions, ionization, relaxation, and dissociation reactions. ZDPlasKin solver was used to predict the thermal and kinetic effects of NRPD on pre-mixed methane/air at the same measured experimental reduced electric field E_N. Boundary conditions for numerical analysis in terms of pressure, temperature, and electrical feeding were same as in Experimental Test 2 (Table 1). The Boltzmann equation solver was coupled with ZDPlaskin to estimate the temporal evolution of neutral radicals and active species. Finally, based on the output of ZDPlasKin, Chemkin was used to predict the flame ignition at high altitude conditions—i.e., low pressure and low temperature. The closed homogenous batch reactor was used for the ignition analysis. The detailed description of numerical procedure and experimental validation of extended version of mechanism has been presented in our previously published articles [22,23].

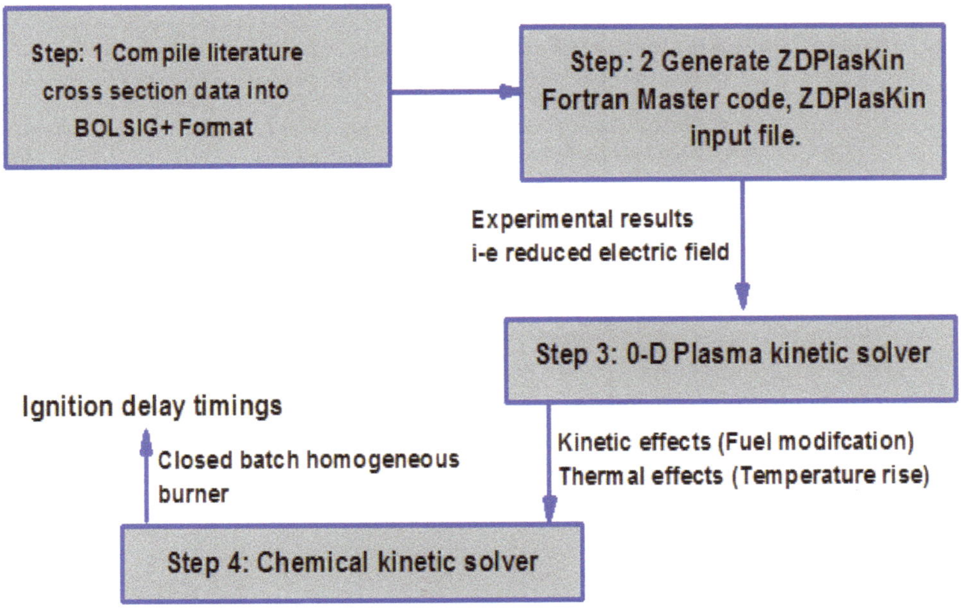

Figure 3. Detailed description of numerical methodology.

4. Results and Discussions

4.1. Electrial Characterization

Table 2 presented the electrical output in terms of electrical power delivered to the non-reactive flow at different plasma actuation conditions; in particular, the electrical power per pulse and the mean electrical power were reported.

Table 2. Electrical output of plasma actuated test cases.

Test Case No.	Voltage Peak (V)	Repetition Rate (Hz)	Pulse Electrical Power (W)	Mean Electrical Power (W)	Uncertainty Pulse Electrical Power (W)
1.	14,152.01	813.6697	644.4663	1.2581	70
2.	16,795.8	755.287	660.782	1.1974	70
3.	166.6829	1196.602	4.2125	0.0121	0.4
4.	189.9815	1059.098	6.2117	0.0158	0.5

It has been noticed that the power delivered to fluid is almost same in Test 1 and 2 however, with the increase of standoff distances, it significantly reduced as in Test 3 and 4. Therefore, it was concluded that controlling of plasma discharge could be possible in two ways, either to control voltage amplitude or to adjust the standoff distance by moving needle in *y*-axis. The time dependent evolution of voltage-current characteristics at different HV amplitude voltages and standoff distances has been shown in Figure 4. The voltage signal was highly variable due to NRPD mode of actuation. The current signals presented several high-amplitude spikes due to the plasma discharges that lead to a fast electrical impedance change within the actuator. The maximum peak voltage was noticed in Test 2 which was selected for numerical analysis. The voltage signal and gap distance between ring and needle electrodes were considered for the prediction of the E_N, because E_N is the major factor to control energy deposition and production of active species during NRPD Energy obtained by electric field is transmitted to the various excitation states via electron impacted reactions. The temporal evolution of E_N of Test 2 is shown in Figure 5, it shows the several peaks during the single pulse discharge. The mean integral value of E_N in the time interval of 10^{-6} was about 200 Td. This value has been previously proved to improve the combustion characteristic time of methane flame [22–24]. Therefore, numerical analysis of ignition characteristics has been performed by considering the electrical output of Test 2

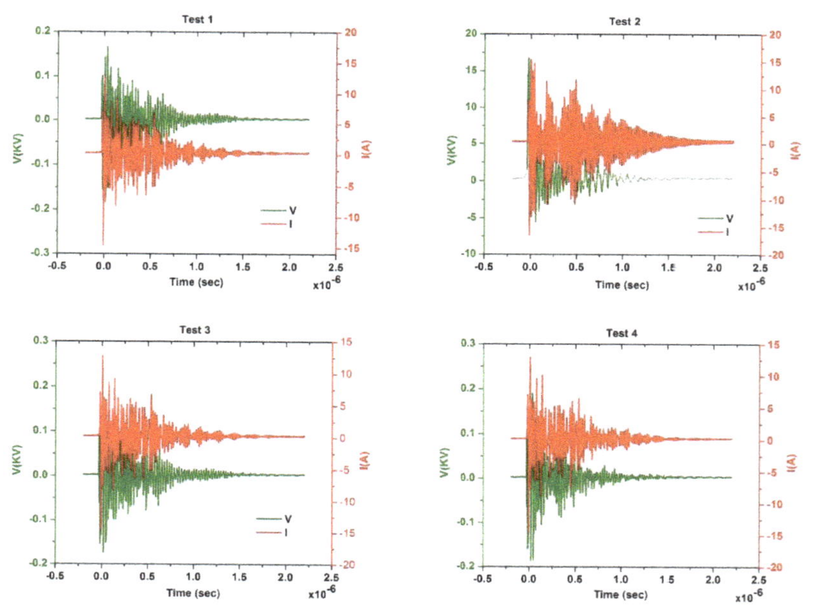

Figure 4. Voltage V(t) and current I(t) characteristics at different percentage of applied voltage and standoff distances.

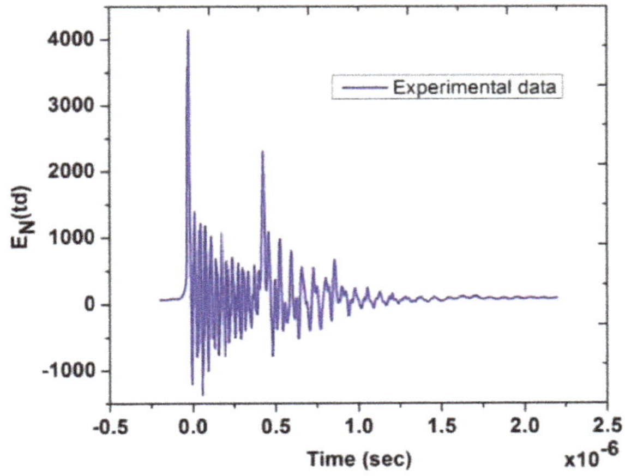

Figure 5. Temporal evolution of experimental E_N of Test 2.

4.2. Flow Visualization

The smoke flow visualizations have been carried out by using a high-resolution camera (MEMRECAM GX-3). An advantage of the flow analysis under quiescent conditions is that it could be possible to better analyze the plasma induced flow due to the momentum transfer, which otherwise would be hidden by effects due to the chemical reactions.

The Plexiglas box was seeded with incense smoke. For assessing the ability of the incense smoke seeding particles to follow the flow field, the particle relaxation time T_p and the Stokes number St were evaluated.

$$T_p = \frac{\rho d^2}{18\mu} \tag{1}$$

The Stokes number is the ratio between the particle residence time and the characteristic time of the carrier fluid corrected with the slip correction factor, C_c.

$$St = \frac{vC_c\rho d^2}{18\mu l} \tag{2}$$

where ρ_p is the particle density, d is the particle diameter, v is a characteristic velocity scale of the flow, μ is the fluid viscosity, l is a characteristic length scale of the flow, and C_c the slip correction factor (or Cunningham's correction factor) [25]. Chinese joss sticks were used for seeding, it was estimated a particle diameter around 0.3 µm and a particle density of 1.06 g/cm^3 [26,27]. It was found that, at standard conditions (293 K and 101 kPa), the slip correction factor was around 1.554 [28,29].

The particle residence time was estimated equal to 0.00004 s and the Stokes number St equal to 0.0018, assuming v of 0.1 m/s (set to mean velocity measured in previous LDV experiments) and l equal to 3.3 cm (inner diameter). According to [30], $St << 0.1$ returns an acceptable flow tracing accuracy with errors below 1%.

Smoke flow visualizations at the downstream region of the exposed electrodes at different HV voltage amplitudes (Tests 1–2) and standoff distances (Test 2, 3, and 4) were performed for the identification of vortical flow pattern generated in quiescent atmospheric air, as shown in Figures 6 and 7. It has been noticed that in Test 1 and Test 2, flow field was uniformly distributed; only one leading vortex was observed on the corners of closed box that could modify the flow recirculation in the burners, with an improvement of the combustion. Moreover, with the increase of the standoff distance, as in Test 3 and Test 4, uniformity and mixing performance were gradually decreased, and the flow was moved to upstream of the closed box.

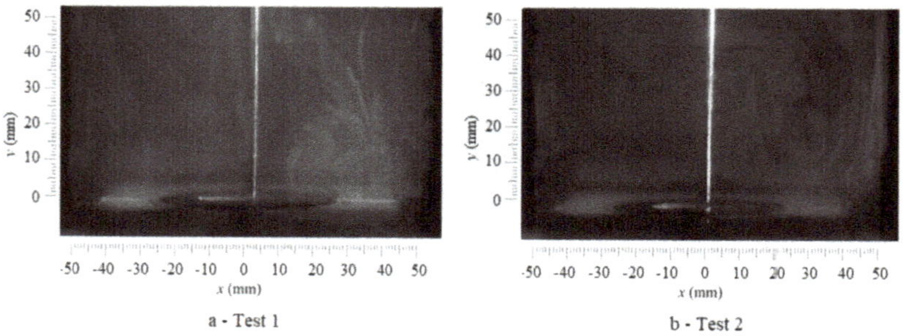

Figure 6. Comparative analysis of smoke flow visualizations at different percentage of amplitudes and fixed standoff distance 0 mm.

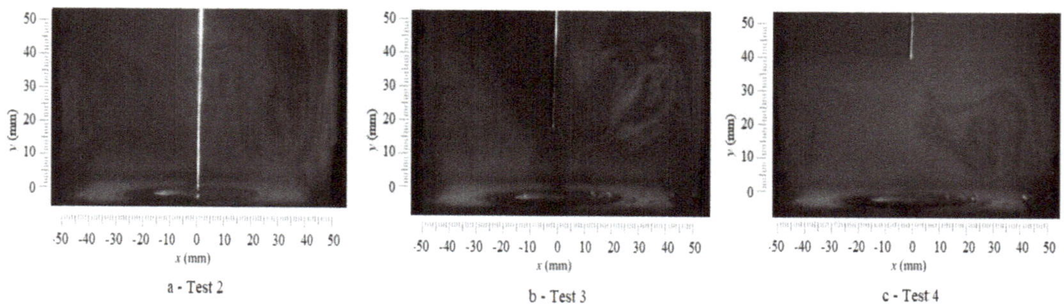

Figure 7. Comparative analysis of smoke flow visualizations at different standoff distances 0 mm, 20 mm, and 40 mm at fixed percentage of amplitude.

Moreover, for a deep analysis of the induced flow vortices and coherent structures, the preliminary investigation of POD analysis of non-reactive flow was performed, which allows the identification of dominant structures from random data. The sign and the intensity of POD eigen structures represent the coherent flow structures. The relative contribution of each mode to the flow field reconstruction is described by their energy. The first POD mode represents the most energetic flow structure, and it is strictly related to the flow variance. The relative mode energy is mainly consumed and distributed by the first three modes in all test cases as shown in Figure 8. Though, the influence of the first mode was relatively high, while POD modes above number 3 are less significant. Hence, only the first three modes were considered for the POD flow analysis. The POD structures at different amplitudes and standoff distances are shown in Figures 9 and 10, respectively. In Test 1, Mode 1, and Mode 2 were more symmetric and they have approximately the same relative energy as shown in Figure 8. However, in Test 2, Mode 1, and Mode 2 present quite different structures and almost 80% of relative energy were captured by Mode 1. Besides this, when the standoff distance was 0 mm, flow disturbances was quite attached to the needle electrode and located on the downstream of closed box as in Test 1 and 2; however, with the increase of standoff distance as in Test 3 and 4 the flow was moved to the corners and upstream of the rectangular box.

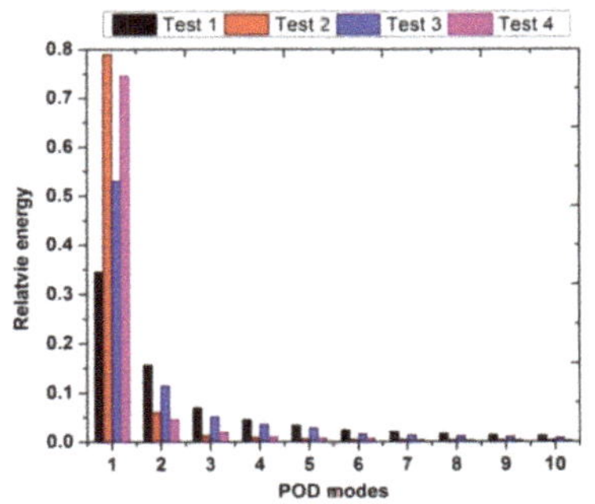

Figure 8. Comparative characterization of relative energy of first 10 modes at different voltage amplitudes and standoff distances.

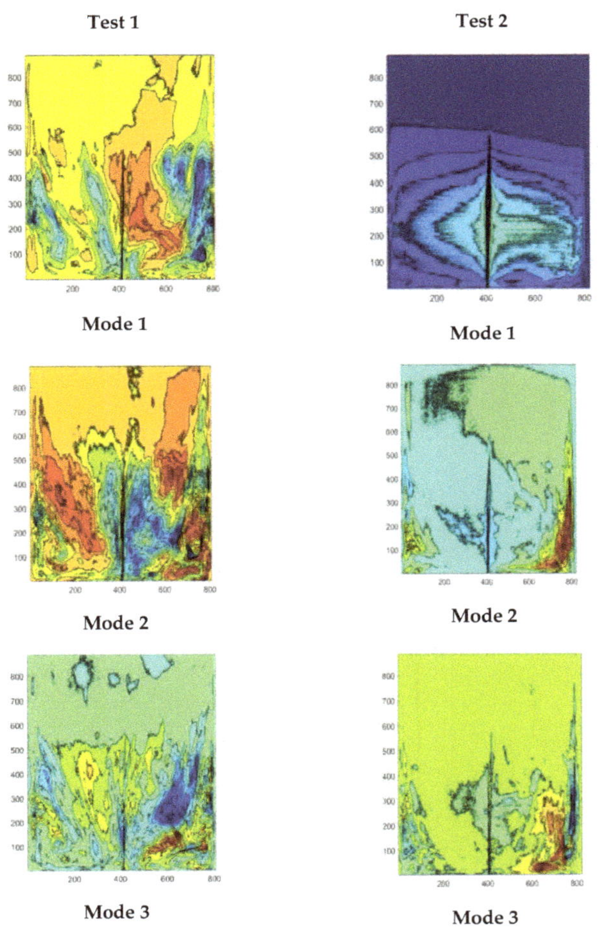

Figure 9. POD analysis of first three modes at different voltage amplitudes and fixed standoff distance of 0 mm.

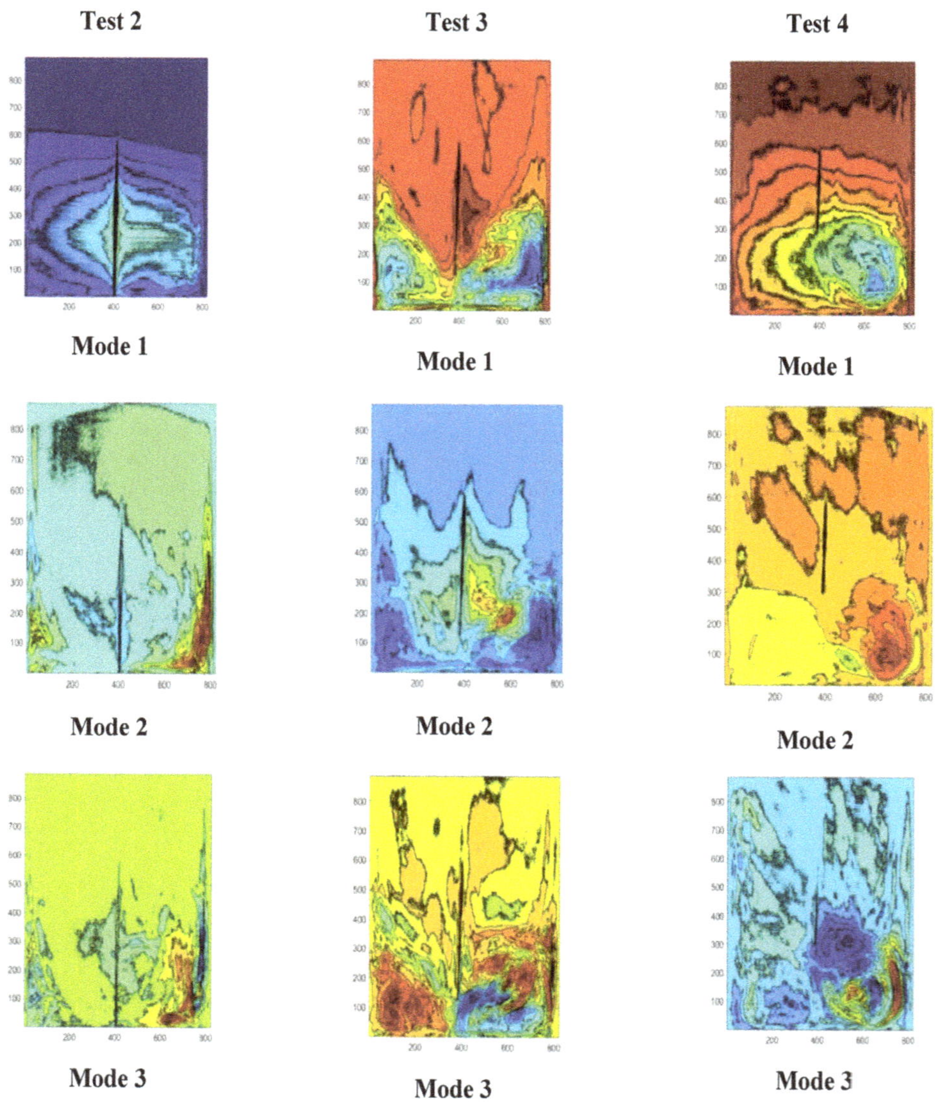

Figure 10. POD analysis of first three modes at different standoff distances of 0 mm, 20 mm, and 40 mm, and fixed voltage amplitude.

4.3. Production of Active Particles

Different from traditional ignition techniques, which depend mainly on gas heating, NTP provides enhancement through kinetics. This happens through collisions between energetic electrons and abundant particles such as diatomic nitrogen and oxygen, finally leading to the generation of radicals that promote ignition. In our previous studies, it has been noticed that the deposition of nanopulsed plasma discharge decomposed the methane/air and improved the concentrations of neutral radicals such as H, CH, OH, CH_3, O, and O_3 by following the electron impact reactions [22–24].

In this paper, numerical analysis of neutral radicals has been performed at different inlets pressure aiming to see particularly the effects on concentration of particles at low pressure. Experimental E_N = 200 Td, inlet temperature equal to 300 K, and initial reaction of $CH_4/N_2/O_2$ were considered. Figure 11 showed the production of free radicals H, CH_3, O, O_3, CH, and OH at three different inlet pressure values. It has been perceived that with the use of plasma discharge active particles were produced; however, by decreasing the inlet pressure, the concentrations of all neutral radicals were also significantly reduced and minimum concentrations were observed at 0.4 bar. This effect could be due to the slow speed of electron impacts reactions and low ionization process at low pressure conditions. However, still the production rate of radicals is applicable to enhance the flame ignition at low pressure as discussed in the next section of ignition process. At a pressure of 1 bar, the peak concentration of all radicals ranges between 10^{-1} and 10^{-2} except CH which lies between 10^{-3} and 10^{-4}. Table 3 is showing the conditions that have been used to simulate the plasma effects.

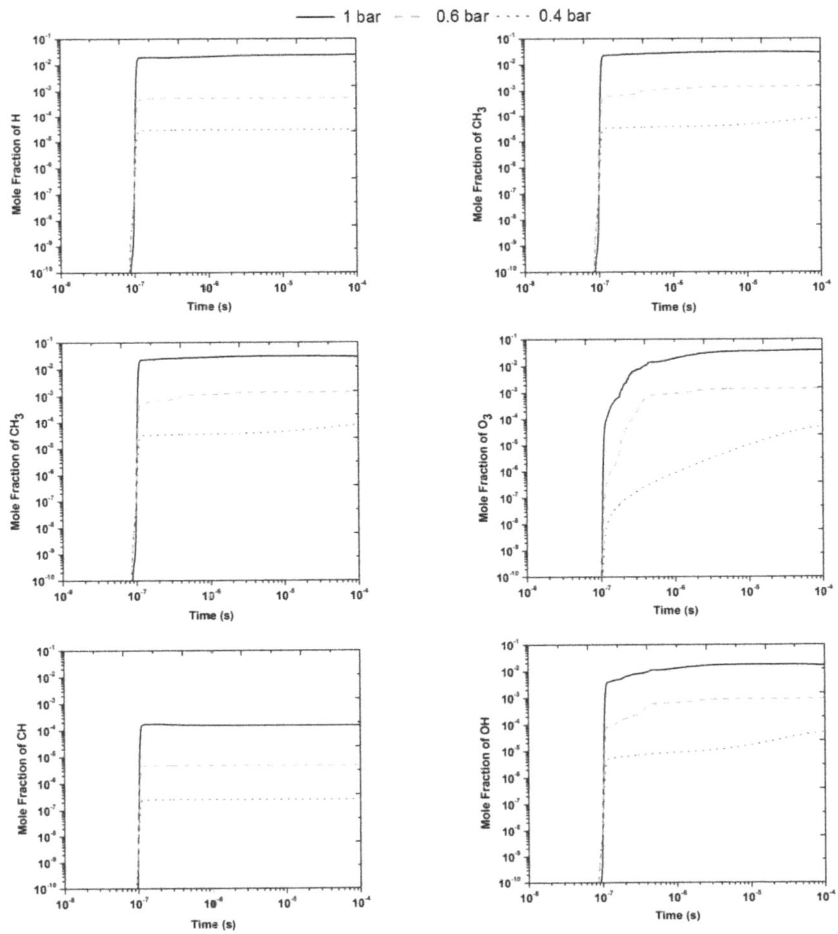

Figure 11. Neutral radicals produced due to plasma actuation at three different inlet pressures.

Table 3. Production of active radicals during the pulse discharge.

Pressure	H	O	OH	CH	CH_3	O_3
1 bar	0.02581	0.0671	0.001818	0.000175622	0.03046	0.05062
0.6 bar	0.000558261	0.00203	0.000944928	4.912×10^{-6}	0.00146	0.00151
0.4 bar	2.96967×10^{-5}	0.000108	6.22008×10^{-5}	2.5848×10^{-7}	8.967×10^{-5}	6.6069×10^{-5}

4.4. Ignition Characteristics at Low Pressure and Temperature Conditions

The impact of needle-ring nanopulsed plasma actuator on the flame ignition was analyzed at low pressure and fixed inlet temperature and plasma actuation conditions. Experiments showed that the transport effects of plasma needle-disk plasma actuator could improve the mixing of the fuel/air mixture, which a possible contribution to the reduction of ignition delay time. ZDPlasKin permitted to estimate the thermal and kinetics effects due to the nanopulsed plasma discharge at the experimentally measured E_N. Finally, based on the thermal and kinetic outputs of ZDPlasKin, in terms of rise of the mixture temperature and production of active species, Chemkin estimated the flame ignition at low pressure and low temperature conditions. A closed homogeneous batch reactor (CHBR) was used to predict the ignition delay time.

The closed homogenous batch reactor has been commonly used for the analysis of ignition delay timings in Chemkin solver. Two important parameters were available in the setting of solver to define the ignition process: maximum concentrations of OH species and temperature inflection point. This means that the computation of ignition delay timings could be predicted by maximum molar fraction of OH species and temperature inflection point (TIP). The TIP is defined as the position where the rate of change of temperature with reference to time is largest.

Figure 12 shows the comparative analysis of ignition delay time of clean case (without plasma) and plasma actuation at different inlet pressures at constant inlet mixture temperature equal to 1400 K and E_N = 200 Td. It was noticed that the time required to achieve the maximum flame temperature with plasma actuation is significantly reduced in comparison with autoignition time (clean case, without plasma actuation). At 1 bar, the ignition time was 1.1×10^{-3} s while it was reduced at 3.5×10^{-5} s in presence of plasma actuation. However, lowering the inlet pressure, the ignition delay time increased for both the clean case (without plasma effect) and plasma assisted case. Results showed that plasma-assisted ignition is more efficient at low pressures. At 0.6 bar flame ignition occurred in case without plasma effects at (0.0048 s) and plasma assisted ignition (0.0022 s), simultaneously, at 0.4 bar it achieved at 0.0063 s and 0.0027 s without and with plasma actuation, respectively. Figure 13 depicts the temporal evolution of maximum molar concentration of OH species at different inlet pressures during the inlet plasma discharge. It has been noticed that maximum OH concentrations were produced at the same time at which temperature inflection point was achieved. At the same pressure, the maximum OH is higher and reached in a shorter time in the case of plasma actuation than in the case without plasma. It is also possible to see the evolution of the OH during the first stage.

Ignition in low-temperature repetitively pulsed nanosecond discharge plasmas occurs mainly due to plasma chemical radical species reactions rather than due to rapid localized heating. In particular, the ignition enhancement is due to the conversion of electron energy into the bond energy of free/active radicals produced during the plasma discharge processes. Then, the bond energy of free/active radicals is transformed into internal energy during the ignition process.

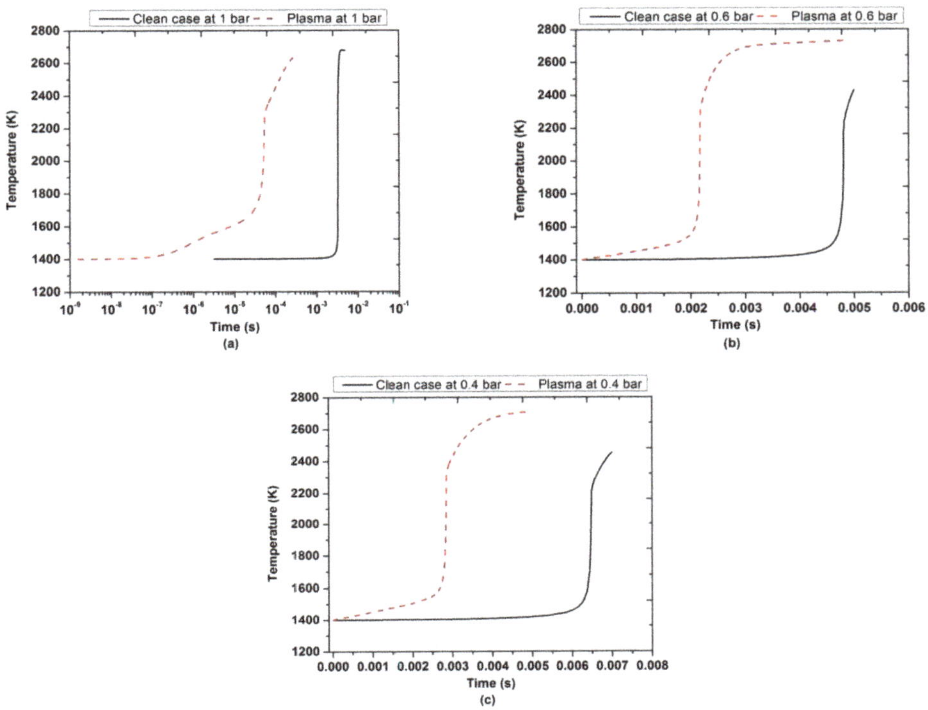

Figure 12. Comparative behavior of ignition characteristics of clean case (without plasma effect, black line) with plasma actuation (red dashed line) at low pressure conditions (**a**) 1 bar, (**b**) 0.6 bar, (**c**) 0.4 bar considering constant inlet temperature 1400 K, E_N 200 Td, and stoichiometric conditions.

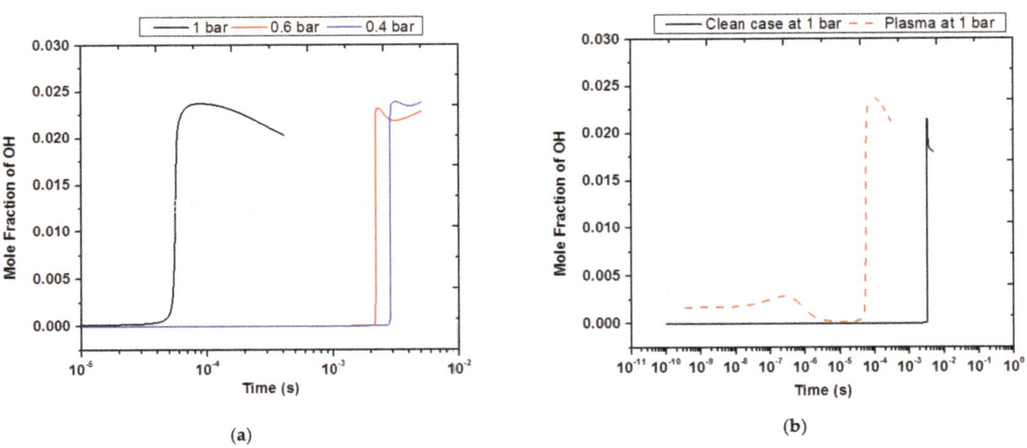

Figure 13. Maximum molar fraction of OH concentration achieved at certain period of time (**a**) different inlet pressures during the plasma discharge; (**b**) with or without the plasma discharge at 1 bar.

Adding to this, flame ignition analysis was also carried out at low inlet mixture temperature by putting constant inlet pressure of 0.4 bar and 0.6 bar. Figure 14 shows the comparative behavior of ignition delay time at different low inlet temperature ranges from 1000 K to 1300 K, investigating the time required to achieve the maximum combustion temperature. It has been noted that, with the increase of inlet temperature, ignition time was reduced, and minimum reduction was observed at 1300 K for both 0.4 bar and 0.6 bar, however, the reduction effect was bit high at 0.6 bar than the 0.4 bar. At a low temperature of 1000 K, ignition times were observed at 0.013 s and 0.010 s for 0.4 and 0.6 bar, respectively. Moreover, the maximum combustion temperature of all initial mixture temperature values was similar around 2700 K for both mixture pressure values. Simultaneously, the parametric study at different inlet temperatures was performed at two different operating pressure values and the results are shown in Figure 15. It has been noticed that the rise of the inlet temperature reduced the influence of the ignition delay time on the operating pressure.

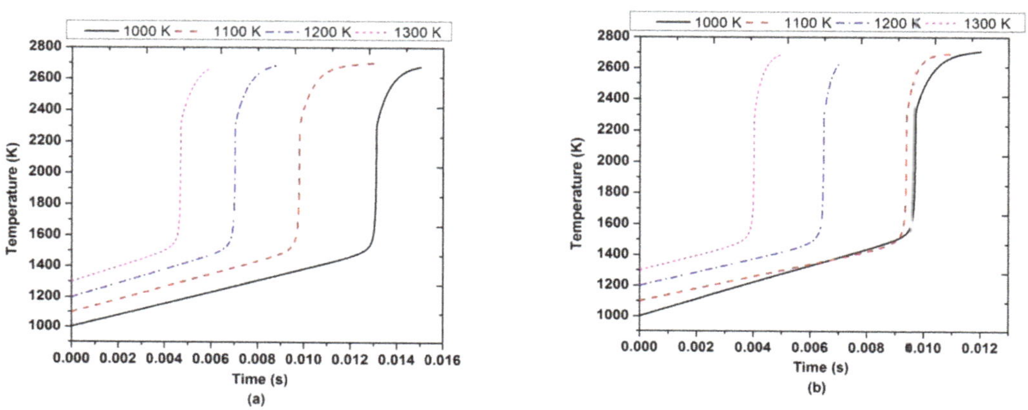

Figure 14. Comparative anlaysis of ignition delay time in presence of PAI at different inlet temperatures from 1000 K to 1300 K at constant low pressure (**a**) 0.4 bar (**b**) 0.6 bar.

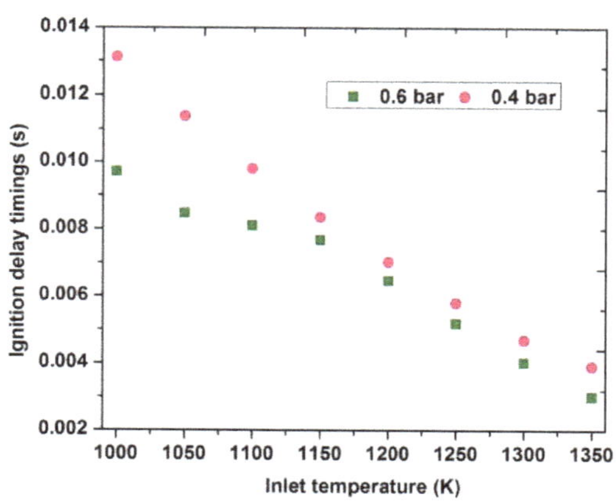

Figure 15. Parametric study of ignition delay time at low inlet temperatures at fixed inlet pressure 0.4 and 0.6 bar in presence of PAI.

5. Conclusions

The present work investigated the experimental and numerical characterization of flow structures and ignition delay time, particularly at high altitude conditions—i.e., low inlet pressure and temperature. Plasma assisted ring-needle plasma actuator was developed and powered by a high-voltage nanosecond pulse generator. The electrical characterization has been performed considering different percentage values of the HV amplitude and different standoff distances. The mean power and optimum value of E_N were experimentally measured and used in the numerical study for the prediction of the ignition delay time. Smoke flow visualizations and the POD analysis showed the transport effects of the plasma discharges, which generated an induced flow in the recirculation zone near the corners of the combustor, with a possible effect on the mixing of the fuel/air mixture. Finally, predictions of the ignition delay time for the case without plasma effects and the plasma assisted ignition cases at different inlet pressure and temperature values were performed, assuming for the plasma effects the experimental E_N equal to 200 Td. It was noticed that the time required to achieve the maximum flame temperature with PAI is significantly reduced in comparison with the autoignition time. Maximum reduction in ignition time was observed at 1 bar (3.5×10^{-5} s) with respect to the case without plasma (1.1×10^{-3} s). However, decreasing the inlet pressure, the ignition delay time of clean and PAI were increased. At 0.6 bar the flame ignition occurred at 0.0022 s in case without plasma and at 0.0048 s for the PAI, while at 0.4 bar it was achieved at 0.0027 s and 0.0063 s without and with PAI, respectively. The flame ignition analysis was also carried out at different inlet temperature values at constant operating pressure, equal to 0.4 bar and 0.6 bar. It has been noted that with the increase of the inlet temperature, the ignition delay time was reduced, and the minimum reduction was observed at 1300 K for both 0.4 bar and 0.6 bar; however, the reduction was less significant at 0.6 bar than at 0.4 bar. At a low temperature of 1000 K, the ignition delay time was equal to 0.013 s and 0.010 s for 0.4 and 0.6 bar, respectively.

Author Contributions: Conceptualization, M.G.D.G.; Methodology, M.G.D.G.; Software, G.M.; Validation, G.M.; Formal analysis, G.M.; Investigation, G.M. and S.B.; Data curation, G.M. and S.B.; Writing—original draft preparation, G.M. and S.B.; Writing—review and editing, M.G.D.G.; Visualization, G.M. and S.B.; Supervision, M.G.D.G.; Project administration, M.G.D.G.; Funding acquisition, M.G.D.G. All authors have read and agreed to the published version of the manuscript.

Funding: The work was supported and funded by the PON R&I 2014–2020 Asse I "Investimenti in Capitale Umano" Azione I.1 "Dottorati Innovativi con caratterizzazione industriale"—Corso di Dottorato in "Ingegneria dei Sistemi Complessi" XXXV ciclo—Università degli Studi del Salento"—Borsa Codice: DOT1312193 no. 3. This project is also received funding from the Clean Sky 2 Joint Undertaking (JU) under the grant agreement no. 831881 (CHAiRLIFT). The JU received support from the European Union's Horizon 2020 research and innovation program and the Clean Sky 2 JU members other than the Union.

Conflicts of Interest: The authors declare no conflict of interest.

References

1. Starikovskiy, A.; Aleksandrov, N. Plasma-Assisted Ignition and Combustion. *Prog. Ener. Comb. Sci.* **2013**, *39*, 61–110. [CrossRef]
2. Ju, Y.; Lefkowitz, J.K.; Reuter, C.B.; Won, S.H.; Yang, X.; Yang, S.; Sun, W.; Jiang, Z.; Chen, Q. Plasma Assisted Low Temperature Combustion. *Plasma Chem. Plasma Process.* **2016**, *36*, 85–105. [CrossRef]
3. Starikovskiy, S.M. Topical review: Plasma-assisted ignition and combustion. *J. Phys. D Appl. Phys.* **2006**, *39*, 265–299. [CrossRef]
4. Ruma, M.; Ahasan, H.; Ranipet, H.B. A Survey of Non-thermal plasma and their generation methods. *Int. J. Renew. Energy Environ. Eng.* **2016**, *4*, 6–12.
5. De Giorgi, M.G.; Pescini, E.; Marra, F.; Ficarella, A. Plasma actuator scaling down to improve its energy conversion efficiency for active flow control in modern turbojet engines compressors. *Appl. Therm. Eng.* **2016**, *106*, 334–350. [CrossRef]
6. Francioso, L.; De Pascali, C.; Pescini, E.; De Giorgi, M.G.; Siciliano, P. Modelling, fabrication and plasma actuator coupling of flexible pressure sensors for flow separation detection and control in aeronautical applications. *J. Phys. D Appl. Phys.* **2016**, *49*, 235201. [CrossRef]
7. Ju, Y.; Sun, W. Plasma Assisted Combustion: Dynamics and Chemistry. *Prog. Energy Combust. Sci.* **2015**, *48*, 21–83. [CrossRef]
8. Tholin, F.; Lacoste, D.A.; Bourdon, A. Influence of Fast-Heating Processes and O Atom Production by a Nanosecond Spark Discharge on the Ignition of a Lean H2-Air Premixed Flame. *Combust. Flame* **2014**, *161*, 1235–1246. [CrossRef]
9. Aleksandrov, N.L.; Kindysheva, S.V.; Kukaev, E.N.; Starikovsjaya, S.M.; Starikovskii, A.Y. Simulation of the Ignition of a Methane-Air Mixture by a High-Voltage Nanosecond Discharge. *Plasma Phys. Rep.* **2009**, *35*, 867–882. [CrossRef]
10. Benard, N.; Braud, P.; Pons, J.; Touchard, G.; Moreau, E. Quasi-steady and unsteady actuation by surface non-thermal plasma discharge for control of a turbulent round air jet. *J. Turbul.* **2007**, *8*, N49. [CrossRef]

11. Bernard, N.; Bonnet, P.J.; Touchard, G.; Moreau, E. Flow control by dielectric barrier discharge actuators: Jet mixing enhancement *AIAA J.* **2008**, *46*, 2293–2305. [CrossRef]
12. Xu, C.; Li, L. Electric effects on the pre-mixing process in DBD assisted combustions. *Appl. Mech. Mater.* **2015**, *722*, 235–238 [CrossRef]
13. Kosarev, I.N.; Aleksandrov, N.L.; Kindysheva, S.V.; Starikovskaia, S.M.; Starikovskii, A.Y. Kinetics of Ignition of Saturated Hydrocarbons by Nonequilibrium Plasma: CH4-Containing Mixtures. *Combust. Flame* **2008**, *154*, 569–586. [CrossRef]
14. Starik, A.M.; Kozlov, V.E.; Titova, N.S. On the influence of singlet oxygen molecules on the speed of flame propagation in methane–air mixture. *Combust. Flame* **2010**, *157*, 313–327. [CrossRef]
15. Uddi, M.; Jiang, N.; Mintusov, E.; Adamovich, I.V.; Lempert, W.R. Atomic oxy- gen measurements in air and air / fuel nanosecond pulse discharges by two photon laser induced fluorescence. *Proc. Combust. Inst.* **2009**, *32*, 929–936. [CrossRef]
16. Pancheshnyi, S.; Eismann, B.; Hagelaar, G.J.M.; Pitchford, L.C. ZDPlaskin Zero-Dimensional Plasma Kinetic Solver 2008. Available online: http://www.zdplaskin.laplace.univ-tlse.fr/ (accessed on 8 September 2021).
17. Lutz, A.E.; Kee, R.J.; Miller, J.A. *SENKIN: A FOR- TRAN Program for Predicting Homogeneous Gas Phase Chemical Kinetics with Sensitivity Analysis*; Report No. SAND87-8248; Sandia National Laboratories: Livermore, CA, USA, 1988.
18. Capitelli, M.; Ferreira, C.M.; Gordiets, B.F.; Osipov, A.I. *Plasma Kinetics in Atmospheric Gases*; Springer: Berlin/Heidelberg, Germany, 2000.
19. Flitti, A.; Pancheshnyi, S. Gas heating in fast pulsed discharges in N_2–O_2 mixtures. *Eur. Phys. J. Appl. Phys.* **2009**, *45*, 21001 [CrossRef]
20. Mao, X.; Rousso, A.; Chen, Q.; Ju, Y. Numerical modeling of ignition enhancement of CH_4/O_2/He mixtures using a hybrid repetitive nanosecond and DC discharge. *Proc. Combust. Inst.* **2019**, *37*, 5545–5552. [CrossRef]
21. Mao, X.; Chen, Q.; Guo, C. Methane pyrolysis with N2/Ar/He diluents in a repetitively-pulsed nanosecond discharge: Kinetics development for plasma assisted combustion and fuel reforming. *Energy Convers. Manag.* **2019**, *200*, 112018. [CrossRef]
22. Fontanarosa, D.; Mehdi, G.; De Giorgi, M.G.; Ficarella, A. Assessment of the impact of nanosecond plasma discharge on the combustion of methane air flames. *E3S Web Conf.* **2020**, *197*, 10001. [CrossRef]
23. Mehdi, G.; Bonuso, S.; De Giorgi, M.G. Effects of Nanosecond Repetitively Pulsed Discharges Timing for Aeroengines Ignition at Low Temperature Conditions by Needle-Ring Plasma Actuator. *Energies* **2021**, *14*, 5814. [CrossRef]
24. Mehdi, G.; De Giorgi, M.G.; Fontanarosa, D.; Bonuso, S.; Ficarella, A. Ozone Production With Plasma Discharge: Comparisons Between Activated Air and Activated Fuel/Air Mixture. In Proceedings of the ASME Turbo Expo 2021: Turbomachinery Technical Conference and Exposition, Virtual, 7–11 June 2021; Volume 3B. Combustion, Fuels, and Emissions V03BT04A036. [CrossRef]
25. Yoon, S.; Ross, J.; Mench, M.; Sharp, K. Gas-phase particle image velocimetry (PIV) for application. *J. Power Sources* **2006**, *160*, 1017–1025. [CrossRef]
26. Cheng, Y.S.; Bechtold, E.; Yu, C.C.; Hung, I.F. Incense smoke: Characterization and dynamics in indoor environments. *Aerosol Sci. Technol.* **1995**, *23*, 271–281. [CrossRef]
27. See, S.W.; Balasubramanian, R.; Joshi, U.M. Physical characteristics of nanoparticles emitted from incense smoke. *Sci. Technol. Adv. Mater.* **2007**, *8*, 25–32. [CrossRef]
28. Hosokawa, M.; Kiyoshi, N.; Naito, M.; Yokoyama, T. *Nanoparticle Technology Handbook*, 1st ed.; Elsevier: Amsterdam, The Netherlands, 2007; ISBN 9780080558028.
29. Hinds, W.C.; Zhu, Y. *Aerosol Technology: Properties, Behaviour, and Measurement of Airborne Particles*; John Willey & Sons, Inc.: Los Angeles, CA, USA, 1998.
30. Tropea, C.; Yarin, A.L.; Foss, J.F. *Springer Handbook of Experimental Fluid Mechanics*; Springer: Berlin, Germany, 2007.

aerospace

MDPI

Article

Initial Assessment of a Fuel Cell—Gas Turbine Hybrid Propulsion Concept

Arne Seitz *, Markus Nickl, Florian Troeltsch and Kathrin Ebner

Bauhaus Luftfahrt e.V., Willy-Messerschmitt-Str. 1, 82024 Taufkirchen, Germany; markus.nickl@bauhaus-luftfahrt.net (M.N.); florian.troeltsch@bauhaus-luftfahrt.net (F.T.); kathrin.ebner@bauhaus-luftfahrt.net (K.E.)

* Correspondence: arne.seitz@bauhaus-luftfahrt.net

Abstract: A fuel cell—gas turbine hybrid propulsion concept is introduced and initially assessed. The concept uses the water mass flow produced by a hydrogen fuel cell in order to improve the efficiency and power output of the gas turbine engine through burner steam injection. Therefore, the fuel cell product water is conditioned through a process of condensation, pressurization and re-vaporization. The vaporization uses the waste heat of the gas turbine exhaust. The functional principles of the system concept are introduced and discussed, and appropriate methodology for an initial concept evaluation is formulated. Essential technology fields are surveyed in brief. The impact of burner steam injection on gas turbine efficiency and sizing is parametrically modelled. Simplified parametric models of the fuel cell system and key components of the water treatment process are presented. Fuel cell stack efficiency and specific power levels are methodically derived from latest experimental studies at the laboratory scale. The overall concept is assessed for a liquid hydrogen fueled short-/medium range aircraft application. Block fuel savings of up to 7.1% are found for an optimum design case based on solid oxide fuel cell technology. The optimum design features a gas turbine water-to-air ratio of 6.1% in cruise and 62% reduced high-level NO_x emissions.

Keywords: fuel cell; gas turbine; hybrid propulsion; hydrogen fuel; aircraft design; heat exchange; water condensation; steam injection; efficiency assessment; NO_x reduction

Citation: Seitz, A.; Nickl, M.; Troeltsch, F.; Ebner, K. Initial Assessment of a Fuel Cell—Gas Turbine Hybrid Propulsion Concept. *Aerospace* **2022**, *9*, 68. https://doi.org/10.3390/aerospace9020068

Academic Editor: Spiros Pantelakis

Received: 17 December 2021
Accepted: 20 January 2022
Published: 26 January 2022

1. Introduction

Driven by the drastic need for decarbonization in aviation, hydrogen as a zero in-flight CO_2 emissions fuel has recently experienced a great revival in aeronautical research and development. Beyond the well-known and demonstrated option of hydrogen combustion in classic gas turbine (GT) engines, significant technological progress in electrochemical conversion systems adds to the dynamic development of hydrogen-based visions for future air transport. However, the application of fuel cells (FCs) in the scheme of propulsion and power for transport category aircraft holds great challenges, and classic GT engines still feature significant advantages in power-specific weight over future FC systems. At the same time, further efficiency improvements and significant NO_x emission reduction will be required for future aero engines. The injection of water in the compression section or the combustion chamber of a GT engine is well-known for its enhancement of cycle-specific work and efficiency, as well as its potential for significant cuts in NO_x emissions.

In this paper, a synergistic concept for a FC–GT hybrid aircraft propulsion system is introduced and initially assessed. A basic functional schematic of concept is presented in Figure 1. The concept utilizes the product water of the FC in order to improve the performance and emission characteristics of the GT engine. The FC product water is first condensed and separated from the FC residual air mass flow, then pressurized and re-evaporated before being injected in the GT engine. This treatment process allows for minimum work effort for the water pressurization—due to the incompressibility of liquid water—while the energy required for its re-evaporation is drawn from the GT exhaust heat.

The electric power produced by the FC is available as a convenient source for on-board electric customers, including aircraft subsystems and possibly decentralized or distributed propulsors. Operational flexibility may be enhanced through an intermediate storage possibility for the liquefied water, in order to decouple the GT steam injection from the FC power setting.

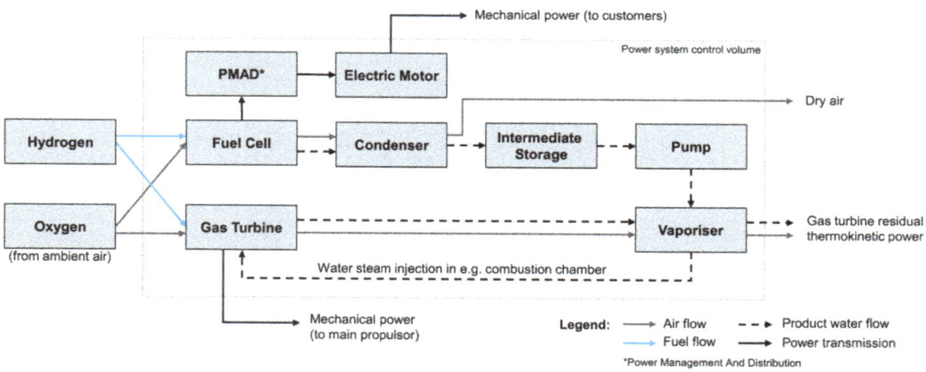

Figure 1. Basic functional schematic of studied FC–GT hybrid propulsion concept.

At first, important developments in the key field of the technology relevant for the studied FC–GT hybrid propulsion system type are reviewed in the paper. Subsequently, the methodological foundation for the conceptual assessment is formulated and discussed, before the results obtained from isolated power system and aircraft-integrated design studies are presented and examined.

2. Review of Key Technologies

The presented study is motivated by and based upon a number of key technological developments in different fields, including hydrogen as an alternative fuel option for aviation, significant advances in FC technology, high-performance heat exchangers (HEXs) and advanced GT technology involving steam injection in the thermodynamic cycle. In this section, relevant aspects of technological progress in those fields are surveyed in brief.

2.1. Hydrogen Fuel in Aviation

Combustion and electrochemical conversion of hydrogen do not create carbon emissions. In a recent well-to-wake analysis of CO_2 equivalent emissions, green hydrogen by far outperformed conventional jet fuel as well as all other considered synthetic fuel options [1]. As a typical intermediate product along many production pathways for alternative hydrocarbon-based fuels, minimum fuel selling prices of liquid hydrogen (LH_2) are likely to be more affordable than most synthetically produced sustainable drop-in fuels [2]. Due to a different stoichiometry of the chemical reaction, the use of hydrogen fuel yields a water production rate that is approximately 2.5 times as high as the one obtained from the combustions of kerosene fuel. The effect of the resulting water vapor emissions on climate change is not yet fully understood—the debate mainly centers around conditions for the formation and avoidance of persistent contrails [3].

For commercial air transport, hydrogen as a fuel poses significant challenges for vehicular design [4] as well as in terms of ground infrastructural prerequisites [5]. The amounts of energy required for typical commercial transport aircraft missions require hydrogen fuel to be stored in its most compact form, i.e., as LH_2. The cryogenic temperature levels associated with LH_2, combined with a persisting low volumetric energy density in comparison to hydrocarbon fuels, require unconventional and complex onboard storage solutions that involve considerable structural mass and aerodynamic drag penalties.

Nevertheless, hydrogen-based aircraft designs have been considered in both theoretical and experimental studies throughout the history of commercial aviation. Famous hydrogen-fuelled flight tests were performed in the 1950s on a modified Martin B-57 "Canberra" and later in the 1980s on the Tupolev Tu-155, a modified version of the commercial Tu-154. Extensive preliminary design studies were performed in the 1970s by Boeing [6] and Lockheed [7,8] under NASA contract as well as during the European Commission (EC) funded CryoPlane project led by Airbus in the early 2000s. Since then, the topic of hydrogen-fuelled aircraft has been continuously investigated as part of academic research, for example at Cranfield University, yielding a series of doctoral theses on hydrogen-fuelled aero-engine and aircraft systems (e.g., [9,10], as well as the ongoing EC-funded project ENABLEH2 [11]).

A multitude of advanced hydrogen-fuelled aircraft concepts have recently been published by research organisations (e.g., NASA's N3-X [12], Bauhaus Luftfahrt's Hyliner2.0 [13], FlyZero [14]) and announced as technology concept planes by the industry (e.g., Airbus ZeroE [15], Do228 FC demonstrator [16]). A broad overview on hydrogen technology, economics and climate impact by 2050 is provided by a study issued by the European "CleanSky" and "Fuel Cells and Hydrogen" Joint Undertakings in 2020 [17]. As a first key technology brick in the context of the present paper, a brief overview of advanced fuel cell concepts is provided in the next section.

2.2. Advanced Fuel Cell Concepts

An FC is an electrochemical energy conversion device, which enables the direct conversion of chemical energy stored in a fuel to electricity. The fuel, typically hydrogen, is oxidized at the anode, while ambient oxygen is reduced at the cathode. In the process, water—and in the case of fuels other than hydrogen—additional by-products are formed and released from the cell (cf. Equation (1)):

$$H_2 + 0.5\, O_2 \rightarrow H_2O. \tag{1}$$

The voltage that could theoretically be drawn from this reaction is 1.23 V at room temperature [18], however, various voltage losses occur in the cell, e.g., due to hindered mass transport, electrical resistance, fuel crossover or sluggish kinetics. Consequently, the practically achievable voltages U_{Cell} are significantly lower, the electrical efficiency η_{Stack} of the cell stack in reference to the heating value per mol of fuel FHV_{mol} (cf. Equation (2)) is diminished and the surplus energy is released as heat:

$$\eta_{Stack} = \frac{2 \cdot F \cdot U_{Cell}}{FHV_{mol}}, \tag{2}$$

where F is the Faraday constant.

In an effort to find FCs applicable to versatile operating environments and application demands, various cell types have been developed. Out of these, the polymer electrolyte FC (PEFC), which is schematically depicted in Figure 2a, is the current state-of-the-art FC for transportation applications due to its comparably high specific power and quick load response. Therefore, most hydrogen-electric aircraft designs based on FC-electric propulsion rely on the PEFC (e.g., research projects Go4Hy2, BALIS, BILBO). This is comprised of a proton-conducting membrane, most commonly a perfluorosulfonic acid (PFSA) polymer that transports the protons formed on the anode over to the cathode while avoiding gas crossover and preventing short-circuiting due to its electrically insulating properties. As these functionalities are only maintained when the membrane's water content is high enough, inlet gases are typically humidified. Generally, water management is crucial for successful PEFC operation in order to balance sufficient humidification on the one hand and the avoidance of excess amounts of (product) water causing pore flooding and mass transport losses on the other. Gas-diffusion layers (GDLs) are employed to aid water removal and gas transport to the catalytically active sites in the catalyst layer (CL).

Catalysts are required to enhance the reaction kinetics of the hydrogen oxidation reaction and the oxygen reduction reaction (HOR and ORR); state-of-the-art high-performance materials are based on the noble metal Pt. It should be briefly noted that the schematics in Figure 2 only depict the core components of FCs. One individual cell would deliver a relatively small voltage as discussed above, which is why typically numerous cells are connected in series through bipolar plates, or so-called interconnects for higher temperature applications, to obtain an FC stack with technically usable output voltage. The stack itself is accompanied by balance-of-plant components such as compressors, humidifiers, thermal management, etc. The entire assembly will, in the following, be termed "FC system".

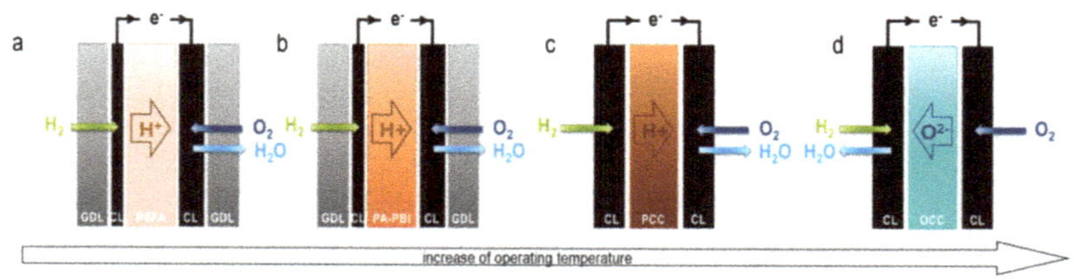

Figure 2. Schematic representation of the FC types considered herein in order of increasing operating temperature: (**a**) PEFC, (**b**) HT-PEFC, (**c**) PCFC, (**d**) SOFC.

The standard operating temperature of the PEFC is between 60 and 90 °C, which is a range in which reaction kinetics employing the above-mentioned catalysts are high enough to yield technically feasible current densities and the membrane is stable and (when humidified) well-conductive, while material degradation is limited. While PEFC systems are continuously further developed to match performance, cost and weight targets for various propulsion applications, they have already reached a high technological maturity and are utilized in commercial applications at the 100 kW scale [19].

Another FC type that has been developed to a high technological readiness level (TRL) is the solid oxide FC (SOFC) [20]. In contrast to the PEFC, the SOFC employs a solid ceramic electrolyte, typically yttria-stabilized zirconia (YSZ), which conducts O^{2-}-ions (cf. OCC, oxygen-conducting ceramic in Figure 2d). In these cells, particularly high electrical efficiency can be achieved as gas crossover is mostly avoided. Furthermore, for sufficient electrolyte conductivity, high operating temperatures (>650 °C and up to 1200 °C) are needed, which enhance the reaction kinetics, minimize activation losses and further improve electrical efficiency, even when fully non-noble metal catalysts are employed. Drawbacks for transportation applications of the SOFC are the brittleness of the ceramic materials used, challenges related to the operating temperature (e.g., unsuitability of metallic interconnects, difficulty to find adequate sealing materials), limited specific power and a slow start up time. Recent lab-scale developments, however, show significant progress of both planar and microtubular designs, reaching stack power densities of up to 17 kW/L and start-up times well below 10 min [20–24].

To overcome the temperature-related struggles described for the low- and high-temperature FCs introduced above, i.e., the need for noble-metal catalysts, humidification issues and difficulty in thermal management of the PEFC on the one hand, and material-related challenges in sealing and interconnecting as well as thermal degradation for the SOFC on the other hand, intermediate temperature approaches are of interest. These can be realized by a change of electrolyte and a concomitant adaption of the catalysts in either the ceramic or the polymeric FC type.

For the former, instead of an OCC, proton-conducting ceramics (PCC) have also been developed. These materials' ion conductivity reaches technically usable values at a significantly lower temperature, starting at around 300 °C [25]. With that, so-called proton-ceramic FCs (PCFCs, cf. Figure 2c) provide a much wider choice of materials for sealing, interconnecting and balance-of-plant components that can sustain the respectable temperature range than SOFCs do [26,27].

When it comes to polymeric electrolytes, a variety of membranes have been proposed, which are mostly phosphoric acid-doped such as PA-PBI (phosphoric acid-doped benzimidazoles) [28]. These cover operating temperatures ranging from 140 to 200 °C (cf. HT-PEFC, Figure 2b) and are proton-conductive without external humidification. Therefore, they alleviate the stringent requirements on the operating conditions that the PFSA membranes pose [29].

It should be noted that the latter two FC types are not at a development stage comparable to the PEFC or the SOFC. Scalability, especially close to the MW-scale as required for aviation, has not yet been shown, neither have standardized accelerated stress tests been systematically performed to assess the durability. Moreover, the bulk of related publications uses the power per electrode area as a means of comparison, whereas stack and system specific power are not specified for the often only lab-scale sized devices. Nevertheless, for the present work, all four cell types have been taken under consideration, aiming at exploring the application potential of each for the proposed hybrid concept for propulsion application in aviation. Forming the complement of the investigated hybrid propulsion concept, the effects of water/steam injection in GT engines are reviewed in brief below.

2.3. Water/Steam Injection in Gas Turbine Engines

The idea of improving the performance of GT engines through water or steam injection has been known at least since the publication of a British patent from 1944 [30]. The performance impact of water injection on the compressors of turbojet engines was analysed by Wilcox and Trout in 1950 [31]. The primary effects obtainable by GT water/steam injection—depending on the injection location and the condition of the injected water—include an increased cycle specific work and enhanced thermodynamic efficiency, as well as the reduction of NO_x emissions during the combustion process. Typical locations for water/steam injection in a GT engine include the air inlet as well as the compression combustion and turbine sections. Inlet and compressor water injection was applied to a number of the first generation, low-bypass ratio turbofan engines in the 1950s in order to enhance engine thrust for an improved take-off performance (e.g., J57 engine on the B-52 and KC-135 aircraft, JT3C engine on Boeing 707, JT9D engine on Boeing 747-100/200 and RR Spey engine on BAC 1–11 aircraft).

The injection of steam into the GT cycle is commonly used in stationary power plant technology (cf. [32–34]) in order to enhance both power output and thermodynamic efficiency. So-called "Steam-Injected Gas turbine" (STIG) power plants, with the steam typically produced using the GT exhaust heat, have been offered as commercial product solutions (cf. [33]) since 1987 [35].

Cheng demonstrated efficiency increases of up to 40% over a comparable simple GT cycle with a simultaneously enhanced power output by 70% for a steam injection rate of approximately 20% [32]. Noticeable efficiency gains may already be expected at much lower steam injection rates [36,37]. A consolidated overview of proposed and implemented GT cycles featuring water/steam injection is provided by Jonsson and Yan [38]. Daggett et al. have shown the significant cuts in NO_x emissions from kerosene combustion due to water injection in the compression section of turbofan engines [39,40]. The emission effects due to the ultra-wet combustion of natural gas and hydrogen have been numerically and experimentally studied by Göke et al., showing significant NO_x reductions for both types of fuel [41,42]. The most recent concept of a "Water-Enhanced Turbofan" (WET) engine, featuring in-flight water-recovery from the GT exhaust flow [43], has been initially assessed by Pouzolz et al. [44] and is currently under technology demonstration [45].

2.4. Advanced Heat Exchangers for Aeronautical Application

High-performance Heat Exchangers (HEXs) play a key role in the investigated propulsion system concept. This includes the HEX applications as part of the water treatment process as well as the HEXs required for the FC thermal management. HEXs are well-known from stationary applications as bulky and heavy components, especially when large amounts of heat need to be transferred. Therefore, HEX applications onboard transport category aircraft today refer exclusively to heat exchange tasks associated with engine auxiliaries and aircraft subsystems, such as air or fuel cooled oil coolers, or as part of the aircraft environmental control system, for instance.

The main heat management of classic GT based propulsion systems is intrinsically alleviated by the fact that most of the waste heat produced by the thermodynamic cycle is rejected at a relatively high temperature level via the core nozzle exit mass flow. This highly efficient heat rejection mechanism, at the same time, represents a significant source of losses in the power plant system. Therefore, significant research and development efforts have been invested over the last three decades in order to tackle the thermal losses in the GT exhaust flow. Common technical concepts include bottoming cycles that take the GT exhaust heat as a source of power, the utilization of the exhaust heat for the generation of steam and GT internal exhaust heat recuperation. For the latter concept, ultra-compact, high temperature resistant air-air HEX components have been developed and demonstrated for aero-engine integration [46,47]. Further advanced HEX configurations have been conceptually elaborated for an improved integration in the nozzle section of aircraft engines [48,49] as well as for secondary fluid recuperation based on multi-fluid HEXs [50]. In parallel, light-weight compact HEX designs for an application as compressor intercoolers have been developed [51,52] in order to increase the achievable fuel optimum cycle pressure ratios. Even more specialized high-performance microchannel HEXs originally developed for intercooling tasks in high-speed propulsion application are available [53].

Further momentum in the development of more advanced HEX technology certainly results from the thermal management demands posed by electrical components used in hybrid electric propulsion and power systems (cf. e.g., [54,55]). The need for rejecting significant amounts of waste heat at low temperature levels is also particularly relevant in the context of low-temperature FC application (cf. Section 2.2 above). Moreover, thermal management systems including various kinds of HEX will be crucial for the use of LH_2 systems for aircraft propulsion, as is reflected in ongoing research [56,57].

Existing empirical data for compact HEX designs are documented in [58,59], allowing the derivation of customized empirical correlations for HEX key performance for propulsion system conceptual design purposes.

2.5. Fuel Cell—Gas Turbine Hybrid Systems

In the quest for high system efficiencies, a number of studies have investigated the possibility of combining FCs with heat engines, typically GTs. While mostly developed for power generation applications in stationery use, these systems have also been proposed as (auxiliary) power units in aircraft and unmanned aerial vehicles [60–62].

Due to their similar operating temperature range, GTs are typically coupled with SOFCs, whereby two general concepts are distinguished: FC topping cycles versus FC bottoming cycles, i.e., energy conversion in the FC occurs upstream or downstream of the turbine [63]. A comprehensive review of past publications related to FC–GT hybrids is beyond the scope of this work and can be found elsewhere (cf. e.g., [60,64,65]). Nevertheless, some noteworthy examples in the context of the present work shall be briefly described in the following.

As mentioned above, one key aspect of the present concept is the use of the FC's product water to enhance the performance of the turbine and benefit from the concomitantly reduced emissions. Utilizing the product water for various purposes in the aircraft, e.g., cabin humidity, grey water or even drinking water, has been proposed in several patents owned by aerospace companies, including Boeing, Airbus and Liebherr Aerospace [66–69]. A direct interaction between an FC and a turbofan engine in order to enhance the engine's efficiency was, to the best of our knowledge, first described in 2016 by General Electric [70].

Amongst other approaches to efficiency enhancement, the author here puts forward the utilization of the FC product water for cooling of the turbofan engine as well as its direct injection into the core air flow path to alter temperature and/or compressibility properties. A suitable conditioning path for the FC product water that is required in order to minimize energy penalties for the water pressurization to relevant GT cycle pressure levels, and to exploit system synergies such as the GT waste heat utilization for the water vaporization, was very recently proposed as part of a patent application by Bauhaus Luftfahrt (BHL) [71].

FC technology as an option for the primary power supply on transport category aircraft, i.e., for the purpose of propulsion, has just very recently picked up momentum, together with the debate on hydrogen as an alternative fuel in aviation (cf. e.g., [72]). A hybrid system employing an SOFC, GTs and lithium ion batteries in order to boost take-off power was conceptually designed and assessed to be initially viable in larger aircraft applications by Collins et al. [73].

Lastly, it has been established that a high-pressure operation (up to 50 bar) could theoretically enhance the overall system efficiencies of FC–GT hybrids [74]. This is based on the fact that, theoretically, FC performance should increase with higher pressure due to an increased Nernst voltage and lower voltage losses [74–77]. Furthermore, the need for auxiliary components of FC–GT hybrid systems could simultaneously be lowered. These pressure levels have, however, not yet been practically realized as state-of-the-art FC technology cannot sustain the mechanical forces induced by these operating conditions. Even to realize operation at intermediate pressures (up to 15 bar [75,77]) the FCs are placed into pressure vessels designed to minimize fluctuations of pressure and consequently the mechanical strain on the FCs components.

3. Concept Evaluation Methodology

In this section, the methodological foundation of the aspired initial concept assessment is formulated and discussed. This includes the definition of system efficiencies and control volumes as well as the specification of key figures of merit for concept performance assessment at power system and aircraft level. Beyond this, the modelling of the power system's key components is presented, the derivation of lab-scale FC technology properties is discussed and a simplified aircraft scaling approach for the integrated concept assessment at the vehicular level is introduced.

3.1. System Definition and Efficiency Formulation

Following the definition of the Thrust Specific Fuel Consumption (TSFC) as fuel flow per unit net thrust produced, the overall efficiency η_{ov} of aero propulsion systems is given by the ratio of fuel supply power P_{supply}, i.e., the product of fuel mass flow $\dot{m}_f$ and its heating value FHV, and the effective propulsive power P_{thrust} (cf. e.g., [78]):

$$\eta_{ov} = \frac{P_{thrust}}{P_{supply}} = \frac{V_0 \cdot F_N}{\dot{m}_f \cdot FHV}, \tag{3}$$

where V_0 represents the flight velocity and F_N denotes the streamtube net thrust.

For the purpose of the present conceptual investigation, the classic breakdown of η_{ov} into the product of the core efficiency η_{co}, the transmission efficiency η_{tr} and the propulsive efficiency η_{pr}, is conveniently reorganized to yield η_{ov} as the product of the effective core engine efficiency $\eta_{co,eff}$ and the effective propulsive device efficiency $\eta_{pd,eff}$ (cf. [79]):

$$\eta_{ov} = \eta_{co} \cdot \eta_{tr} \cdot \eta_{pr} = \eta_{co,eff} \cdot \eta_{pd,eff}, \tag{4}$$

where $\eta_{pd,eff}$ relates the propulsive power demand in terms of F_N and V_0 to the effective core engine exit power $P_{co,eff}$, and $\eta_{co,eff}$ forms the ratio of $P_{co,eff}$ and P_{supply} [79]:

$$\eta_{pd,eff} = \frac{V_0 \cdot F_N}{P_{co,eff}}, \quad \eta_{co,eff} = \frac{P_{co,eff}}{\dot{m}_f \cdot FHV}. \tag{5}$$

In GT power plants, $P_{co,eff}$ is composed by the free shaft power extracted from the LPT in order to drive the propulsor $P_{co,sht}$ and the residual excess power in the core flow at the LPT exit (Station 5) $P_{co,res}$ (cf. [79]):

$$P_{co,eff} = P_{co,sht} + P_{co,res} \,. \tag{6}$$

For turbofan engines $P_{co,sht}$ and $P_{co,res}$ are thermodynamically expressed as (cf. [79]):

$$P_{co,sht,GT} = \dot{m}_{co} \cdot \Delta h_{LPT,free} = \frac{P_{Fan,o}}{\eta_{mech,LP}} = \frac{\dot{m}_{bp} \cdot \Delta h_{Fan,o}}{\eta_{mech,LP}} \,, \tag{7}$$

$$P_{co,res,GT} = \dot{m}_{co} \cdot \left(\Delta h_{is,5 \to amb} - \frac{V_0^2}{2} \right) , \tag{8}$$

where $\Delta h_{LPT,free}$ denotes the effective specific free work of the LPT and $\Delta h_{Fan,o}$ is the effective specific work of the outer fan. The term $\Delta h_{is,5 \to amb}$ represents the ideal residual work remaining after the turbine expansion process. The core and bypass mass flows are indicated by $\dot{m}_{co}$ and $\dot{m}_{bp}$, respectively, while $\eta_{mech,LP}$ denotes the low pressure spool mechanical efficiency. The efficiency-optimal ratio between $P_{co,sht,GT}$ and $P_{co,res,GT}$ can be determined based on ideality conditions for turbofan engines [80] as well as for turboprop engines [81].

In the case of a combined power system, such as the considered FC–GT hybrid, $P_{co,eff}$ equals the sum of the effective core engine exit powers for the installed GT engines, $P_{co,eff,GT}$, and the correspondingly equivalent effective power output of the FC power train, $P_{co,eff,FC}$:

$$P_{co,eff} = P_{co,eff,GT} + P_{co,eff,FC} \,. \tag{9}$$

The primary power output of the FC power train is the shaft power delivered by the electric motor, $P_{sht,eMot}$, which is driven by the FC system (FCS) electric power output. It is correlated to the fuel power supplied to the FC, namely $\left(\dot{m}_f \cdot FHV\right)_{FC}$, via the chain of electric component efficiencies:

$$P_{sht,eMot} = \left(\dot{m}_f \cdot FHV\right)_{FC} \cdot \eta_{FCS} \cdot \eta_{PMAD} \cdot \eta_{eMot} , \tag{10}$$

where η_{FCS} represents the FC system electric efficiency, η_{PMAD} refers to the efficiency of the Power Management And Distribution (PMAD) system and η_{eMot} denotes the efficiency of the controlled electric motor.

Beside its (primary) electric power output, an FC delivers thermo-kinetic power via the stack outflow. The usefulness of this thermo-kinetic power strongly depends on the FC type and operating conditions, namely temperature and pressure. In case of high temperature and/or high pressure FCs, the stack outflow is typically expanded through a turbine in order to provide power for system internal customers such as the inlet air compressor. The residual thermo-kinetic power at the exit of the FC system, $P_{co,res,FC}$, eventually adds to the effective electric shaft power delivery:

$$P_{co,eff,FC} = P_{sht,eMot} + P_{co,res,FC} . \tag{11}$$

As part of the present power system concept, the FC exit mass flow is routed through a condenser HEX in order to extract liquid water from it. While the extracted liquid water is pressurized and piped to the GT, the residual of the FC exit mass flow $\dot{m}_{FC,res,dry}$ and its remaining ideal work potential at the condenser exit $\Delta h_{is|Cond,ex,dry \to amb}$ define $P_{co,res,FC}$ analogously to Equation (8):

$$P_{co,res,FC} = \dot{m}_{FC,res,dry} \cdot \left(\Delta h_{is|Cond,ex,dry \to amb} - \frac{V_0^2}{2} \right) . \tag{12}$$

Applying the definition of $\eta_{co,eff}$ (cf. Equation (5))to the FC power train yields:

$$\eta_{co,eff,FC} = \frac{P_{co,eff,FC}}{P_{supply,FC}} = \frac{P_{sht,eMot} + P_{co,res,FC}}{\left(\dot{m}_f \cdot FHV\right)_{FC}} \tag{13}$$

and for the GT, respectively:

$$\eta_{co,eff,GT} = \frac{P_{co,sht,GT} + P_{co,res,GT}}{\left(\dot{m}_f \cdot FHV\right)_{GT}}. \tag{14}$$

With this, the effective core efficiency for a FC–GT hybrid power system can be formulated as a power weighted superimposition of the GT and FC power train:

$$\eta_{co,eff} = \frac{P_{co,eff}}{P_{supply}} = \frac{P_{co,eff,GT} + P_{co,eff,FC}}{\left(\dot{m}_f \cdot FHV\right)_{GT} + \left(\dot{m}_f \cdot FHV\right)_{FC}}. \tag{15}$$

While for GT engines, the lower heating value of the supplied fuel is used, it should be noted that in the case of low temperature FCs, the water contained in the product mass flow at the fuel exit might be condensed, i.e., in a liquid state, therefore requiring the application of the higher heating value of the fuel.

As can be seen from the equations above, the specific control volume definition of the effective core efficiency $\eta_{co,eff}$ ensures a direct consistency of the main power outputs of both parts of the FC–GT hybrid power system, namely the free shaft power available from the GT LPT(s), $P_{co,sht,GT}$, and the shaft power delivered by the FC electric power train(s) $P_{sht,eMot}$, while also catering for the residual powers that are available for the conversion to propulsive thrust.

A key design and performance descriptor for FC–GT hybrid power systems is the power split parameter PS, which can be defined as follows:

$$PS = \frac{P_{sht,emot}}{P_{co,eff}} = \frac{P_{co,eff,FC}}{P_{co,eff,GT} + P_{co,eff,FC}}. \tag{16}$$

As can be seen from Equation (16), the PS parameter expresses the share of the FC power train output relative to the total power system output, based on the $\eta_{co,eff}$ control volume.

For the fuel burn assessment, the Breguet-Coffin equation in integral form is solved for consumed fuel mass m_f as a function of aircraft instantaneous gross weight $m_{A/C,end}$ at the end of a considered range segment ΔR:

$$m_f = m_{A/C,end} \cdot \left(e^{\frac{\Delta R \cdot g}{FHV \cdot L/D \cdot \eta_{ov}}} - 1\right), \tag{17}$$

where L/D refers to the aircraft's aerodynamic efficiency, namely the ratio of lift-to-drag, and g represents the gravity constant. FHV is the effective fuel heating value of the hybrid power system, averaged according to the supply power split between the GT and FC power train parts. For all three parameters, FHV, L/D and η_{ov}, representative value for the considered flight segment should be used when evaluating Equation (17).

3.2. Mapping of Key Power System Component Properties

In order to facilitate an initial performance and emissions assessment of the investigated power system concept, the efficiency and mass properties of all key components are estimated using suitably simplified methods. All thermodynamic data for mass flows of semi-ideal gases are based on NASA's Chemical Equilibrium with Applications (CEA) database [82], while all thermodynamic data for air-water mixtures are read from a real gas database based on the IAPWS IF-97 standard [83]. The used hydrogen-specific thermodynamic data also refer to [82]. Important assumptions and model paradigms for the individual power system components are documented in the following.

3.2.1. Fuel Cell

The FC as a system includes the cell stack as well as a variety of auxiliary systems, including balance of plant and thermal management systems in often complex arrangements. The objective of the FC modelling in this paper is an appropriately simplified representation of aspects relevant for the hybrid power system concept. This includes the water production rate obtainable from the FC, but also the mapping of the FC system efficiency η_{FCS} (cf. Equation (10)). Moreover, for the estimation of the overall system mass, the thermal household requires suitable consideration.

The FC system efficiency η_{FCS} describes the ratio of net electric power delivered by the FC system $P_{FCS,el}$ to the system power supply via fuel flow, i.e., the product of fuel mass flow required for the cell redox reaction and its heating value $(\dot{m}\cdot FHV)_{H_2}$:

$$\eta_{FCS} = \frac{P_{FCS,el}}{(\dot{m}\cdot FHV)_{H_2}} = \frac{P_{Stack,el} - P_{Aux,el}}{(\dot{m}\cdot FHV)_{H_2}}, \quad (18)$$

where $P_{FCS,el}$ equals the stack electric power output $P_{Stack,el}$, reduced by all electric power demands P_{Aux} of the FC auxiliary systems including balance of plant and thermal management. It should be noted that, given the considered FC operating conditions in the present study, the lower heating value LHV is used for all FC efficiency evaluations.

Now, with η_{Stack} (cf. Equation (2)) rewritten as:

$$\eta_{Stack} = \frac{P_{Stack,el}}{(\dot{m}\cdot FHV)_{H_2}}. \quad (19)$$

η_{FCS} can be expressed in a factorized form:

$$\eta_{FCS} = \eta_{Stack}\cdot f_{\eta,BoP,TMS}, \quad (20)$$

with an efficiency factor $f_{\eta,BoP,TMS}$ capturing all auxiliary electric power demands within the FC system being defined as:

$$f_{\eta,BoP,TMS} = 1 - \frac{P_{Aux,el}}{P_{Stack,el}}. \quad (21)$$

Typical customers of $P_{Aux,el}$ amongst the FC auxiliary systems mainly focus on the reactant supply units, i.e., pumps and compressors, and electric controllers. It should be noted that the residual effective excess power of the FC power train $P_{co,res,FC}$ (cf. Equation (12)) is conservatively neglected when evaluating $\eta_{co,eff,FC}$ acc. to Equation (13).

While for the present study, fuel supply pressure is assumed to be provided through an active regulation of the LH_2 fuel tank delivery pressure, the FC inlet air supply requires active pressurization. This is realized by a turbo compressor. For low temperature FCs without sufficient thermo-kinetic power at the exit of a FC stack, the inlet air compressor requires an electric drive. The correspondingly required power for the compression of inlet air mass flow $\dot{m}_{air}$ to the stack operating pressure is calculated as follows:

$$P_{comp,air} = \frac{\dot{m}_{air}\cdot\bar{c}_p\cdot T_{in}}{\eta_{Comp,air}}\cdot\left(\left(\frac{p_{out}}{p_{in}}\right)^{\frac{\gamma-1}{\gamma}} - 1\right), \quad (22)$$

where $\bar{c}_p$ and γ are the representative isobaric specific heat and isentropic exponents for the air compression process. The compressor inlet total pressure p_{in} results from the freestream total pressure with an inlet pressure ratio of 99% assumed for the study. The compressor outlet total pressure p_{out} corresponds to the FC operating pressure. The efficiency $\eta_{Comp,air}$

includes the isentropic efficiency of the turbo compressor, the mechanical efficiency of its shaft and bearing system and its electric drive unit. It should be noted that in the case of high temperature and/or high pressure FCs, the stack outflow is typically expanded through a turbine in order to drive the inlet air compressor, thereby reducing $P_{comp,air}$ to zero.

In order to obtain the value of $P_{Aux,el}$, and thus $f_{\eta,BoP,TMS}$, the power requirements of all customers need to be added together. Beside $P_{comp,air}$, this also includes the power required in order to drive the liquid water pump within the product water treatment process (see Section 3.2.2). The electric power demand for the FC electric controller and auxiliary pumps and blowers within the balance of plant is assumed to be 3% of $P_{Stack,\ el}$.

The water mass flow $\dot{m}_{Stack,H2O}$ produced by an FC can be expressed as a function of $P_{Stack,el}$, η_{Stack}, the heating value per mol of fuel FHV_{mol} and the product water's molar Mass M_{H2O}:

$$\dot{m}_{Stack,H2O} = \frac{P_{Stack,el} \cdot M_{H2O}}{\eta_{Stack} \cdot FHV_{mol}}. \tag{23}$$

A parametric study of Equation (23) for a hydrogen FC is presented in Figure 3 below. As can be seen, the expectable water mass flows for typical stack efficiencies range from approximately 0.10 kg/s to 0.18 kg/s per MW of stack electric power.

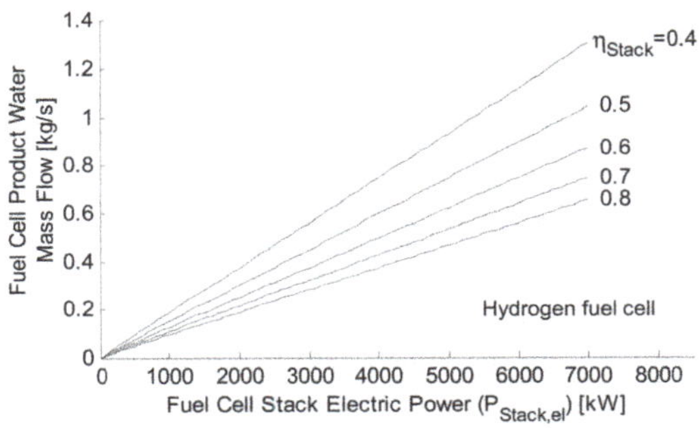

Figure 3. Product water mass flow obtainable from hydrogen FC.

All FC stacks considered in the study are assumed to be technically operated at an excess supply of reactants. While the excess air is simply passed through the stack, the unused fuel is assumed to be recirculated to the stack inlet without leakage.

While for the efficiency formulation above only the net mass flow rate of fuel that is effectively participating in the cell redox reaction was relevant, the thermal balance requires a more careful consideration of the actual mass flow rate of the reactants. In order to simplify the modelling of this, the stack is embedded in an auxiliary control volume that internalizes the recycling of unused fuel as well as any potentially required preheating of the fuel prior to entering the stack. A schematic representation of the stack and the auxiliary control volume referred to as "FC module" is presented in Figure 4.

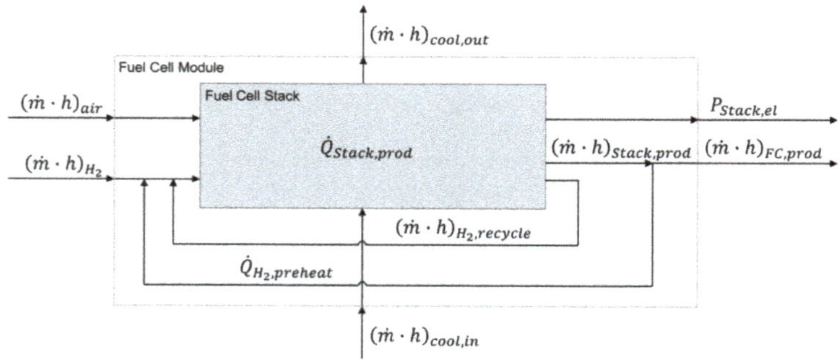

Figure 4. Schematic representation of FC stack including surrounding auxiliary control volume for thermal balance mapping.

The thermal balance of the FC module in Figure 4 is given by the heat production rate due to the redox reaction in stack $\dot{Q}_{Stack,prod}$, the net enthalpy flow across the FC due to its reactants net mass inflows, i.e., fuel and air supply, and its product mass outflow $\Delta\dot{H}_{FC}$ and the residual waste heat to be removed from the cell $\dot{Q}_{FC,cool}$. The stack internal heat source term can be calculated from the stack efficiency:

$$\dot{Q}_{Stack,prod} = \left(\dot{m}\cdot FHV\right)_{H_2} - P_{Stack,el} = P_{Stack,el}\cdot\left(\frac{1}{\eta_{Stack}} - 1\right). \tag{24}$$

The heat transfer terms result from the mass flows entering and exiting the control volume of the FC module:

$$\Delta\dot{H}_{FC} = \left(\dot{m}\cdot h\right)_{H_2} + \left(\dot{m}\cdot h\right)_{air} - \left(\dot{m}\cdot h\right)_{FC,prod}, \tag{25}$$

$$\dot{Q}_{FC,cool} = \left(\dot{m}\cdot h\right)_{cool,in} - \left(\dot{m}\cdot h\right)_{cool,out}. \tag{26}$$

As discussed above, $\left(\dot{m}\cdot h\right)_{air}$ represents the enthalpy of the gross stack inlet air flow, while $\left(\dot{m}\cdot h\right)_{H_2}$ refers to the enthalpy of the net fuel inflow. For a stationary equilibrium stack operating temperature, the following balance must be reached:

$$\Delta\dot{H}_{FC} + \dot{Q}_{Stack,prod} + \dot{Q}_{FC,cool} = 0. \tag{27}$$

while the FC heat production rate $\dot{Q}_{FC,prod}$ is always positive, the sign and absolute value of $\Delta\dot{H}_{FC}$ depend on the inflow temperatures of the reactants, the stack operating temperature and the effective operating stoichiometry, i.e., the excess air supply ratio.

In a case where $\Delta\dot{H}_{FC}$ is negative and its value exceeds $\dot{Q}_{Stack,prod}$ this would mean residual waste heat to be removed from the FC becomes negative, i.e., the cell needs to be heated in order to maintain its operating temperature. While this may be less relevant for low temperature FCs, this may often be the case for highly efficient high-temperature FCs. High temperature cell operation also increases the demand for a preheating of the reactants prior to entering the stack in order to reduce local thermal stresses. While preheating of the pre-compressed inlet air is often not applicable, fuel preheating is often required. In practice, both requirements are tackled by utilising the heat contained in the stack's product exit mass flow to preheat the fuel inlet mass flow [18,60].

In order to emulate this for the present study, heat is transferred from the stack product mass flow to the fuel inflow via an FC internal HEX until $\Delta\dot{H}_{FC}$ equals $\dot{Q}_{Stack,prod}$, in cases where $\dot{Q}_{FC,cool}$ would otherwise be <0. As a result in these cases, the temperature of the product mass flow leaving the FC is reduced relative to the stack operating temperature. The total pressure of the product outflow is assumed to be sufficiently close to the stack operating pressure.

The hydrogen fuel is preheated from the fuel supply temperature at the FC entry. Prior to entering the control volume of the FC, the hydrogen fuel mass flow from the LH_2 storage in the fuel tank is assumed to have absorbed a certain amount of heat due to its function as a cryogenic heat sink, e.g., used for the electric power train components as well as non-perfect thermal insulation of the fuel transmission system. A fuel supply temperature to the FC of 200 K is assumed for the present study.

3.2.2. Water Treatment Process

In order to allow for an efficient injection of the FC product water as superheated steam in the GT combustion chamber, the water first needs to be extracted from the FC product outflow, then pressurized in liquid form and re-vaporized. The basic thermodynamic implications of such a water treatment process are illustrated in Figure 5 based on a temperature entropy diagram for water. The water treatment process examples illustrated in the figure refer to the water released by a typical PEMFC and a typical SOFC, respectively.

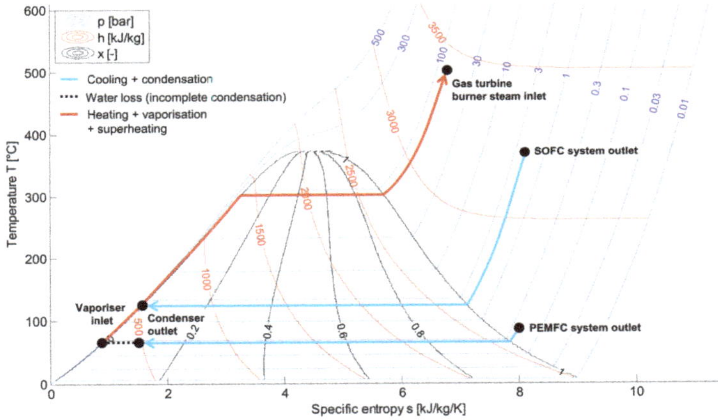

Figure 5. Temperature-entropy diagram for water with annotation of the water treatment processes for typical PEMFC and SOFC options.

An SOFC releases its product water at the anode (cf. Figure 2) while all other considered FC types release the water at the cathode, together with the excess mass flow from the cells' oxidant supply. With the hydrogen fuel recirculated inside the FC, this means a pure water condenser for SOFCs. For all other FC types, the condenser sees a mixture of the FC product water mass flow and the residual air fraction of FC oxidant supply. Consequently, the partial pressure of the PEMFC's product water in Figure 5 is significantly lower than that of the SOFC, resulting in a larger removal of heat required for the condensation process. At the same time, the condensation process of the PEMFC product water takes place at a lower temperature. The precooling of the FC product mass flow before condensation starts obviously requires more effort for the SOFC example due to the higher release temperature at the FC system exit.

For all studies presented in this paper, a liquid recovery of $\geq$90% of the FC product water can be assumed. The corresponding heat exchanger design modelling is discussed in Section 3.2.4. The liquefied fraction of the FC product water is subsequently pressurized by a water pump to a pressure level suitably above the total pressure in the GT combustion chamber and is then piped to the gas turbine core exhaust locations where the high-pressure water is reheated, vaporized and superheated using the residual GT exhaust heat. The power input required for the water pressurization is expressed as follows:

$$P_{Pump} = \frac{\dot{m}_{H_2O}}{\rho_{H_2O,liquid}} \cdot \frac{P_{out} - P_{in}}{\eta_{Pump}}. \quad (28)$$

where $\dot{m}_{H_2O}$ is the liquid water mass flow through the pump and $\rho_{H_2O,liquid}$ refers to the corresponding fluid density. The pump system efficiency η_{Pump} includes the hydraulic efficiency of the pump as well as the involved mechanical, electric drive and power system efficiencies A typical pump-delivered pressure level p_{out} would be 80 bar. The water inlet pressure level to the pump system, P_{in}, is defined by the FC operating pressure reduced by all pressure losses along the water treatment process between the FC stack and the pump inlet. Assuming P_{in} of 1.5 bar and η_{Pump} to be 0.5, the resultant water mass flow specific power input to the water pump system yields 15.7 kW/(kg/s). Extracted from the FC system this equals less than 0.2% of its electric power output. This low pumping power requirement is also reflected the water T-s diagram (cf. Figure 5) where the condenser outlet conditions (in case of full water recovery) are in immediate proximity of the vaporizer inlet conditions.

The mass of the water pump is calculated from a reference pump given in [61] using water mass flow-based proportional mass scaling. The mass of the required water piping system is derived in analogy to the aircraft fuel system. As a reference point, fuel system mass and throughput characteristics similar to an Airbus A320 aircraft are assumed. Piping system mass is then estimated via mass flow scaling using the liquid water throughput. An optional water tank acting as an intermediate reservoir of liquid water for increased operational flexibility of the overall system is not yet considered in the modelling.

3.2.3. Gas Turbine Engine

For the present study, the GT water injection is assumed to be realized via superheated steam injection into the combustion chamber. The resulting immediate effects on the thermodynamic cycle include an additional turbine mass flow due to injected water/steam and an increased mass flow-specific heat capacity in turbine, as well as a reduced compressor mass flow for a given power demand. A key descriptor for the implications of GT water/steam injection is the water-air-ratio (*WAR*), being defined as the ratio of steam mass flow $\dot{m}_{Steam}$ to compressor delivery mass flow at the combustor inlet $\dot{m}_{31}$, in the present study:

$$WAR = \frac{\dot{m}_{Steam}}{\dot{m}_{31}}. \quad (29)$$

The increased heat capacity in the turbine section for different levels of temperature and *WAR* is displayed in Figure 6a, while Figure 6b shows the reduction of core compressor mass flow against increasing steam injection.

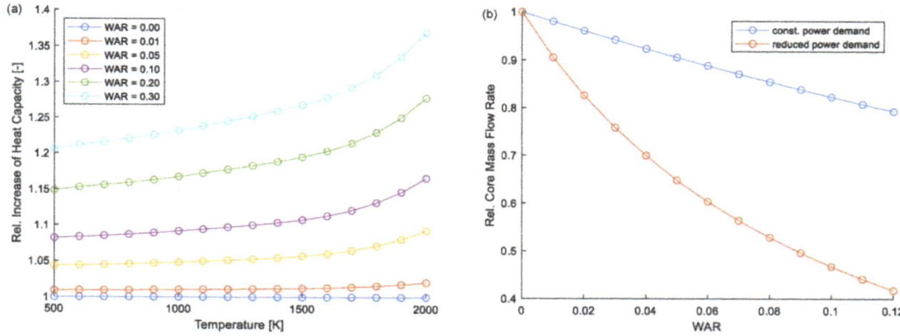

Figure 6. Impact of steam injection in GT combustion chamber on (**a**) the specific heat capacity of turbine working fluid and (**b**) the compressor mass flow rate.

As can be seen from the figure above, the specific work potential enhancement due to change of thermodynamic fluid properties is significant, e.g., approximately 25% at $WAR = 0.3$ for a typical average temperature level in the turbine section. At the same time, the core compressor mass flow reduces at a steep rate of approximately 1.8% per percent WAR increase, if a constant cycle output power is prescribed. This over-proportional compressor mass flow decrease results from the combined effect of the increased specific heat capacity in turbine flow and the added steam mass flow acting in the turbine section only. In this case, the required cycle output power is additionally reduced by the power produced through water supplying FC and the core compressor mass flow is reduced by more than 50% at $WAR = 0.1$.

Assuming the steam production from the liquefied FC product water mass flow to be performed using heat from the GT exhaust mass flow, the presence of the HEX in the core engine exhaust section will cause a pressure loss as a function of the amount of heat to be exchanged and as a trade-off with HEX size and mass.

In order to appropriately capture these main effects on the GT's effective core efficiency, a simplified formulation approach is used for the present study. Accordingly, all the primary effects due to the steam injection are combined in the efficiency factor $f_{\eta,co,eff,WAR}$. The GT effective core efficiency $\eta_{co,eff,GT}$ is obtained when multiplying the steam injection efficiency factor with a baseline efficiency value $\eta_{co,eff,base}$:

$$\eta_{co,eff,GT} = f_{\eta,co,eff,WAR} \cdot \eta_{co,eff,base}\left(P_{co,eff,GT}\right). \tag{30}$$

It should be noted that the baseline efficiency $\eta_{co,eff,base}$ will strongly depend on GT size effects. In order to also allow for technological sensitivity, the modelling of $\eta_{co,eff,base}$ is factorised into two parameters, and a technology factor $f_{\eta,co,eff,tech}$ is used to emulate the $\eta_{co,eff,GT}$ impact of advanced technology status and a true reference efficiency $\eta_{co,eff,ref}$, which is modelled as a function of $P_{co,eff,GT}$:

$$\eta_{co,eff,base} = f_{\eta,co,eff,tech} \cdot \eta_{co,eff,ref}\left(P_{co,eff,GT}\right). \tag{31}$$

For $\eta_{co,eff,ref}$, a simple data fitting based on BHL-internal models of existing aircraft GT engines of different power classes typical for the year 2000 in service technology is employed in this paper:

$$\eta_{co,eff,ref} = 0.5089 - 5.0654 \cdot \left(P_{co,eff,GT}[MW] + 3.8\right)^{-1.9}. \tag{32}$$

Based on its regression input data set, the validity of the $\eta_{co,eff,ref}$ correlation ranges from $P_{co,eff,GT} \approx 1.5$ MW up to $P_{co,eff,GT} \approx 30$ MW in cruise. Throughout the studies presented in this paper, a value of 1.2 is chosen for $f_{\eta,co,eff,tech}$ in order to represent an appropriately advanced technology level.

For the $f_{\eta,co,eff,WAR}$ factor, a simple heuristic based on published data and in-house performance synthesis computations was derived. The data basis for the heuristic law is visualized in Figure 7 below.

Roumeliotis and Mathioudakis [37] investigated the effects of water injection on GTs more generally, taking varying injection locations and conditions into account. For water steam injection in the combustion chamber, a potential for core efficiency enhancement of approximately 4% per 1% WAR is stated for low WAR injection rates and constant turbine inlet temperature operation. In Figure 7, data points for 2% and 4% WAR are depicted. Jonsson and Yan [38] collected data for various GT water/steam injection options. Again, a relevant data point derived from efficiency values provided for steam injection in the combustion chamber is annotated in Figure 7. Finally, Cheng and Nelson predicted a potential for an efficiency increase of up to 43% at WAR slightly above 20% while using exhaust gas recuperation for steam generation.

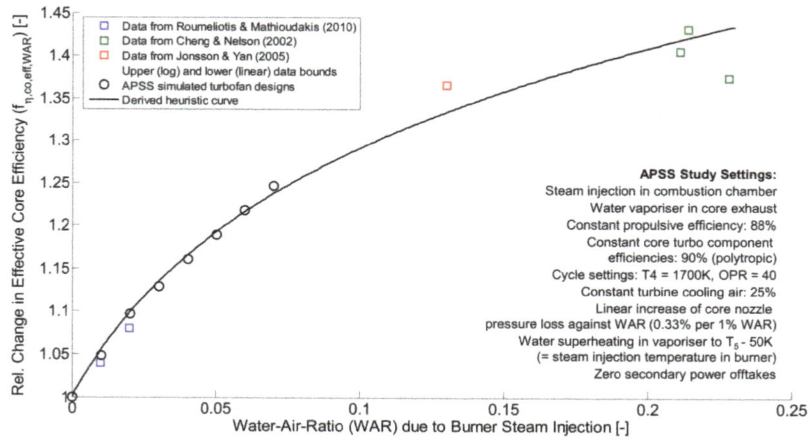

Figure 7. Semi-empirical derivation of heuristic correlation for GT effective core efficiency improvement versus water-to-air ratio due to burner steam injection.

In order to supplement the data obtained from the literature, a number of GT cycle design points were computed for the present study using BHL's in-house "Aircraft Propulsion System Simulation" (APSS) framework (cf. e.g., [84–86]). For these in-house simulated data points (black circles in Figure 7), steam injection in the combustion chamber of a 2-spool turbofan engine architecture was modelled using constant core turbo component polytropic efficiencies of 90%, a cycle featuring a burner exit temperature T_4 of 1700 K at an Overall Pressure Ratio (OPR) of 40 and a constant turbine cooling air supply of 25% of the high pressure compressor exit mass flow. Secondary customer offtakes including power and bleed air were set to zero. The vaporization and superheating of the recovery FC product water mass flow were modelled via a turbine exit duct-mounted vaporizer assuming a water heating up to LPT exit temperature T_5-50 K (=steam injection temperature in burner). To reflect the increasing heating effort required for the water vaporization and superheating as the water to GT exhaust air flow ratio grows, a linear correlation of core nozzle pressure loss against *WAR* was assumed that yields 5% at *WAR* = 0.15. For a given ideal nozzle exit velocity ratio according to Gasparovic [80], this effectively equals an increased back pressure for the cycle, hence reduced turbine expansion capacity. As such, the efficiency implication of the water vaporizer behind the LPT is effectively mapped to the effective core efficiency of the GT.

The heuristic correlation for $f_{\eta,co,eff,WAR}$ derived from the data collected in Figure 7 yields:

$$f_{\eta,co,eff,WAR} = 1 + 0.21 \cdot \ln(30 \cdot WAR + 1). \tag{33}$$

As can be seen in Figure 7, the correlation is supported by reference data up to *WAR* values of approximately 20%. As a first indication for the correlation's confidence level, the logarithmic upper and linear lower bounds of the reference data basis are indicated as dashed lines in the figure.

The impact of steam injection in the GT combustion chamber on cruise NO_x emissions is estimated based on the experimental results for ultra-wet hydrogen combustion published by Göke et al. in their Figure 8 [42]. Therefore, quadratic data fittings for the NO_x concentrations measured in the combustion products under water-to-air ratios between 0% and 30% were produced at relevant combustion temperature levels. For a given *WAR*, an average NO_x concentration can be calculated from the set of data fittings. Absolute GT NO_x emissions obtained as the product of NO_x concentration and actual fuel flow can then be related to the NO_x emissions at *WAR* = 0 in order to evaluate the NO_x reduction potentials.

Assuming zero NO_x emissions from the FC system, GT-produced NO_x represents the total NO_x emissions of the propulsion system.

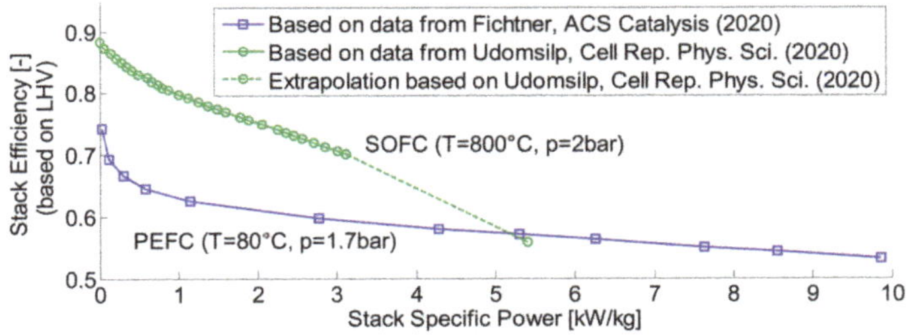

Figure 8. Performance characteristics of selected lab-scale PEMFC and SOFC technology concepts, calculated based on the assumptions specified in Appendices A.2–A.4.

3.2.4. Heat Exchangers

While for the efficiency of the FC module internal HEX used for the preheating of fuel (cf. Section 3.2.1), a prescribed value of 90% is assumed, the HEXs involved in the water treatment process require a more detailed consideration: The first one along the process is a HEX (condenser) required for water recovery from the FC product mass flow. It should be noted that the composition and thermal conditions of the relevant part of the FC outflow strongly depend on the FC type (cf. Section 3.2.2).

The condenser is modelled to be operated with freestream air as a coolant. The freestream mass flow rate is tailored to ensure a sufficient pinch point temperature difference between hot and cold HEX sides of 40 K at design point conditions (FL350, M0.78, ISA+10). The HEX performance calculation is therefore divided into two steps:

- temperature reduction down to saturation conditions; and,
- subsequent further cooling for latent heat transfer during phase change.

A second HEX (vaporizer) uses the GT exhaust gas mass flow for the vaporization and superheating of the pressurized water. The heat transfer of this HEX is segmented in three parts:

- heating the liquid water to boiling temperature;
- water vaporization; and,
- subsequent superheating of the water steam.

For sufficient thermodynamic conservatism in the studies presented in Section 4, steam superheating was limited to $T_5 - 50$ K. Both HEXs are modelled as counter flow HEXs with assumptions on pressure losses as summarized in Table 1.

Table 1. Summary of key assumptions taken for the initial estimation of HEX properties within the water treatment process.

HEX	Property	Value
Condenser	Pressure Ratio (cold side)	0.95
	Pressure Ratio (hot side)	0.96
	Pinch Point Temperature Distance	40 K
Vaporizer	Pressure Ratio (cold side)	0.9 *
	Pressure Ratio (hot side)	f(*WAR*)
	Superheating Limitation	$T_5 - 50$K

* compensated by water pump with very low effort due to incompressibility.

HEX efficiency η_{HEX} for the condenser and vaporizer components is defined by the ratio of realized temperature difference between inlet to outlet and the theoretically maximum temperature difference between HEX hot side and cold side inlet:

$$\eta_{HEX,Cond} = \frac{T_{Cond,in,h} - T_{Cond,out,h}}{T_{Cond,in,h} - T_{Cond,in,c}}, \quad \eta_{HEX,Vap} = \frac{T_{Vap,out,c} - T_{Vap,in,c}}{T_{Vap,in,h} - T_{Vap,in,c}}. \tag{34}$$

HEX mass estimation is based on a semi-empirical method for air-air HEX acc. to Grieb [59]. Therefore, data for mass flow-specific matrix weights of exhaust recuperators versus HEX efficiency η_{HEX} provided in [59] were fitted and scaled by a constant factor representing GT exhaust recuperator HEX installation weight effects $f_{m,HEX,inst}$, yielding the following correlation:

$$m_{HEX} = f_{m,HEX,inst} \cdot 82 \cdot \dot{m}_{air} \cdot (\eta_{HEX} + 0.2)^{8.1}, \tag{35}$$

where $\dot{m}_{air}$ is the representative air mass flow through the HEX matrix. Taking into account the specific matrix weight data used for the regression, the validity of Equation (34) ranges from $\eta_{HEX} \approx 0.6$ up to $\eta_{HEX} \approx 0.9$. For all HEX weight estimates in the present paper, a value of 1.6 is assumed for $f_{m,HEX,inst}$ (cf. also [59]).

In order to emulate the sizing of the multi-fluid, multi-phase HEXs involved in the water treatment process, the correlation in Equation (35) is applied using an equivalent air mass flow derived from the actually required heat capacity flows in the condenser and vaporiser HEXs. The respective equivalent air mass flows are tailored to facilitate the same amounts of heat transfer at the same temperature differences as the real multi-fluid, phase-changing condensers and vaporizers.

3.2.5. Electric Power Train

The electric power train represents the link between the net electric output power of the FC system and the mechanical shaft power absorbed by the targeted onboard customers. As such, it is constituted by a Power Management And Distribution (PMAD) system and a dedicated number of controlled electric motors. In the studies presented in this paper, as a first approach, the FC electric power is used to assist in driving the installed propulsive devices on the aircraft. Owing to the presence of the onboard LH_2 fuel storage as a convenient low-temperature heat sink, the employment of cryogenic technologies for key components of the electric power train seems warranted. This is reflected in a PMAD system efficiency η_{PMAD} of 99%, as well as the assumption of a High Temperature Superconducting (HTS) electric motor design paradigm also featuring 99% controlled efficiency (cf. e.g., [78]). Correspondingly, an advanced specific power target of 10 kW/kg is assumed for the overall electric power train.

3.2.6. Propulsive Devices

The propulsive device represents the assembly of the ducted fan, including the fan module, the fan drive gear system and the nacelle system. For all propulsive devices in the study, an effective propulsive device efficiency of 75% during typical cruise is assumed. This corresponds to a classic propulsive efficiency of approximately 88%, representing an advanced low specific thrust fan design.

Propulsive device mass is modelled based on a prescribed thrust-to-weight ratio in cruise. For the study, a cruise thrust-to-weight ratio of 1.7 is assumed, which is approximately 3 times the cruise thrust-to-weight ratio estimated for the advanced turbofan-type power plant systems.

3.3. Derivation of Lab-Scale Fuel Cell Technology Properties

In an attempt to evaluate the application potential of several FC types as introduced in Section 2.2 for the hybrid-concept presented herein, a number of recent publications on FC developments were taken into consideration [21,23,24,29,73,87–90]. Note that the focus lies on power performance only—other considerations of employing an FC in an aircraft (e.g., vibration tolerance, transient performance, etc.) go beyond the scope of this evaluation (cf. Appendix A.1).

As the polarization curves of novel FC developments are typically quoted per electrode area (i.e., in W/cm^2), a number of assumptions (e.g., on thickness and weight share of cell components) were made in order to estimate the respective specific power in kW/kg as required for an assessment of the presented hybrid concept. For each FC type, a value for a conservative and a more aggressive estimation is specified in Table A1.

All corresponding assumptions, technical details and references are given in Appendix A.2 to Appendix A.4. After recalculating the polarization curve in terms of stack specific power by normalizing with the derived peak power tabulated in Appendix A.5, the two best performing candidates were identified. Figure 8 displays the so-calculated performance characteristics of both cases based on the aggressive assumptions introduced in Appendix A.

For a given electric power output requirement and a specified operating point along the polarization curve, FC stack mass m_{Stack} can be calculated from the data provided in Figure 8. In order to obtain the total mass of the FC system m_{FCS} including its balance of plant and auxiliaries, the following correlation is used:

$$m_{FCS} = m_{Stack} + m_{Stack} \cdot \left(f_{m,BoP,TMS} + f_{m,BoP,Aircomp} + f_{m,BoP,Res}\right), \tag{36}$$

where the additive mass factors $f_{m,BoP}$ indicate typical masses of key component groups of the balance of plant in relation to the stack mass. A synopsis of the factors used for the estimation of the FC system masses associated with the PEMFC and SOFC technology cases in Figure 8 is provided in Table 2.

Table 2. Synopsis of FC system balance of plant additive mass factors PEMFC and SOFC technology cases.

Additive Mass Factors Relative to Stack	PEFC	SOFC
Thermal management ($f_{m,BoP,TMS}$)	1.61 *	0.39
Air compression system ($f_{m,BoP,Aircomp}$)	0.12	0.11
Other ($f_{m,BoP,Res}$)	0.16 *	0.21 **

* Based on HyPoint "Technical White Paper" [91]. ** Based on Tornabene et al. [61].

For the FC system mass calculation, three main component groups of the balance of plants and auxiliaries are distinguished: for the FC thermal management ($f_{m,BoP,TMS}$), the air compression system ($f_{m,BoP,Aircomp}$) and the residual of all other components ($f_{m,BoP,Res}$). As can be seen from Table 2, the values of the individual additive mass factors strongly depend on the FC type and its typical operating conditions, namely temperature and pressure. In particular, the significant additional mass connected to the thermal management system of the low-temperature FC technology option, namely the PEMFC, should be noted.

3.4. Simplified Aircraft Scaling Approach

Aircraft conceptual design and sizing for the integrated power system assessment studies presented in Section 4 of the paper is realized using a simplified discipline-oriented set of analysis modules based on Seitz [92]. Starting from the definition of the configurational setup of the aircraft and the integrated propulsion system, the design synthesis procedure includes

- the basic geometric description of the aircraft and its components primarily relevant for the study;
- the estimation of key aircraft component masses based on the underlying simplified geometric descriptions;
- the computation of aircraft aerodynamics during high-speed operation using the pre-calculated geometric and weight properties; and,

- the analysis of aircraft design mission performance based on the system weights, the aerodynamic properties and the energy and propulsion system efficiency characteristics.

During aircraft sizing, these disciplinary modules are consecutively evaluated. The required feedback correlations during the iterative aircraft scaling procedure are handled using a gradient-free iteration strategy as described and validated in [92]. Further verification and validation of the aircraft scaling approach were performed by Seitz and Engelmann [93]. For the present study, the pre-existing set of methods was supplemented in order to appropriately facilitate the aspired design studies for an LH_2-fuelled short-/medium range aircraft, with the cryogenic fuel storage accommodated in the fuselage. In the following, key relevant modelling aspects are discussed along the main disciplinary modules.

3.4.1. Geometry

Lifting surfaces are represented by simple trapezoidal planforms. Wing reference area scaling is performed for a prescribed wing loading at maximum landing weight in order to ensure appropriate low speed performance without detailed modelling of high-lift performance. Wing sweep is determined based on the simple sweep theory, i.e., employing cosine correction of the design cruise Mach number, applied to the aerodynamic center line at 25% chord length. Here, the effective design incidence Mach number for the wing airfoils is used as in input. Wing taper ratio is balanced to approximate elliptical spanwise lift distribution as a function of aircraft design cruise Mach number, wing aspect ratio and sweep angle according to Torenbeek [94]. Tail plane sizes are mapped as functions of wing reference area through prescribed volume coefficients. The wetted areas of lifting surfaces are determined based on the exposed regions of the component planform, translated using an empirical correlation given by McCormick [95]. Fuselage dimensioning in terms of diameter and overall length is performed under consideration of the design payload capacity and volumetric requirement associated with the accommodation of the onboard LH_2 fuel storage. The assumed LH_2 tank shape features spherically shaped heads that enclose an optional cylindrical center section. For the presented studies, an inner tank volume utilization of 87% is assumed, which corresponds to a 1.2 bar filling pressure and a venting pressure of 4 bar [9]. The corresponding gravimetric density of LH_2 is read from [9]. An additional allowance of 4% for trapped and unusable fuel in the tank is taken into account. A volumetric ratio of the tank walls to the inner storage volume of 40% is assumed based on results presented by Winnefeld et al. [96]. In summary, this yields a volumetric efficiency of the LH_2 storage, i.e., ratio of design mission fuel volume to the external displacement volume of the LH_2 tank in the fuselage, of 64%. During parametric studies with constant design passenger payload, fuselage diameter is retained constant, setting an upper limit to the allowable outer tank diameter. Hence, variations in required LH_2 tank size are accommodated via stretching or shrinking the length of the cylindrical fuselage center section.

3.4.2. Weights

The mapping of aircraft component masses is based on a combination of suitable textbook methods. The selection of the employed handbook methods for the individual components is made in order to secure appropriate functional sensitivity for essential physical effects connected to the aspired conceptual design study at aircraft level. A compact overview of the weight prediction methods for key aircraft components is given as follows: Wing and fuselage structural masses are calculated according to [97]. For LH_2 tank mass estimation a constant gravimetric tank efficiency as defined by Verstraete [9] of 50% is assumed, i.e., the tank's structural mass equals the stored fuel mass. The mass of the LH_2-specific onboard fuel system is considered as a constant fraction (2.5%) of aircraft Maximum Take-Off Weight (MTOW). All other airframe components, systems and equipment masses are summarized in a residual share of MTOW. Optimized mass and balance for aircraft with fuselage stored LH_2 tanks is neglected, in the first instance.

3.4.3. Aerodynamics

The mapping of aircraft aerodynamics is based on handbook methods (cf. [95,98–100]). A simple symmetric polar approach is used with the induced drag share modelled based on the wing's geometric properties using a correlation for Oswald's efficiency factor given by [100]. Skin friction and form drag for the wing and fuselage components are determined according to [95]. For wing aerodynamic mapping, the Mean Aerodynamic Chord (MAC) is treated as representative. The flow around the fuselage is assumed to be fully turbulent. The relative chordwise transition point from laminar to turbulent flow along the wing's MAC is prescribed by a free input parameter. The skin friction and form drags of all other aircraft components, including the empennage and propulsion system nacelles, are directly geared to the wing's drag using a constant factor. The wave drag of the overall aircraft is prescribed as a constant share of its total drag. The additional drag due to the rejection of water condenser as well as the FC waste heat is estimated as a delta in drag coefficient based on a change in drag counts per rejected heat load given by Pratt et al. [101]. Trim drag (typically 1–2% of aircraft total drag) is not explicitly modelled, but is considered as part of the induced drag share.

3.4.4. Performance

The performance mapping of the study aircraft is based on the performance formulation presented in Section 3.1. For the mission, fuel burn calculation during aircraft sizing is evaluated using the Breguet-Coffin equation as provided in Equation (17). Mission trip fuel is computed solely based on the performance properties for a typical cruise point. As such, the transversal flight phases are averaged as part of the considered cruise segment. Mission block fuel is obtained by adding a mass fraction of 1% of aircraft MTOW for LTO cycle and taxi fuel to the calculated trip fuel mass. For the computation design loaded fuel, the design stage length is increased by an additional 800 nmi as a convenient means of incorporating the required contingencies and reserves.

4. Concept Evaluation Results

In this section, the evaluation results for the investigated hybrid propulsion and power system concept are presented. Therefore, initial design studies at the isolated power system level and at the integrated vehicular level are introduced and discussed, before the impact of two selected scenarios of lab-scale FC technology on the potential aircraft design fuel and in-flight NO_x emission savings is assessed. To round off, more advanced integration options for the investigated technology concept are discussed in brief. The power system layout considered throughout the initial assessment features one FC system driving an electrically powered fan and two under-wing installed turbofan-type GT power plants. Specific aircraft-level integration implications of a third propulsive device, such as geometric integration aspects, changed structural load paths and local aerodynamic interference drag effects, as well as center of gravity shifts, control surface sizing impacts and changed OEI climb gradients during aircraft sizing, are not taken account in the first instance. Instead, the shaft power absorbed, the thrust produced and the mass added by the electrically driven fan is simply added to the propulsive devices of the turbofan power plants, thereby emulating a twin-engine aircraft configuration. All system and component sizing during the concept evaluation studies presented in the following are purely based on representative cruise conditions with typical margins built in for key high-power and low-speed operating points.

4.1. Initial Power System Design Study

In order to gauge the basic behavior of key design properties of the investigated hybrid power system concept, a variation of the design power split PS_{des} between the FC based and the GT based branches of the overall power train (cf. Equation (16)) is studied in this section. The study is conducted at typical cruise conditions, with the components of the two parallel branches of the power system and the water treatment process sized for a range of PS_{des} between 0 (all GT powered) and 0.7 (70% FC, 30% GT powered), assuming a constant $P_{co,eff}$ of 10 MW for the overall system. All water recovered from the fuel cell exit is immediately applied to the GT cycle, i.e., no intermediate water storage is considered. For the study, an advanced FC system operating at 80 °C and 2 bar is assumed featuring a specific power of 2 kW/kg at a constant efficiency of 60% in cruise. The FC system efficiency includes a constant efficiency factor for auxiliary electric power customers assumed to be 0.9 in the first instance, meaning 10% of the stack electric power output would be absorbed by the balance of plant. The water treatment process is sized for a 95% recovery of FC product water. Electric power train and propulsive device settings refer to Sections 3.2.5 and 3.2.6, respectively. The primary figures of merit targeted in the study include the power system total effective core efficiency $\eta_{co,eff}$ and the overall mass of the power system plus propulsive devices. The results obtained from the study are presented in Figure 9, including an annotation of key relevant settings.

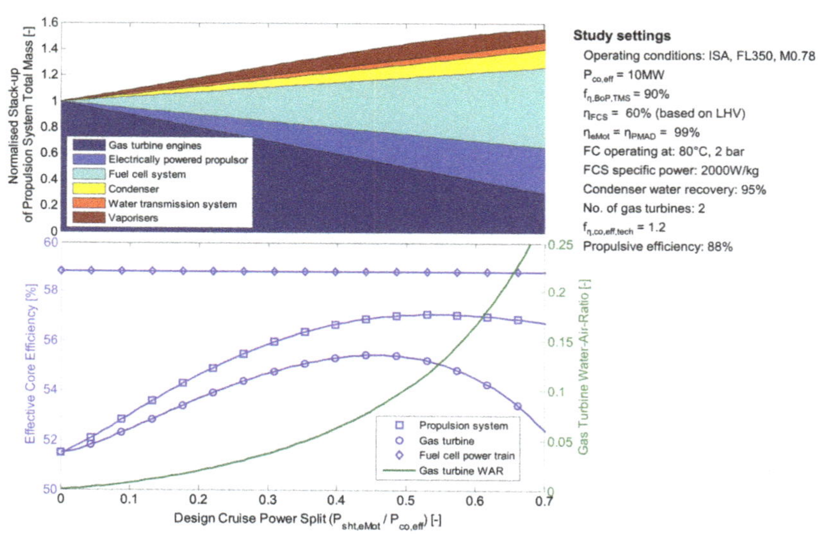

Figure 9. Effective core efficiency and propulsion system mass breakdown results of initial power system design study.

Inspection of the effective core efficiency trends displayed in Figure 9 reveals how the effective core efficiency of the overall power system results from the power split-based superimposition of the efficiency trending behaviors of its two parallel branches. While the $\eta_{co,eff,FC}$ appears constant against PS_{des}, $\eta_{co,eff,GT}$ is strongly affected by both the increasing water-air-ratio and reducing size as effective core excess power is incrementally shifted over to the FC branch via the PS_{des} descriptor. Given the decreasing positive gradient of $f_{\eta,co,eff,WAR}$ as WAR incrementally rises (cf. Equation (33)) and the simultaneously increasing negative gradient of $\eta_{co,eff,ref}$ as $P_{co,eff,GT}$ diminishes (cf. Equation (32)), a maximum of $\eta_{co,eff,GT}$ is formed at PS_{des} = 0.45. In consequence, an optimum of the complete power system effective core efficiency $\eta_{co,eff}$ is obtained in the middle ground between $\eta_{co,eff,GT}$ and $\eta_{co,eff,FC}$. In the studied case with $\eta_{co,eff,FC}$ exceeding the maximum value of $\eta_{co,eff,GT}$, $\eta_{co,eff}$ for the complete power system occurs at a larger design power split than for the GT branch alone, i.e., PS_{des} = 0.56.

Figure 9 also shows the trending behaviour of the overall propulsion system normalized mass stack-up as PS_{des} is varied. Starting from the pure turbofan baseline arrangement at PS_{des} = 0, the GT masses reduce constantly as PS_{des} increases. The simultaneously rising masses of the FC power train and water treatment process components, however, lead to a monotonic increase of the overall propulsion system mass. The linear behavior of the GT, FC and the electrically powered propulsor (propulsive device including electric drive motor and PMAD system) results from the constant specific power assumptions for the individual components. The study behaviour of the HEX component masses is mainly driven by linear scaling effects of the involved HEX fluid mass flows, such as the linear increase in FC product water mass flow as PS_{des} is raised. A synopsis of relevant design performance and mass data for the key design cases of the study is provided in Table 3.

Table 3. Summary of power system performance and mass properties for selected design cases.

Property	Unit	$PS_{des} = 0$	$PS_{des} = 0.45$ [1]	$PS_{des} = 0.56$ [2]
$\eta_{co,eff}$	[-]	0.515	0.569	0.571
$\eta_{co,eff,GT}$	[-]	0.515	0.554	0.550
$\eta_{co,eff,FC}$	[-]	0.588	0.588	0.588
η_{Stack}	[-]	0.667	0.667	0.667
$P_{Stack,el}$	[kW]	0	5100	6300
$\dot{Q}_{Stack,prod}$	[kW]	0	2550	3150
$\dot{Q}_{FC,cool}$	[kW]	0	2042	2522
$P_{sht,eMot}$	[kW]	0	4499	5557
$\dot{m}_{Stack,H2O}$	[kg/s]	0.000	0.569	0.703
$\dot{Q}_{Cond}$	[kW]	0	1884	2327
$\eta_{HEX.Cond}$	[-]	0.581	0.581	0.581
P_{Pump}	[kW]	0.0	8.5	10.5
$\dot{Q}_{Vap}$ (both engines)	[kW]	0	1093	1270
$\eta_{HEX,Vap}$	[-]	0.871	0.882	0.889
$P_{co,eff,GT}$ (per engine)	[MW]	5.000	2.751	2.221
$\dot{m}_{Steam}$	[kg/s]	0.000	0.540	0.668
WAR	[%]	0.0	8.0	13.3
$f_{\eta,co,eff,WAR}$	[-]	1.000	1.257	1.338
$\eta_{co,eff,ref}$	[-]	0.515	0.441	0.411
GT mass [3] (both engines)	[kg]	5863	3228	2607
Electric propulsor mass	[kg]	0	1330	1643
FC system mass	[kg]	0	2295	2835
Condenser mass	[kg]	0	529	654
Water transmission system mass	[kg]	0	208	257
Vaporiser mass (both engines)	[kg]	0	650	716
Overall propulsion system mass	[kg]	5863	8240	8711

[1] $\eta_{co,eff,GT}$ = maximum; [2] $\eta_{co,eff}$ = maximum; [3] including propulsive devices.

Table 3 shows that for the $\eta_{co,eff}$ optimum design at $PS_{des} = 0.56$, the total mass of the overall propulsion system is increased by 49% over the baseline design purely featuring turbofan engines. The corresponding relative improvement $\eta_{co,eff}$ amounts to 11% with a WAR in the GT cycle of 13.3%. The water treatment in this case is sized for an FC product water mass flow of approximately 0.7 kg/s, yielding a share of 19% of the overall propulsion system mass. The pumping power required for the pressurisation of the water mass flow is only 0.17% of the electric power output of the FC stack. It can be seen that the heat transfer required for the water condensation $\dot{Q}_{HEX,Cond}$ clearly exceeds the heat transfer during steam production $\dot{Q}_{HEX,Vap}$.

The ratio between the heat produced by the FC stack $\dot{Q}_{Stack,prod}$ and its residual waste heat to be removed by the thermal management system $\dot{Q}_{FC,cool}$ indicates that in the study case, approximately 20% of $\dot{Q}_{Stack,prod}$ is rejected from the FC through the net balance of its reactants and product enthalpy flows $\cdot\dot{H}_{FC}$.

4.2. Initial Aircraft-Integrated Assessment

In order to obtain an initial aircraft-integrated fuel burn assessment of the FC–GT hybrid power system concept, the previously presented propulsion system study is transferred to the vehicular level. As a relevant and convenient technology application case, a generic LH_2-fuelled short-/medium-range transport task is selected featuring a design payload capacity of 180 passengers on a 3000 nmi stage length. Key characteristics of the corresponding advanced baseline aircraft are listed in Table 4 below.

Table 4. Key characteristics of baseline aircraft for aircraft-integrated concept assessment studies.

Property	Value
Design Range	3000 nmi
Fuel Type	LH_2
Cruise Condition	ISA, M0.78, FL350
Lift/Drag	18.0
Fuel Heating Value	120.9 MJ/kg (LHV)
TSFC *	4.83 g/kN/s
$\eta_{co,eff}$	53%
$\eta_{pd,eff}$	75%
OEW	49,000 kg
Payload	18,000 kg
Design Loaded Fuel	6300 kg
MTOW	73,300 kg
MTOW/$S_{ref,Wing}$	650 kg/m^2

* Corresponds to 13.5 g/kN/s based on FHV 43 MJ/kg.

The baseline aircraft characterised in Table 4 is powered by turbofan-type GT engines without hybridization. The efficiency properties given in the table, namely Lift/Drag, TSFC and propulsion system efficiency figures, refer to typical (mid-)cruise conditions. The additional structural and systems masses intrinsic to the use of LH_2 as the onboard fuel are reflected in the OEW which is considerably higher than for a kerosene fueled aircraft with similar MTOW. The effective MTOW mass shares for the baseline aircraft's airframe structure and for its systems and equipment are 30% and 15%, respectively. The baseline aircraft's increased fuselage size and corresponding drag characteristics due to the internal LH_2 fuel storage are additionally reflected in the aircraft's aerodynamic efficiency, which might otherwise be interpreted as relatively low for a highly advanced short-/medium-range aircraft.

Similarly to the power system level study, the design power split parameter PS_{des} in cruise is investigated as part of the aircraft integrated sizing study targeting design mission block fuel as the central figure of merit. In order to ensure first order consistency against the investigated aircraft design variations, a number of basic aircraft scaling laws are followed throughout the study: The aircraft lifting surfaces are scaled at a constant maximum loading. Correspondingly, the maximum loading of the wing, i.e., aircraft MTOW per wing reference area $S_{ref,Wing}$, is retained at a constant of 650 kg/m^2 in order to ensure appropriate low-speed aerodynamic performance. Propulsion system scaling is performed based on the $P_{co,eff}$ design requirement obtained from the aircraft during cruise. Identical design specific thrust settings are adopted for all aircraft designs. The scaling implications of the onboard LH_2 infrastructure refer to the description provided in Section 3.4.

Together with the PS_{des} parameter, the study includes sensitivities for key FC system design and performance figures, namely FC system efficiency $\eta_{FCS,cr}$ and specific power $SP_{FCS,cr}$ in cruise. For the study, again, a low temperature FC type is assumed, operating at a temperature of 80°C and pressure of 2 bar. An FC product water recovery of 90% in the condenser is assumed. The results of the aircraft-integrated sizing study in terms of design mission block fuel are summarised in Figure 10, where Figure 10a shows the $\eta_{FCS,cr}$ and Figure 10b displays the sensitivity with regard to $SP_{FCS,cr}$.

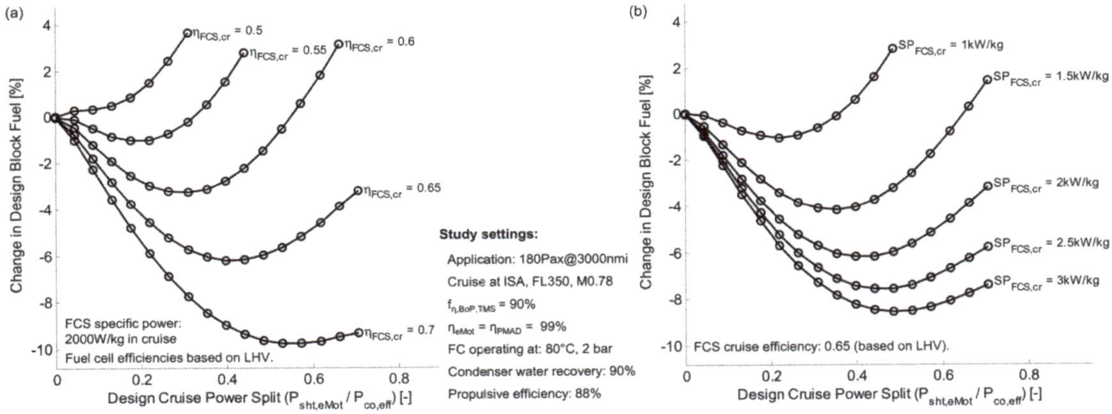

Figure 10. Design block fuel results of initial aircraft-integrated sizing study: (**a**) sensitivity to FC system efficiency, (**b**) sensitivity to FC system specific power.

As expected, the block fuel saving potential of the FC–GT hybrid propulsion concept increases with both, improving FC system efficiency and FC system specific power (cf. Figure 10). Design fuel savings may generally be expected when $\eta_{FCS,cr}$ and $SP_{FCS,cr}$ lie above 60% and 1500 W/kg, respectively. Above these values, improving FC efficiency and specific power continuously enhances the potential for fuel burn reduction while simultaneously shifting the fuel-optimum design cruise power splits to higher values. It should be noted, however, that the lever for further improvement through additional enhancement of $SP_{FCS,cr}$ progressively diminishes as the FC system mass share of the aircraft gross weight declines. Further improvements in FC system efficiency, in contrast, may open the potential for significant further fuel savings. For a selected representative scenario of $\eta_{FCS,cr}$ and $SP_{FCS,cr}$ from the performed study, a more detailed view is presented in Figure 11 below.

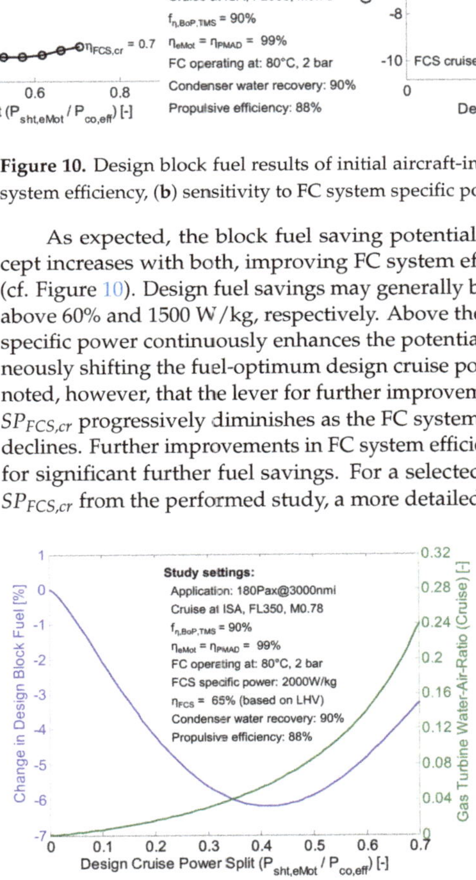

Figure 11. Design block fuel and GT water-to-air ratio trends for selected aircraft design study scenario.

Beside the characteristic design block fuel trend versus design cruise power split, Figure 11 also displays the associated nonlinear increase in GT water-to-air ratio as design power is shifted more and more to the FC branch of the overall power system. In the region of low block fuel designs, GT *WAR* notably ranges from 5% to 10%. The corresponding implications can be exemplified for the fuel optimum design at $PS_{des,cr} \approx 0.42$ as highlighted in Table 5.

Table 5. Summary of key characteristics for block optimum aircraft from Figure 11.

Property	Value [%]
Design cruise power split $PS_{des,cr}$	41.5
GT *WAR*	5.72
Change in $\eta_{co,eff}$ *	12.8
Change in aircraft c_{D0} due to system waste heat rejection *	2.26
Change in propulsion system mass *	35.0
Change in aircraft OEW *	6.40
Change in aircraft MTOW *	3.78
Change in design block fuel *	−6.18
Change in GT fuel flow *	−44.1
Change in NO_x emissions *	−61.2

* relative to baseline aircraft ($PS_{des,cr}$ = 0).

As can be seen from the table above, the selected fuel optimum aircraft design from Figure 11 features a $\eta_{co,eff}$ improvement of 12.8% at an increase of the overall propulsion system mass of 35% compared to the non-hybrid baseline aircraft design. The aerodynamic drag penalty calculated for the propulsion system heat rejection expressed in terms of the aircraft zero-drag coefficient c_{D0} yields approximately 2.3%. In summary, a design block reduction of 6.2% is obtained. Together with the GT fuel flow reduction of 44.1% at the given $PS_{des,cr}$ the *WAR* in the GT engines of approximately 5.7% leads to a calculated NO_x emission reduction of approximately 61%.

4.3. Introduction of Lab-Scale Fuel Cell Technology

In an important final step of the initial assessment of the investigated FC–GT hybrid propulsion concept, the generic assumptions regarding FC design characteristics from the previous studies are replaced by lab-scale experimental data of pre-selected low- and high-temperature FC technology concepts taken from the technology survey presented in Section 2.2 (corresponding assumptions in Appendix A). Therefore, suitable cruise operating points are selected from the performance curves of the PEMFC and SOFC stacks characterised in Section 3.3 (cf. Figure 8). In the first instance, cruise operation of both the stack technologies is considered to be at 40% of maximum loading, thereby ensuring adequate power reserves for high-power low-speed operations such as take-off and go-around (cf. [78]). Nevertheless, it should be noted that the present study is a purely stationary consideration, i.e., no transient implications are taken into account in the first instance. In order to meet rapid transient response time requirements, the SOFC-based hybrid power system would be expected to necessitate, for example, an additional high specific power battery and/or enhanced operational characteristics of the involved GT engines.

The cruise operating points deliver both the design specific powers and the efficiencies of PEMFC and SOFC stacks. The balance of plant efficiency factors $f_{\eta,BoP,TMS}$ for both FC system cases are predicted as described in Section 3.2.1. The FC system masses are calculated from the stack specific powers at cruise conditions (cf. Figure 8) using the additive mass factors for the balance of plant items provided in Table 2. A summary of the resulting FC cruise efficiency and specific power metrics at stack and system level is provided in Table 6 below.

Table 6. Cruise efficiency and specific power characteristics calculated for selected PEMFC and SOFC technology cases.

FC Cruise Property	Unit	PEMFC Case	SOFC Case
η_{Stack}	[-]	0.58 *	0.74 *
$f_{\eta,BoP,TMS}$	[-]	0.77	0.97
η_{FCS}	[-]	0.45	0.72
P/W_{Stack}	[kW/kg]	3.94 *	2.16 *
P/W_{FCS}	[kW/kg]	1.03	1.26

* at 40% peak power, based on Figure 8.

The SOFC technology case in the table above features a higher system efficiency η_{FCS} than the PEMFC concept. This is rooted in both, a higher stack efficiency and lower losses associated with the balance of plant systems, as the air supply compressor is driven by an exhaust turbine. With regard to the calculated stack specific powers P/W_{Stack} the PEMFC case shows clear advantages over the SOFC. The significant masses added by the balance of plant, especially by the thermal management system, however, diminish this advantage of the low temperature technology case, when benchmarking the specific powers at FC system level P/W_{FCS}. The resulting aircraft design block fuel trends of both FC technology cases is plotted versus the $PS_{des,cr}$ parameter in Figure 12.

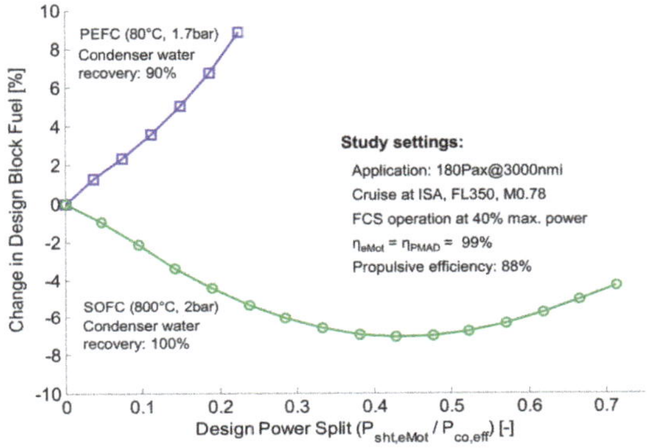

Figure 12. Aircraft design block fuel trends for selected PEMFC and SOFC technology cases versus design cruise power split.

While hybridization based on the selected PEMFC technology case does not lead to a design fuel reduction, Figure 12 displays a clear opportunity for fuel savings for the studied SOFC case. This is primarily due to the higher system efficiency, but also due to the slightly lower masses at FC system level and for the HEXs within the water treatment process. As a result of this, a full FC product water recovery is numerically achieved in contrast to a best and balanced 90% recovery in the PEMFC case. A brief characterization of the optimum fuel design for the SOFC technology case is given in Table 7.

Table 7. Summary of key characteristics of block fuel optimum aircraft design based on selected SOFC technology case.

Property	Value [%]
Design cruise power split $PS_{des,cr}$	42.8
GT WAR	6.11
Change in $\eta_{co,eff}$ *	18.3
Change in GT fuel flow *	−44.4
Change in propulsion system mass *	64.9
Change in aircraft MTOW *	8.70
Change in design block fuel *	−7.07
Change in NO_x emissions *	−62.4

* relative to baseline aircraft ($PS_{des,cr} = 0$).

Minimum design block fuel for the SOFC case is obtained at $PS_{des,cr} \approx 0.43$, equivalent to an improvement of approximately 7.1% over the turbofan powered baseline aircraft. The corresponding cruise NO_x emission benefit of the optimum hybrid design is calculated to approximately 62%.

4.4. Discussion of Advanced Integration Options

The initial concept assessment studies presented earlier in this section were performed for a most simplistic implementation option for the proposed FC–GT hybrid power system. However, a number of more advanced conceptual implementation options are very well conceivable that could feature an even more synergistic integration of technologies at the aircraft level. A few important aspects with regard to advanced technology integration opportunities based on the studied hybrid power system are briefly discussed here.

Depending on the effectively achievable design power split values for the hybrid power system, there are multiple potential onboard customers for the FC system electric power output. These include the electric subsystems of the aircraft as well as distributed propulsion systems and advanced aircraft annexed technologies such as wing tip propellers for vortex drag improvements, wing hybrid laminar flow control or the possibility of emission free ground operation.

The presence of an onboard cryogenic heat sink due to the LH_2 fuel storage certainly is a key enabler for the consideration of superconducting technology as part of the fuel cell based electric power train. Beyond this, the inevitable boil-off from the LH_2 storage might be an attractive source of power for independent ground operations. While a purely LH_2-based fuel supply appears most attractive for the proposed hybrid propulsion concepts, a dual-fuel configuration is also conceivable with the FC operated on hydrogen while the GT engines remain fueled by a sustainable drop-in fuel.

The FC system itself could be strategically decentralized through distributed stacks and structurally embedded HEXs, thereby enhancing structural load alleviation of the dry wing design in case of LH_2-fuelled aircraft and tapping the potential for efficient heat rejection through aircraft external surfaces without dedicated radiator devices [102,103].

From the investigation of lab-scale FC technologies, the benefits of high-temperature SOFC technology for the use in the presented hybrid have been clearly identified in this assessment focusing on specific power performance only. The key issue of SOFC operational limitations due to very slow transient responses is certainly alleviated through the combination with GT engines in a hybrid concept when compared to a fully SOFC based propulsion system. However, further work will be required to investigate the transient operational implications of FC–GT hybrid system designs with a special focus on SOFC technology options.

An intermediate storage capability for the liquefied FC product water will facilitate new opportunities for optimally tailored system performance characteristics within the entire flight envelope, and for minimized local NO_x emissions.

The GT performance benefits obtainable from a given fuel product water mass flow depend on the engine size and the specific work capacity of the underlying cycle. For an increased specific work capacity of the baseline cycle, a predefined water mass flow translated to a higher water-to-air ratio in the cycle. Therefore, the hybrid combination of advanced FC technology with high specific work GT cycles, for example through topping cycles such as that realized in a Composite Cycle Engine [84,104,105] seems to be most beneficial. First steps towards the evaluation of hydrogen-fueled Composite Cycle Engine concepts are presented in [106].

A particularly synergistic technology configuration becomes obvious when combining the proposed LH_2-fuelled FC–GT hybrid power system with fuselage boundary layer ingesting propulsion as implemented by the recently proven Propulsive Fuselage Concept [107].

5. Conclusions and Further Work

In the present paper, a FC–GT hybrid propulsion concept was introduced and initially assessed. The studied concept uses the by-product of one central system element, i.e., the water mass flow released by a hydrogen FC while producing electric power, in order to improve the efficiency and power output of another key system element, namely a GT engine enhanced by a steam-injected thermodynamic cycle. The FC product water is conditioned through a process of condensation, pressurization and re-vaporization prior to its GT injection. The water vaporization is performed using the waste heat in the GT core exhaust.

The concept assessment was performed for an LH_2 fueled short-/medium range transport task featuring a design of 180 passengers on a 3000 nmi stage length. The assessment included design studies for the isolated propulsion system as well as at the integrated aircraft level. Therefore, appropriate methodology was formulated, and parametric modelling of all key components was performed. As a key power system parameter, the impact of the design power split ratio, i.e., the ratio between the power delivered by the FC branch and the total power delivery of the hybrid power system, was investigated.

At the power system level, an increasing design power split ratio at constant total power output has two main effects on the efficiency of the involved GT engines: (1) The FC product water mass flow increases proportionately with the increasing power output of the FC stack. Together with the simultaneously reducing size and power output target of the GT engines, the ratio of injected water mass flow to the GT air mass flow, i.e., the water-to-air-ratio, rises non-linearly and causes significant enhancements of the cycle efficiency. (2) The shrinking GT size diminishes the associated baseline efficiency. The resulting GT efficiency trending behavior against the design power split ratio, when superimposed with the more constant efficiency scaling behavior of the FC, forms a distinct efficiency optimum of the hybrid power system sized for a total power output of 10 MW. The pumping power required for the pressurization of the water mass flow is lower than 0.2% of the electric power output of the FC stack, and thus has only a minor impact on power system efficiency.

Despite monotonic increases in overall propulsion mass due to the components of the FC and water treatment process as design power split is shifted more to the FC branch, the optimal behavior of the power system efficiency basically translates also to the vehicular level. For the studied aircraft application case, design fuel savings were found as soon as the efficiency of the FC system in cruise reached 60% and higher at simultaneous cruise-specific powers of 1.5 kW/kg or above.

These findings were completed by the investigation of two distinct study cases based on selected lab-scale PEMFC and SOFC technology, which showed that a propulsion system hybridization using the selected PEMFC technology case does not lead to design fuel reductions for the air transport task considered. In contrast, the studied SOFC technology case displayed a clear opportunity for design fuel savings. This was primarily due to the higher system efficiency for the SOFC case in cruise when compared to the selected PEMFC concept. It should be noted, however, that no transient operation was considered, which represents one of the operational advantages of the PEFC.

Specifically, minimum design block fuel for the SOFC case was identified at a design power split ratio of approximately 43%, yielding an improvement of 7.1% over the turbofan powered baseline aircraft. The corresponding water-to-air ratio of approximately 6.1% in the GT cycle and the simultaneous design fuel flow reduction of 44% in the GT combustion chambers translated to a cruise NO_x emission benefit of approximately 62%.

The conceptual design evaluation studies presented in this paper were performed for a most simplistic architectural implementation of the proposed hybrid propulsion concept, featuring two under-wing podded turbofan engines and an electric ducted fan driven by a central FC power train. Specific aircraft-level integration implications of a third propulsive device were not taken into account. All water recovered from the FC exit was immediately applied to the GT cycle, i.e., no intermediate water storage was considered. The system and component sizing was mainly based on representative cruise conditions without detailed consideration of key high-power and low-speed operating points. Therefore, further work needs to exhaustively analyse all relevant operating conditions within the design and sizing of the hybrid propulsion system concept. This particularly includes the consideration of the aircraft's overall thermal household and abnormal operating modes, e.g., under partial system failure, as well as optimized operations featuring an intermediate storage of the liquefied FC product water, which will need to be explored and reflected in order to obtain best and balanced design solutions.

For a broader evaluation of the efficiency and emission-saving potentials associated with the presented concept, further studies should also include synergistically annexed technologies combined in consistent overall aircraft technology configurations. Therefore, the compatibility of the FC product water utilization in advanced high specific power heat engine cycles needs to be explored. Among various potential customers for the electric power produced by the FC branch of the hybrid power system, the fuselage boundary layer ingesting propulsor of a propulsive fuselage concept aircraft certainly represents a particularly attractive option.

To summarize, the performed initial assessment demonstrates the potential of future SOFC technology synergistically combined with GT engines for application in transport aircraft propulsion and power systems. Exploiting the synergies of the key components within the presented hybrid concept enables promising opportunities for relevant emission reductions and, consequently, a reduced climate impact of aviation.

Author Contributions: Conceptualization: A.S., M.N. and F.T.; Methodology: A.S., M.N. and K.E.; Validation: A.S., M.N. and K.E.; Formal analysis: A.S., M.N. and K.E.; Investigation: A.S. and M.N.; Data curation: A.S., M.N. and K.E.; Writing—original draft preparation: A.S., K.E., M.N. and F.T.; Writing—review and editing: A.S., K.E., M.N. and F.T.; Visualization: A.S., M.N. and K.E. All authors have read and agreed to the published version of the manuscript.

Funding: This research received no external funding.

Conflicts of Interest: The authors declare no conflict of interest.

Nomenclature

Abbreviation/Acronym	**Description**
ACARE	Advisory Council for Aviation Research and innovation in Europe
APSS	Aircraft Propulsion System Simulation
BHL	Bauhaus Luftfahrt
CEA	Chemical Equilibrium with Applications
CL	Catalyst Layer
CO_2	Carbon Dioxide
Eq	Equation
FC	Fuel Cell
FCS	Fuel Cell System
FHV	Fuel Heating Value
GDL	Gas Diffusion Layer
GT	Gas Turbine
GTF	Geared TurboFan
H_2O	Water
HEX	Heat EXchanger
HOR	Hydrogen Oxidation Reaction
HT	High Temperature
IAPWS	International Association for the Properties of Water and Steam
LH_2	Liquid Hydrogen
LHV	Lower Heating Value
LPT	Low Pressure Turbine
LTO	Landing and Take-Off
MTOW	Maximum Take-Off Weight
NO_x	Nitrogen Oxides
OCC	Oxygen Conducting Ceramic
OEW	Operative Empty Weight
OPR	Overall Pressure Ratio
ORR	Oxygen Reduction Reaction
PA-PBI	Phosphoric Acid-Doped Benzimidazoles
PAX	Passengers
PCC	Proton Conducting Ceramic

PCFC		Proton-Ceramic Fuel Cell
PEFC		Polymer Electrolyte Fuel Cell
PEMFC		Proton-Exchange Membrane Fuel Cell
PFSA		Perfluorosulfonic Acid
PMAD		Power Management and Distribution
PS		Power Split
SOFC		Solid Oxide Fuel Cell
SP		Specific Power
STIG		Steam-Injected Gas Turbine
TMS		Thermal Management System
TRL		Technology Readiness Level
TSFC		Thrust Specific Fuel Consumption
WAR		Water-Air-Ratio
WET		Water-Enhanced Turbofan
YSZ		Yttria-Stabilized Zirconia
Symbol	**Unit**	**Description**
$\bar{c}_p$	$\frac{J}{kg\cdot K}$	Representative specific heat capacity
F	$\frac{C}{mol}$	Faraday constant
f_m	-	Mass scaling factor
f_{tech}	-	Technology factor
f_η	-	Efficiency factor
FHV	$\frac{J}{kg}$	Fuel Heating Value
FL	$100\,feet$	Flight Level
F_N	N	Streamtube Net Thrust
g	$\frac{m}{s^2}$	Gravity constant
H	J	Enthalpy
h	$\frac{J}{kg}$	Specific enthalpy
ISA	K	International Standard Atmosphere
L/D	-	Aerodynamic Lift-to-Drag ratio
m	kg	Mass
$\dot{m}$	$\frac{kg}{s}$	Mass flow
M	-	Mach number
P	W	Power
$\dot{Q}$	$\frac{J}{s}$	Heat flux
s	$\frac{J}{kg\cdot K}$	Specific enthropy
T	K	Temperature
U	V	Voltage
V	$\frac{m}{s}$	Velocity
x	-	Water steam fraction
η	-	Efficiency
γ	-	Isentropic exponent
ρ	$\frac{kg}{m^3}$	Density
Indices/Subscripts		
31		Gas turbine thermodynamic station at burner inlet
5		Gas turbine thermodynamic station at low pressure turbine exit
amb		Ambient conditions
aux		Auxiliary
base		Baseline
BoP		Balance of Plant
bp		Bypass
c		Cold
co		Core
Comp		Compressor
Cond		Condenser
cr		Cruise
des		Design
eff		Effective
el		Electric
eMot		Electric Motor

Fan,o	Outer fan
h	Hot
is	Isentropic
in	At inlet
liquid	In liquid state
LP	Low Pressure shaft system
LPT	Low Pressure Turbine
mech	Mechanical
out	At outlet
ov	Overall
pr	Propulsive
prod	Product
Pump	Pump system
ref	Reference
res	Residual
sht	Shaft
Stack	FC Stack
Steam	Water in gaseous state
tech	Technological Advanced
tr	Transmission
Vap	VapoPropulsion D1 & M33 D5riser

Appendix A. Explanatory Notes on Lab-Scale FC Technology

Appendix A.1. Assumptions on FC Specifications

In the following, the assumptions that have been taken to estimate the relevant FC specifications (most importantly the specific power of the stack) are described in detail. It should be pointed out that these numbers are merely indicative, as power performance is mostly only described per electrode area (W/cm^2), while exact material densities/porosities and detailed device configurations are often left unclear. Furthermore, the performance characterization has been executed on lab-scale sized setups and the recorded power performance may not translate to large-scale systems. Nevertheless, in an attempt to evaluate their potential future power performance in the proposed hybrid propulsion system, we have given these estimations our best effort. Further, it should be noted that there are other considerations beyond power performance when employing an FC in an aircraft (e.g., durability, tolerance to vibration, etc.) which are, however, beyond the scope of this work, which focuses on power performance.

Appendix A.2. PEFC & HT-PEFC

As these two FC types' components closely resemble one another, similar assumptions have been taken when it comes to the weight of the non-active layers: For the GDLs, an average thickness and density of the commercially available GDLs at Freudenberg [108] was used for calculation, while for the bipolar plates as major stack weight contributors two approaches were calculated. Both consider metallic bipolar plates using coated SUS304 as the main material, however, in the conservative approach 0.3 mm thickness is assumed, while in an aggressive approach the thickness is set to 0.1 mm, as shown, for example, in [109]. As an example of a high-performance laboratory-scale PEFC, a cell featuring a surface-tuned catalyst, as presented by Fichtner et al. [87], is used for calculation. For the HT-PEFCs, a cell operated at 120 °C [88] as well as an ion-pair polymer based approach operated at 240 °C [29] are considered. Furthermore, values of laboratory-scale prototypes of a commercial supplier are given as a reference in [91]. As additional hardware for, e.g., compression is needed in the stack in addition to the cell block, a 30% reduction of cell block specific power is assumed for stack-specific power.

Appendix A.3. PCFC

While there have been several reports of PCFC performance reaching power densities beyond 0.5 W/cm^2, we will take [89] as a primary reference for this cell type, as, on the one hand, outstanding performance at low temperature is demonstrated and, on the other hand, materials are processed using a scalable method that would allow for large cell areas. The temperature dependence of the cell performance has also been investigated in detail. Furthermore, details of layer thicknesses and materials are specified, enabling estimations of the cell specific power density. To do so, the peak electrode power density is divided by the bulk material weight (calculated from the layer thicknesses specified and corresponding bulk material densities). The weight-share of non-active components (sealing, interconnects) is factored in (a) a conservative estimation by using the values provided by [73], where the interconnects are responsible for ca. 66% of the total stack weight, while sealing and insulation amounts to ca 6%, or (b) an aggressive estimation considering ultrathin 0.1 mm stainless steel for forming the interconnects instead [110].

Appendix A.4. SOFC

For SOFC performance, several options are considered: the first one is a planar design operated at 0.785 V and 750 °C within a pressure vessel made from a composite material for which the relevant parameters are all specified in the publication [73]. To calculate the stack specific power or the microtubular design as presented by [21], the following assumptions are taken: (1) extensive sealings are not needed due to the intrinsic tight tubular design, (2) the void space due to the tubes is ca. 50% as estimated from the SEM images provided, (3) interconnects are assumed to be ceramic and to make up ca. 20% of the total mass. Lastly, an anode supported SOFC published by Udomsilp et al. [23]. As details on layer thickness and materials were specified in the latter publication and in [111], the specific power density could again be estimated in an a) conservative or b) aggressive manner as described above.

Appendix A.5. Summary of Lab-Scale FC Performance Characteristics

Table A1. Summary of the performance characteristics of lab-stage FC developments. Values in grey italic font have been calculated/estimated using the assumptions described above, while values in standard font have been taken directly from the cited publications.

Cell Type	Estimation Type	Electrode Power Density +	Stack Specific Power *+x	Operating Pressure	Operating Temperature	Notes	Ref.
		W/cm²	kW/kg	Bar	°C	-	-
PEFC	conservative	1.33	*4.20 (6.01)* x	1.7	80	H_2/O_2	[87]
PEFC	aggressive	1.33	*9.86 (14.09)* x	1.7	80	H_2/O_2	[87]
HT-PEFC	-	~0.7	3	n.a.	160		[91]
HT-PEFC	conservative	0.512	*1.55 (2.22)* x	1 (100 sscm H_2, 200 sscm O_2)	120	H_2/O_2	[88]
HT-PEFC	aggressive	0.512	*2.77 (3.96)* x	1 (100 sscm H_2, 200 sscm O_2)	120	H_2/O_2	[88]
HT-PEFC	conservative	0.91	*2.58 (3.69)* x	1.47	240	H_2/air	[29]
HT-PEFC	aggressive	0.91	*6.2 (8.86)* x	1.47	240	H_2/air	[29]
PCFC	conservative	0.535	*0.33*	n.a.	500	Planar	[89]
PCFC	conservative	1.61	*0.99*	n.a.	650	Planar	[89]
PCFC	aggressive	0.535	*0.63*	n.a.	500	Planar	[89]
PCFC	aggressive	1.61	*1.90*	n.a.	650	Planar	[89]
PCFC	conservative	0.548	*0.15*	n.a.	500	Planar	[90]
PCFC	conservative	1.4	*0.38*	n.a.	650	Planar	[90]
PCFC	aggressive	0.548	*0.38*	n.a.	500	Planar	[90]
PCFC	aggressive	1.4	*0.97*	n.a.	650	Planar	[90]
SOFC	-	n.a.	1.88 (at 0.785 V)	5.07	750	Planar, including pressure vessel	[73]
SOFC	-	n.a.	4.6 (at 0.785 V)	5.07	750	Monolithic, incl. pressure vessel	[73]
SOFC	-	n.a.	2.5	n.a.	850	Planar	[24]
SOFC		1.27	*2.7–4.7*	n.a. (30 sscm)	800	Micro-monolithic, allows fast startup	[21]
SOFC	conservative	3.13	*4.2*	n.a. (1000 sscm H_2, 2000 sscm O_2)	800	Anode supported	[23]
SOFC	conservative	1.96	*2.6*	n.a. (1000 sscm H_2, 2000 sscm O_2)	650	Anode supported	[23]
SOFC	aggressive	3.13	*5.4*	n.a. (1000 sscm H_2, 2000 sscm O_2)	800	Anode supported	[23]
SOFC	aggressive	1.96	*3.4*	n.a. (1000 sscm H_2, 2000 sscm O_2)	650	Anode supported	[23]

+ peak power unless stated otherwise. * cell block components only. x for (HT-)PEFCs, additional hardware e.g., for compression is needed in the stack. This is estimated to result in a ca. 30% reduced stack specific power. The values in parenthesis refer to the corresponding specific power of the cell block.

References

1. Penke, C.; Falter, C.; Batteiger, V. Pathways and environmental assessment for the introduction of renewable hydrogen into the aviation sector. In *Progress in Life Cycle Assessment 2019*; Springer: Cham, Switzerland, 2021.
2. Sizmann, A.; Roth, A.; Batteiger, V. Renewable jet fuel: Upcoming technology options. In Proceedings of the 4th International Symposium of Bauhaus Luftfahrt, Taufkirchen near Munich, Munich, Germany, 8 May 2019.
3. Lee, D.S.; Fahey, D.W.; Skowron, A.; Allen, M.R.; Burkhardt, U.; Chen, Q.; Doherty, S.J.; Freeman, S.; Forster, P.M.; Fuglestvedt, J.; et al. The contribution of global aviation to anthropogenic climate forcing for 2000 to 2018. *Atmos. Environ.* **2021**, *244*, 1–29. [CrossRef] [PubMed]
4. Rao, A.G.; Yin, F.; Werij, H. Energy Transition in Aviation: The Role of Cryogenic Fuels. *Aerospace* **2020**, *7*, 181. [CrossRef]
5. Hoelzen, J.; Silberhorn, D.; Zill, T.; Bensmann, B.; Hanke-Rauschenbach, R. Hydrogen-powered aviation and its reliance on green hydrogen infrastructure—Review and research gaps. *Int. J. Hydrog. Energy* **2021**, *244*. [CrossRef]
6. Boeing Preliminary Design Department. *An Exploratory Study to Determine the Integrated Technological Air Transportation System Ground Requirements if Liquid-Hydrogen-Fueled Subsonic, Long-Haul Civil Air Transports*; NASA Contractor Report No. 2699; National Aeronautics and Space Administration: Washington, DC, USA, September 1976.
7. Brewer, G.D. *LH2 Airport Requirements Study*; NASA Contractor Report No. 2700; NASA: Burbank, CA, USA, 1976; p. 91520.
8. Brewer, G.D. *Hydrogen Aircraft Technology*, 1st ed.; CRC Press: Boca Raton, FL, USA, 1991; ISBN 0-8493-5838-8.
9. Verstraete, D. The Potential of Liquid Hydrogen for Long Range Aircraft Propulsion. Ph.D. Thesis, Cranfield University, Cranfield, UK, 2009.
10. Svensson, F. Potential of Reducing the Environmental Impact of Civil Subsonic Aviation by Using Liquid Hydrogen. Ph.D. Thesis, Cranfield University, Cranfield, UK, 2005.
11. Sethi, B. Enabling Cryogenic Hydrogen-Based CO_2-Free Air Transport. *Hambg. Aerosp. Lect. Ser.* **2021**. [CrossRef]
12. Kim, H.D.; Felder, J.L.; Tong, M.T.; Armstrong, M. Revolutionary Aeropropulsion Concept for Sustainable Aviation: Turboelectric Distributed Propulsion. In Proceedings of the 21th ISABE Conference, Busan, Korea, 9–13 September 2013. ISABE-2013-1719.
13. Troeltsch, F.M.; Engelmann, M.; Scholz, A.E.; Peter, F.; Kaiser, J.; Hornung, M. Hydrogen Powered Long Haul Aircraft with Minimized Climate Impact. In Proceedings of the AIAA AVIATION 2020 FORUM, Virtual Event, 15–19 June 2020; American Institute of Aeronautics and Astronautics: Reston, VI, USA, 2020. ISBN 978-1-62410-598-2.
14. The Engineer. FlyZero Concept Promises Carbon-Free Flights. 2021. Available online: https://www.theengineer.co.uk/concept-aircraft-ati-flyzero-liquid-hydrogen/ (accessed on 14 December 2021).
15. Airbus. Airbus Reveals New Zero-Emission Concept Aircraft: Airbus Press Release. Available online: https://www.airbus.com/en/newsroom/press-releases/2020-09-airbus-reveals-new-zero-emission-concept-aircraft (accessed on 15 December 2021).
16. Perry, D. DLR Prepares Do 228 for Fuel Cell Conversion Alongside MTU. Available online: https://www.flightglobal.com/aerospace/dlr-prepares-do-228-for-fuel-cell-conversion-alongside-mtu/146587.article (accessed on 15 December 2021).
17. McKinsey & Company. Hydrogen-Powered Aviation: A Fact-Based Study of the Hydrogen Technology, Economics, and Climate Impact by 2050. 2020. Available online: https://www.fch.europa.eu/sites/default/files/FCH%20Docs/20200507_Hydrogen%20Powered%20Aviation%20report_FINAL%20web%20%28ID%208706035%29.pdf (accessed on 8 July 2020).
18. Larminie, J.; Dicks, A. *Fuel Cell Systems Explained*; John Wiley & Sons, Ltd.: West Sussex, UK, 2003; ISBN 9781118878330.
19. Hyzon Motors. Third Party Tests Confirm Hyzon Motors' New Liquid-Cooled Fuel Cell Stack Leads the World in Power Density. Available online: https://hyzonmotors.com/third-party-tests-confirm-hyzon-motors-leads-fuel-cell-power-density/ (accessed on 8 July 2021).
20. Boldrin, P.; Brandon, N.P. Progress and outlook for solid oxide fuel cells for transportation applications. *Nat. Catal.* **2019**, *2*, 571–577. [CrossRef]
21. Li, T.; Heenan, T.M.M.; Rabuni, M.F.; Wang, B.; Farandos, N.M.; Kelsall, G.H.; Matras, D.; Tan, C.; Lu, X.; Jacques, S.D.M.; et al. Design of next-generation ceramic fuel cells and real-time characterization with synchrotron X-ray diffraction computed tomography. *Nat. Commun.* **2019**, *10*, 1497. [CrossRef]
22. Wachsman, E.D.; Lee, K.T. Lowering the temperature of solid oxide fuel cells. *Science* **2011**, *334*, 935–939. [CrossRef]
23. Udomsilp, D.; Rechberger, J.; Neubauer, R.; Bischof, C.; Thaler, F.; Schafbauer, W.; Menzler, N.H.; De Haart, L.G.J.; Nenning, A.; Opitz, A.K.; et al. Metal-Supported Solid Oxide Fuel Cells with Exceptionally High Power Density for Range Extender Systems. *Cell Rep. Phys. Sci.* **2020**, *1*, 100072. [CrossRef]
24. NASA. High Power Density Solid Oxide Fuel Cell. Available online: https://technology.nasa.gov/patent/LEW-TOPS-120 (accessed on 15 March 2021).
25. Gao, J.; Meng, Y.; Duffy, J.H.; Brinkman, K.S. Low Temperature Protonic Ceramic Fuel Cells through Interfacial Engineering of Nanocrystalline $BaCe_{0.7}Zr_{0.1}Y_{0.1}Yb_{0.1}O_{3-\delta}$ (BCZYYb) Electrolytes. *Adv. Energy Sustain. Res.* **2021**, *2*, 2100098. [CrossRef]
26. Hossain, S.; Abdalla, A.M.; Jamain, S.N.B.; Zaini, J.H.; Azad, A.K. A review on proton conducting electrolytes for clean energy and intermediate temperature-solid oxide fuel cells. *Renew. Sustain. Energy Rev.* **2017**, *79*, 750–764. [CrossRef]
27. Le, L.Q.; Hernandez, C.H.; Rodriguez, M.H.; Zhu, L.; Duan, C.; Ding, H.; O'Hayre, R.P.; Sullivan, N.P. Proton-conducting ceramic fuel cells: Scale up and stack integration. *J. Power Sources* **2021**, *482*, 228868. [CrossRef]
28. Haider, R.; Wen, Y.; Ma, Z.-F.; Wilkinson, D.P.; Zhang, L.; Yuan, X.; Song, S.; Zhang, J. High temperature proton exchange membrane fuel cells: Progress in advanced materials and key technologies. *Chem. Soc. Rev.* **2021**, *50*, 1138–1187. [CrossRef] [PubMed]

29. Gittleman, C.S.; Jia, H.; De Castro, E.S.; Chisholm, C.R.I.; Kim, Y.S. Proton conductors for heavy-duty vehicle fuel cells. *Joule* **2021**, *45*, 57. [CrossRef]
30. Fairey Aviation Co. Ltd. Improvements in or relating to power plants for jet propelled aircraft. British Patent GB 583341A, 19 October 1944.
31. Wilcox, E.C.; Trout, A.M. Analysis of Thrust Augmentation of Turbojet Engines by Water Injection at Compressor Inlet Including Charts for Calculating Compression with Water Injection. NASA Report No. 1006. Available online: https://ntrs.nasa.gov/citations/19930092063 (accessed on 16 December 2021).
32. Cheng, D.Y.; Nelson, A.L.C. The Chronological Development of the Cheng Cycle Steam Injected Gas Turbine During the Past 25 Years. In Proceedings of the Volume 2: ASME Turbo Expo 2002, Parts A and B: Power for Land, Sea, and Air, Amsterdam, The Netherlands, 3–6 June 2002; pp. 421–428, ISBN 0-7918-3607-X.
33. Badeer, G.H. *GE Aeroderivative Gas Turbines—Design and Operating Features*; GE Power Systems: Evendale, OH, USA, 2000.
34. Bahrami, S.; Ghaffari, A.; Genrup, M.; Thern, M. Performance Comparison between Steam Injected Gas Turbine and Combined Cycle during Frequency Drops. *Energies* **2015**, *8*, 7582–7592. [CrossRef]
35. Cheng, D.Y. The Distinction Between the Cheng and STIG Cycles. In Proceedings of the ASME Turbo Expo 2006: Power for Land, Sea, and Air, Volume 4: Cycle Innovations; Electric Power; Industrial and Cogeneration; Manufacturing Materials and Metallurgy, Barcelona, Spain, 8–11 May 2006; pp. 101–116, ISBN 0-7918-4239-8.
36. Daggett, D.L. Water Injection Feasibility for Boeing 747 Aircraft. NASA Contractor Report 2005-213656; 2005. Available online: https://ntrs.nasa.gov/citations/20060004124 (accessed on 16 December 2021).
37. Roumeliotis, I.; Mathioudakis, K. Evaluation of water injection effect on compressor and engine performance and operability. *Appl. Energy* **2010**, *87*, 1207–1216. [CrossRef]
38. Jonsson, M.; Yan, J. Humidified gas turbines—A review of proposed and implemented cycles. *Energy* **2005**, *30*, 1013–1078. [CrossRef]
39. Daggett, D.L.; Hendricks, R.C. Water Misting and Injection of Commercial Aircraft Engines to Reduce Airport NOx. NASA Contractor Report 2004-212957; 2004. Available online: https://ntrs.nasa.gov/citations/20040035576 (accessed on 16 December 2021).
40. Daggett, D.L.; Hendricks, R.C.; Fucke, L.; Eames, D.J.H. *Water Injection on Commercial Aircraft to Reduce Airport Nitrogen Oxides*; NASA Technical Memorandum 2010-213179; NASA: Cleveland, OH, USA, 2010. Available online: https://ntrs.nasa.gov/citations/20100015629 (accessed on 16 December 2021).
41. Göke, S. Ultra Wet Combustion. Ph.D. Thesis, Technische Universität Berlin, Berlin, Germany, 2012.
42. Göke, S.; Füri, M.; Bourque, G.; Bobusch, B.; Göckeler, K.; Krüger, O.; Schimek, S.; Terhaar, S.; Paschereit, C.O. Influence of steam dilution on the combustion of natural gas and hydrogen in premixed and rich-quench-lean combustors. *Fuel Processing Technol.* **2013**, *107*, 14–22. [CrossRef]
43. Klingels, H. Reducing contrails during operation of aircraft. German Patent DE 10 2018 203 159 A1, 2 March 2018.
44. Pouzolz, R.; Schmitz, O.; Klingels, H. Evaluation of the Climate Impact Reduction Potential of the Water-Enhanced Turbofan (WET) Concept. *Aerospace* **2021**, *8*, 59. [CrossRef]
45. Schmitz, O.; Kaiser, S.; Klingels, H.; Kufner, P.; Obermüller, M.; Henke, M.; Zanger, J.; Grimm, F.; Schuldt, S.; Marcellan, A.; et al. Aero Engine Concepts Beyond 2030: Part 3—Experimental Demonstration of Technological Feasibility. *J. Eng. Gas Turbines Power* **2021**, *143*, 4678. [CrossRef]
46. Schoenenborn, H.; Ebert, E.; Simon, B.; Storm, P. Thermomechanical Design of a Heat Exchanger for a Recuperative Aeroengine. *J. Eng. Gas Turbines Power* **2006**, *128*, 736–744. [CrossRef]
47. Wilfert, G. CLEAN—Validation of a High Efficient Low NOx core, a High Speed Turbine and an Integration of a Recuperator in an Environmantal Friendly Engine Concept. Aeronautics Days. In Proceedings of the 41st AIAA/ASME/SAE/ASEE Joint Propulsion Conference & Exhibit, Vienna, Austria, 20 June 2006.
48. Misirlis, D.; Vlahostergios, Z.; Flouros, M.; Salpingidou, C.; Donnerhack, S.; Goulas, A.; Yakinthos, K. Optimization of Heat Exchangers for Intercooled Recuperated Aero Engines. *Aerospace* **2017**, *4*, 14. [CrossRef]
49. Vlahostergios, Z.; Misirlis, D.; Flouros, M.; Donnerhack, S.; Yakinthos, K. Efforts to improve aero engine performance through the optimal design of heat recuperation systems targeting fuel consumption and pollutant emissions reduction. In Proceedings of the 12th European Conference on Turbomachinery Fluid Dynamics and Thermodynamics, European Turbomachinery Society, Stockholm, Sweden, 3–7 April 2017.
50. Grönstedt, T.; Xisto, C.; Sethi, V.; Rolt, A.; Roas, N.G.; Seitz, A.; Misirlis, D.; Whurr, J.; Tantot, N.; Dietz, M.; et al. Conceptual design of ultra-efficient cores for mid-century aircraft turbine engines. In Proceedings of the 24th ISABE Conference, Canberra, Australia, 22–27 September 2019. ISABE-2019-24335.
51. Zhao, X.; Tokarev, M.; Adi Hartono, E.; Chernoray, V.; Grönstedt, T. Experimental Validation of the Aerodynamic Characteristics of an Aero-engine Intercooler. *J. Eng. Gas Turbines Power* **2017**, *139*, 11701. [CrossRef]
52. NEWAC. *NEWAC Publishable Final Activity Report*; NEWAC-ART-DEL-FAR-R1.0; Grant agreement ID: 30876; NEWAC Consortium; European Commission Directorate-General for Research and Innovation: Brussels, Belgium, 8 September 2011. Available online: http://trimis.ec.europa.eu/sites/default/files/project/documents/20121029_130736_70767_Publishable_Final_Activity_Report.pdf (accessed on 16 December 2021).
53. Evans, K. Thermal Management Technologies to Unlock Zero Emission Aviation. In Proceedings of the RAeS Alternative Propulsion Systems Conference, Virtual Event, 1–2 December 2021.

54. Kellermann, H.; Lüdemann, M.; Pohl, M.; Hornung, M. Design and Optimization of Ram Air–Based Thermal Management Systems for Hybrid-Electric Aircraft. *Aerospace* **2021**, *8*, 3. [CrossRef]
55. Kellermann, H.; Fuhrmann, S.; Hornung, M. Design of a Battery Cooling System for Hybrid Electric Aircraft. In Proceedings of the AIAA Aviation 2021 Forum, American Institute of Aeronautics and Astronautics, Virtual Event, 2–6 August 2021; ISBN 978-1-62410-610-1.
56. Grönstedt, T. Opportunities for heat management in hydrogen fueled aircraft. In Proceedings of the UTIAS 7th International Workshop on Aviation and Climate Change, University of Toronto Institute for Aerospace Studies, Virtual Event, 20 May 2021.
57. Xisto, C.; Abedi, H.; Jonsson, I.; Grönstedt, T. Integration of cryogenic hydrogen and propulsion system for commercial aviation In Proceedings of the 9th EASN Conference on Innovation in Aviation and Space, Athens, Greece, 3–6 September 2019.
58. Kays, W.M.; London, A.L. *Compact Heat Exchangers*, 3rd ed.; Krieger: Malabar, FL, USA, 1998; ISBN 9781575240602.
59. Grieb, H. *Projektierung von Turboflugtriebwerken*; Birkhäuser Verlag: Berlin, Germany, 2004; ISBN 978-3-0348-7938-5.
60. Fernandes, M.D.; De Andrade, S.P.; Bistritzki, V.N.; Fonseca, R.M.; Zacarias, L.G.; Gonçalves, H.N.C.; De Castro, A.F.; Domingues, R.Z.; Matencio, T. SOFC-APU systems for aircraft: A review. *Int. J. Hydrog. Energy* **2018**, *43*, 16311–16333. [CrossRef]
61. Tornabene, R.; Wang, X.-y.; Steffen, C.J.; Freeh, J.E. Development of Parametric Mass and Volume Models for an Aerospace SOFC/Gas Turbine Hybrid System. 2005. Available online: https://ntrs.nasa.gov/search.jsp?R=20050203084 (accessed on 11 June 2019).
62. Whyatt, G.A.; Chick, L.A. *Electrical Generation for More-Electric Aircraft Using Solid Oxide Fuel Cells*; PNNL-21382; Pacific Northwest National Lab. (PNNL): Richland, WA, USA, 2012.
63. McLarty, D.; Brouwer, J.; Samuelsen, S. Fuel cell–gas turbine hybrid system design part II: Dynamics and control. *J. Power Sources* **2014**, *254*, 126–136. [CrossRef]
64. Azizi, M.A.; Brouwer, J. Progress in solid oxide fuel cell-gas turbine hybrid power systems: System design and analysis, transient operation, controls and optimization. *Appl. Energy* **2018**, *215*, 237–289. [CrossRef]
65. Azizi, M.A. Solid Oxide Fuel Cell-Gas Turbine Hybrid Power Systems: Energy Analysis, Control Assessments, Fluid Dynamics Analysis and Dynamic Modeling for Stationary and Transportation Applications. Ph.D. Thesis, University of California, Irvine, Irvine, CA, USA, 2018.
66. Hayashi, M.; Saito, H. Explosion-proof equipment of fuel supply section. Japanese Patent JP 2001338660 A, 31 May 2000.
67. Hoffjann, C.; Heinrich, H.-J. Process for the recovery and distribution of water generated on board an aircraft, land based vehicle and/or watercraft. European Patent EP 1354856 B1, 9 April 2003.
68. Metzler, D. System for water reclamation from an exhaust gas flow of a fuel cell of an aircraft. US Patent US 20060029849 A1, 18 July 2005.
69. Huber, J.M.; Lindstrom, J.D.; Daggett, D.L.; Friend, M.G. Solid oxide fuel cell as auxiliary power source installation in transport aircraft. US Patent US 6641084 B1, 4 November 2002.
70. El Hacin Sennoun, M. Hybrid Propulsion System for a Gas Turbine Engine Including a Fuel Cell. US Patent US 10774741 B2, 15 September 2020.
71. Nickl, M.; Seitz, A.; Troeltsch, F. Energieumwandlungsvorrichtung, Fahrzeug und Verfahren zum Betrieb einer Energieumwandlungsvorrichtung. German Patent Application DE 10 2020 107 905 A1, 23 March 2020.
72. Summary of Key Messages Delivered at the Technical Workshop "Hydrogen-Powered Aviation Research and Innovation", Clean Sky 2 Joint Undertaking (CS 2 JU) and Fuel Cells and Hydrogen 2 Joint Undertaking (FCH 2 JU), Brussels, Belgium, 6 May 2021. Available online: https://www.fch.europa.eu/sites/default/files/20210526_CA-CH_Workshop_key_messages.pdf (accessed on 16 December 2021).
73. Collins, J.M.; McLarty, D. All-electric commercial aviation with solid oxide fuel cell-gas turbine-battery hybrids. *Appl. Energy* **2020**, *265*, 114787. [CrossRef]
74. Yi, Y.; Rao, A.D.; Brouwer, J.; Samuelsen, G.S. Analysis and optimization of a solid oxide fuel cell and intercooled gas turbine (SOFC–ICGT) hybrid cycle. *J. Power Sources* **2004**, *132*, 77–85. [CrossRef]
75. Freeh, J.E.; Pratt, J.W.; Brouwer, J. Development of a Solid-Oxide Fuel Cell/ Gas Turbine Hybrid System Model for Aerospace Applications. In Proceedings of the ASME Turbo Expo 2004 Power for Land, Sea, Air, American Society of Mechanical Engineers, Vienna, Austria, 14–17 July 2004. GT2004-53616.
76. Jensen, S.H.; Graves, C.; Chen, M.; Hansen, J.B.; Sun, X. Characterization of a Planar Solid Oxide Cell Stack Operated at Elevated Pressure. *J. Electrochem. Soc.* **2016**, *163*, F1596–F1604. [CrossRef]
77. Jensen, S.H.; Sun, X.; Ebbesen, S.D.; Chen, M. Pressurized Operation of a Planar Solid Oxide Cell Stack. *Fuel Cells* **2016**, *16*, 205–218. [CrossRef]
78. Seitz, A.; Schmitz, O.; Isikveren, A.T.; Hornung, M. Electrically Powered Propulsion: Comparison and Contrast to Gas Turbines. In Proceedings of the 61. Deutscher Luft- und Raumfahrtkongress (DLRK), Deutsche Gesellschaft für Luft- und Raumfahrt. Berlin, Germany, 10–12 September 2012.
79. Seitz, A.; Habermann, A.L.; Van Sluis, M. Optimality Considerations for Propulsive Fuselage Power Savings. *Proc. Inst. Mech. Eng. Part. G J. Aerosp. Eng.* **2021**, *235*, 22–39. [CrossRef]
80. Gašparović, N. Das Zweistromtriebwerk bei optimaler und nicht-optimaler Auslegung. *Ing.-Wes.* **1976**, *42*, 157–168. [CrossRef]
81. Trout, A.M.; Hall, E.W. Method for Determining Optimum Division of Power between Jet and Propeller for Maximum Thrust Power of a Turbine-Propeller Engine. NASA Technical Note No. 2178; 1950. Available online: https://ntrs.nasa.gov/citations/19930082829 (accessed on 16 December 2021).

82. Gordon, S.; Mcbride, B.J. Computer Program for Calculation of Complex Chemical Equilibrium Compositions and Applications. Part 1: Analysis. NASA Reference Publication 1995-0013764; 1995. Available online: https://ntrs.nasa.gov/citations/19950013764 (accessed on 16 December 2021).
83. Wagner, W.; Cooper, J.R.; Dittmann, A.; Kijima, J.; Kretzschmar, H.-J.; Kruse, A.; Mareš, R.; Oguchi, K.; Sato, H.; Stöcker, I.; et al. The IAPWS Industrial Formulation 1997 for the Thermodynamic Properties of Water and Steam. *J. Eng. Gas. Turbines Power* **2000**, *122*, 150–184. [CrossRef]
84. Kaiser, S.; Seitz, A.; Donnerhack, S.; Lundbladh, A. Composite Cycle Engine Concept with Hectopressure Ratio. *J. Propuls. Power* **2016**, *32*, 1413–1421. [CrossRef]
85. Bijewitz, J.; Seitz, A.; Hornung, M. Power Plant Pre-Design Exploration for a Turbo-Electric Propulsive Fuselage Concept. In Proceedings of the Joint Propulsion Conference, Cincinnati, OH, USA, 9–11 July 2018; ISBN 978-1-62410-570-8.
86. Seitz, A.; Nickl, M.; Stroh, A.; Vratny, P.C. Conceptual study of a mechanically integrated parallel hybrid electric turbofan. *Proc. Inst. Mech. Eng. Part. G J. Aerosp. Eng.* **2018**, *232*, 2688–2712. [CrossRef]
87. Fichtner, J.; Watzele, S.; Garlyyev, B.; Kluge, R.M.; Haimerl, F.; El-Sayed, H.A.; Li, W.-J.; Maillard, F.M.; Dubau, L.; Chattot, R.; et al. Tailoring the Oxygen Reduction Activity of Pt Nanoparticles through Surface Defects: A Simple Top-Down Approach. *ACS Catal.* **2020**, *10*, 3131–3142. [CrossRef]
88. Luo, F.; Zhang, Q.; Yang, Z.; Guo, L.; Yu, X.; Qu, K.; Ling, Y.; Yang, J.; Cai, W. Fabrication of Stable and Well-connected Proton Path in Catalyst Layer for High Temperature Polymer Electrolyte Fuel Cells. *ChemCatChem* **2018**, *10*, 5314–5322. [CrossRef]
89. An, H.; Lee, H.-W.; Kim, B.-K.; Son, J.-W.; Yoon, K.J.; Kim, H.; Shin, D.; Ji, H.-I.; Lee, J.-H. A 5×5 cm^2 protonic ceramic fuel cell with a power density of 1.3 W cm^{-2} at 600 °C. *Nat. Energy* **2018**, *3*, 870–875. [CrossRef]
90. Choi, S.; Kucharczyk, C.J.; Liang, Y.; Zhang, X.; Takeuchi, I.; Ji, H.-I.; Haile, S.M. Exceptional power density and stability at intermediate temperatures in protonic ceramic fuel cells. *Nat. Energy* **2018**, *3*, 202–210. [CrossRef]
91. Hypoint. Technical White Paper. 2021. Available online: https://docsend.com/view/t9aw2mk (accessed on 15 December 2021).
92. Seitz, A. Advanced Methods for Propulsion System Integration in Aircraft Conceptual Design. Ph.D. Thesis, Technische Universität München, Munich, Germany, 2012.
93. Seitz, A.; Engelmann, M. *D1.03—Multi-Disciplinary Evaluation of the Optimized Propulsive Fuselage Aircraft*; Public Project Deliverable D1.03, Grant Agreement No. 723242; CENTRELINE Consortium; European Commission Directorate-General for Research and Innovation: Brussels, Belgium, 2020.
94. Torenbeek, E. *Synthesis of Subsonic Airplane Design: An Introduction to the Preliminary Design of Subsonic General Aviation and Transport Aircraft, with Emphasis on Layout, Aerodynamic Design, Propulsion and Performance*; Springer: Dordrecht, The Netherlands, 1982; ISBN 9789048182732.
95. McCormick, B.W. *Aerodynamics, Aeronautics, and Flight Mechanics*, 2nd ed.; Wiley: Hoboken, NJ, USA, 1995; ISBN 0471110876.
96. Winnefeld, C.; Kadyk, T.; Bensmann, B.; Krewer, U.; Hanke-Rauschenbach, R. Modelling and Designing Cryogenic Hydrogen Tanks for Future Aircraft Applications. *Energies* **2018**, *11*, 105. [CrossRef]
97. *Arbeitskreis Masseanalyse (MA)*; Luftfahrttechnisches Handbuch; LTH-Koordinierungsstelle (IABG mbH): Ottobrunn, Germany, 2015.
98. Raymer, D. *Aircraft Design: A Conceptual Approach*, 6th ed.; American Institute of Aeronautics and Astronautics, Inc.: Washington, DC, USA, 2018; ISBN 978-1-62410-490-9.
99. Roskam, J. *Airplain Design Part. I: Preliminary Sizing of Airplanes*, 4th ed.; DARcorporation: Lawrence, KS, USA, 1997; ISBN 9781884885426.
100. Howe, D. *Aircraft Conceptual Design Synthesis*; Professional Engineering Publishing: London, UK, 2010; ISBN 1860583016.
101. Pratt, J.W.; Klebanoff, L.E.; Munoz-Ramos, K.; Akhil, A.A.; Curgus, D.B.; Schenkman, B.L. Proton exchange membrane fuel cells for electrical power generation on-board commercial airplanes. *Appl. Energy* **2013**, *101*, 776–796. [CrossRef]
102. Kellermann, H.; Habermann, A.L.; Vratny, P.C.; Hornung, M. Assessment of fuel as alternative heat sink for future aircraft. *Appl. Therm. Eng.* **2020**, *170*, 114985. [CrossRef]
103. Kellermann, H.; Habermann, A.L.; Hornung, M. Assessment of Aircraft Surface Heat Exchanger Potential. *Aerospace* **2020**, *7*, 1. [CrossRef]
104. Kaiser, S.; Kellermann, H.; Nickl, M.; Seitz, A. A Composite Cycle Engine Concept for Year 2050. In Proceedings of the 31st Congress of the International Council of the Aeronautical Sciences, Belo Horizonte, MG, Brazil, 9–14 September 2018; ISBN 9783932182884.
105. Kaiser, S. Multidisciplinary Design of Aeronautical Composite Cycle Engines. Ph.D. Thesis, Technische Universität München, Munich, Germany, 2020.
106. Nickl, M.; Winter, F.; Gümmer, V. Piston Engine Modelling for Hydrogen Fueled Composite Cycle Engines. *IOP Conf. Ser. Mater. Sci. Eng.* **2022**. accepted for publication.
107. Seitz, A.; Habermann, A.; Peter, F.; Troeltsch, F.; Castillo Pardo, A.; Della Corte, B.; Van Sluis, M.; Goraj, S.; Kowalski, M.; Zhao, X.; et al. Proof of Concept Study for Fuselage Boundary Layer Ingesting Propulsion. *Aerosp. J.* **2020**, *10*, 16. [CrossRef]
108. Freudenberg. Fuel Cell Components/Gas Diffusion Layers. Available online: https://fuelcellcomponents.freudenberg-pm.com/Products/gas-diffusion-layers (accessed on 8 April 2021).
109. Song, Y.; Zhang, C.; Ling, C.-Y.; Han, M.; Yong, R.-Y.; Sun, D.; Chen, J. Review on current research of materials, fabrication and application for bipolar plate in proton exchange membrane fuel cell. *Int. J. Hydrog. Energy* **2020**, *45*, 29832–29847. [CrossRef]

110. fuelcellmaterials. Available online: https://fuelcellmaterials.com/ (accessed on 16 July 2021).
111. Haydn, M.; Ortner, K.; Franco, T.; Uhlenbruck, S.; Menzler, N.H.; Stöver, D.; Bräuer, G.; Venskutonis, A.; Sigl, L.S.; Buchkremer, H.-P.; et al. Multi-layer thin-film electrolytes for metal supported solid oxide fuel cells. *Mater. Today* **2014**, *256*, 52–60. [CrossRef]

aerospace

MDPI

Article

Application of Probabilistic Set-Based Design Exploration on the Energy Management of a Hybrid-Electric Aircraft

Andrea Spinelli *, Hossein Balaghi Enalou, Bahareh Zaghari, Timoleon Kipouros and Panagiotis Laskaridis

School of Aerospace, Transport and Manufacturing, Cranfield University, Cranfield MK43 0AL, UK; hb.enalou@cranfield.ac.uk (H.B.E.); bahareh.zaghari@cranfield.ac.uk (B.Z.); t.kipouros@cranfield.ac.uk (T.K.); p.Laskaridis@cranfield.ac.uk (P.L.)

* Correspondence: andrea.spinelli@cranfield.ac.uk

Abstract: The energy management strategy of a hybrid-electric aircraft is coupled with the design of the propulsion system itself. A new design space exploration methodology based on Set-Based Design is introduced to analyse the effects of different strategies on the fuel consumption, NO_x and take-off mass. Probabilities are used to evaluate and discard areas of the design space not capable of satisfying the constraints and requirements, saving computational time corresponding to an average of 75%. The study is carried on a 50-seater regional turboprop with a parallel hybrid-electric architecture. The strategies are modelled as piecewise linear functions of the degree of hybridisation and are applied to different mission phases to explore how the strategy complexity and the number of hybridised segments can influence the behaviour of the system. The results indicate that the complexity of the parametrisation does not affect the trade-off between fuel consumption and NO_x emissions. On the contrary, a significant trade-off is identified on which phases are hybridised. That is, the least fuel consumption is obtained only by hybridising the longest mission phase, while less NO_x emissions are generated if more phases are hybridised. Finally, the maximum take-off mass was investigated as a parameter, and the impact to the trade-off between the objectives was analysed. Three energy management strategies were suggested from these findings, which achieved a reduction to the fuel consumption of up to 10% and a reduction to NO_x emissions of up to 15%.

Keywords: hybrid-electric aircraft; hybrid-electric propulsion; energy management strategy; set-based design; design space exploration; optimisation

Citation: Spinelli, A.; Enalou, H.B.; Zaghari, B.; Kipouros, T.; Laskaridis, P. Application of Probabilistic Set-Based Design Exploration on the Energy Management of a Hybrid-Electric Aircraft. *Aerospace* **2022**, *9*, 147. https://doi.org/10.3390/aerospace9030147

Academic Editors: Andreas Strohmayer and Konstantinos Kontis

Received: 17 January 2022
Accepted: 3 March 2022
Published: 8 March 2022

Publisher's Note: MDPI stays neutral with regard to jurisdictional claims in published maps and institutional affiliations.

1. Introduction

Electrification has been at the forefront of sustainable aviation research for the last 10 years. Projects such as the European Union's FlightPath 2050 seek to reduce the CO_2 and NO_x emissions of the commercial fleet by 75% and 90%, respectively, compared to the best-of-class in 2000 [1]. These goals fall within a global trend of aviation de-carbonification measures proposed by the ICAO [2]. One type of electrified aircraft concept is the hybrid-electric aircraft (HEA), where the power required for propulsion is supplied by more than one type of energy source, usually fuel and batteries [3,4]. Hybrid propulsion is proposed as a stepping stone towards full electrification, as future electric technologies are expected to be inadequate for achieving enough range to be competitive with conventional commercial aircraft [5].

Unlike other propulsion systems, HEAs have multiple energy sources and power distribution paths. An Energy Management Strategy (EMS) can be introduced to find the optimal use of the available energy for a defined mission. EMS is a function of the degree of hybridisation (DoH) [6] of the power over time, where DoH is the ratio of the power that is supplied by the electric source over the overall power required by the aircraft at that point in time. This single parameter is sufficient if there are only two power paths. Otherwise, multiple non-dimensional parameters have to be specified for each path, such as how much energy is provided by batteries over the total electrical power production if the system has multiple electrical power sources [3,4].

The problem of specifying an optimal EMS is complicated by how unpredictable its operational conditions can be. The HEA can operate between any airports within its maximum range and service ceiling. The approaches to solve this problem are divided into two families: offline methods and online methods [7,8]. The offline approaches find the optimal strategy by simulating the vehicle's operational life and tuning the parameters of the EMS. Some examples include heuristic rules [9], fuzzy control [10] and global optimisation algorithms such as dynamic programming [11]. On the contrary, online methods tune the EMS during the vehicle operation, such as Pontryagin's Maximum Principle [12].

However, few studies have been published where the EMS and HEPS sizing are coupled [13–16]. Indeed, the EMS of an HEA has to address the change in power requirements from the change in aircraft mass during its operation, which is not present in HE land vehicles. In this respect, two opposite approaches are generally adopted. The first one consists of considering a fixed energy management strategy, usually a uniform degree of hybridisation over cruise and climb, and the size of the aircraft according to the energy requirements This approach is usually selected to study and compare different HE architectures and can be found in Pornet [13] and Zamboni [14]. The other approach is to iteratively couple the energy management optimisation with the system sizing. Trawick [15] implemented a global optimisation methodology where dynamic programming is used to construct the optimal EMS and sequentially size the system. This approach is more flexible since it does not assume any analytical form of the power schedule but calculates the optimal schedule by starting from the end state of the aircraft and searching backwards the optimal power setting that minimises the objective function. The drawback is the high memory and computational cost, as the algorithm has to evaluate $N \times M^2$ states when searching for the optimal path, assuming a single-setting discretised M times over N time steps.

The presented work tries to implement a new design space exploration methodology to study the coupling between the EMS and the HEPS. For this reason we selected an architecture that is well-studied in the literature [17–25] and is the most mature within the FUTPRINT50 project. Regarding the modelling of the EMS, an intermediate approach is adopted where the class of possible DoH functions are restricted within the set of polynomial functions. Then, the design space methodology searches for the parameters that minimise fuel consumption and NO_x emissions. At each iteration, the evaluation function sizes the mass of required fuel and batteries, taking into account the power–mass coupling. The goal is to identify an optimal energy management strategy for a given mission and value of maximum take-off mass.

The problem of finding the EMS that satisfies the requirements and constraints of the system under design is an engineering design problem. Hence, the design space exploration and optimisation techniques are required. A significant challenge in this field is handling the complex interactions and feedbacks between multiple subsystems [26].

Several approaches have been developed, which can be divided into two families: an iterative or point-based approach and a convergent or parallel-based approach. In the first type, a candidate design or solution is selected from an initial trade-off analysis and refined with several iterations with optimisation tools and higher fidelity analysis. This approach is the traditional process of conceptual, preliminary and detail design, often called the "waterfall model" in software engineering. Examples of this approach can be found in Ullman [27] and Pahl [28]. In the context of design space exploration, the Margin Value Method [29] is a point-based approach that, from a starting configuration, identifies excessive margins, any parameter that sub-optimally satisfies the constraints and evaluates the possible improvement by quantifying the trade-off between robustness and performance degradation. Guenov [30] presents a Margin Allocation method to dynamically assess the effect of margins on performance, other margins and constraint satisfaction, so that the designer can understand the interactions between the system under design, its components and the overall requirements.

Conversely, the second type considers and evolves many candidate designs in parallel, discarding the unfeasible, undesirable or non-robust ones. Design methodologies that fall within this category are the Method of Controlled Convergence of Pugh [31], the Design–Build–Test cycle of Wheelwright [32] and Set-Based Design (SBD) [33,34].

This paper focuses on Set-Based Design, whose principle is to generate as many configurations and delay critical design decisions as much as possible [35]. The aim is to avoid problems that often surface at advanced stages of the design process, which would require a major redesign or reversal of early decisions, prompting an increased cost both in terms of time and resources [36]. Set-Based Design has been recently investigated for the problem of designing products under flexible requirements [37,38], and incorporate resilience to unexpected modes of operation [39]. Furthermore, the large sweeping of the design space combined with a multi-disciplinary model makes SBD principles suitable for design space exploration and trade-off activities of complex systems [40]. SBD has also been applied in decision analysis, where it has been combined with Value of Information methods, enabling robust design selection [41,42].

The framework developed for this study combines set-based design with multi-objective optimisation. Most optimisation methods are point-based in nature, where single designs are optimised; hence, several authors have attempted to extend these for design sets.

Some authors use a multi-objective optimiser to search for the optimal set or family of solutions without exploring the individual designs. Sets are evaluated based on abstract quantities such as general optimality, variability, robustness [43], hypervolume size and imprecision [44]. Trade-off studies are performed. Some examples of this general approach are the Set-Based Concept method [43,45], the Set Swarm Optimisation [46] and the Set-Based Genetic Algorithm [44].

In contrast, others use SBD principles to reduce the searchable design space before introducing the optimiser. These approaches evaluate sets by mapping their input parameters to the optimisation objectives [47]. The final result expresses multiple individual designs that belong to the surviving sets.

Some examples are the Set-Based MDO of Hannapel [48], the Technology Charaterization Models method [47] and the ADOPT framework [49]. The last one was chosen as it allows us to study the interaction between the input parameters and the design requirements. Furthermore, it allows us to identify and reject areas of the design space which are unable to satisfy the top-level requirements. However, the major drawback of this framework is the necessity of expertise on the problem under design for constructing if–else expert rules capable of filtering out unfeasible and undesirable configurations. Otherwise, even with a moderate number of parameters, the combinatorial explosion makes the entire methodology impractical to use. In fact, one major challenge in SBD methodologies is identifying the sets or subspaces out of all the possible combinations of discretised inputs that satisfy the requirements and discarding those that do not. Previous approaches can be divided into three main categories:

- Algorithmic approaches which classify the sets by propagating constraints. Examples include Constraint Satisfaction solvers [50,51] and Fuzzy Set theory [52].
- Machine Learning classification which constructs a decision boundary based on samples responses. Examples include Bayesian network classifiers [53], Support Vector Machines [47] and k-Means Clustering [54].
- Decision-based approaches which model the restriction of the design space as a transition between states. Examples include Markov Decision Processes [55] and Reinforcement Learning [56].

These approaches offer a deterministic classification of desirability of each design space subset. In the methodology presented in this paper the quantification of the uncertainty of the mapping procedure is considered. A Gaussian Process Regression was adopted to evaluate the capability of each set to satisfy the requirements and constraints. While this mathematical tool falls within the family of Machine Learning approaches, it provides an estimation of the variance of its response [57], hence enabling a probabilistic assessment of the design space [58].

While the application of probability theory has been widely used for the propagation of uncertainties on system performance [59,60], robust optimisation [61] and sensitivity analysis [62], it has been rarely used to assess the uncertainty of top level design requirements satisfaction. The previous use of probabilities in the context of top-level requirements satisfaction has been found recently in Guenov [30] and Di Bianchi [63]. Both authors use probabilities to estimate the ability of a design to satisfy the constraints, given some uncertainty: specifically on the margin for Guenov and on the constraint values for Di Bianchi. We seek to explore this approach and apply it in the context of Set-Based Design Optimisation.

In this paper, Section 2 describes the modeling and methodologies adopted for this study with Section 2.1, focusing on the new design exploration framework, while Sections 2.2 and 2.3 describe the parametrisation of the energy management schedule and the overall sizing and analysis of the hybrid-electric propulsion system. Section 3 presents the baseline aircraft, the sizing mission, the selected hybrid-electric propulsion architecture and the matrix of experiments performed in this study. Section 4 presents the results of these experiments, both from the perspective of the methodology (Section 4.1) and the energy management scheduling problem (Sections 4.2–4.4). These are discussed in Section 5.1 and Sections 5.2–5.4, respectively. Finally, three families of energy management strategies were recommended in Section 5.5, and the findings are summarised in Section 6.

2. Materials and Methods

2.1. Design Exploration Methodology

The present methodology implements and extends the previous work by Georgiades [49], which employs a two-step process to explore and refine the design space with a convergent approach. Here, the principles of Set-Based Design are implemented with the enumeration of all possible combinations of discrete input parameter levels. These design "subspaces" are evaluated and discarded according to a set of feasibility and desirability rules, which encode the expert knowledge of the system under design. Then, an optimisation algorithm is used to search the surviving subspaces for the best design points. This new information is used by the designer to further discard areas of the design space and continue the convergence towards a final set of candidate solutions.

This method is improved by replacing the feasibility and desirability rules with a statistical response approach, which seeks to estimate the probability to satisfy the optimisation constraints and, if known, to bound the objectives. This second type of constraint is a "soft constraint" defined as a minimum desired outcome from the optimisation. For instance, if the designer desires the model to have no more than a value f for the objective to be minimised F, the framework allows us to introduce the constraint $F(x) < f$ to discard the areas of the design space that, even if minimised, will not meet this desired objective. This approach allows us to replace the desirability rules used previously [58]. Figure 1 presents the flowchart of the improved methodology.

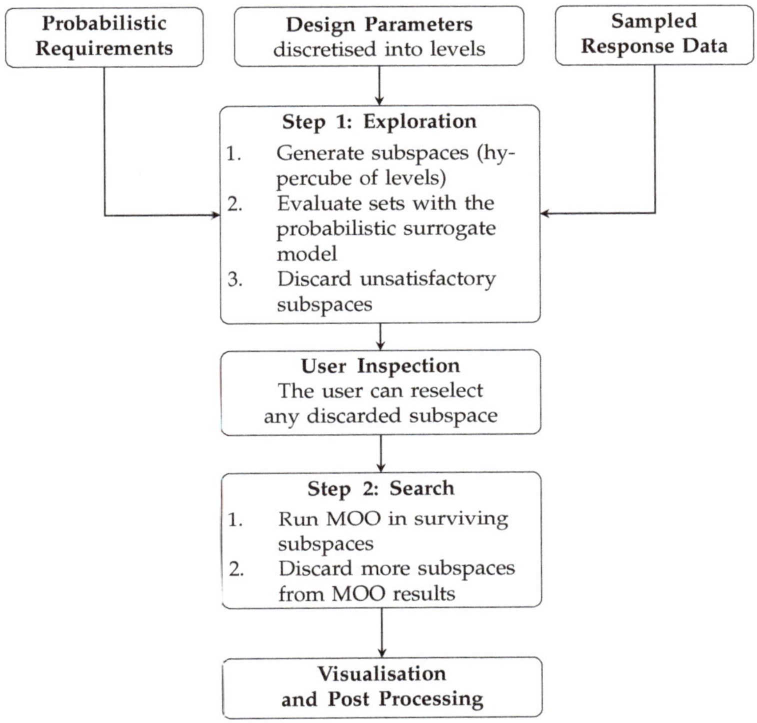

Figure 1. Flowchart of the proposed methodology.

The exploration step, where areas of the design space are marked as discarded and ignored, is necessary to avoid the combinatorial explosion of a traditional full search while allowing for alternatives that could be less performing but more robust. Equation (1) defines the number of total subspaces given n_{par} parameters, where each k-th parameter has l_k levels. For example, two parameters such as climb DOH and cruise DOH (both ranging between 0 and 1), divided into 4 levels each would generate 16 subspaces such as [(0–0.25), (0–0.25)] (level 0 of both parameters) or [(0.5–0.75), (0.25–0.5)] (level 1 and level 2). Table 1 shows how rapidly the number of subspaces can grow even with a moderate number of discrete input parameters.

$$N_{subspace} = \prod_{k=0}^{n_{par}} l_k \tag{1}$$

Table 1. Total number of subspaces for different numbers of parameters and levels.

n_{par}	$l = 2$	$l = 3$	$l = 4$	$l = 5$
2	4	9	16	25
3	8	27	64	125
4	16	81	256	625
5	32	273	1024	3125
6	64	729	4096	15,625

The evaluation of each subspace is carried out with a statistical surrogate model trained for each of the constraints, returning its response and its standard deviation. Gaussian processes, reliably used in surrogate modeling [64], are employed which have the advantage of returning the statistical momentum of a Gaussian distribution as output for its trained inputs. This allows us to calculate the probability over a single sample using the Gaussian cumulative distribution function Φ, as described next. The implementation used is the Gaussian Process Regressor algorithm of `scikit-learn` Python library [65].

Let y_i be a quantity of interest from the model under design $F(X)$, which is constrained by the inequality $y_i < \overline{g_i}$. Then, the target probabilities for this constraint $\overline{P_i}$ are defined such that:

$$P(y_i < \overline{g_i}) \geq \overline{P_i} \tag{2}$$

Next, the probability to satisfy this constraint for a single sample in the design space defined by the vector of parameters $X_k = \{x_1, x_1, ..., x_{n_{par}}\}$, has to be calculated. First, a Gaussian Process $y_i \sim \mathcal{GP}(X)$ is trained, from which the mean response μ_i^k and its variance σ_i^k can be calculated. Then, the probability of Equation (2) for sample X_k is found:

$$P^k(y_i < \overline{g_i}) = \Phi\left(\frac{\overline{g_i} - \mu_i^k}{\sigma_i^k}\right) \tag{3}$$

The Equation (3) can be used on a single sample only. To estimate the probability of satisfying the constraint $y_i < \overline{g_i}$ over a subspace, a large number of points within the subspace are sampled using a Latin Hypercube [66] scheme, then the number of points which satisfy the inequality of Equation (2) are counted over the total number of samples Effectively the conditional probability of satisfying the constraint over a single subspace is calculated with a Monte Carlo approach. Finally, if more constraints are present, the total probability of the subspace is obtained by multiplying all the individual conditioned probabilities together, assuming they are all conditionally independent from each other If the total probability is less than a threshold set by the designer, the subspace is marked as discarded. In summary, the algorithm used for the exploration step is presented in Algorithm 1, with the respective flowchart shown in Figure 2.

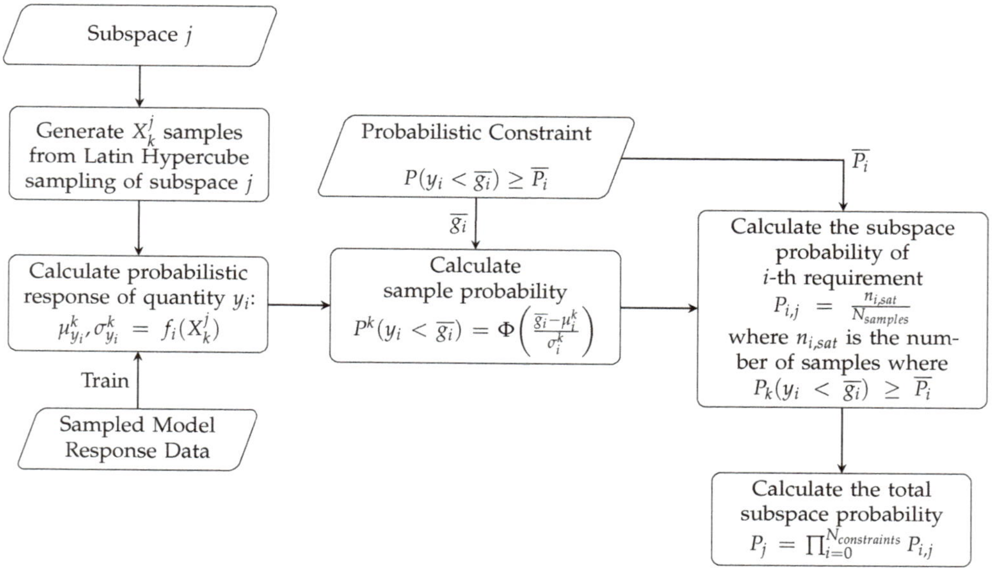

Figure 2. Flowchart of the exploration step.

Once every subspace has been evaluated, those with a probability under a user-specified threshold are marked for discard and are not evaluated in the second step, i.e., the search step. Those that survive are converted into an optimisation problem where the levels of the parameters become variable bounds. Objectives and constraints are the same from the exploration step, except for the "soft constraints" used previously. The results from the exploration step are stored in a .csv file, which can be inspected by the designer. If desired, discarded subspaces can be manually re-introduced before running the search step.

Algorithm 1 Exploration Algorithm

$X_{samples} \leftarrow LHS(Parameters)$
$y_{samples} \leftarrow F(X_{samples})$ ▷ Generate Samples to Train the Surrogate Models
for y_i in $N_{constraints}$ **do**
  Train $y_i \sim \mathcal{GP}_i(X)$
  $f_i \leftarrow \mathcal{GP}_i(X)$
end for
for Subspace j **do**
  $X_{samples} \leftarrow LHS(Subspace_j)$
  for y_i in $N_{constraints}$ **do**
    $n_{i,sat} \leftarrow 0$ ▷ Variable to count how many samples satisfy Equation (2)
    for $X_k = \{x_1, x_2, ..., x_{N_{par}}\}$ in $X_{samples}$ **do**
      $\mu_i^k, \sigma_i^k \leftarrow f_i(X_k)$
      $P_k^i \leftarrow \Phi\left(\frac{\overline{g_i}-\mu_i^k}{\sigma_i^k}\right)$ ▷ $\overline{g_i}$ from the constraint $y_i < \overline{g_i}$
      if $P_k^i > \overline{P_i}$ **then**
        $n_{i,sat} \leftarrow n_{i,sat} + 1$
      end if
    end for
    $P_{i,j} = \frac{n_{i,sat}}{N_{samples}}$ ▷ Probability of i-th constraint of j-th subspace
  end for
  $P_j \leftarrow \prod_{i=0}^{N_{constraints}} P_{i,j}$
  if $P_j < \overline{P}_{threshold}$ **then**
    Mark Subspace j as discarded
  end if
end for

Optimisation is performed using the Python library pymoo [67], which implements several optimisation algorithms based on heuristic methods. In the context of design exploration, optimality is sought as a mechanism to undestand the impact of the parameters over the objectives and its trade-offs, rather than identifying the single best design. With this goal in mind, the population-based Universal Non-dominated Sorting Genetic Algorithm III (U-NSGA-III) was selected, which is efficient in both single, multi- and many-objective problems [68]. It is chosen as it is a gradient-free method, granting flexibility in the evaluation function. The drawback of population methods, however, is the large amount of evaluations of the objective function; hence, modeling should be as computationally cheap as possible. Surrogate methods such as Radial Basis Function interpolation [69], Kriging [64] and, more recently, Artifical Neural networks [70], can be introduced to replace costly simulations. Indeed, RBF interpolation is used for calculating the fuel flow of the gas turbine, as explained in Section 2.3.

After all optimisations have been run, results are stored in a .csv format for visualisation and post-processing. An interactive web-based visualisation tool is being developed alongside the described methodology. The multi-dimensionality of the output data requires multiple types of graphs where data can be selected and visualised on multiple axes. A convenient form is the parallel coordinates plot [71], which can be augmented if combined with scatter plots [72].

2.2. Energy Management Modeling

Energy management strategies are defined as a continuous piecewise linear degree of hybridisation over the entire mission, with values ranging from 0 to 1. DoH is set to zero when a mission phase is powered only by the gas turbine. Otherwise, the DoH is defined with a combination of two parameters: the degree of hybridisation itself h and its relative distance position x across the mission phase. These two quantities are defined as follows:

$$\begin{aligned} h_i &= \frac{P_{re}}{P_{rt}} \in [0,1] \\ x_i &= \frac{d}{L} \in [0,1] \end{aligned} \tag{4}$$

where P_{rt} is the total power (in Watts) required to sustain flight across the mission phase, of which P_{re} is the amount to be provided by electric propulsion, d is the length of the phase spanned over the total phase length L (in kilometers). Within each phase, a number of points are defined, where h is set and linearly interpolated between them. This approach allows modeling both constant, linear or piecewise linear energy management strategies. Figure 3 shows some possible hybridisation strategies over a single mission phase. If a single or two h values are provided, their respective positions x are omitted, as the code automatically assumes a constant or linear interpolation over the phase.

Figure 3. Examples of energy management definitions over a single mission phase.

2.3. Propulsion System Modeling

A low-fidelity simulation code was developed to size the mass of the batteries with a power flow approach [14], which allows for fast function evaluations but with sufficient accuracy to enable the study of these trade-offs. Each mission phase is divided in a set of steps, where each flight parameter (Mass, L/D, V, V_z, etc.) is assumed constant. For each step, the required flight power is calculated after the propeller as:

$$P_{rt} = gV\frac{M}{L/D} + gV_zM \tag{5}$$

where g is the acceleration of gravity, V is flight velocity, V_z the climb/descent rate and M is the current aircraft mass.

This power is divided between the gas turbine and the electric motors depending on the degree of hybridisation h_i (Equation (4)) at that position in the mission.

The energy sources of the hybrid propulsive system are the gas turbine and the battery. The total electric power, which has to be provided by the battery to fly the mission is calculated by applying the efficiencies of each component in the electric power chain. The value of each efficiency is assumed constant for the entire mission. The battery is sized by dividing this power by its energy density.

Required fuel, instead, is calculated by estimating the fuel flow from table of gas turbine data with a radial-basis function interpolation and multiplying it by the step time. These data have been previously computed with the in-house gas turbine performance tool TURBOMATCH [73], with a span of all possible combinations of altitude, Mach and required shaft power from the selected mission. The step adopted for altitude, Mach number and power are 300 m, 0.1 and 100 kW, respectively. The maximum power for each combination of altitude and Mach number is limited by the maximum allowable TET (Turbine Entry Temperature).

While the required fuel, which is consumed during the mission, can be used as an indicator of CO_2 emissions, NO_x emissions had to be estimated by using the Boeing FuelFlow2 method [74]. The data required to model the turboprop EI_{NO_x} were collected from Filippone and Bojdo [75]. The model does not take into account the presence of a Thermal Management System (TMS) to cool the electronic equipment.

3. Test Case Definitions

3.1. Aircraft Model Assumptions

The goal of the model for the presented numerical experiments is to capture the interaction between the EMS, the HEPS and the aircraft. In particular, the variables of interest of the aircraft are its mass M and its aerodynamic performance L/D.

The selected aircraft is a retrofit of the ATR 42-600, where the original Pratt & Whitney Canada PW-127 gas turbine is kept and electric propulsion is introduced alongside. The battery cells are assumed to be replaceable and, much like fuel, limited to the amount needed to fly a specific mission.

As explained in Section 2.3, the mass of both fuel and batteries is computed iteratively, until the required power to fly the mission matches the power that the fuel and batteries can provide. In this computation, the payload mass and the operating empty mass are assumed constant, as well as the mass of the electrical equipment (motors, power electronics, cabling). Finally, structural re-sizing of the airframe is not considered in this study either. Therefore, the maximum take-off mass is constrained to 20,000 kg, which reflects to a 7.5% increment over the ATR 42-600 nominal maximum take-off weight [76]. Table 2 presents the aircraft data used in this study.

Table 2. Aircraft Properties.

Maximum Take-Off Mass	20,000 kg
Operating Empty Mass	11,550 kg
Payload Mass	5000 kg
L/D Climbout	10.5
L/D Climb/Descent	16
L/D Cruise	14.5
L/D Final	7.5

3.2. Mission

A single mission is selected between all the presented test cases. This mission is the maximum range reference flight mission defined by FUTPRINT50 [77], whose profile is shown in Figure 4. It is divided into two stages, a main flight stage of 800 km of distance and a flight to an alternate airport stage of 95 km with a 30 min holding pattern. Average climb rate is 996 ft/min (5.059 m/s), while cruise speed is 268 kt (137.78 m/s).

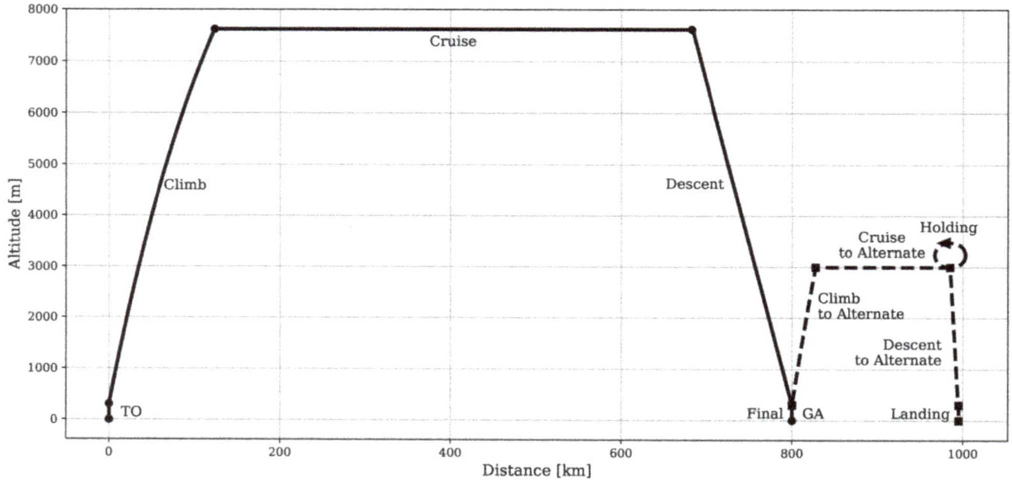

Figure 4. Mission Profile.

3.3. Propulsion System Model and Assumptions

Figure 5 presents the structure of the selected parallel hybrid propulsion system. The power supplied to the propeller is generated by the gas turbine and the electric motor, through a planetary gearbox. The assumed properties of the engines, electrical components and the propeller are presented in Tables 3 and 4.

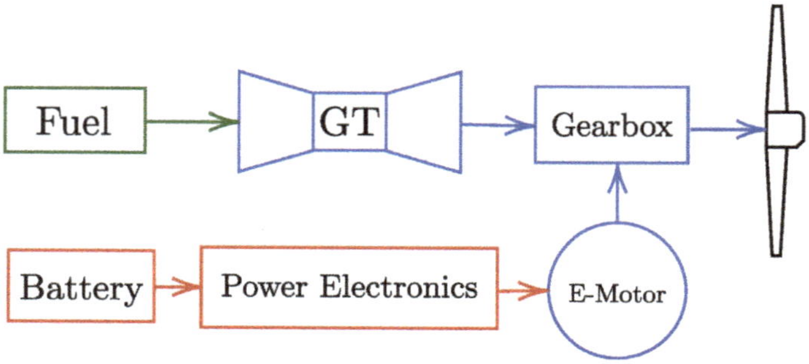

Figure 5. Propulsion system configuration.

Table 3. Propulsion system parameters.

Propeller Efficiency (Take-Off)	0.64
Propeller Efficiency (Climb)	0.73
Propeller Efficiency (Cruise)	0.86
Propeller Efficiency (Other)	0.8
Gearbox Efficiency	0.99
Electric Motor Efficiency	0.95
Power Electronics Efficiency	0.94
Cabling Efficiency	0.995
Battery Efficiency	0.95
Battery Energy Density [Wh/kg]	500

Table 4. Engine specifications at ISA-SLS.

OPR	14.9	HPT cooling flow to core flow ratio	0.12
Mass flow (kg/s)	8.4	LPT cooling flow to core flow ratio	0.05
TET (K)	1250	LPC efficiency	0.87
Net power (kW)	1698.6	HPC efficiency	0.88
Residual thrust (N)	712.1	HPT efficiency	0.89
Efficiency	0.35	LPT efficiency	0.89
SP (MWsec/kg)	0.20	Combustor pressure loss	0.05

The propulsion system model ignores the effects of auxiliary components such as the thermal management system (TMS). This aspect was not included as it is secondary to the interaction between EMS and HEPS. Furthermore, the efficiency of the components is assumed constant, except for the gas turbine.

The selected gas turbine engine is thermodynamically similar to PW127 used on the ATR-72-600. The engine is a three-spool gas turbine, where the gas generator is comprised of a high-pressure turbine (HPT), which drives a high pressure compressor (HPC), and a low pressure turbine (LPT), which drives a low pressure compressor (LPC). The free turbine (FT) drives the propeller through the gearbox, as shown in Figure 5. The same baseline engine is kept for different studies by de-rating the engine for different DoHs. This is simply performed by running the engine at a lower operating point which does not have any impact on the operability of the compressors. By de-rating the engine, the operating point of the compressors moves along their normal operating line.

The battery energy density figure was selected by considering a 2035 technological scenario [78].

The efficiencies of the electrical components are estimated as the average of the mostly adopted ones for power electronics and electrical machines, as found in the literature [79]. These values may vary depending on the design requirements, such as the type of the machine and its specific power [80]. However, for the model in this study, some of the efficiency values were chosen based on experience. Similarly, the propeller efficiency for different mission segments is calculated from in-house simulations.

Finally, for the purpose of this study, battery recharging during flight has been ignored, as it is not the focus. Therefore, electrical power always flows from the battery to the propeller.

3.4. Energy Management Experiments

Given the flexibility of the energy management modeling, it is possible to define as many parameters as desired to describe one or more hybridised mission phases. Therefore, a matrix of experiments has been defined by combining different hybridised mission phases with different energy management strategies. Each experiment is identified by a number for ease of reference during discussion in Table 5, plus an acronym composed by the hybridised mission phase names and its parametrisation type. For instance, Experiment 6, which has Climb and Cruise hybridised phases and a Linear DoH parametrisation, would be referred to as CC-L.

Table 5. Definition of Experiments.

Hybridised Mission Phases (In Order)	Uniform DoH	Linear DoH	Two Segments DoH	Three Segments DoH
Cruise	ex. 1 C-U	ex. 5 C-L	ex. 9 C-2S	ex. 11 C-3S
Climb, Cruise	ex. 2 CC-U	ex. 6 CC-L	ex. 10 CC-2S	ex. 12 CC-3S
TO, Climb, Cruise, Final	ex. 3 TCC-U	ex. 7 [a] TCC-L	-	-
Climb, Cruise, Climb to Alternate, Cruise to Alternate	ex. 4 CCA-U	ex. 8 CCA-L	-	-

[a] DoH for TO and Final phases are kept uniform as they are very short.

All experiments share the same optimisation problem formulated in Equation (6) and have DoH bounded between 0.1 and 1. P_{sat} is the target satisfaction probability of the constraint, as it was defined in Equation (2) If a hybridised phase is defined with more than three points, additional constraints are added to guarantee the x position of the inner points do not cross each other (i.e., $x_2 > x_1, x_3 > x_2$, etc.). Descent was excluded from hybridisation since battery charging is ignored (see Section 3.3), and the engine is assumed to be set to idle in this mission phase.

$$\begin{aligned} \underset{X}{\text{minimise}} \quad & M_{fuel}, M_{NO_x} \\ \text{subject to} \quad & M_{takeoff} \leq 20,000 \text{ kg } (P_{sat} \geq 0.5) \end{aligned} \tag{6}$$

For the purpose of discussing the results in Section 4, these experiments are grouped into test cases. However, given the large amount of results, the most relevant ones for each discussed test case were selected and presented. These cases are as follows:

- **Parametrisation Complexity**: Study the effects of increasing the complexity of the energy management parametrisation. The experiments considered are 1, 2, 5, 6, 9, 10, 11 and 12.
- **Mission Hybridisation**: Understand which and how many mission phases should be hybridised, and the effects of their inclusion in the energy management system. The experiments considered are 1, 2, 3, 4, 5, 6, 7 and 8.
- **Take-Off Mass Constraint**: Understand the impact the take-off mass constraint has over the performance of the propulsive system. The experiments considered are 6 and 10.

The baseline values of the reference aircraft (ATR-42-600) are calculated with the methods presented in Section 2.3 and shown in Table 6. These quantities are used to calculate the improvement of the presented energy management strategies and normalise the scatter plots.

Table 6. Baseline reference values.

Take-Off Mass	17,568	kg
Burned Fuel (M_f)	1018.22	kg
NO_x Emissions (M_{NO_x})	11.179	kg

4. Results

4.1. *Effectiveness of Design Space Exploration Methodology*

This section presents the computational cost for running each experiment and the effectiveness of the probabilistic constraints and estimates how much time was saved with the presented methodology. All the calculations presented in this study were performed on a 16-core machine (Two Intel E5-2620 v4 CPUs) and a total of 128 GB of memory.

Tables 7 and 8 present the total number of design subspaces, the discarded percentile and the time required to execute the entire calculation for each experiment.

Table 7. Number of discarded subspaces.

Experiment	$Level^{N_{par}}$	N_{par}	$N_{subspaces}$	% Discarded
C-U	4^1	4	4	50.0%
CC-U	4^2	4	16	75.0%
TCC-U	$4^2, 3^2$	7	144	79.2%
CCA-U	4^4	4	256	94.5%
C-L	4^2	2	16	62.5%
CC-L	4^4	4	256	85.1%
TCC-L	3^6	6	729	90.1%
CCA-L	3^8	8	6561	98.9%
C-2S	3^4	4	81	61.7%
CC-2S	3^8	8	6561	90.8%
C-3S	3^6	6	729	84.2%
CC-3S	3^{12}	12	531,441	99.9%

Table 8. Breakdown of the total computational time.

Experiment	Sampling Time (s)	Training Time (s)	Exploring Time (s)	Search Time (s)	Total Run Time (h)
C-U	2.644	0.025	0.357	166.30	0.047
CC-U	3.543	0.037	0.018	863.16	0.241
TCC-U	9.512	0.193	1.166	6552.60	1.823
CCA-U	4.529	0.058	1.174	3611.16	1.005
C-L	9.058	0.058	0.019	1245.18	0.348
CC-L	9.603	0.073	0.857	9878.86	2.747
TCC-L	10.745	0.127	0.673	16,188.48	4.500
CCA-L	10.204	0.025	24.367	19,518.74	5.431
C-2S	9.560	0.162	0.460	6754.59	1.879
CC-2S	11.005	0.042	25.138	135,967.70	37.779
C-3S	7.357	0.064	3.329	25,144.75	6.988
CC-3S	9.673	0.171	5819.547	54,580.23	16.780

Table 9 compares the actual computational time with the estimated one if the framework is searched in every subspace, ignoring the results from the exploration phase. Even with the overhead of the sampling, Gaussian process training and the probabilistic exploration, the total time required by the framework is less than would have been needed to search each subspace. The amount of saved time, however, is directly proportional to the number of discarded subspaces. The best practice is to have relaxed probabilistic constraints and fewer parameters for a first run, then increase the probabilistic constraints to drive the exploration towards the desired response. For instance, after identifying where the Pareto front is, the user can add a probabilistic soft constraint on the objectives such that the framework discards those areas of the design space far from the global Pareto front. This approach is consistent with the "convergent approach" to design, of which Set-Based design is an implementation.

Table 9. Comparison between the actual computational time and the estimated one without the probabilistic exploration.

Experiment	Total Run Time (ch)	Estimated Runtime without Discarded Subspaces (ch)	% Saved Time
C-U	0.75	1.48	49.09%
CC-U	3.85	15.35	74.90%
TCC-U	29.17	138.01	78.86%
CCA-U	16.08	293.48	94.52%
C-L	5.57	14.76	62.22%
CC-L	43.95	295.79	85.14%
TCC-L	72.00	728.48	90.12%
CCA-L	86.90	7796.80	98.89%
C-2S	30.07	78.44	61.67%
CC-2S	604.46	6553.42	90.78%
C-3S	111.80	708.43	84.22%
CC-3S	268.49	530,519.84	99.95%

$$\begin{aligned} M_{fuel} &\leq 940\,\text{kg} \quad (P_{sat} \geq 0.5) \\ M_{NO_x} &\leq 10.25\,\text{kg} \quad (P_{sat} \geq 0.5) \end{aligned} \tag{7}$$

In addition to the experiments presented in Table 5, an additional experiment was performed by adding soft constraints over the objectives to the Climb-Cruise with Two Segments case (experiment 10). These additional soft constraints, shown in Equation (7), were selected from the inspection of the Pareto fronts (See Sections 4.2 and 4.3) to further discard subspaces that most likely do not contain global optimum points.

Figures 6 and 7 present the data points of the original case and the one with additional constraints, respectively. These results show the probabilistic constraints are effective at limiting the search to those subspaces within the desired constraints, especially hard ones such as the take-off mass constraint. Figure 8 presents the objective space of the two cases overlapped, indicating the soft constraints were capable of restricting the search away from areas with low objective improvement. However, the soft constraints did not strongly limit near the Pareto front as it was required, with some optimal points above the constraints Secondly, the relatively low total probability in Figure 7 suggests that the system under design does not have the same sensibility to each constraint.

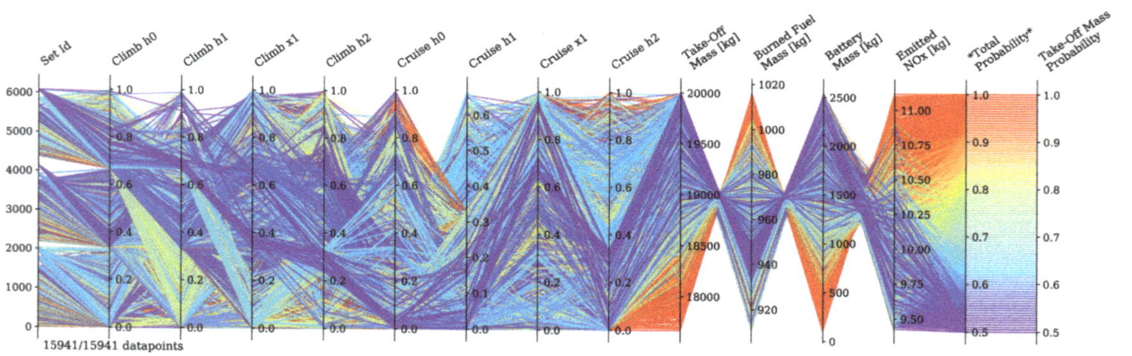

Figure 6. Data of Experiment CC-2S.

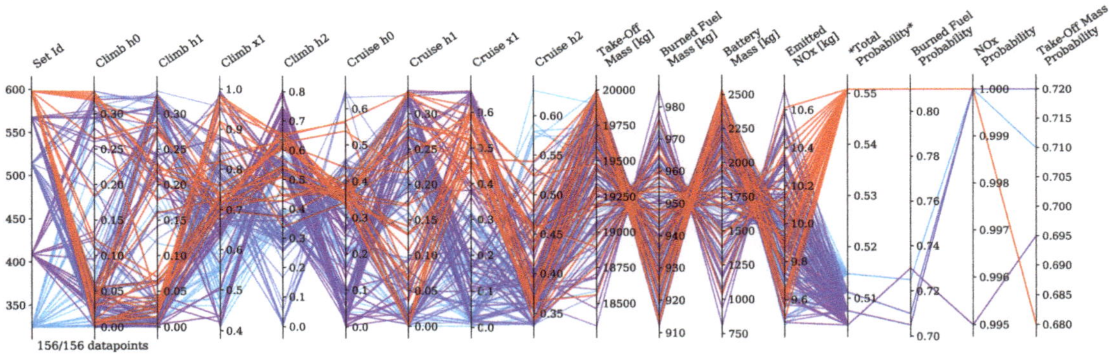

Figure 7. Data of Experiment CC-2S with additional soft constraints.

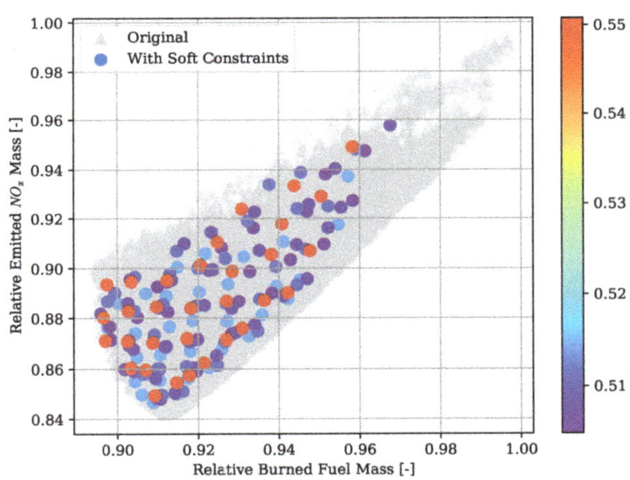

Figure 8. Objective space of data from Figures 6 and 7.

4.2. Parametrisation Complexity

The experiments selected for this study have two subdivisions based on the number of hybridised phases. The first set presents a single hybridised phase, cruise, while the second set hybridises both climb and cruise.

Figure 9a presents the results of only those experiments with cruise as hybridised. There is no Pareto front between M_f and M_{NO_x}. Instead, the constant and linear parametrisations overlap with a single optimal point, whereas the two-segment and three-segment strategies have two close points. By tracing the shape of each energy management strategy (Figure 10), it is observed that they all converge towards the constant parametrisation. In particular, the three-segment strategies degenerate into a two-segment one by having their first point overlapping the beginning of the mission phase.

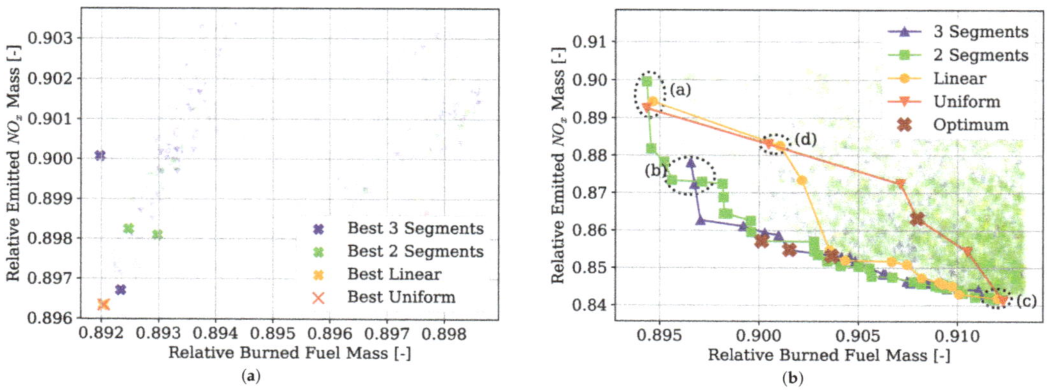

Figure 9. Effects of parametrisation complexity: (**a**) Optimal points Objective space of cruise-only hybridised strategies; (**b**) Pareto fronts of climb and cruise hybridisation strategies.

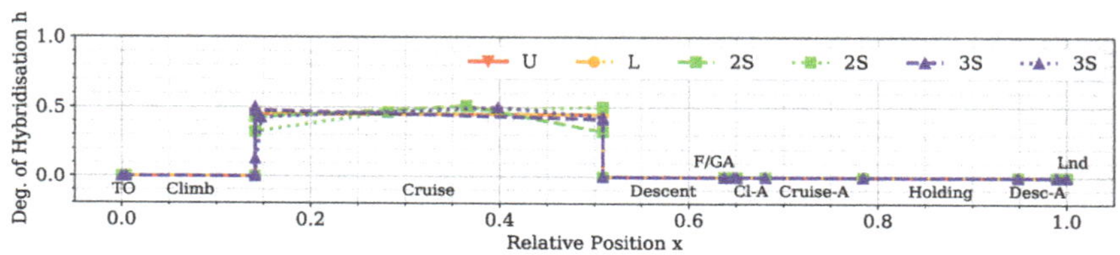

Figure 10. Hybridisation strategies of the optimal points of Figure 9a.

Figure 9b presents the results of those experiments where climb and cruise are hybridised. Contrary to the results in Figure 9a, here, a Pareto front is present. The more complex parametrisations have more points and include some strategies which dominate the simpler ones. Figure 11 shows the optimal energy management strategies found, corresponding to the marked points in Figure 9b. Points which are close to each other in the objective space present very similar topologies, regardless of parametrisation. This is particularly true for the extremities of the Pareto fronts (Figure 11a,b) and the highlighted points marked as "b" and "d" (Figure 11c). In Figure 11d, the optimiser found a linearly increasing *h* strategy for the climb phase, which allows these points to dominate the original uniform strategy. Details of these optimal points are presented in Table A1.

These result indicate that complex parametrisations beyond two segments do not produce significant improvements over the objectives. According to analysis, linear and two-segment strategies are flexible enough to identify the Pareto front in the objective space. However, it should be noted that the optimal points of the two-segment type present a 2% extra reduction in NO_x emissions to the linear counterparts towards the upper-left area of the front (where the burned fuel mass is lower). This indicates that the topology and parametrisation of the energy management strategy affects more the emissions of NO_x than the fuel consumption (and by extent, the CO_2 emissions).

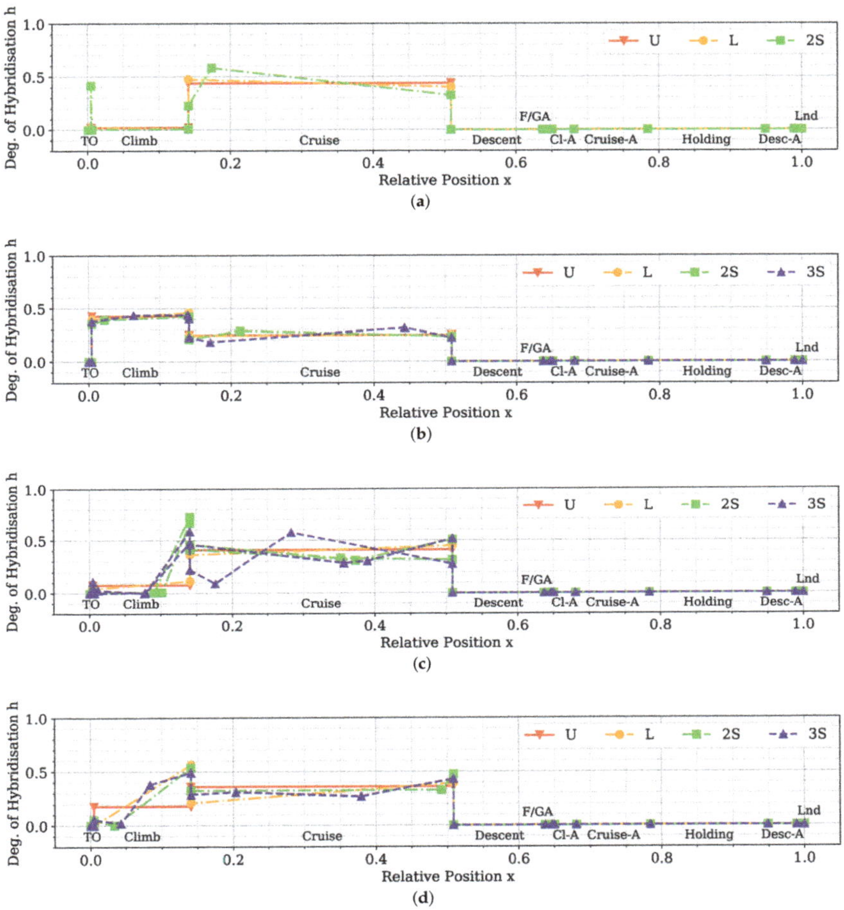

Figure 11. Optimal energy management strategies of Figure 9b: (**a**) Selection a; (**b**) Selection c; (**c**) Selections b and d; (**d**) Point closest to origin, marked with "X" in Figure 9b.

4.3. Mission Hybridisation

This group of experiments has two subsets depending on the type of parametrisation used in the energy management definition. The first set has uniform parametrisation (Figure 12a), while the second one is linear (Figure 12b). The details of the strategies of the most notable optimal points can be found in appendix (Figures A5 and A6).

The results in the objective space indicate that hybridising shorter phases, such as take-off and final approach, bring no significant change in total emissions or fuel consumption. Indeed, in Figure 12a,b, the discrepancy between the Pareto front of Climb–Cruise missions and TO–Climb–Cruise–Land is of 0.5% in burned fuel mass and 1% in NO_x emissions. However, as it will be argued in Section 5, the hybridisation of low-altitude operations are fundamental to reducing pollution and noise near airports and the communities living close by.

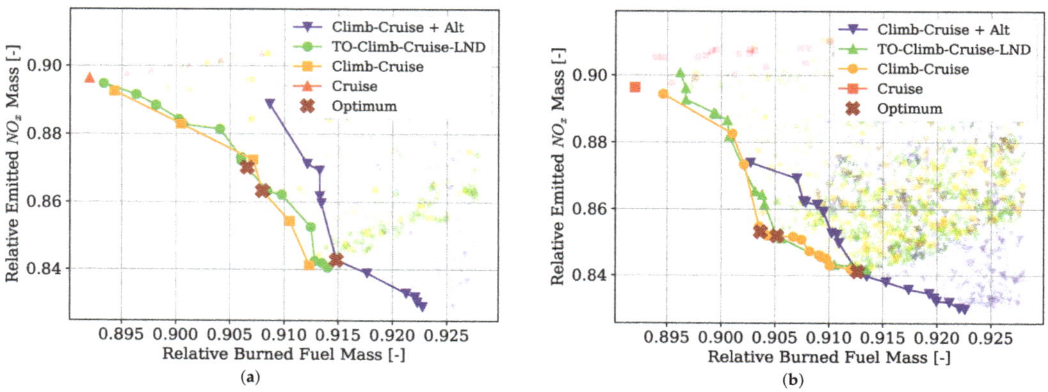

Figure 12. Effects of mission hybridisation: (**a**) Pareto fronts of uniform strategies with different hybridised mission phases; (**b**) Pareto fronts of linear strategies with different hybridised mission phases.

Furthermore, the hybridisation of the alternate portion of the mission, alongside the main one, produced a shift towards less NO_x emissions at the cost of higher fuel consumption. Strategies that hybridise only the main portion of the mission dominate those that hybridise both the main and alternate parts.

Table 10 reports the detailed results for each of the identified optimal points of Figure 12a. In each Pareto front, there is a trade-off between the battery mass and fuel mass, while the overall efficiency of the propulsion system η_{total} and the NO_x emissions are seemingly correlated. Furthermore, if the overall efficiency is greater, a reduction in NO_x is present but a reduction in CO_2 (fuel consumption) is not, η_{total} is defined as:

$$\eta_{total} = \frac{P_{tr}}{P_{total\ source}} \tag{8}$$

where P_{tr} is the total required power to fly the aircraft, and $P_{total\ source}$ is the total power provided by the gas turbine and battery before efficiency.

Table 10. Detailed results of the points in Figure 12a.

Point	TOM [kg]	M_{fuel} [kg]	$M_{battery}$ [kg]	M_{NO_x} [kg]	η_{total}
Baseline	17,568	1018.22	0	11.179	0.299
C-U, $min(M_f, M_{NO_x})$	19,998	908.30	2539.75	10.021	0.316
CC-U, $min(M_f)$	19,998	910.63	2537.65	9.978	0.316
CC-U, $min(M_f, M_{NO_x})$	19,992	924.52	2518.23	9.651	0.319
CC-U, $min(M_{NO_x})$	19,999	928.93	2520.33	9.405	0.326
TCC-U, $min(M_f)$	19,998	909.63	2538.67	10.004	0.320
TCC-U, $min(M_f, M_{NO_x})$	19,995	923.11	2522.46	9.728	0.322
TCC-U, $min(M_{NO_x})$	19,999	930.69	2519.03	9.398	0.332
CCA-U, $min(M_f)$	19,984	925.21	2508.95	9.937	0.328
CCA-U, $min(M_f, M_{NO_x})$	19,998	931.51	2516.56	9.423	0.328
CCA-U, $min(M_{NO_x})$	19,999	939.59	2510.19	9.269	0.342

The comparison between Figure 12a,b indicates that, despite the higher amount of parameters and degrees of freedom from the linear energy management strategies, there is no difference in the trends from the number and types of hybridised mission phases. It can be concluded that the parametrisation of the hybridisation strategy does not affect the behavior in the objective space when a different number of hybridised phases are compared

4.4. Maximum Take-Off Mass

Figure 13 presents a family of Pareto fronts using data from experiments 6 and 10, where each front is the optimal energy management strategy available for a given take-off mass constraint. The fronts tend to shrink and collapse into a single point as the maximum take-off mass limit is reduced.

Figure 14 shows the energy management strategy, which lies halfway between each Pareto front for each constraint level (more visualised strategies can be found in Figure A7). It is evident that the DoH is greater the higher is the maximum take-off mass limit, since it allows for more mass battery capacity.

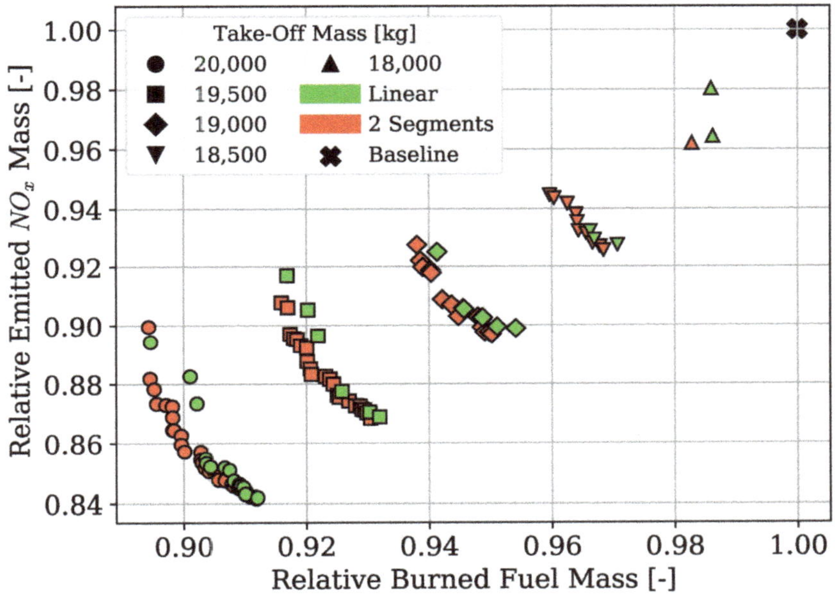

Figure 13. Pareto fronts for different Take-off Mass constraints.

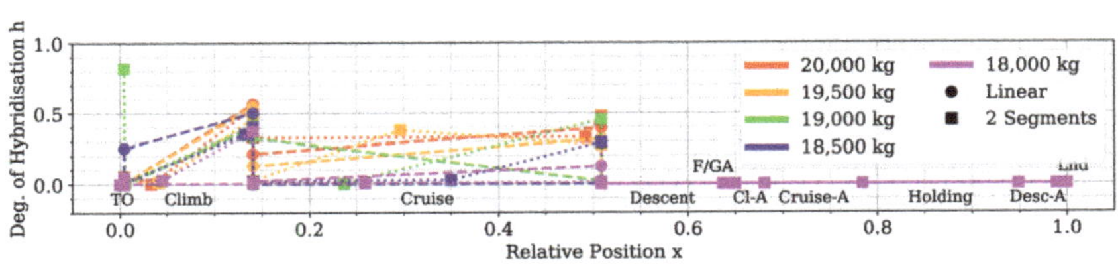

Figure 14. Energy management strategies for different take-off mass constraint levels.

The shape of the presented energy management strategies is similar for both types of functions (linear and two-segment piece-wise linear) for each type of optimal point. As presented in Section 4.2, the added freedom of an intermediate point does not produce significantly different shapes than the linear counterpart. Instead, the trade-off between hybridising climb or cruise is present in each take-off mass constraint level and corresponding to the trade-off between NO_x emissions and fuel consumption.

5. Discussion

5.1. Design Space Exploration Capabilities

This study was carried out with the design space exploration methodology presented in Section 2.1. This framework applies Set-Based Design principles to optimisation by enumerating end evaluating many design subspaces and discarding the ones unable to satisfy the requirements set by the designer. A probabilistic approach to perform this evaluation was introduced, where the probability of a subspace to satisfy a set of inequalities expressing the requirements was estimated by the use of machine learning and sampling. The optimiser is then introduced in the surviving subspaces to identify the local Pareto front. Using a thorough exploration rather than an optimiser in the entire design space allows us to analyse the interaction between input parameters, objectives and constraints, independently of the global optimum. Additionally, a whole family of possible solutions are generated, which can be used if the global constraints are more conservative than what was initially set. For instance, the results presented in Section 4.4 show the effect of the take-off mass constraint over the objective space and, consequently, on the input parameters. From the perspective of performance, the data in Section 4.1 demonstrate that the computational time of the presented framework is less than introducing the optimiser alone in each subspace, even though the exploration phase sweeps all the subspaces.

Finally, it was shown in Figure 6 that the subspace probability constraint and the value itself do correlate. Therefore, the exploration phase effectively discards areas of the design space which would not satisfy the hard constraints, as it was previously demonstrated [58].

The addition of soft constraints for restricting the search near the global Pareto front (Figure 7) resulted in less surviving subspaces and a more restricted objective space (Figure 8). While the least optimal solutions were effectively discarded, the restriction was not as hard as could be expected, with a "spillover" of possible solutions outside of the constraints of Equation (7).

While this effect requires further investigation and testing, several causes can be suggested: the number of discrete levels of the input parameters, the number of sampled points in each subspace, the sensibility of the model responses to the input parameters and the satisfaction probability. Indeed, if the subspaces are too coarse, they might cover areas that include both globally optimal solutions and the least desirable solutions. These subspaces, therefore, include the probabilistic constraint boundary and, if not adequately sampled, are discarded, eliminating some possible optimal points. Conversely, if they are kept, the sub-optimal solutions are retained, and spillover is visible in the objective space. Model sensibility would also play a role in the shape and steepness of the probabilistic constraint boundary. However, unlike the two other parameters, it cannot be controlled. Finally, the satisfaction probability P_{sat} determines what the threshold for a single sample to be accepted is, and therefore, how much the hyper-volume of a design subspace is capable of satisfying the probabilistic constraints.

In summary, the inclusion of soft constraints in the exploration phase defines an "exploration–exploitation" problem, where the continuous parameter discretisation, the sampling and the satisfaction probability determine if the accepted design space is restricted to the most desirable solutions (exploitation) or relaxed to accept sub-optimal solutions (exploration).

5.2. Effects of Parametrisation Complexity

The results presented in Section 4.2 describe the interaction between the topology and parametrisation complexity of an energy management strategy and the effect on fuel consumption and NO_x emissions. It was concluded that parametrisations no more complex than linear and two-segment piecewise linear solutions are enough to find the Pareto front over the fuel consumption and emissions objectives. However, the linear parametrisation was not able to capture some of the solutions between the extremities of the Pareto front. Instead, the two-segment piecewise linear solutions explored the upper-left region of the Pareto front, dominating the linear solutions (Figure 9b). Finally, when close to the Pareto front, the strategies that had more parameters presented the same topological shape as their simpler counterparts.

Additionally, the comparison between the experiments of Figures 9a and 12a demonstrates that at least two mission phases are required to obtain a Pareto front in the two objectives. The trade-off is present when cruise and climb are hybridised, and hybridising cruise minimises fuel consumption over NO_x emissions, and vice-versa for the climb phase.

The explanation of this phenomenon is related to the amount of power required in these phases: The climb is a power-intensive process where usually the maximum cycle temperature of the gas turbine is defined, which is the combustion chamber exhaust temperature. This is correlated with NO_x emissions, as high temperatures dissociate O_2 to produce oxygen radicals that react with N_2 to produce nitrous oxides [81]. Hybridising climb allows us to reduce the required power from the gas turbine and therefore lower its combustion temperature and NO_x emissions. However, because the climb is a shorter phase than cruise and the maximum amount of battery mass is limited by the maximum take-off mass, the cruise will be less hybridised, and more fuel will be consumed in this phase. The opposite happens when the cruise is prioritised: More battery mass is allocated to provide energy during the cruise, but in climb, NO_x emissions will be higher as the gas turbine has to provide more power.

As shown in Section 5.3, the total burned fuel is minimised when the battery mass is mainly allocated to the longest mission phases, especially cruise.

5.3. Effects of Mission Hybridisation

From the results in Figure 12a,b, it can be concluded that the longer phases of the mission, when hybridised, have the highest impact on reducing fuel consumption and NO_x emissions. Even if take-off is energy-intensive, shorter phases' hybridisation did not significantly improve the Pareto front. Nevertheless, the hybridisation of these short phases at low altitudes, such as take-off and landing, should be taken into consideration as they provide a reduction in noise and pollution at airports and nearby communities. While noise was not modelled in this study, as the focus was on gaseous emissions (CO_2 and NO_x), airport pollution and noise are two of the major drivers of current hybrid-electric aircraft research [2,77].

However, the strategy that minimises the fuel consumption the most, and therefore the CO_2 emissions, is the one where the longest mission phase, cruise, is hybridised. If climbing, take-off, landing and the climb and cruise to alternate phases are added, the fuel consumption increases despite having more of the mission hybridised and less NO_x emissions. Descent is not hybridised in all the experiments since battery recharging is ignored (see Sections 3.3 and 3.4). As explained in Section 3.2, Table 10 demonstrates this behavior clearly. Because the maximum take-off mass is limited, and therefore the amount of mass to be split between batteries and fuel is limited, when more mission phases are included in the hybridisation, more of the battery energy has to be distributed. Thus, the degree of hybridisation is lower for the longest mission phases (see Figure A5 for details). The consequence is an overall increment in the efficiency of the propulsive system and reduction in NO_x, as, on average, more of the required power is provided by the more chain-efficient electric subsystem and the gas turbine operates at a lower power setting.

However, more fuel will be consumed in those long phases due to the lower degree of hybridisation. This trade-off is visible when comparing the Pareto front of Climb–Cruise–Alternates to the original Climb–Cruise only hybridisation (Figure 12a,b). Interestingly, this trade-off behaviour is largely independent of the type of energy-management parametrisation used.

5.4. Effects of Maximum Take-Off Mass

In Section 5.3, it was discussed how the trade-off between fuel consumption and NO_x emissions is caused by the allocation of the limited amount of electrical energy stored in the batteries, which is proportional to the maximum amount of battery mass that can be allowed within the maximum take-off limit of the aircraft.

The results from Section 4.4 further demonstrate the effects of the take-off mass limitation over the aircraft performance. Figure 13 demonstrates how, as the maximum take-off mass is reduced, the Pareto front between the two objectives is smaller up to collapse into a single optimal point. The available mass for batteries is so small that only one of the two phases can be hybridised. As is seen in Figure A7, the lower is the maximum take-off mass limit, the lower the average degree of hybridisation is during the mission.

Given the trend, under this model, it might assumed that increasing the maximum take-off mass limit of the aircraft would always reduce the fuel consumption and the emissions overall. However, the model presented in Section 2.3 assumes a constant value of empty operating mass (OEM), which is a major simplification adopted, as this study did not increase the maximum take-off mass over the reference aircraft too much. A more detailed aircraft model should include an estimation of the empty mass due to the structural sizing of the airframe for carrying the additional battery mass and the additional weight due to the electrical system having to manage more electrical power. Therefore, this trend is expected to have diminishing returns, as more batteries would require a heavier aircraft with higher production costs, requiring more power to fly.

5.5. Recommended Strategies

According to the discussion of the results four groups of optimum strategies were identified. Two for the Climb–Cruise with Linear parametrisation and two for the Climb-Cruise with Two Segments parametrisation case. These selections of groups of solutions are shown along the trade-off in Figure 15 and also in the corresponding Parallel Coordinates plots in Figures 16 and 17. These selections highlight the effects of the trade-off between the two objectives.

The shapes of each respective energy management strategy are shown in Figures 18 and 19, respectively. Average trends were identified from these selected points, shown in Figures 18 and 19 with a black dashed line. These average strategies have been simulated and drawn in the objective space along their original samples (see the crosses in Figure 15).

While some of the averaged strategies do not fall within the sampled group due to the non-linearity of the simulation, these points are nonetheless close to the original selection and follow their respective trends within 1% of discrepancy for both objectives.

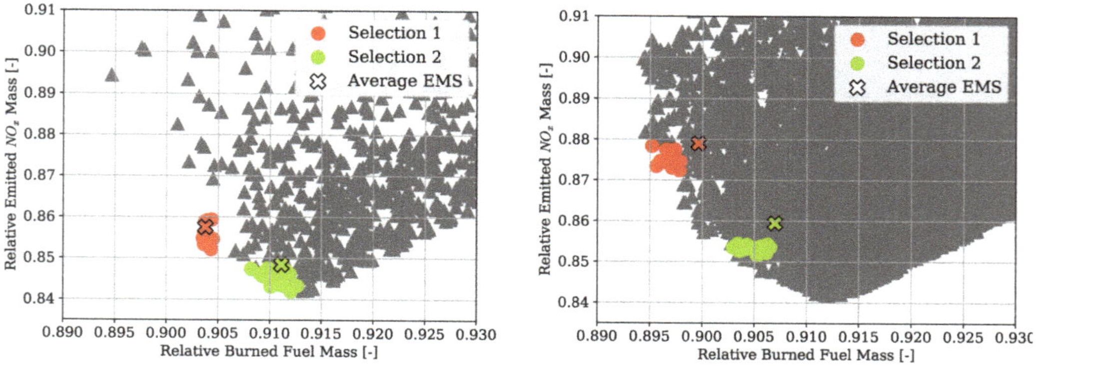

(**a**) Experiment 6 (**b**) Experiment 10

Figure 15. Points selected for the recommended strategies.

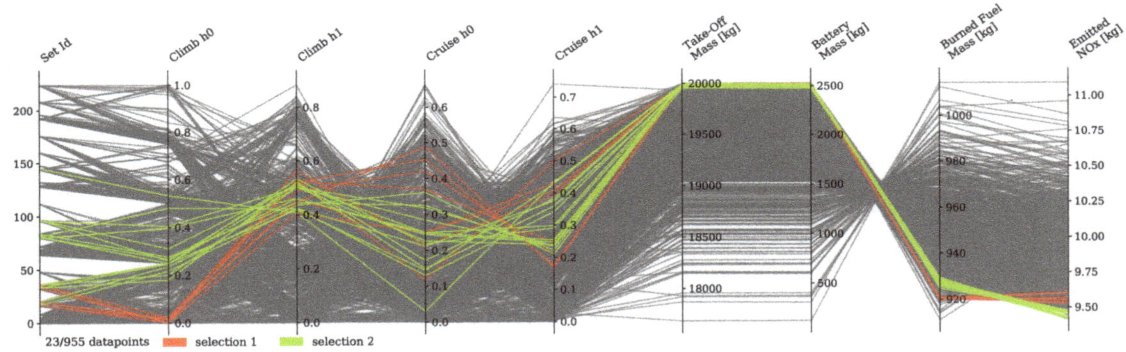

Figure 16. Parallel coordinates visualisation of Experiment 6.

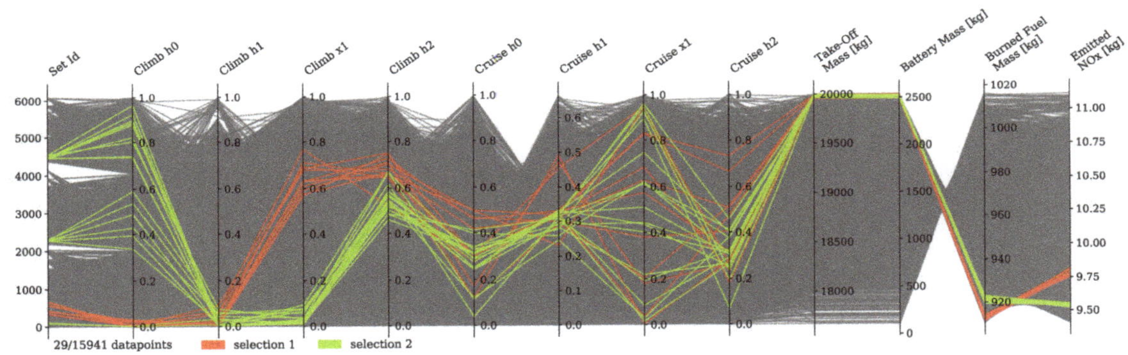

Figure 17. Parallel coordinates visualisation of Experiment 10.

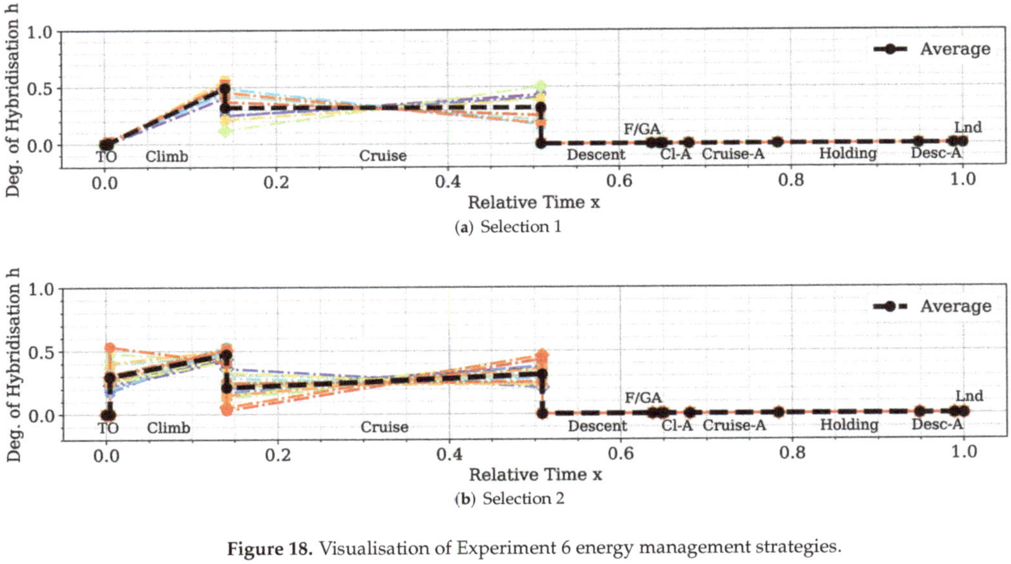

(**a**) Selection 1

(**b**) Selection 2

Figure 18. Visualisation of Experiment 6 energy management strategies.

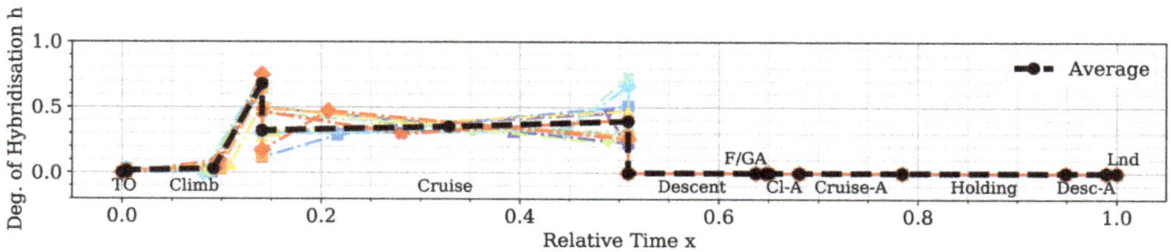

(**a**) Selection 1

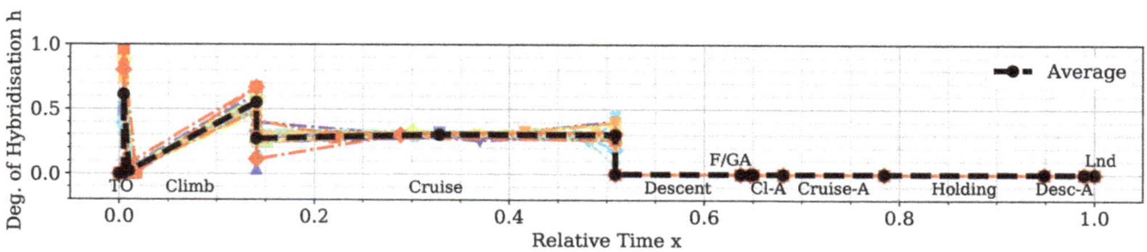

(**b**) Selection 2

Figure 19. Visualisation of Experiment 10 energy management strategies.

Three possible general shapes for the energy management strategy are indicated. As evident from Figure 15, the points of Selection 1 of Experiment 6 and Selection 2 of Experiment 10 fall within the same objective area. Their respective energy management strategies are similar, except for an impulse at the beginning of the climb phase (Figure 19b), which is an artefact of the optimisation algorithm. Another point of note is the steep climb after a late introduction of electric power in the climb phase of Selection 1 of Experiment 10 (Figure 19a). This behaviour is what allows the two-segment strategies to expand the Pareto front in this region and dominate linear strategies. The improvement in M_{fuel} is the consequence of the increased utilisation of batteries during the cruise phase (see Section 5.3). Finally, the proposed general shapes in detail are:

1. Linearly increasing DoH in climb, starting from zero, and constant DoH in cruise (Figures 18a and 19b). This strategy is balanced between the two objectives and represents a good compromise.
2. Late deployment of electric power in climb, with a steep linear increment in DoH and moderately increasing DoH in cruise (Figure 19a). This strategy prioritises the reduction of fuel consumption over the emissions of NO_x, as more battery capacity is allocated in cruise.
3. Moderately increasing DoH in climb, starting from a non-zero value, and moderately increasing DoH in cruise (Figure 18b). This strategy prioritises the reduction of NO_x over fuel consumption, as more battery capacity is allocated in climb.

No specific degree of the hybridisation values is provided for these strategies because, as discussed in Section 5.4, the amount of battery mass available determines the average degree of hybridisation. These DoH values of these strategies are determined by the 20,000 kg take-off mass limit that has been adopted in this study. Nonetheless, it is possible to summarise two general trends. First, the shape of the DoH functions are either constant, increasing or a mixture of the two. Secondly, the trade-off between NO_x emissions and burned fuel is directly correlated with the hybridisation of climb or cruise. Finally, the results indicate that the best parametrisation is the two segments piece-wise linear one, since it is capable to cover the entire Pareto front with the least number of parameters.

6. Conclusions

Hybrid-electric vehicles are characterised by providing propulsive power from more than a single energy source. Finding the optimal energy schedule is a common issue that affects hybrid-electric aircraft, albeit with the added complexity of take-off mass limitations. This study presented the problem of finding the optimal energy management scheduling for a hybrid-electric aircraft combined with its battery sizing. An ATR-42 class aircraft retrofitted with a hybrid-electric parallel propulsion system was modelled, and a piecewise linear parametrisation of the degree of hybridisation was introduced for flexibly defining any energy management strategy. A new design space exploration methodology based on the principles of Set-Based Design was introduced, enhanced by a probabilistic approach, which is capable of discarding design subspaces unable to satisfy the requirements set by the designer. This methodology allowed the optimisation algorithm to focus on areas of the design space without constraint violations and, when specified, closer to the global Pareto front. Indeed, the presented test cases had a large portion of the design space unable to satisfy the constraints. The framework was capable of identifying it and discarding it, reducing the number of configurations to evaluate by 75% on average. Furthermore, the optimisation results obtained with this approach allowed us to reveal the behaviour of the system. Nevertheless, a set of tuning hyperparameters of this framework was identified, the granularity of the subspaces and the minimum satisfaction probability, which affect the number of possible results. Tuning studies will be performed to benchmark the proposed methodology and define best practice hyperparameters. Several experiments were performed by changing the complexity of the parametrisation and the number of hybridised mission phases, from which several lessons were learned. The complexity of the energy management parametrisation has diminishing returns on the objectives. The piecewise two-segment topology was found to be enough for identifying the Pareto front. Introducing more phases did not improve the objective significantly to justify the increased computational cost. Furthermore, points in the same area of the objective space have similar energy management strategies, regardless of how complex and flexible their parametrisation is. Indeed, the parametrisation itself was not responsible for the trade-off between the objectives.

Instead, the length and number of the hybridised mission phases affected the shape and presence of a Pareto front. The results clearly showed that, while the longest mission phases contributed the most to the reduction of fuel consumption, if most mission phases were hybridised, the NO_x emissions would go down together with an increment in the overall propulsive efficiency. Moreover, the maximum take-off mass limit affects how much trade-off is present between the objectives. These phenomena can be explained in the limited amount of available electric energy due to the limited amount of battery mass. If this energy is distributed over more mission phases, on average, less gas turbine power is used, and therefore, less NO_x is emitted as it runs at a lower combustion temperature. However, the DoH would be lower on the longer mission phases, requiring most fuel burned. Indeed, if the electric energy is used only on the longest mission phase, cruise, the mission fuel consumption is minimised.

Lastly, the maximum take-off mass determines how much battery mass is distributed over the mission phases. While this might suggest increasing this limit indefinitely, the increment in the battery mass and the higher amount of electric power to manage would greatly impact the aircraft's structural and systems mass, increasing the aircraft's cost and complexity and ultimately reducing the fuel efficiency benefits.

This aspect was not included in this study but will be included in future work. Furthermore, from this study, the thermal management aspect of the hybrid-electric propulsion system was ignored, which is one of the current challenges in the design and production of HEA. Future studies will introduce the sizing of the thermal management system (TMS) from the electric system lost energy and its effects on aircraft mass and available power.

Finally, a critical aspect in the study and development of HE and Electric propulsion is the uncertainty of future technology. In this study, a fixed value of battery energy density was selected for a 2050 scenario. However, future studies will introduce an uncertainty into the presented methodology for studying different technological scenarios and their impact on the design parameters and requirements at the same time.

Author Contributions: Conceptualization, T.K. and A.S.; methodology, A.S., T.K. and H.B.E.; software, A.S. and H.B.E.; validation, A.S. and H.B.E.; formal analysis, A.S. and T.K.; investigation, A.S. and H.B.E.; resources, T.K.; data curation, A.S.; writing—original draft preparation, A.S. and H.B.E.; writing—review and editing, A.S., H.B.E., B.Z. and T.K.; visualization, A.S. and T.K.; supervision, T.K. and P.L.; project administration, T.K.; funding acquisition, T.K. and P.L. All authors have read and agreed to the published version of the manuscript.

Funding: This project has received funding from the European Union's Horizon 2020 Research and Innovation programme under Grant Agreement No 875551.

Institutional Review Board Statement: Not applicable.

Informed Consent Statement: Not applicable.

Data Availability Statement: Results not available, raw data are cited in text.

Acknowledgments: We thank all the researchers of Cranfield University working in the FUTPRINT50 project. We would also like to thank the reviewers for their constructive comments which helped to improve the quality of the manuscript.

Conflicts of Interest: The authors declare no conflict of interest.

Abbreviations

The following abbreviations are used in this manuscript:

C	Cruise
CC	Climb and Cruise
CCA	Climb and Cruise and their respective alternates
Cl-A	Climb in Flight to Alternate
Cruise-A	Cruise in Flight to Alternate
Desc-A	Descent in Flight to Alternate
DP	Dynamic Programming
DoH	Degree of Hybridisation
EMS	Energy Management System
EVE	Electrically Variable Engine
FGA	Final and Go-Around
FT	Free Turbine
ICAO	International Civil Aviation Organization
ISA	International Standard Atmosphere
SLS	Sea Level Static
MOO	Multi-Objective Optimisation
MTOM	Maximum Take-Off Mass
HE	Hybrid-Electric
HEA	Hybrid-Electric Aircraft
HEPS	Hybrid-Electric Propulsion System
HPC	High-Pressure Compressor
HPT	High-Pressure Turbine
L	Linear DoH Segment
Lnd	Land
LPC	Low-Pressure Compressor
LPT	Low-Pressure Turbine
OEM	Operating Empty Mass
OPR	Overall Pressure Ratio
RBF	Radial Basis Function
TET	Turbine Entry Temperature
TOM	Take-Off Mass
TCC	Take-Off, Climb, Cruise and Landing

TMS	Thermal Management System
TO	Take-Off
TOM	Take-Off Mass
SBD	Set-Based Design
U	Uniform DoH Segment
U-NSGA	Universal Non-dominated Sorting Genetic Algorithm
2S	Two Linear DoH Segments
3S	Three Linear DoH Segments

Appendix A

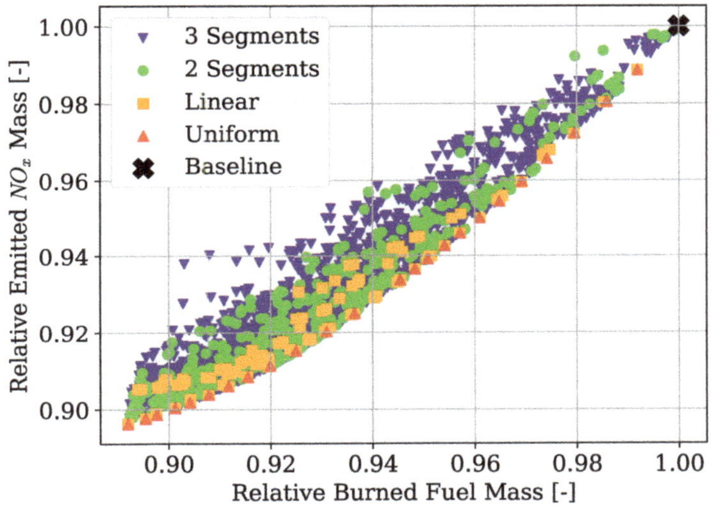

Figure A1. Objective space of cruise-only hybridised strategies.

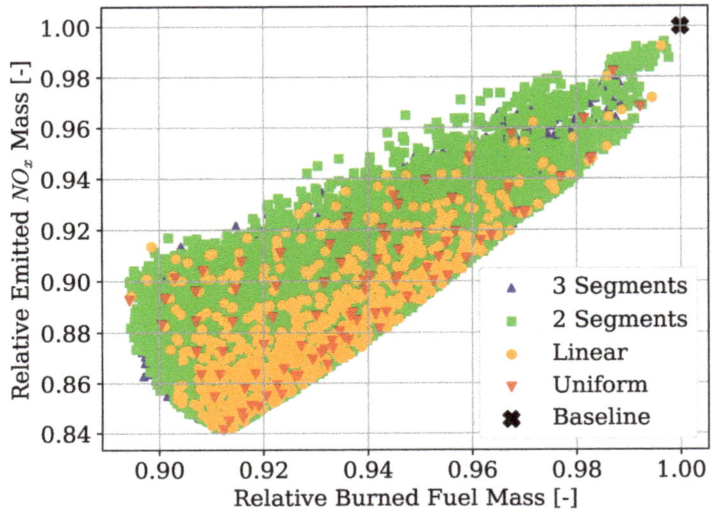

Figure A2. Objective space of Climb and cruise hybridisation strategies.

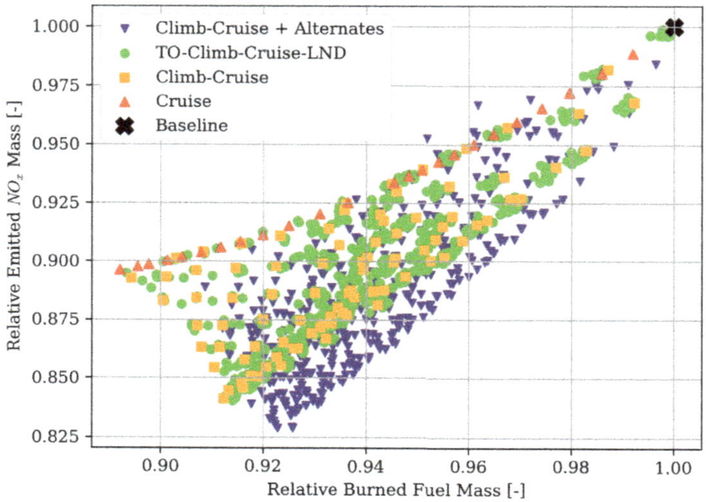

Figure A3. Objective space of Uniform strategies with different hybridised mission phases.

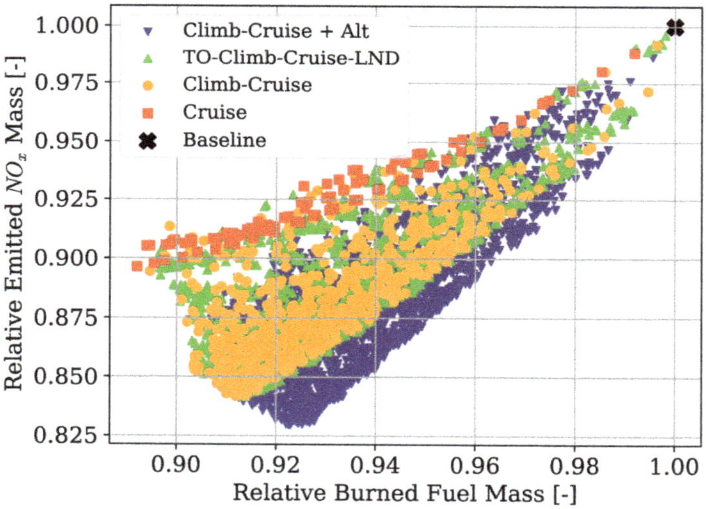

Figure A4. Objective space of Linear strategies with different hybridised mission phases.

Table A1. Detailed results of the optimal points where Climb and Cruise are hybridised. (Section 5.2).

Point	TOM [kg]	M_{fuel} [kg]	$M_{battery}$ [kg]	M_{NO_x} [kg]	η_{total}
Baseline	17,568	1018.22	0	11.179	0.299
CC-U, $min(M_f)$	19,998	910.63	2537.65	9.978	0.316
CC-U, $min(M_f, M_{NO_x})$	19,992	924.52	2518.23	9.651	0.319
CC-U, $min(M_{NO_x})$	19,999	928.93	2520.33	9.405	0.326
CC-L, $min(M_f)$	19,981	910.94	2520.08	10.053	0.316
CC-L, $min(M_f, M_{NO_x})$	19,994	920.12	2524.12	9.594	0.322
CC-L, $min(M_{NO_x})$	19,998	928.64	2519.55	9.467	0.326
CC-2S, $min(M_f)$	19,997	910.63	2536.66	10.112	0.316
CC-2S, $min(M_f, M_{NO_x})$	19,999	915.15	2590.40	9.588	0.321
CC-3S, $min(M_{NO_x})$	19,999	928.46	2520.66	9.464	0.325
CC-3S, $min(M_f)$	19,997	910.63	2536.66	10.112	0.316
CC-3S, $min(M_f, M_{NO_x})$	19,999	915.15	2590.40	9.588	0.321
CC-3S, $min(M_{NO_x})$	19,999	928.46	2520.66	9.464	0.325

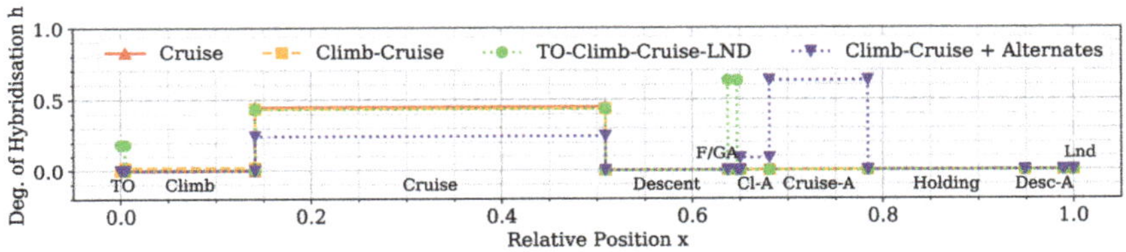

(a) Minimum M_f

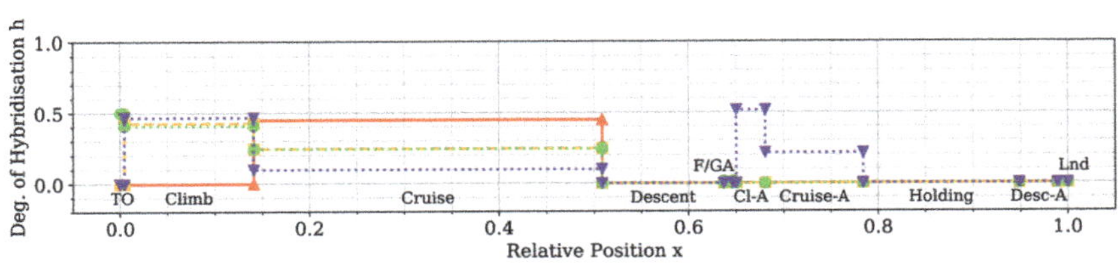

(b) Minimum NO_x

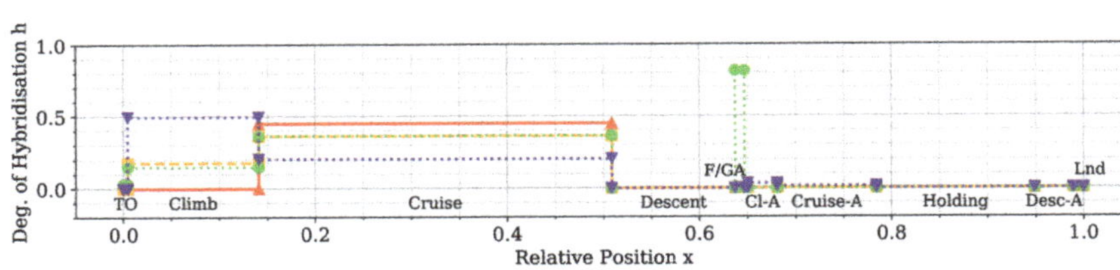

(c) Point closest to origin, marked with "X"

Figure A5. Optimal energy management strategies of points in Figure 12a.

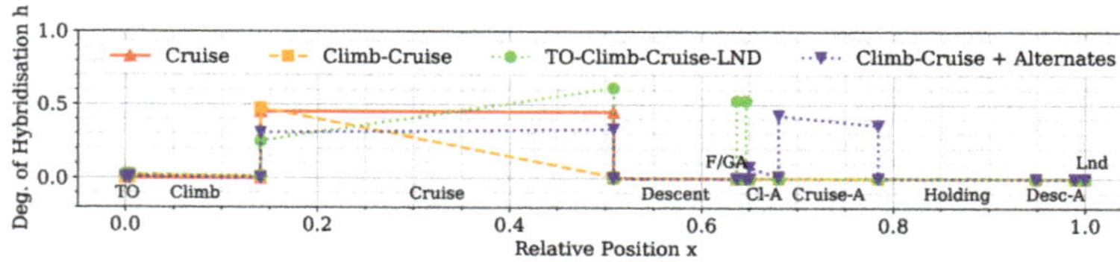

(a) Minimum M_f

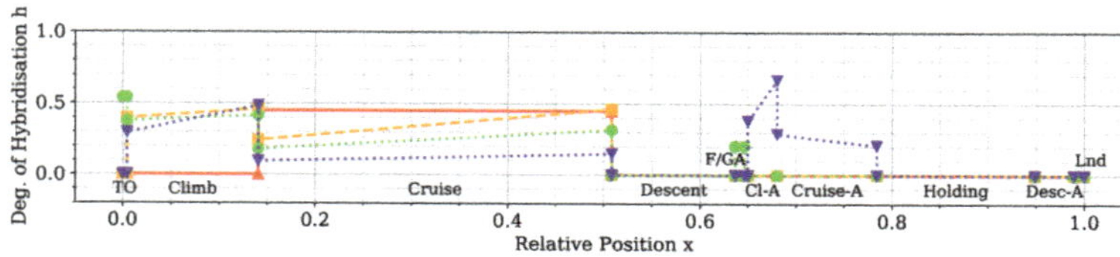

(b) Minimum NO_x

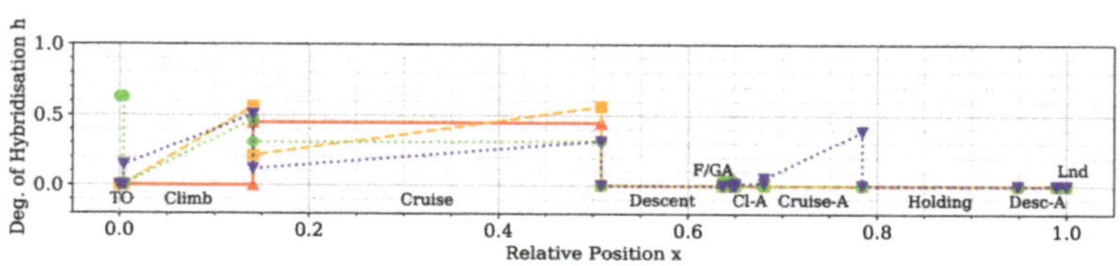

(c) Point closest to origin, marked with "X"

Figure A6. Test 4 optimal energy management strategies.

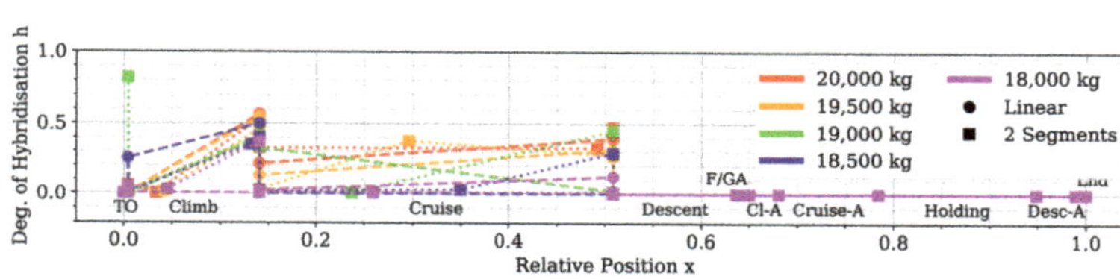

(a) Point closest to origin

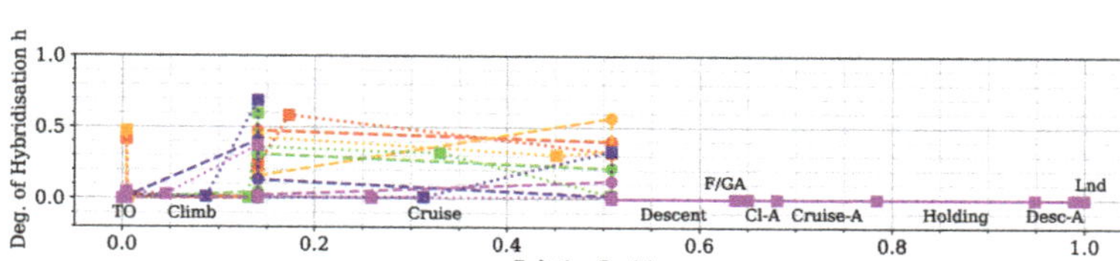

(b) Minimum Burned Fuel

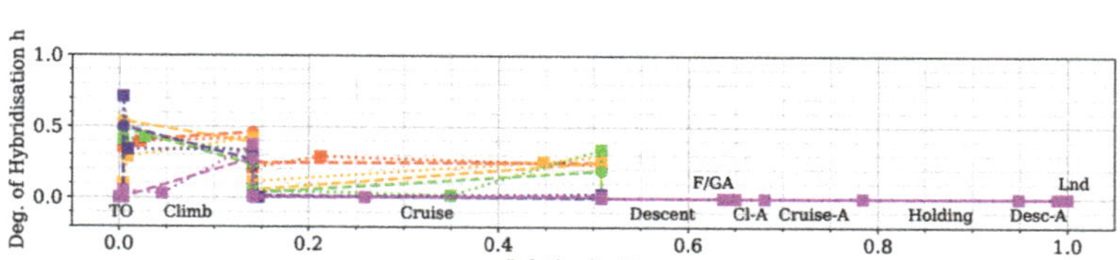

(c) Minimum NO_x emissions

Figure A7. Optimal energy management strategies for different take-off mass constraint levels.

References

1. European Commission. Flightpath 2050: Europe's Vision for Aviation-Report of the High Level Group on Aviation Research. In *Report of the High Level Group on Aviation Research*; European Commission: Luxembourg, 2011. Available online: https://www.acare4europe.org/sites/acare4europe.org/files/document/Flightpath2050_Final.png (accessed on 17 November 2021).
2. Zaporozhets, O.; Isaienko, V.; Synylo, K. Trends on Current and Forecasted Aircraft Hybrid Electric Architectures and Their Impact on Environment. *Energy* **2020**, *211*, 118814. [CrossRef]
3. Brelje, B.J.; Martins, J.R.R.A. Electric, Hybrid, and Turboelectric Fixed-Wing Aircraft: A Review of Concepts, Models, and Design Approaches. *Prog. Aerosp. Sci.* **2019**, *104*, 1–19. [CrossRef]
4. Sahoo, S.; Zhao, X.; Kyprianidis, K. A Review of Concepts, Benefits, and Challenges for Future Electrical Propulsion-Based Aircraft. *Aerospace* **2020**, *7*, 44. [CrossRef]
5. Epstein, A.H.; O'Flarity, S.M. Considerations for Reducing Aviation's CO_2 with Aircraft Electric Propulsion. *J. Propuls. Power* **2019**, *35*, 572–582. [CrossRef]
6. Isikveren, A.T.; Kaiser, S.; Pornet, C.; Vratny, P.C. Pre-Design Strategies and Sizing Techniques for Dual-Energy Aircraft. *Aircr. Eng. Aerosp. Technol.* **2014**, *86*, 525–542. [CrossRef]
7. Zhang, F.; Wang, L.; Coskun, S.; Pang, H.; Cui, Y.; Xi, J. Energy Management Strategies for Hybrid Electric Vehicles: Review, Classification, Comparison, and Outlook. *Energies* **2020**, *13*, 3352. [CrossRef]
8. Xie, Y.; Savvarisal, A.; Tsourdos, A.; Zhang, D.; Gu, J. Review of Hybrid Electric Powered Aircraft, Its Conceptual Design and Energy Management Methodologies. *Chin. J. Aeronaut.* **2021**, *34*, 432–450. [CrossRef]
9. Hung, J.Y.; Gonzalez, L.F. On Parallel Hybrid-Electric Propulsion System for Unmanned Aerial Vehicles. *Prog. Aerosp. Sci.* **2012**, *51*, 1–17. [CrossRef]
10. Karunarathne, L.; Economou, J.T.; Knowles, K. Adaptive Neuro Fuzzy Inference System-Based Intelligent Power Management Strategies for Fuel Cell/Battery Driven Unmanned Electric Aerial Vehicle. *Proc. Inst. Mech. Eng. Part G J. Aerosp. Eng.* **2010**, *224*, 77–88. [CrossRef]
11. Pinto Leite, J.P.S.; Voskuijl, M. Optimal Energy Management for Hybrid-Electric Aircraft. *Aircr. Eng. Aerosp. Technol.* **2020**, *92*, 851–861. [CrossRef]
12. Misley, A.A.; D'Arpino, M.; Ramesh, P.; Canova, M. A Real-Time Energy Management Strategy for Hybrid Electric Aircraft Propulsion Systems. In Proceedings of the AIAA Propulsion and Energy 2021 Forum, Online, 9–12 August 2021; American Institute of Aeronautics and Astronautics: Reston, VA, USA, 2021. [CrossRef]
13. Pornet, C.; Gologan, C.; Vratny, P.C.; Seitz, A.; Schmitz, O.; Isikveren, A.T.; Hornung, M. Methodology for Sizing and Performance Assessment of Hybrid Energy Aircraft. *J. Aircraft* **2015**, *52*, 341–352. [CrossRef]
14. Zamboni, J.; Vos, R.; Emeneth, M.; Schneegans, A. A Method for the Conceptual Design of Hybrid Electric Aircraft. In Proceedings of the AIAA Scitech 2019 Forum, San Diego, CA, USA, 7–11 January 2019; American Institute of Aeronautics and Astronautics: Reston, VA, USA, 2019. [CrossRef]
15. Trawick, D.R. *A Methodology for the Determination of Optimal Operational Schedules of Hybrid Electric Architectures*; Georgia Institute of Technology: Atlanta, GA, USA, 2018.
16. Trainelli, L.; Salucci, F.; Rossi, N.; Riboldi, C.E.D.; Rolando, A. Preliminary Sizing and Energy Management of Serial Hybrid-Electric. In Proceedings of the XXV AIDAA National Congress, Italian Association of Aeronautics and Astronautics, Rome, Italy, 9–12 September 2019.
17. Bradley, M.K.; Droney, C.K. *Subsonic Ultra Green Aircraft Research: Phase 2—Volume II— Hybrid Electric Design Exploration*; NASA Langely Research Center: Hampton, VA, USA, 2015.
18. Lents, C.E.; Hardin, L.W.; Rheaume, J.; Kohlman, L. Parallel Hybrid Gas-Electric Geared Turbofan Engine Conceptual Design and Benefits Analysis. In Proceedings of the 52nd AIAA/SAE/ASEE Joint Propulsion Conference, Salt Lake City, UT, USA, 25–27 July 2016; American Institute of Aeronautics and Astronautics: Reston, VA, USA, 2016. [CrossRef]
19. Lents, C.E.; Hardin, L.W. Fuel Burn and Energy Consumption Reductions of a Single-Aisle Class Parallel Hybrid Propulsion System. In Proceedings of the AIAA Propulsion and Energy 2019 Forum, Indianapolis, IN, USA, 22–24 August 2019; American Institute of Aeronautics and Astronautics: Reston, VA, USA, 2019. [CrossRef]
20. Rheaume, J.M.; MacDonald, M.; Lents, C.E. Commercial Hybrid Electric Aircraft Thermal Management System Design, Simulation, and Operation Improvements. In Proceedings of the AIAA Propulsion and Energy 2019 Forum, Indianapolis, IN, USA, 22–24 August 2019; American Institute of Aeronautics and Astronautics: Reston, VA, USA, 2019. [CrossRef]
21. Trawick, D.; Perullo, C.; Armstrong, M.; Snyder, D.; Tai, J.C.; Mavris, D.N. Development and Application of GT-HEAT for the Electrically Variable Engine(TM) Design. In Proceedings of the 55th AIAA Aerospace Sciences Meeting, Grapevine, TX, USA, 9–13 January 2017; American Institute of Aeronautics and Astronautics: Reston VA, USA, 2017. [CrossRef]
22. Gladin, J.C.; Trawick, D.; Mavris, D.N.; Armstrong, M.J.; Bevis, D.; Klein, K. Fundamentals of Parallel Hybrid Turbofan Mission Analysis with Application to the Electrically Variable Engine. In Proceedings of the 2018 AIAA/IEEE Electric Aircraft Technologies Symposium, Cincinnati, OH, USA, 12–14 July 2018; American Institute of Aeronautics and Astronautics: Reston VA, USA, 2018. [CrossRef]

23. Perullo, C.; Trawick, D.; Armstrong, M.; Tai, J.C.; Mavris, D.N. Cycle Selection and Sizing of a Single-Aisle Transport with the Electrically Variable Engine(TM) (EVE) for Fleet Level Fuel Optimization. In Proceedings of the 55th AIAA Aerospace Sciences Meeting, Grapevine, TX, USA, 9–13 January 2017; American Institute of Aeronautics and Astronautics: Reston VA, USA, 2017. [CrossRef]
24. Pornet, C.; Kaiser, S.; Isikveren, A.T.; Hornung, M. Integrated Fuel-Battery Hybrid for a Narrow-Body Sized Transport Aircraft. *Aircr. Eng. Aerosp. Technol. Int. J.* **2014**, *86*, 568–574. [CrossRef]
25. Xin, Z.; Sahoo, S.; Kyprianidis, K.; Sumsurooah, S.; Valente, G.; Rashed, M.; Vakil, G.; Hill, C.I.; Jacob, C.; Gobbin, A. A Framework for Optimization of Hybrid Aircraft. In Proceedings of the ASME Turbo Expo 2019: Turbomachinery Technical Conference and Exposition, Phoenix, AZ, USA, 17–21 June 2019; American Society of Mechanical Engineers (ASME): New York, NY, USA, 2019; Volume 3. [CrossRef]
26. Nunez, M.; Datta, V.C.; Molina-Cristobal, A.; Guenov, M.; Riaz, A. Enabling Exploration in the Conceptual Design and Optimisation of Complex Systems. *J. Eng. Des.* **2012**, *23*, 852–875. [CrossRef]
27. Ullman, D. G. *The Mechanical Design Process*; David Ullman LLC: Independence, OR, USA, 2017; ISBN 978-0-387-21507-5.
28. Pahl, G.; Beitz, W.; Feldhussen, J.; Karl-Heinrich G. *Engineering Design. A Systematic Approach*; Springer: London, UK, 2007; ISBN 978-1-4471-6025-0.
29. Brahma, A.; Wynn, D.C. Margin Value Method for Engineering Design Improvement. *Res. Eng. Design* **2020**, *31*, 353–381. [CrossRef]
30. Guenov, M.D.; Chen, X.; Molina-Cristóbal, A.; Riaz, A.; van Heerden, A.S.J.; Padulo, M. Margin Allocation and Tradeoff in Complex Systems Design and Optimization. *AIAA J.* **2018**, *56*, 2887–2902. [CrossRef]
31. Pugh, S. *Total Design: Integrated Methods for Successful Product Engineering*; Prentice Hall: Wokingham, UK, 1990; ISBN 978-0-201-41639-8.
32. Wheelwright, S.C.; Clark, K.B. *Revolutionizing Product Development: Quantum Leaps in Speed, Efficiency, and Quality*; Simon and Schuster: New York, NY, USA, 1992; ISBN 978-0-02-905515-1.
33. Sobek II, D.; Ward, A.; Liker, J. Toyota's Principles of Set-Based Concurrent Engineering. *MIT Sloan Manag. Rev.* **1999**, *40*, 67-83.
34. Singer, D.J.; Doerry, N.; Buckley, M.E. What Is Set-Based Design? *Naval Eng. J.* **2009**, *121*, 31–43. [CrossRef]
35. Shallcross, N.; Parnell, G.S.; Pohl, E.; Specking, E. Set-Based Design: The State-of-Practice and Research Opportunities. *Syst. Eng.* **2020**, *23*, 1–22. [CrossRef]
36. Small, C.; Parnell, G.S.; Pohl, E.; Goerger, S.R.; Cilli, M.; Specking, E. Demonstrating Set-Based Design Techniques: An Unmanned Aerial Vehicle Case Study. *J. Def. Model. Simul.* **2020**, *17*, 339–355. [CrossRef]
37. Al Handawi, K.; Andersson, P.; Panarotto, M.; Isaksson, O.; Kokkolaras, M. Scalable Set-Based Design Optimization and Remanufacturing for Meeting Changing Requirements. *J. Mech. Des.* **2020**, *143*, 21702 . [CrossRef]
38. Inoue, M.; Suzuki, W.; Yamada, S.; Aoyama, K. A Universal Design Method That Considers Variability in User Requirements: A Case Study of Mechanical Pencil Design. *J. Adv. Mech. Des. Syst. Manuf.* **2021**, *15*, JAMDSM0022. [CrossRef]
39. Wade, Z.; Parnell, G.S.; Goerger, S.R.; Pohl, E.; Specking, E. Designing Engineered Resilient Systems Using Set-Based Design. In *Proceedings of the Systems Engineering in Context*; Adams, S., Beling, P.A., Lambert, J.H., Scherer, W.T., Fleming, C.H., Eds.; Springer International Publishing: Cham, Switzerland, 2019; pp. 111–122._10. [CrossRef]
40. Specking, E.; Shallcross, N.; Parnell, G.S.; Pohl, E. Quantitative Set-Based Design to Inform Design Teams. *Appl. Sci.* **2021**, *11*, 1239. [CrossRef]
41. Shallcross, N.J.; Parnell, G.S.; Pohl, E.; Goerger, S.R. A Value of Information Methodology for Multiobjective Decisions in Quantitative Set-Based Design. *Syst. Eng.* **2021**, *24*, 409–424. [CrossRef]
42. Shallcross, N.J.; Parnell, G.S.; Pohl, E.; Goerger, S.R. Using Value of Information in Quantitative Set-Based Design. *Syst. Eng.* **2021**, *24*, 439–455. [CrossRef]
43. Avigad, G.; Moshaiov, A. Set-Based Concept Selection in Multi-Objective Problems Involving Delayed Decisions. *J. Eng. Des.* **2010**, *21*, 619–646. [CrossRef]
44. Gong, D.; Sun, J.; Miao, Z. A Set-Based Genetic Algorithm for Interval Many-Objective Optimization Problems. *IEEE Trans. Evol. Comput.* **2018**, *22*, 47–60. [CrossRef]
45. Moshaiov, A.; Snir, A.; Samina, B. Concept-Based Evolutionary Exploration of Design Spaces by a Resolution-Relaxation-Pareto Approach. In Proceedings of the 2015 IEEE Congress on Evolutionary Computation (CEC), Sendai, Japan, 25–28 May 2015; pp. 1845–1852. [CrossRef]
46. Veenhuis, C.B. A Set-Based Particle Swarm Optimization Method. In *Proceedings of the Parallel Problem Solving from Nature—PPSN X*; Rudolph, G., Jansen, T., Beume, N., Lucas, S., Poloni, C., Eds.; Springer: Berlin/Heidelberg, Germany, 2008; pp. 971–980. [CrossRef]
47. Parker, R.R.; Malak, R.J., Jr. Technology Characterization Models and Their Use in Designing Complex Systems. In Proceedings of the ASME Design Engineering Technical Conference, Washington, DC, USA, 28–31 August 2011; pp. 1063–1075. [CrossRef]
48. Hannapel, S.; Vlahopoulos, N. Implementation of Set-Based Design in Multidisciplinary Design Optimization. *Struct. Multidisc. Optim.* **2014**, *50*, 101–112. [CrossRef]
49. Georgiades, A.; Sharma, S.; Kipouros, T.; Savill, M. ADOPT: An Augmented Set-Based Design Framework with Optimisation. *Des. Sci.* **2019**, *5*, e4. [CrossRef]

50. Finch, W.W.; Ward, A.C. A Set-Based System for Eliminating Infeasible Designs. In Proceedings of the ASME 1997 Design Engineering Technical Conferences, Sacramento, CA, USA, 14–17 September 1997; American Society of Mechanical Engineers: New York, NY, USA, 1997. [CrossRef]
51. Yannou, B.; Yvars, P.-A.; Hoyle, C.; Chen, W. Set-Based Design by Simulation of Usage Scenario Coverage. *J. Eng. Des.* **2013**, *24*, 575–603. [CrossRef]
52. Venter, G.; Haftka, R.T. Using Response Surface Approximations in Fuzzy Set Based Design Optimization. *Struct. Optim.* **1999**, *18*, 218–227. [CrossRef]
53. Shahan, D.W.; Seepersad, C.C. Bayesian Network Classifiers for Set-Based Collaborative Design. *J. Mech. Design* **2012**, *134*. [CrossRef]
54. Han, H.; Chang, S.; Kim, H. Multiple Target Exploration Approach for Design Exploration Using a Swarm Intelligence and Clustering. *J. Mech. Des.* **2019**, *141*, 091401. [CrossRef]
55. Miller, S.W.; Yukish, M.A.; Simpson, T.W. Design as a Sequential Decision Process. *Struct. Multidisc. Optim.* **2018**, *57*, 305–324. [CrossRef]
56. Chhabra, J.P.S.; Warn, G.P. A Method for Model Selection Using Reinforcement Learning When Viewing Design as a Sequential Decision Process. *Struct. Multidisc. Optim.* **2019**, *59*, 1521–1542. [CrossRef]
57. Rasmussen, C.E.; Williams, C.K.I. *Gaussian Processes for Machine Learning*; Adaptive Computation and Machine Learning Series; MIT Press: Cambridge, MA, USA, 2005; ISBN 978-0-262-18253-9.
58. Spinelli, A.; Anderson, L.; Enalou, H.B.; Zaghari, B.; Kipouros, T.; Laskaridis, P. Application of Probabilistic Principles to Set-Based Design for the Optimisation of a Hybrid-Electric Propulsion System. *IOP Conf. Ser. Mater. Sci. Eng.* **2022**, *1226*, 12064. [CrossRef]
59. Smith, R.C. *Uncertainty Quantification: Theory, Implementation, and Applications*; SIAM: Philadelphia, PA, USA, 2013; ISBN 978-1-61197-321-1.
60. Cunha, A.; Nasser, R.; Sampaio, R.; Lopes, H.; Breitman, K. Uncertainty Quantification through the Monte Carlo Method in a Cloud Computing Setting. *Comput. Phys. Commun.* **2014**, *185*, 1355–1363. [CrossRef]
61. Yao, W.; Chen, X.; Luo, W.; van Tooren, M.; Guo, J. Review of Uncertainty-Based Multidisciplinary Design Optimization Methods for Aerospace Vehicles. *Prog. Aerosp. Sci.* **2011**, *47*, 450–479. [CrossRef]
62. Palar, P.S.; Zuhal, L.R.; Shimoyama, K.; Tsuchiya, T. Global Sensitivity Analysis via Multi-Fidelity Polynomial Chaos Expansion. *Reliab. Eng. Syst. Saf.* **2018**, *170*, 175–190. [CrossRef]
63. Bianchi, D.H.B.D.; Sêcco, N.R.; Silvestre, F.J. A Framework for Enhanced Decision-Making in Aircraft Conceptual Design Optimisation under Uncertainty. *Aeronaut. J.* **2021**, *125*, 777–806. [CrossRef]
64. Oliver, M.A.; Webster, R. Kriging: A Method of Interpolation for Geographical Information Systems. *Int. J. Geogr. Inf. Sys.* **1990**, *4*, 313–332. [CrossRef]
65. Pedregosa, F.; Varoquaux, G.; Gramfort, A.; Michel, V.; Thirion, B.; Grisel, O.; Blondel, M.; Prettenhofer, P.; Weiss, R.; Dubourg, V.; et al. Scikit-Learn: Machine Learning in Python. *J. Mach. Learn. Res.* **2011**, *12*, 2825–2830.
66. Helton, J.C.; Davis, F.J. Latin Hypercube Sampling and the Propagation of Uncertainty in Analyses of Complex Systems. *Reliab. Eng. Syst. Saf.* **2003**, *81*, 23–69. [CrossRef]
67. Blank, J.; Deb, K. Pymoo: Multi-Objective Optimization in Python. *IEEE Access* **2020**, *8*, 89497–89509. [CrossRef]
68. Seada, H.; Deb, K. A Unified Evolutionary Optimization Procedure for Single, Multiple, and Many Objectives. *IEEE Trans. Evol. Comput.* **2016**, *20*, 358–369. [CrossRef]
69. Regis, R.G. Particle Swarm with Radial Basis Function Surrogates for Expensive Black-Box Optimization. *J. Comput. Sci.* **2014**, *5*, 12–23. [CrossRef]
70. Pagliuca, G.; Kipouros, T.; Savill, M.A. Surrogate Modelling for Wing Planform Multidisciplinary Optimisation Using Model-Based Engineering. *Int. J. Aerosp. Eng.* **2019**, *2019*, e4327481. [CrossRef]
71. Inselberg, A. *Parallel Coordinates: Visual Multidimensional Geometry and Its Applications*; Springer: New York, NY, USA, 2009; ISBN 978-0-387-21507-5.
72. Kipouros, T.; Inselberg, A.; Parks, G.; Savill, A.M. Parallel Coordinates in Computational Engineering Design. In Proceedings of the 54th AIAA/ASME/ASCE/AHS/ASC Structures, Structural Dynamics, and Materials Conference, Boston, MA, USA, 8–11 April 2013; American Institute of Aeronautics and Astronautics: Reston VA, USA, 2013. [CrossRef]
73. Nikolaidis, T.; Jafari, S.; Bosak, D.; Pilidis, P. Exchange Rate Analysis for Ultra High Bypass Ratio Geared Turbofan Engines. *Appl. Sci.* **2020**, *10*, 7945. [CrossRef]
74. DuBois, D.; Paynter, G. Fuel Flow Method2 for Estimating Aircraft Emissions. *SAE Trans. J. Aerosp.* **2006**, *115*, 1–14. [CrossRef]
75. Filippone, A.; Bojdo, N. Statistical Model for Gas Turbine Engines Exhaust Emissions. *Transp. Res. Part Transp. Environ.* **2018**, *59*, 451–463. [CrossRef]
76. ATR-42-600 Factsheet. Available online: https://www.atr-aircraft.com/wp-content/uploads/2020/07/Factsheets_-_ATR_42-600.png (accessed on 21 February 2022).
77. Eisenhut, D.; Moebs, N.; Windels, E.; Bergmann, D.; Geiß, I.; Reis, R.; Strohmayer, A. Aircraft Requirements for Sustainable Regional Aviation. *Aerospace* **2021**, *8*, 61. [CrossRef]
78. Edström, K.; Dominko, R.; Fichtner, M.; Otuszewski, T.; Perraud, S.; Punckt, C.; Tarascon, J.-M.; Vegge, T.; Winter, M.; Dominko, R. BATTERY 2030+ Roadmap. 2020. Available online: https://battery2030.eu/wp-content/uploads/2021/08/c_860904-l_1-k_roadmap-27-march.png (accessed on 7 December 2021)

79. Ojeda-Rodríguez, Á.; González-Vizuete, P.; Bernal-Méndez, J.; Martín-Prats, M.A. A Survey on Bidirectional DC/DC Power Converter Topologies for the Future Hybrid and All Electric Aircrafts. *Energies* **2020**, *13*, 4883. [CrossRef]
80. Sirimanna, S.; Thanatheepan, B.; Lee, D.; Agrawal, S.; Yu, Y.; Wang, Y.; Anderson, A.; Banerjee, A.; Haran, K. Comparison of Electrified Aircraft Propulsion Drive Systems with Different Electric Motor Topologies. *J. Propuls. Power* **2021**, *37*, 733–747. [CrossRef]
81. Wulff, A.; Hourmouziadis, J. Technology Review of Aeroengine Pollutant Emissions. *Aerosp. Sci. Technol.* **1997**, *1*, 557–572. [CrossRef]

aerospace

MDPI

Article

Wing Structural Model for Overall Aircraft Design of Distributed Electric Propulsion General Aviation and Regional Aircraft

Raquel Alonso Castilla †, Florent Lutz *,†, Joël Jézégou and Emmanuel Bénard

ISAE-SUPAERO, University of Toulouse, 31000 Toulouse, France;
Raquel.ALONSO-CASTILLA@student.isae-supaero.fr (R.A.C.); Joel.JEZEGOU@isae-supaero.fr (J.J.);
Emmanuel.BENARD@isae-supaero.fr (E.B.)
* Correspondence: Florent.LUTZ2@isae-supaero.fr
† These authors contributed equally to this work.

Abstract: In the context of reducing the environmental footprint of tomorrow's aviation, Distributed Electric Propulsion (DEP) has become an increasingly interesting concept. With the strong coupling between disciplines that this technology brings forth, multiple benefits are expected for the overall aircraft design. These interests have been observed not only in the aerodynamic properties of the aircraft but also in the structural design. However, current statistical models used in conceptual design have shown limitations regarding the benefits and challenges coming from these new design trends. As for other methods, they are either not adapted for use in a conceptual design phase or do not cover CS-23 category aircraft. This paper details a semi-analytical methodology compliant with the performance-based certification criteria presented by the European Union Aviation Safety Agency (EASA) to predict the structural mass breakdown of a wing. This makes the method applicable to any aircraft regulated by EASA CS-23. Results have been validated with the conventional twin-engine aircraft Beechcraft 76, the innovative NASA X-57 Maxwell concept using DEP, and the commuter aircraft Beechcraft 1900.

Keywords: aircraft design; distributed propulsion; general aviation; regional aircraft; structural mass; wing design; aviation; MDAO; FAST-OAD; innovative aircraft architecture

Citation: Alonso Castilla, R.; Lutz, F.; Jézégou, J.; Bénard, E. Wing Structural Model for Overall Aircraft Design of Distributed Electric Propulsion General Aviation and Regional Aircraft. *Aerospace* **2022**, *9*, 5. https://doi.org/10.3390/aerospace9010005

Academic Editors: Spiros Pantelakis, Andreas Strohmayer and Liberata Guadagno

Received: 26 November 2021
Accepted: 16 December 2021
Published: 22 December 2021

1. Introduction

The goals of reducing the environmental footprint and decreasing the noise levels determine the trends of future aviation. In this sense, distributed electric propulsion (DEP) aircraft represent a potential key approach for meeting the overall objectives without compromising costs and safety requirements. The concept of blowing air at high speed over the wing generates a relative speed of the air with respect to the aircraft. As a result, increasing the number of propellers demands less thrust to reach the minimum speed during take-off, and it increases the lift coefficient and requires a smaller wing surface [1]. From a structural point of view, the incorporation of DEP along the span leads to the alleviation of the bending moment and the reduction of the wing structural mass [2]. Therefore, DEP affects the mass breakdown of the wing, and its assessment is one of the main challenges to be faced in the weight analysis.

The design of a new aircraft, either innovative or conventional, generally starts by a conceptual design (see Figure 1) in which the preliminary assessment of a variety of possible configurations is studied. This phase, if conducted properly, allows us to rule out unfeasible layouts and identify key features [3]. This stage is but one of the iterations of the aircraft design process and is likely to be conducted several times for each tested configuration. Additionally, many new aircraft concepts are taking shape, which creates the need for a way to quickly and reliably assess their likeliness to fly in conceptual and preliminary design steps. FAST-OAD-GA is an overall aircraft design (OAD) code that aims to fulfill the need for general aviation (GA) aircraft. As part of the undertaking of adapting FAST-OAD-GA for novel configuration, the necessity for an improved wing mass estimation method was identified.

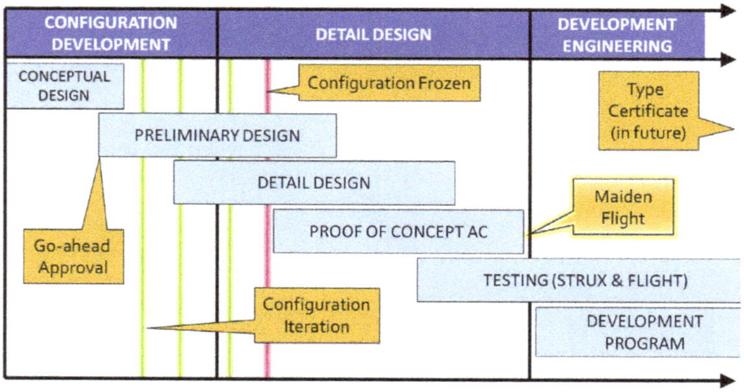

Figure 1. Aircraft design process for a typical GA aircraft. Reprinted from Ref [3].

A literature review has thus been conducted to assess the relevancy of existing wing mass estimation methods for conceptual and preliminary aircraft sizing. Three types of methods have been identified based on their fidelity level: empirical, physics-based, and finite-element methods.

Empirical models such as the one presented by Raymer [4] and Nicolai [5] are particularly well suited for the conceptual design stage and are commonly used. Their computation time is virtually instantaneous, and the level of detail required to apply the formulae matches what is usually found at this stage of aircraft design. Such methods are, however, strongly dependent on the database used to establish them, and with the lack of information on innovative designs, they would fail to capture the aforementioned effects [6]. More refined methods such as the one presented in NASA's FLOPS [7] are capable of encapsulating the design features of unconventional aircraft configurations. This is the case for high-aspect-ratio wings, strut-braced wings, or aeroelastic tailoring. Nevertheless, the aerodynamic loads and the engine relief effect are only used in part of the computation of the wing weight, and they must be specified by the user.

On the opposite end of the complexity spectrum, one finds finite element methods. Some of them, such as the one presented in [8], have been shown to be able to estimate the wing mass with satisfactory results. However, such methods present computational times that are not acceptable for conceptual design phase. Furthermore, the required inputs and precision level of those models are usually not known at this stage of the design. The final limiting step is the complexity of such codes or software that makes their integration efficient in an overall aircraft design software complex.

Finally, existing analytical and physics-based methods have been reviewed. These combine the ease of integration with flexibility and relatively low computational time. Nonetheless, they are often proprietary software or embedded in an already existing OAD tool. Other approaches such as the one developed in [9] or [10] present the appropriate level of fidelity and are described with enough precision that they can be implemented to estimate a basic mass breakdown of the wing. However, these methodologies do not consider the loads cases and structural requirements prescribed by EASA CS-23, and they are more oriented towards commercial transport aircraft. Nonetheless, these have served as the basis to build the methodology presented in this paper.

This work focuses on developing and validating a semi-analytical approach to predict the structural mass breakdown of a wing for a usage in a preliminary aircraft design tool. The choice of a semi-analytical model was made to combine the robustness of the method, the ease of the integration, and the low computational time. This model shall be applicable to aeroplanes in EASA CS-23 category and be adapted to both conventional and distributed propulsion airplanes. The methodology developed by [10] was adapted to consider both the relevant CS-23 loads certification requirements and the effects of DEP. After a validation of the developed method on three aircraft covered by the CS-23 requirements, parametric studies were conducted to investigate different design choices. This methodology was then implemented in FAST-OAD-GA, which brought its own set of challenges, chief among which was the concern for the computational time and opportunities such as benefiting from the software framework.

2. Materials and Methods

2.1. FAST-OAD-GA (Future Aircraft Sizing Tool—Overall Aircraft Design—General Aviation)

FAST-OAD-GA is an open-source plugin of FAST-OAD, an open-source aircraft sizing and optimization tool developed by ISAE-SUPAERO and ONERA based on the OpenMDAO framework [11]. The links to the code are available in the Supplementary Material section. A FAST-OAD run consists in the assembly of several OpenMDAO components with inputs and outputs that represents the application of an aircraft sizing methodology. Moreover, FAST-OAD allows us to define specific problems by using a configuration file. This file provides an easy way to switch between models or to remove/add some modules. Coupled with OpenMDAO's multidisciplinary design analysis capabilities, it allows for a preliminary estimation of the viability of possible aircraft configurations. Currently, FAST-OAD-GA is bundled with statistical models for mass estimation targeted at conventional airplane architectures. However, in the extension of this software to cover innovative aircraft designs, this was targeted as a limiting factor, which prompted the creation of the presented methodology.

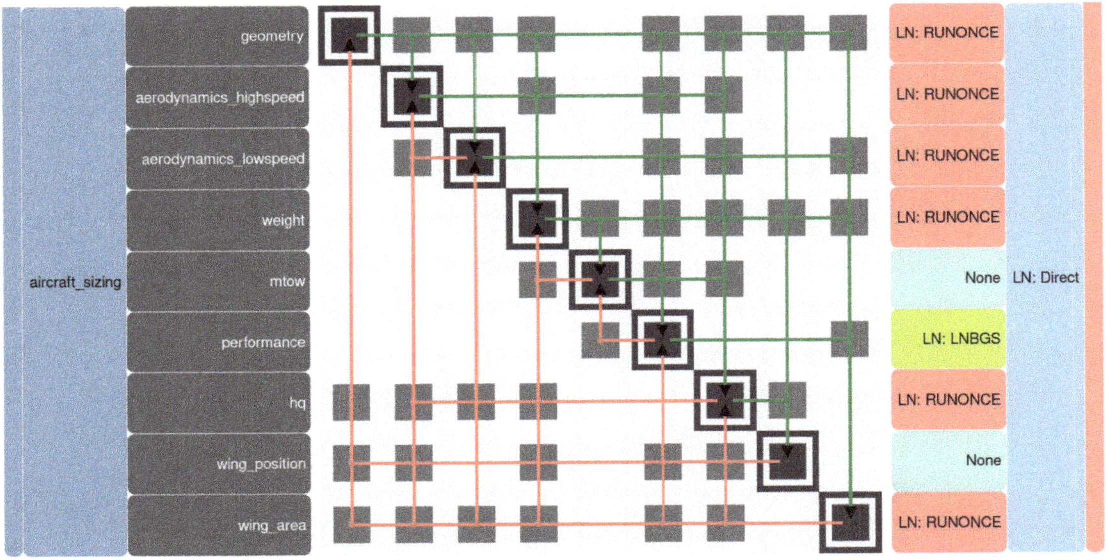

Figure 2. Representation of the typical FAST-OAD-GA process as an N2 diagram.

A representation of the typical FAST-OAD-GA sizing process is presented in Figure 2 in the format of an N2 diagram. Each of the colored lines on the bottom-left side of the diagonal represents a loop in the process, while the one on the upper-right are forward computation. Henceforth, the outermost loop, i.e., the loop on the wing area, will be referred to as an iteration. Figure 2 shows that each of the modules (gray blocks in the main diagonal) can be used several times in each iteration. More specifically, the weight module that computes the overall empty weight (OEW) is inside 3 loops in each iteration, meaning it is run at least three times. The OEW is currently computed by summation of the weights of every component of the aircraft broken down into categories adapted from [12]. The wing corresponds to one of the main components that contributes to the OEW. The work presented in this paper consists in replacing the current way of computing the wing weight by the semi-analytical methodology mentioned in the introduction to improve the precision of the sizing process while compromising the computational time. Figure 2 also illustrates that this model will be used in a complex environment with a methodology at various levels of fidelity, making it unnecessary to have a model with a level of detail in the inputs greater than what the rest of the process can provide, which was one of the reasons why a finite element method was not retained.

Furthermore, Figure 2 shows that many modules are run before the computation of the mass breakdown. This means that since the outputs of some components can be the inputs for others, it is easy to reuse the results of previous models without requiring their computational time. An example of a module in this situation is the computation of the aerodynamic performance of the aircraft, which contains the computation of the lift distribution along the span.

In order to be able to use a real lift distribution as part of the computation of the aerodynamics loads, FAST-OAD-GA relies on OpenVSP through an interface that allows an automated processing of the computation and results. It is an open-source software released under NASA Open Source Agreement as a parametric geometry tool. One of OpenVSP's tools is VSPAERO which allows aerodynamic analysis based on either a vortex lattice or a panel method. OpenVSP is well suited in early aircraft design stage because of its low computational time and ability to run various types of analyses. Another advantage of using this software is that it allows the simulation of the wing-propeller interaction due to the induced velocity through the use of an actuator disk model [13], which is a key feature to capture the effect of DEP in the aerodynamic loads.

2.2. Wing Mass Model

For the purpose of initial design, the wing is assumed to be trapezoidal, with N engines distributed along the wingspan and divided into a primary and secondary structure. The leading and trailing edge devices such as ailerons, flaps, and slats form the secondary structure of the wing. The primary structure is modeled as a wing-box, represented in Figure 3 with a miscellaneous weight to account for the joints, cut-outs, and connections. The wing-box structure is then broken into different components. Each of these components is assumed to withstand only one type of stress:

- The spar web withstands the shear force;
- The spar caps supports the bending mometn;
- The skin reacts to the torsion moment;
- The ribs bear the buckling loads.

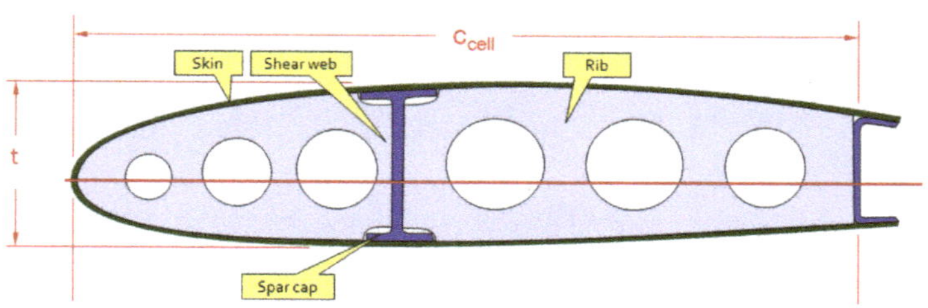

Figure 3. Wing-box model. Reprinted from Ref [3].

This set of assumptions allows a simplified analytical estimation of their individual mass based on É. Roux's model [10], which is only valid for isotropic materials. The use of composites is accounted for in the miscellaneous weight through the FCOMP factor [7], ranging from 0 for no composites to 1 for its maximum utilization.

The choice was made to ultimately not consider buckling for the sizing of the ribs. Indeed, this method was elaborated to be used as part of a conceptual design tool and does not substitute a full load analysis, which means we had to rule out some failure scenarios such as buckling.

In addition to the previous model, Sui An [14] shows in her studies the impact of flutter on the wing mass when using distributed propulsion. She proves that the flutter constraint increases the optimum mass by 6–14%. Therefore, the current method will approximate such effects in the applicable cases through a mean coefficient of 1.10.

2.3. Methodology Description

The following methodology is applicable to airplanes regulated by EASA CS-23 certification specification. Requirements from CS-23 at amendment 4 [15] are considered throughout this paper as an acceptable means of compliance to amendment 5, the latest applicable issue written in a performance-based language. The goal is to compute for each part of the wing the quantity of material necessary to support the stresses induced by the load cases prescribed in the certification requirements. Figure 4 describes the iterative process developed for the evaluation of the mass breakdown of the wing, which begins with an initial estimation of the total wing weight for the computation of the structural loads and then follows the steps presented hereafter.

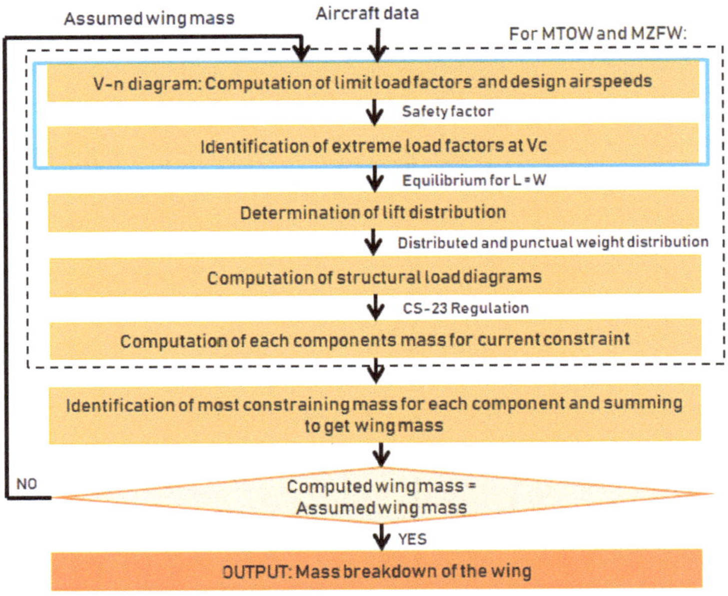

Figure 4. Flow diagram of the method.

1. Load factors: CS 23.321 and CS 23.333 [15] state that the aircraft must comply with the load requirements for any combination of load factor and airspeed within the limits of the flight envelope. This envelope results from the application of the maneuver and gust criteria defined for every weight and flight altitude computed according to CS 23.335, CS 23.337, and CS 23.341. To simplify the process, two constraining weight cases for the wing structure at the typical cruise altitude were selected: the aircraft

at MTOW and with minimum fuel in the wings [15]. The choice of considering only the cruise altitude and cruise equivalent airspeed V_c was made since these conditions lead to the highest gust load factor. In addition to that, according to CS 23.305, the wing should also demonstrate compliance to the strength requirements at ultimate loads. These are calculated by applying a safety factor to the maximum/minimum load factors obtained for the aforementioned conditions.

2. Lift distribution: The computation of the lift distribution was separated into two parts: the clean lift, which is the lift without accounting for engine wake and the slipstream lift, meaning the additional lift generated by the engine wake. The clean lift distribution is obtained from OpenVSP and was assumed to be of constant shape along the wingspan. It is then scaled to match the lift coefficient required to sustain the extreme load factors. In the computation of the clean lift, the tail down force necessary to maintain the aircraft level was also considered, leading to a 5% increase assumed to balance the weight in cruise (L = W) [12]. The slipstream lift distribution is computed using OpenVSP through a simple actuator-disk engine model to capture the effects of DEP on the aerodynamic loads. For the computation of the slipstream effect it was considered that two types of propellers could be installed on the wing: the High-Lift Propeller (HLP) [16] to provide extra lift at low speed and the Wingtip Propeller (WTP) to provide thrust during cruise and reduce induced drag. In cruise, the HLPs are folded, but at the climb cruise transition, they are still activated and working close to their maximum operating speed. At this time of flight, the potential risk of whirl flutter require setting the torque at the minimum level with a consequent reduction of power [17]. These few seconds of blown wing under cruise conditions are taken into account in the wing sizing process under a steady approximation to ensure conservative results. This lift distribution is then multiplied by the chord at the corresponding station along the span to obtain the aerodynamic loads at each point of the wing. The results of the computation of the lift load for the X-57 are shown in Figure 5.

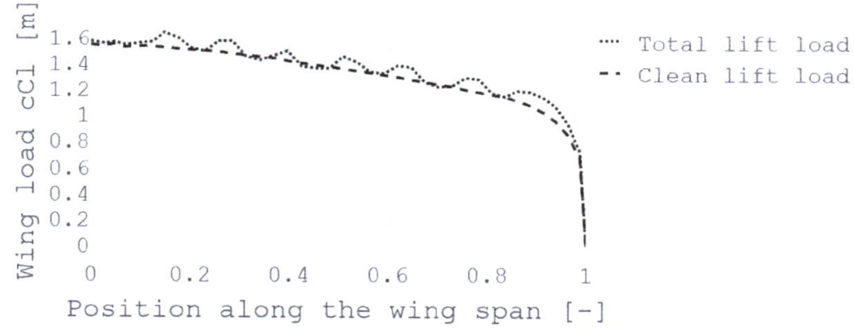

Figure 5. Lift load distribution along the dimensionless wing span.

3. Structural load diagrams: five types of loads acting on the wing are considered: the aerodynamic lift load previously obtained, the distribution of the wing and fuel weights (proportional to the local chord), the punctual weight of the engines and the weight of the landing gear. However, when the fuel is placed in the same position as an engine or landing gear housing, its weight is decreased at this section to account for the reduced space available. The shear force diagram is then obtained by the numerical integration of the distributed loads. Similarly, the bending moment diagram can be constructed with the numerical integral of the shear diagram. Pursuing the flexibility of this methodology for hybrid or DEP configurations, the fuel weight is only considered in the applicable cases.
4. Mass breakdown of the wing: the mass of each structural component of the wing is proportional to the quantity of material necessary to support the loads, defined as

the density of the material ρ_m multiplied by the volume of the part. The following computations will be done for the two previously mentioned aircraft masses: MTOW and minimum fuel in the wing. Consequently, two values of the mass will be obtained for each element, with the maximum being retained. Hereunder, a non-dimensional reference coordinate system will be used, where the Y axis ($Y = y/(b/2)$) is defined positive to the right side of the aircraft, with b the span.

- Web mass in shear M_{web}: the spar web is a vertical panel designed to react, under assumptions of the presented methodology, to all the shear force of the wing $T(y)$. CS 23.305 stress criteria is applied, which states that the stress on the web must be lower than the maximum shear stress of the material $|\tau_{web}| \leq \tau_{max}$. So, the mass of the web for the whole wing is calculated integrating its section area S_{web} along its elastic axis η which sweep angle is ϕ_e:

$$M_{web} = 2\rho_m \frac{b/2}{cos\phi_e} \int_0^1 S_{web} dY \quad ; \quad S_{web} = \frac{|T_{max}|}{\tau_{max}} \tag{1}$$

- Spar caps mass in bending M_{caps}: the lower and upper spar caps of the main spar react to the bending moment $M_f(y)$. Similarly to the web sizing, the CS 23.305 stress criteria imposes that the stress on the spar caps cannot surpass the maximum normal stress of the material, $\left|\sigma_{fl}\right| \leq |\sigma_{max}|$, neither in traction ($\sigma_{max_t} > 0$) nor in compression ($\sigma_{max_c} < 0$). In general, the resultant shear force of the wing is in the upwards direction. Hence, the upper spar cap will work under compression, while the lower spar cap will be subjected to traction. The consequence is that the upper spar cap will face higher risk of failure under buckling than the lower spar cap [10]. With those considerations, the mass of both spar caps can be calculated with Equation (2), where $h(Y)$ is the distance between the center of gravity of the spar caps retrieved from [10] and S_E, S_I are the cross-section areas of the upper and lower spar caps, respectively.

$$M_{caps} = 2\rho_m \frac{b/2}{cos\phi_e} \left[\int_0^1 S_E dY + \int_0^1 S_I dY \right] \tag{2}$$

$$\begin{cases} S_I = \frac{1}{h(Y)} max(\frac{Mf(y)^+}{\sigma max_t}, \frac{Mf(y)^-}{\sigma max_c}) \\ S_E = \frac{1}{h(Y)} max(\frac{-Mf(y)^+}{\sigma max_c}, \frac{-Mf(y)^-}{\sigma max_t}) \end{cases} \tag{3}$$

- Skin mass in torsion M_{skin}: The skin is a thin panel of a thickness assumed constant that provides the necessary strength to bear the torsion moment. This assumption makes sense from a manufacturing point of view, where varying the thickness for wings of this size represents additional manufacturing complexity that is not desirable for general aviation manufacturers. Besides having a varying skin thickness did not significantly improve the results in [10], hence our choice to keep the wing skin thickness constant. The regulation imposes two sizing criteria: CS 23.455 that limits the deformation of the wing in rolling conditions and one that corresponds to the maximum admissible stress at ultimate loads stated in the CS 23.305. Hence, the skin mass will be determined by the most demanding requirement. In previous versions of the model [18], both criteria were computed for various aircraft. The results revealed that the deformation dimensioning also meets the stress requirements. Consequently, only the CS 23.455 rule was implemented in the current edition of the code. Then, as presented in Equation (4), the skin mass is obtained by integrating the product of the skin thickness, e and the perimeter of the wing-box, l. The objective of this criterion is to guarantee the efficiency of the ailerons control by limiting the roll speed at the design airspeeds p_A, p_C, p_D. When ailerons are deflected, this norm restricts the wing deformation that changes the angle of attack $\Delta\alpha$. This $\Delta\alpha$ can be approximated by the variation of the twist angle due to the torsion moment produced by the ailerons. The skin weight will then be computed based on the necessary thickness to react that torsion moment. For further detail on the calculation of e, p_A, p_C, p_D the user is encouraged to see [10] (pp. 84–104).

$$M_{skin} = 2\rho_m \int_{\frac{R_{fus}}{cos\phi_e}}^{\frac{b/2}{cos\phi_e}} e(\eta) l d\eta \quad ; \quad \begin{cases} p_C \geq p_A \\ 3p_D \geq p_A \end{cases} \tag{4}$$

The mass of the ribs M_{ribs} and miscellaneous $M_{miscellaneous}$ items are obtained empirically according to the expressions proposed by É. Roux [10] and NASA FLOPS [7], respectively. As for the secondary structure, Torenbeek [2] states that it contributes to around a 25% of the total mass of the wing. Therefore, its weight will be empirically obtained with this proportion. Finally, the total mass of the wing will be the sum of all its components:

$$\begin{gathered} M_{primary} = M_{web} + M_{caps} + M_{skin} + M_{ribs} + M_{miscellaneous} \\ M_{wing} = M_{primary} + M_{secondary} \end{gathered} \tag{5}$$

5. Iteration: the method finishes by comparing the computed mass of the wing with the one initially estimated. If these values are different, the tool iterates on the solution with the computed value until the error is lower than 0.1%. This ensures the obtained wing weight is close enough to the structural weight used to compute relief effects.

2.4. Methodology Validation

Validation of the proposed methodology and related models is performed through an assessment for three aircraft use-cases: the Beechcraft 76 as a conventional twin-engine airplane, the innovative NASA Maxwell X-57 as a DEP concept, and the Beechcraft 1900 as a commuter aircraft. In this way, the robustness and flexibility of the method can be proved through a realistic range of weights and sizes of aircraft. Input data used for those use-cases are detailed in Appendix A. The structural safety factor for the Maxwell X-57 was set to 1.8 according to the airworthiness approach set for the Mod III version of the aircraft [19].

2.5. Parametric Analysis

Two parametric studies were conducted to show the importance of considering the impact of DEP and engine placement on the wing sizing. For the analysis described in Sections 2.5.1 and 2.5.2, the aircraft we will use as a reference is the Maxwell X-57. For the study in Section 2.5.3, we will use the Beechcraft 76. As previously introduced, the weight of the engines and the wing-propeller interaction modify the structural loads with subsequent effects on wing mass. For a better comprehension, each of these effects is presented and analysed independently.

2.5.1. Effect of the Engines Weight Distribution

The investigation presented in this section aims to analyse exclusively the engines mass effect on the wing sizing. The complete study of the effects of a change in the propulsion system is an Overall Aircraft Design problematic that is not addressed in this paper. The results of this study, however, can be seen as the first iteration of the global OAD process. For such purpose, the quantity of high-lift propellers per semi-wing has been varied from 1 to 6, and the following assumptions have been retained:

- The high-lift engines are evenly spaced along the wingspan and it is assumed that the distance between the last HLP and the WTP was greater than the distance between two HLP. The position used for this study is detailed in Table 1.

- The maximum power and mass of the engines are proportionally scaled with the number of HLP. The propeller disc size is also adjusted to have a constant power coefficient. The baseline is defined by the 12 high-lift motors mounted in the X-57 Maxwell, with a propeller diameter of 0.58 m, a shaft power of 10.5 kW and a mass of 6.8 kg per engine.
- The slipstream effect of HLP is not considered for this study. This analysis is focused on the engines weight and not on the aerodynamics though the two parameters will be coupled in an overall aircraft analysis.
- The WTP remains the same in every simulation. This enables to maintain the main source of thrust of the aircraft to counteract the drag.

Table 1. HLP position considered in the parametric study on the effect of HLP weight (in % of the semi-span) per half-wing.

N_{eng}	Position (s) of the HLP (s)						Position of the WTP
2 (1 + 1)	0.33	-	-	-	-	-	1.0
3 (2 + 1)	0.25	0.5	-	-	-	-	1.0
4 (3 + 1)	0.2	0.4	0.6	-	-	-	1.0
5 (4 + 1)	0.17	0.35	0.5	0.65	-	-	1.0
6 (5 + 1)	0.143	0.286	0.429	0.572	0.715	-	1.0
7 (6 + 1)	0.188	0.3075	0.4269	0.5462	0.667	0.785	1.0

2.5.2. Effect of Slipstream Consideration

The second parametric study shows the sensitivity of the model to the slipstream effect in cruise conditions. This analysis is made to evaluate the impact of the choice made in this paper to consider that HLP were still active in the cruise conditions. The original configuration of the X-57 Maxwell with 14 engines is used. In order to achieve this objective, the mass of the wing has been obtained with and without adding the slipstream lift.

2.5.3. Effect of Engine Placement

The third parametric study aims at analysing the influence of the position of the engine on the overall wing mass for a conventional design. The baseline configuration for this study is the Beechcraft 76. The position of the engine varies between the root and the tip of the wing. For a correct interpretation of the conclusions of this study, it is necessary to mention that, for the computation where the engine is located closer to the fuselage centerline than the reference position, results are physically irrelevant. The baseline engine location is indeed already a lower limit based on the propeller clearance, which can be specified as a design constraint. The results will nevertheless be included as an indicative basis.

3. Results

3.1. Methodology Validation

For the purpose of validation of the methodology, the outputs of the use-cases have been evaluated against data available for Beechcraft 76, Beechcraft 1900 and NASA X-57 Maxwell with DEP. For the latter, the weight of the wing was directly available in the literature [20]. For the Beechcraft 76 and the Beechcraft 1900, as no real values are publicly available, the comparison is made with a traditional statistical method [4,5,12]. Said statistical methods are applicable to normal category aircraft and commuter aircraft and were established based on database containing aircraft which technological level is close to that of the Beechcraft 76 and 1900. They are written in the Appendix B.

A summary of the results of the validation study is presented in Table 2.

For the Beechcraft 76 use-case, the deviation of the wing mass obtained with the presented semi-analytical model compared to the reference value is less than 1%. The actual value of the wing mass is unknown; however, this result shows that the newly established methodology gives the same results as a classical empirical formula. This prompted the authors to compare the result with an empirical methodology [5]. In this case, the difference computed was of around 6%, which confirms that no significant precision is lost when using the semi-analytical approach.

Table 2. Outputs of the presented methodology applied on the Beechcraft 76, the X-57 Maxwell, and the Beechcraft 1900.

Estimation Method	Beechcraft 76		X-57 Maxwell		Beechcraft 1900	
	Wing Mass (kg)	Error (%)	Wing Mass (kg)	Error (%)	Wing Mass (kg)	Error (%)
Developed model	182.2	0.4	174.3	4.5	722.8	0.7
Statistical model	183.0 [4]	Reference	131.8 [5]	20.1	717.5 [12]	Reference
Real value	Unknown	-	166.7 [20]	Reference	Unknown	-

The same conclusion can be drawn from the validation of the method against the commuter Beechcraft 1900. The empirical formula used to compare the results obtained for this aircraft is selected from [12], which suggests the use of commercial transport aircraft models as the most appropriate choice to describe turbo-propeller aircraft with a maximum take-off mass over 5670 kg, as is the case for this aircraft.

For the NASA X-57 Maxwell use-case, the empirical formula from the literature gives results with an error of around 20%. This can partly be explained by the fact that the correlation was established on aircraft whose wing loading was fairly low because of the flaps technology used. Given that DEP allows one to reach a lift coefficient of over 4.0 in certain configurations [21], this leads to small wings but with heavy structure. By contrast, the semi-analytical methodology provides results that are much closer to the real wing mass. The reason behind this is that it actually takes into account the structure required to withstand the loads compliant with CS-23 requirements.

In addition to the total wing mass, the presented methodology also provides a further breakdown of the wing structural components shown in Table 3.

From this table, a noticeable remark can be formulated about the change in the proportional importance of the spar caps on the total wing weight. Indeed, for the X-57 Maxwell they take a much larger share. This can be explained by the increase in the wing aspect ratio (AR), which is equal to 15 for X-57 Maxwell, compared to 8 for the Beechcraft 76 and 10.8 for the Beechcraft 1900. The AR is one of the main factors in the computation of the bending moment [3], and since the bending moment is the defining factor for the spar caps computation, this explains such an increase.

Table 3. Detailed mass breakdown (absolute and share of total wing weight) of the Beechcraft 76, the X-57 Maxwell, and the Beechcraft 1900.

Wing Structural Component	Beechcraft 76	X-57 Maxwell	Beechcraft 1900
Web	2.235 kg (1.2%)	2.409 kg (1.4%)	15.538 kg (2.1%)
Lower spar cap	10.889 kg (6.0%)	20.549 kg (11.8%)	64.643 kg (8.8%)
Upper spar cap	14.548 kg (8.0%)	27.455 kg (15.7%)	86.365 kg (11.8%)
Skin	49.73 kg (27.3%)	68.501 kg (39.3%)	267.065 kg (36.0%)
Ribs	22.128 kg (12.1%)	3.206 kg (1.8%)	47.643 kg (6.6%)
Misc	37.113 kg (20.4%)	8.64 kg (5.0%)	70.863 kg (9.7%)
Primary structure	136.643 kg (75.0%)	130.76 kg (75.0%)	542.117 kg (75.0%)
Secondary structure	45.548 kg (25.0%)	43.587 kg (25.0 %)	180.705 kg (25.0%)
Total	182.191 kg (100.0%)	174.347 kg (100.0%)	722.821 kg (100.0%)

3.2. Parametric Studies

3.2.1. Effect of the Engines Weight Distribution

The goal of this study is to see if the presented methodology was capable of replicating the relief in the bending moment that comes from the distribution of engines along the wing. The results are presented in Table 4 and plotted in Figure 6, where the dashed lined is the reference weight for the X-57 Maxwell.

Table 4. Mass breakdown [kg] of the X-57 Maxwell wing for different number of engines on the wing.

N_{eng}	M_{web}	M_{caps}	M_{skin}	M_{ribs}	M_{misc}	$M_{secondary}$	M_{total}
2 (1 + 1)	2.34	47.10	68.50	3.21	8.64	43.26	173.05
3 (2 + 1)	2.32	46.40				43.02	172.10
4 (3 + 1)	2.31	45.95				42.87	171.48
5 (4 + 1)	2.31	45.66				42.77	171.08
6 (5 + 1)	2.30	45.37				42.67	170.69
7 (6 + 1)	2.28	44.58				42.40	169.6

The results show an alleviation effect on the total wing mass of around 3.5 kg for an increasing number of engines. It can be observed that the webs, spar caps, and secondary structure masses are the only values that change. As explained before, the ribs and miscellaneous weights are computed through empirical methods. These techniques do not take into account the loads or the number of engines, which justifies that no change is observed. As for the skin, its mass only depends on the characteristic speed of the aircraft. Therefore, considering this study's assumptions of constant MTOW and no slipstream effects, the skin weight does not vary with the number of engines.

Based on these observations and the hypothesis made regarding the secondary structure mass, which will account for 25% of the total wing mass, the weight variation is due to a change in the primary structure. This is consistent, since the primary mass is mainly obtained as the weight of the material required to withstand the loads. Since the weight of the engines relieves the aerodynamic forces to react to the structure of the wing, a difference in the engine weight and position will impact the structural weight.

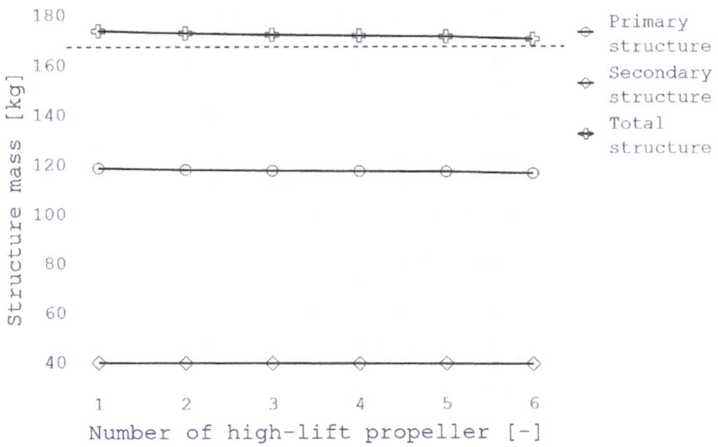

Figure 6. Effect of the number of engines on the overall wing mass.

For a deeper analysis into the reason for this lightening of the primary structure mass, Figure 7 plots the contribution of each structural component to the primary structure mass. This bar chart was plotted based on the results obtained for the two limit cases with only one HLP and with six HLP.

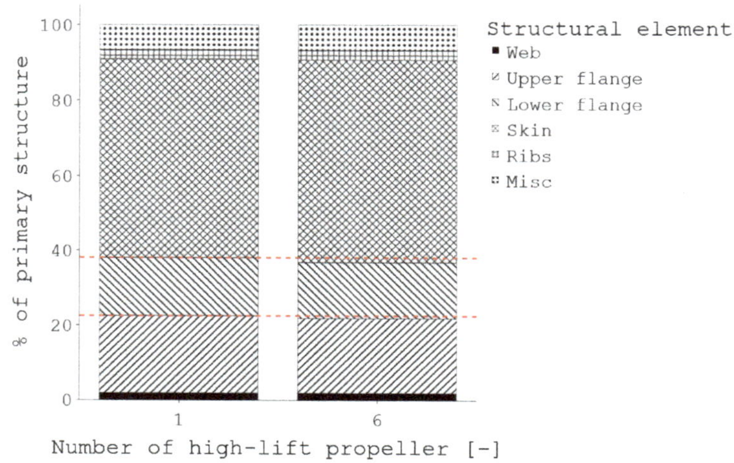

Figure 7. Effect of the number of engines on the primary structure mass.

As seen in Table 4 and Figure 7, the change in the primary structure mass comes from spar caps and web. In particular, the shear force seen by the web and the bending moment borne by the spar caps are lower. Thus, the structures will be lighter. Similar effects were observed in the literature [22], albeit on bigger aircraft, which reinforces the idea that this model can capture the structural benefits that distributed propulsion can bring forth.

3.2.2. Effect of Slipstream Consideration

This study investigates the impacts of the slipstream effects on the wing sizing. The results are presented in Table 5.

Table 5. Mass breakdown (kg) of the X-57 Maxwell wing with and without slipstream effects considered.

<table>
<tr><th></th><th>M_{web}</th><th>M_{caps}</th><th>M_{skin}</th><th>M_{ribs}</th><th>M_{misc}</th><th>$M_{secondary}$</th><th>M_{total}</th></tr>
<tr><td>w/slipstream</td><td>2.41</td><td>48.04</td><td rowspan="2">68.50</td><td rowspan="2">3.21</td><td rowspan="2">8.64</td><td>43.59</td><td>174.35</td></tr>
<tr><td>w/o slipstream</td><td>2.28</td><td>44.58</td><td>42.40</td><td>169.6</td></tr>
</table>

The results reveal a sensitivity of almost 5 kg ($\simeq$3%), resulting in a heavier wing when the induced velocity of the HLP is considered. The air blown by the engines on the wing increases the aerodynamic loads that the wing structure reacts to. Consequently, reinforced and heavier web and spar caps are required.

On the other hand, the current analysis highlights an important point. Depending on the power considered for the HLPs, the slipstream can heavily constrain the wing sizing. In the current approach, a minimum thrust is set, since the study is performed immediately at the end of the climbing phase at V_c speed. However, in more detailed design stages, the take-off phase will have to be analyzed with full engine power. In this case, the slipstream effect will significantly exceed the actual 3%, challenging the choice of using V_c as the sizing point. The point of design was chosen based on the maximum load factor obtained within the maneuver and gust envelopes. However, for commuter-category aircraft, the certification specification requires us to also consider the point of maximum gust from the gust envelope [15].

The V-n diagram we obtained in our study on the Beechcraft 1900 is shown in Figure 8, in which point A corresponds to the point we considered and point B corresponds to the maneuvering point. At this point, even if the load factor is no greater than at cruise, the speed is lower and thus the slipstream effects are greater. As a result, this might lead to a less homogeneous aerodynamic loads and a need for stronger primary structure.

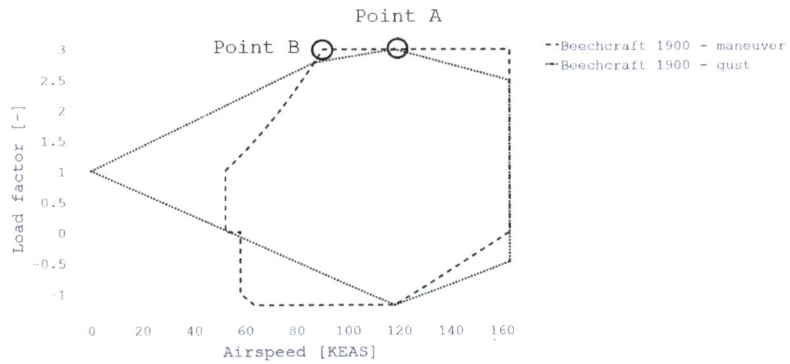

Figure 8. V-n diagram obtained by FAST-OAD-GA for the Beechcraft 1900, a commuter type aircraft.

3.3. *Effect of Engine Placement*

This study focuses on the effect of changing the location of a sole engine along the wing span for a twin-engine aircraft. The reference case is the Beechcraft 76. The results are presented in Figure 9.

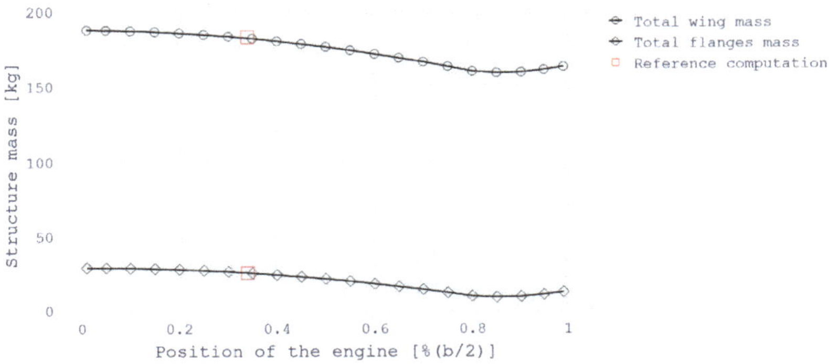

Figure 9. Effect of the location of the engine on the wing and caps mass.

The results show that the further the engine is from the fuselage, the lighter the spar caps are and thus the lighter the wing mass. This echoes results presented in Section 3.2.1 for the study of distributing the propulsion over the whole wing. The amplitude of the changes, however, is much more significant, around 16% of the baseline weight. These results lead to the same conclusion developed in [22]: moving the thrust generation on the outer part can have a positive impact on the wings' structural weight. Nevertheless, the last points of the figure may be too optimistic, as the model currently does not take into account any structural reinforcement that would be needed to withstand such a mass at the wingtip. It also omits the case of hard landing, which might constrain the stiffness of the wing.

Placing the engine outboard can have significant positive or negative impacts on the overall aircraft design. On one hand, studies on the aerodynamic effect of placing an engine at the tip have shown that clear reduction of the induced drag have been obtained [21,23]. On the other hand, the closer the engines are to the wingtips, the higher the roll inertia of the aircraft will be. This will result in either a decreased roll rate or in larger aileron, which means more complexity and weight. It can also affect the sizing of the vertical tail plane in case of an engine failure during low speed conditions. A rough estimation of the vertical area needed based on the one- engine-inoperative certification requirements was made using FAST-OAD-GA. The results show that from the baseline configuration to the case where the engine is at the tip, the required area is more than doubled. This illustrates that while wing weight savings may be obtained, the overall empty weight might increase, highlighting the importance of coupling this method with an overall aircraft design process.

3.4. Computational Time

One of the main objectives of the newly developed methodology was that, once implemented in an overall aircraft sizing tool, its computational time was to remain acceptable for conceptual and preliminary design. Numerous experiments were conducted on each use-cases configuration, and the time to converge was registered. The average results presented in Table 6 were obtained.

Table 6. Comparison of the computational time between the newly developed methodology and statistical formula.

	Semi-Analytical Model	Statistical Model
Beechcraft 76	76 ms	Virtually instantaneous
X-57 Maxwell	155 ms	
Beechcraft 1900	87 ms	

As expected, the newly developed methodology takes more times than the statistical formula suggests. However, the time is fairly low compared to some other modules of FAST-OAD-GA (mainly propeller performances and aerodynamic computations). With the default sizing process, the computation of the OEW (where the developed method is used) is entered three times in each iteration. Knowing that it takes 10 to 15 iterations to converge, the new model only adds around 7 s to the overall process. Compared to the 5–10 min needed for the whole FAST-OAD-GA process to converge, the time added by the new methodology is deemed acceptable. It is also worth noting that there is a strong dependency of the computational time on the initial estimated wing mass. During all those studies, the initial point was taken as the result provided by a statistical formula [4]. This partly explains why the process was much faster on a traditional aircraft for which both results are fairly close. By using the real wing mass of the X-57 Maxwell as the initial value, the time necessary to converge on the wing mass was decreased to about 120 ms. There is, however, a difficulty in repeatedly predicting an initial wing mass that is close to the results. This implies that for unusual wing configurations, the process might take longer to converge than for a conventional design.

4. Discussion

The newly developed methodology has proved to be accurate for a conventional twin-engine aircraft, an innovative DEP concept and a commuter aircraft, and it has provided valuable estimations at preliminary aircraft design stage. However, results must be considered carefully. Indeed, even if FAST-OAD-GA allows those values to be changed, a certain number of the assumptions retained in this paper for the structure of the wing (e.g., height of the spar caps, wingbox depth) are based on commonly used values, which may be a source for errors and would require attention from aircraft designers. Some of the computations could also be further improved, namely the ribs' mass, which is currently obtained with an assumed constant spacing. This could be improved by considering, for instance, that the ribs are sized to prevent the wing from buckling [10,24].

For the skin weight, early results in conventional design seemed to show that the deformation criterion was more constraining than the stress one. However, the use of leading edge propellers placed further from the leading edge may reverse this common tendency. This position has been identified as a major design parameter when considering the sizing of a wing with distributed propulsion [16]. It also plays a major role in the computation of the torsion moment and has an impact on the aerostructural properties of the wing by moving the section's center of gravity forward [25].

On the topic of aero-structural interaction, it was considered that for every aircraft using distributed propulsion, an additional 10% should be added to the wing mass to prevent flutter. This percentage was the mean value of several experiments on different DEP configurations [14]. Further improvements could be added to the presented methodology by running tailored aero-structural analysis to better estimate this coefficient. This, however, would come with an increased computational time and would be an optional module in FAST-OAD-GA.

5. Conclusions

In this paper, and in the context of the continuous improvement of FAST-OAD-GA, a methodology has been developed to estimate the structural mass breakdown of a wing that while taking into account the effect of distributed propulsion. This was done by adapting a semi-analytical methodology to take into account a realistic load distribution. This new method was specifically designed for aircraft covered by the EASA CS-23 certification specification and was adapted for use in the conceptual and preliminary design stages. Good agreement with actual available aircraft data or previous methodology was shown, and documented effects could be replicated, which justifies its use inside the mass estimation step of the overall aircraft design process. The novelty of this method comes not only from the adaptation of the models to conform to CS-23 and to enable the consideration of DEP, but also from the work that has been done regarding its ease of integration inside an overall aircraft sizing tool. Thus, the present work enables a preliminary consideration of the structural synergies brought forth by DEP. In addition to that and thanks to FAST-OAD-GA's API, this model can be used in any Python-based OAD tools.

Supplementary Materials: FAST-OAD is available online at https://github.com/fast-aircraft-design/FAST-OAD, accessed on 26 November 2021; FAST-OAD-GA is available online at https://github.com/supaero-aircraft-design/FAST-GA, accessed on 26 November 2021.

Author Contributions: Conceptualization E.B., J.J. and R.A.C.; methodology, R.A.C. and F.L.; software, R.A.C. and F.L.; validation, R.A.C. and F.L.; investigation, R.A.C. and F.L.; data curation, F.L.; writing—original draft preparation, R.A.C., F.L., E.B. and J.J.; writing—review and editing, F.L., R.A.C. and J.J.; visualization, F.L.; supervision, J.J. and E.B. All authors have read and agreed to the published version of the manuscript.

Funding: The presented work is supported by Daher through the research chair ISAAR "Innovative Solutions for Aviation Architecture & Regulation".

Institutional Review Board Statement: Not applicable.

Informed Consent Statement: Not applicable.

Data Availability Statement: The data presented in this study are available in the FAST-OAD-GA repository in Supplementary Materials and in the article.

Acknowledgments: The authors acknowledge the support of Daher through the research chair ISAAR "Innovative Solutions for Aviation Architecture & Regulation".

Conflicts of Interest: The authors declare no conflict of interest.

Abbreviations

The following abbreviations are used in this manuscript:

AR	Aspect Ratio
DEP	Distributed Electric Propulsion
EASA	European Union Aviation Safety Agency
FAST	Future Aircraft Sizing Tool
GA	General Aviation
HLP	High-Lift Propeller
MTOW	Maximum Take-Off Weight
OAD	Overall Aircraft Design
OEW	Operating Empty Weight
OEI	One Engine Inoperative
WTP	WingTip Propeller

Appendix A. Statistical Formula Used for Methodology Validation

This formula taken from [4] estimates the wing mass based on the wing area S_w in ft^2, the fuel weight inside the wing W_{fw} in lb, the aspect ratio A, the sweep angle Λ at 25% MAC in rad, the dynamic pressure at cruise q in lb/ft^2, the taper ratio λ, the thickness-to-chord ratio t/c, the ultimate load factor N_z, and the flight design gross weight W_{dg} in lb.

$$W_{wing} = 0.036 S_w^{0.758} W_{fw}^{0.0035} \left(\frac{A}{cos^2\Lambda}\right)^{0.6} q^{0.006} \lambda^{0.04} \left(\frac{100t/c}{cos\Lambda}\right)^{-0.3} (N_z W_{dg})^{0.49} \tag{A1}$$

The formula from [5] computes the wing mass based on the ultimate load factor N_z, the flight design gross weight W_{dg} in lb, the aspect ratio A, the sweep angle Λ at 25% MAC in rad, the wing area S_w in ft^2, the taper ratio λ, the thickness-to-chord ratio t/c, and the maximum level airspeed at sea level V_H in KEAS.

$$W_{wing} = 96.948 \left[\left(\frac{N_z W_{dg}}{10^5}\right)^{0.65} \left(\frac{A}{cos^2\Lambda}\right)^{0.57} \left(\frac{S_W}{100}\right)^{0.61} \left(\frac{1+\lambda}{2t/c}\right)^{0.36} \sqrt{1+\frac{V_H}{500}} \right]^{0.993} \tag{A2}$$

The formula from [12] used for the Beechcraft 1900 uses the maximum zero fuel weight W_{MZF} in lb, the span b in ft, the sweep angle Λ_{50} at 50% MAC in rad, the ultimate load factor N_z, the wing area S_w in ft^2, and the maximum thickness at wing root t_r in ft.

$$W_{wing} = 0.0017 W_{MZF} \left(\frac{b}{cos\Lambda_{50}}\right)^{0.75} \left[1+\sqrt{\frac{6.3cos\Lambda_{50}}{b}}\right] N_z^{0.55} \left(\frac{bS_W}{t_r W_{MZF} cos\Lambda_{50}}\right)^{0.30} \tag{A3}$$

Appendix B. Reference Technical Data for the X-57 Maxwell and the Beechcraft 76

Table A1. Aircraft weights.

Parameter	Unit	X-57 Maxwell	Beechcraft 76	Beechcraft 1900
MTOW	kg	1360	1769	7688
MZFW	kg	1360	1381.7	6804
Wing mass	kg	166.7	-	-
Design fuel	kg	0	387.3	884

Table A2. Aircraft geometry.

Parameter	Unit	X-57 Maxwell	Beechcraft 76	Beechcraft 1900
Fuselage width	m	1.22	1.20	1.38
Wing span	m	9.66	11.58	17.67
Aspect ratio	-	15	7.98	10.84
Wing surface	m^2	6.2	16.8	28.8
Root chord	m	0.74	1.48	2.22
Root thickness	m	0.088	0.21	0.4
Tip chord	m	0.52	1.42	0.93
Leading edge sweep	deg	1.9	0.0	0.0
Aileron chord	%c	25	26	0.28
Aileron span	%b/2	22	32	0.38
Aileron max deflection	deg	22	20	20
Airfoil thickness ratio	-	0.12	0.15	0.18
Ribs step	m	0.6	0.6	0.6

Table A3. Aircraft aerodynamics.

Parameter	Unit	X-57 Maxwell	Beechcraft 76	Beechcraft 1900
Cruise altitude	ft	8000	8000	20,000
Cl_α	rad^{-1}	6.1	5.55	7.0
Cl_{max}	-	4.2	1.97	2.10

Table A4. Aircraft design airspeeds (EAS).

Parameter	Unit	X-57 Maxwell	Beechcraft 76	Beechcraft 1900
V_c	$m{\cdot}s^{-1}$	78.19	78.63	118.27
V_d	$m{\cdot}s^{-1}$	97.7	110.0	163.02
V_a	$m{\cdot}s^{-1}$	58.17	64.75	90.38

Table A5. Propulsion.

Parameter	Unit	X-57 Maxwell	Beechcraft 76	Beechcraft 1900
Cruise engines	n°	2	2	2
HLP engines	n°	12	0	0
Cruise engine power	kW	60.0	130.0	953
HLP power	kW	10.5	-	-
Cruise engines weight	kg	53.1	175.0	325.4
HLP weight	kg	6.8	-	-
Cruise propeller dia	m	1.5	1.93	2.78
HLP dia	m	0.58	-	-

Table A6. Mission.

Parameter	Unit	X-57 Maxwell	Beechcraft 76	Beechcraft 1900
q	lb/ft^2	61.48	72.44	-
V_H	KEAS	150	153	-
N_z	-	6.12	5.7	4.5

References

1. Moore, K.R.; Ning, A. Distributed electric propulsion effects on existing aircraft through multidisciplinary optimization. In Proceedings of the 2018 AIAA/ASCE/AHS/ASC Structures, Structural Dynamics, and Materials Conference, Kissimmee, FL, USA, 8–12 January 2018; p. 1652.
2. Ko, A.; Schetz, J.A.; Mason, W.H. Assessment of the potential advantages of distributed-propulsion for aircraft. In Proceedings of the XVI International Symposium on Air Breathing Engines (ISABE), Cleveland, OH, USA, 31 August–5 September 2003.
3. Gudmundsson, S. *General Aviation Aircraft Design: Applied Methods and Procedures*; Butterworth-Heinemann: Oxford, UK; Waltham, MA, USA, 2013.
4. Raymer, D.P. *Aircraft Design: A Conceptual Approach*; AIAA Education Series; American Institute of Aeronautics and Astronautics: Reston, VA, USA, 2012.
5. Nicolai, L.M.; Carichner, G.E. *Fundamentals of Aircraft and Airship Design, Volume 1—Aircraft Design*; American Institute of Aeronautics and Astronautics: Reston, VA, USA, 2010.
6. Dorbath, F. A Flexible Wing Modeling and Physical Mass Estimation System for Early Aircraft Design Stages. Ph.D. Thesis, Technische Universität Hamburg, Hamburg, Germany, 2014.
7. Wells, D.P.; Horvath, B.L.; McCullers, L.A. The Flight Optimization System Weights Estimation Method. 2017. Available online: https://ntrs.nasa.gov/api/citations/20170005851/downloads/20170005851.pdf (accessed on 17 November 2021).
8. Bindolino, G.; Ghiringhelli, G.; Ricci, S.; Terraneo, M. Multilevel structural optimization for preliminary wing-box weight estimation. *J. Aircr.* **2010**, *47*, 475–489. [CrossRef]
9. Torenbeek, E. *Development and Application of a Comprehensive, Design-Sensitive Weight Prediction Method for Wing Structures of Transport Category Aircraft*; Delft University of Technology, Faculty of Aerospace Engineering, Report LR-693; Delft University of Technology: Delft, The Netherlands, 1992.
10. Roux, E. Modèle de Masse Voilure: Avions de Transport Civil. Ph.D. Thesis, SupAéro-ONERA, Toulouse, France, 2006.
11. David, C.; Delbecq, S.; Defoort, S.; Schmollgruber, P.; Benard, E.; Pommier-Budinger, V. From FAST to FAST-OAD: An open source framework for rapid Overall Aircraft Design. In *IOP Conference Series: Materials Science and Engineering*; IOP Publishing: Bristol, UK, 2021; Volume 1024, p. 012062.
12. Roskam, J. *Airplane Design: Part 5—Component Weight Estimation*; DARcorporation: Lawrence, KS, USA, 1985.
13. Litherland, B.; Rieth, K. *VSP Aircraft Analysis User Manual*; NASA: Washington, DC, USA, 2014.
14. An, S. Aeroelastic Design of a Lightweight Distributed Electric Propulsion Aircraft with Flutter and Strength Requirements. Ph.D. Thesis, Georgia Institute of Technology, Atlanta, GA, USA, 2015.
15. EASA. Certification Specifications and Acceptable Means of Compliance for Normal, Utility, Aerobatic, and Commuter Category Aeroplanes CS-23 Amendment 4. July 2015. Available online: https://www.easa.europa.eu/sites/default/files/dfu/CS-23%20Amendment%204.pdf (accessed on 17 November 2021).
16. Patterson, M.D. Conceptual Design of High-Lift Propeller Systems for Small Electric Aircraft. Ph.D. Thesis, Georgia Institute of Technology, Atlanta, GA, USA, 2016.
17. Borer, N.K.; Patterson, M.D. X-57 High-Lift Propeller Control Schedule Development. In Proceedings of the AIAA AVIATION 2020 FORUM, Virtual Event, 15–19 June 2020; p. 3091.
18. Ruscio, J.P.; Jezegou, J.; Benard, E.; Gomez Pacheco, A.; Laonet, P.; Alonso Castilla, R. Hybrid electric distributed propulsion overall aircraft design process and models for general aviation (FAST GA). In *IOP Conference Series: Materials Science and Engineering*; IOP Publishing: Bristol, UK, 2021; Volume 1024, p. 012072.
19. Li, W. Overview of the X-57 Structural Requirements, Modifications, and Airworthiness. 2019. Available online: https://ntrs.nasa.gov/api/citations/20190026541/downloads/20190026541.pdf (accessed on 17 November 2021).
20. Moore, M.D. *Distributed Electric Propulsion (DEP) Aircraft*; NASA Langley Research Center: Hampton, VA, USA, 2012.
21. Deere, K.A.; Viken, J.K.; Viken, S.; Carter, M.B.; Wiese, M.; Farr, N. Computational analysis of a wing designed for the X-57 distributed electric propulsion aircraft. In Proceedings of the 35th AIAA Applied Aerodynamics Conference, Denver, CO, USA, 5–9 June 2017; p. 3923.
22. Habermann, A.L. Effects of Distributed Propulsion on Wing Mass in Aircraft Conceptual Design. In Proceedings of the AIAA AVIATION 2020 FORUM, Virtual Event, 15–19 June 2020; p. 2625.
23. Sinnige, T.; van Arnhem, N.; Stokkermans, T.C.; Eitelberg, G.; Veldhuis, L.L. Wingtip-mounted propellers: Aerodynamic analysis of interaction effects and comparison with conventional layout. *J. Airc.* **2019**, *56*, 295–312. [CrossRef]
24. Niu, C.; Niu, M.C. *Airframe Structural Design: Practical Design Information and Data on Aircraft Structures*; Conmilit Press: Hong-Kong, China, 1988.
25. Megson, T.H.G. *Aircraft Structures for Engineering Students*; Butterworth-Heinemann: Oxford, UK; Waltham, MA, USA, 2016.

aerospace

MDPI

Article

Multi-Fidelity Optimization of a Composite Airliner Wing Subject to Structural and Aeroelastic Constraints

Angelos Kafkas [1,*,†], Spyridon Kilimtzidis [2,†], Athanasios Kotzakolios [2,†], Vassilis Kostopoulos [2,†] and George Lampeas [1,†]

1 Laboratory of Technology and Strength of Materials, Mechanical Engineering and Aeronautics Department, University of Patras, Rio Campus, 26500 Patras, Greece
2 Applied Mechanics Laboratory, Mechanical Engineering and Aeronautics Department, University of Patras, Rio Campus, 26500 Patras, Greece
* Correspondence: angel.kafkas@upnet.gr
† These authors contributed equally to this work.

Abstract: Efficient optimization is a prerequisite to realize the full potential of an aeronautical structure. The success of an optimization framework is predominately influenced by the ability to capture all relevant physics. Furthermore, high computational efficiency allows a greater number of runs during the design optimization process to support decision-making. The efficiency can be improved by the selection of highly optimized algorithms and by reducing the dimensionality of the optimization problem by formulating it using a finite number of significant parameters. A plethora of variable-fidelity tools, dictated by each design stage, are commonly used, ranging from costly high-fidelity to low-cost, low-fidelity methods. Unfortunately, despite rapid solution times, an optimization framework utilizing low-fidelity tools does not necessarily capture the physical problem accurately. At the same time, high-fidelity solution methods incur a very high computational cost. Aiming to bridge the gap and combine the best of both worlds, a multi-fidelity optimization framework was constructed in this research paper. In our approach, the low-fidelity modules and especially the equivalent-plate methodology structural representation, capable of drastically reducing the associated computational time, form the backbone of the optimization framework and a MIDACO optimizer is tasked with providing an initial optimized design. The higher fidelity modules are then employed to explore possible further gains in performance. The developed framework was applied to a benchmark airliner wing. As demonstrated, reasonable mass reduction was obtained for a current state of the art configuration.

Keywords: wing design optimization; structural optimization; composite materials; computational fluid dynamics; multi-fidelity optimization

Citation: Kafkas, A.; Kilimtzidis, S.; Kotzakolios, A.; Kostopoulos, V.; Lampeas, G. Multi-Fidelity Optimization of a Composite Airliner Wing Subject to Structural and Aeroelastic Constraints. *Aerospace* **2021**, *8*, 398. https://doi.org/10.3390/aerospace8120398

Academic Editors: Spiros Pantelakis, Andreas Strohmayer and Liberata Guadagno

Received: 17 October 2021
Accepted: 12 December 2021
Published: 15 December 2021

1. Introduction

Ever since the early years of aviation, the design, analysis, and optimization of aircraft structures has been receiving ever-increasing attention from the scientific community and has, therefore, been subjected to extensive research studies. As a result, a wide variety of multi-fidelity computational tools, with application to a plethora of analysis disciplines, has emerged. During an aircraft design process, consisting mainly of the conceptual, preliminary and detailed design stages [1], the various computational tools are designated to a corresponding design stage mainly based on the computational time, the modeling complexity and the ability to capture the physical phenomena present. Early on, low-fidelity models, associated with fast turnaround times, yet incapable of modeling higher order phenomena, are employed, along with empirical knowledge, to steer the design towards optimality. As the design knowledge on the current candidate configuration matures, higher fidelity tools are employed, aiming to replicate the relevant phenomena with greater accuracy, albeit at an elevated and often prohibitive computational cost. Throughout the stages, all relevant tools are coupled with optimization algorithms in order to gain further knowledge regarding the design space, as well as to obtain the corresponding optimized solution.

A key enabler to the mass reduction trend of aeronautical structures is the rise of composite materials in the aeronautics field. Lighter yet stiffer configurations have emerged with improved static and dynamic aeroelastic response through tailoring of their properties [2]. One of the main challenges, nevertheless, remains the optimization of such structures, since the introduction of composite materials induces further computational and algorithmic complexity. On one hand, ply angles and thicknesses can be expressed as design variables, resulting into a more complex and irregular design space, but with facilitated algorithmic applicability. On the other hand, the use of lamination parameters for optimization purposes has been extensively studied over the past few years. Miki and Sugiyama [3] was among the first to generate optimum designs of laminated composite plates for required in-plane and maximum bending stiffness, buckling strength and natural frequency utilizing lamination parameters. Fukunaga et al. [4] explored the effect of bend-twist coupling on the fundamental frequency of symmetric laminated plates, indicating that this type of coupling reduces the fundamental frequencies. The optimal laminate configuration that maximizes the fundamental frequencies was also obtained. In his work, Liu et al. [5] maximized the buckling load of composite materials panels using flexural parameters and compared its results with a stacking sequence optimization design generated via a genetic algorithm, indicating a close correlation between the two methods. In later works, efforts were directed towards closure of the up-to-then incomplete feasible design space of the lamination parameters [6], and the aeroelastic tailoring of regular and variable stiffness composite materials wings [7,8], as well as the stiffness optimization subject to aeroelastic constraints [9]. Macquart et al. [10], as well as Bordogna et al. [11], extended the capabilities of the state-of-the art lamination parameters optimization algorithms, introducing blending constraints in order to guarantee a certain degree of ply continuity inside a variable stiffness composite wing.

It becomes evident that various disciplines must be involved in the design optimization process of a modern airliner composite wing, the main contributors being the structural and aerodynamic analysis, since they significantly influence the performance and safety of wing structures. Additional disciplines may be included on a case by case basis. Regarding the structural representation, Finite Element Analysis (FEA) models are commonly used at the preliminary and detailed design stages, with reduced order, equivalent beam or plate models aiming to provide greater insight into the candidate configurations, as well as reduce the design space at an early stage of the design process. The equivalent plate methodology (EPM), often preferred due to its inherent facilitated coupling capabilities with concurrent aerodynamic and optimization codes, initially developed by Giles in a series of research papers [12–14], was proven to be a robust and accurate enough numerical tool for the structural analysis of aircraft wings, as the comparison with a 3D FEA wing model showed. Later on, Livne et al. [15] included the EPM in a multidisciplinary structural, aerodynamics and control analysis and optimization framework of a composite aircraft wing, and went on to also include transverse shear effects in the EPM formulation [16]. The large deformation non-linear static and dynamic behavior of wings via the EPM formulation was also studied by Livne and Navarro [17], with good correlation up to relatively large levels of loading being obtained. In their work, Kapania and Liu [18] introduced Legendre polynomials in the EPM in order to avoid the numerical ill-conditioning problem often accompanied by the usage of simple polynomials as trial functions. The accuracy and efficiency of the method was demonstrated via a series of static, as well as free, vibration loading scenarios. Through a series of works, Krishnamurthy and Eldred [19], Krishnamurthy and Tsai [20], Krishnamurthy [21] developed an optimization framework for obtaining an equivalent plate by minimizing an error function accounting for the displacements and frequencies between the two numerical models under consideration. The methodology was initially tested in plates with and without discrete damage and was later used to evaluate the static and dynamic structural response of a baseline aircraft wing. In each case, an optimum thickness and concentrated mass distribution for the equivalent plate model was obtained. A scaling down methodology was also developed and put into test via a series of numerical tests. The developed methodology indicated that the static response of the wing structure was accurately predicted, with the equivalent plate model being capable of reproducing the first five frequencies of the wing structure within five percent. Over the last few years, Na and Shin [22], along with Henson and Wang [23], further enhanced the current EPM capabilities to account for control surfaces, as well as for tow steered composite wing skins, respectively.

On the other hand, implementation of dedicated computational analysis tools for the relevant disciplines into optimization frameworks is of paramount importance. Pioneering research was conducted by Triplett [24] and Love and Bohlman [25], where one of the earliest multidisciplinary design and optimization tool, aeroelastic Tailoring and Structural Optimization (TSO), was developed. The Rayleigh-Ritz EPM for the structural model, coupled with the Doublet Lattice Method were combined to optimize the thickness distribution and laminate orientations of a fighter wing subject to strength and flutter velocity constraints. Haftka [26] developed an automated procedure for the design and optimization of composite wings subject to strength and flutter constraints. A major conclusion drawn from this work was that the flutter speed may not be a continuous function of the structural stiffness; hence, in optimization under flutter constraint, the constraint should be formulated in terms of other, continuous parameters relative to the flutter phenomenon. Haftka [27] also coupled a lifting-line theory-based aerodynamics model with a wing-box FEA model to perform aeroelastic analysis and optimization of a wing under stress and drag constraints. Weight versus drag trade-off studies for aluminium and composite wings were conducted. It was found that the composite wings were associated with lower mass and drag but their increased flexibility resulted in a larger variation in drag in the Pareto front. One of the earliest aerostructural optimization analysis was performed by Grossman et al. [28], where, utilizing lifting-line aerodynamic models, along with beam equations for the structural representation, the optimum shape, and structural configuration of a sailplane wing was obtained. In a similar but more complex fashion, by means of numerical tools, Grossman et al. [29] optimized a transport aircraft wing, while in parallel developing methodologies for the calculation of the sensitivity derivatives, as well as a sequential approximate optimization module.

While optimization procedures using low-fidelity components for the representation of the structural and aerodynamic problem have to date offered major gains in the overall performance of wing structures, they have inherent limitations. The limitations result from the main assumptions of such methods; for aerodynamics, non-linearities are typically neglected and simplified wetted geometries are used. On the structural side, the assumption of linear behavior, together with simplified FEA models, sees widespread use in optimization. Increasing the fidelity of the aerodynamic and structural solvers can influence the optimization results [30] and represents a step beyond the current state-of-the-art [31,32]. For the coupled aeroelastic problem, linear approaches, both in terms of structure and aerodynamics represented in the frequency domain, are commonly used. Studying the dynamic aerostructural response with such formulations should be limited to linear or weakly non-linear cases due to the underlying assumptions. The dynamic response of wings that operate in the transonic region, as well as cases of boundary layer-shockwave interaction and flutter from flow separation (common in rotor blades), requires a high-fidelity representation of the aerodynamics to be accurately captured [33–35]. Even when the aeroelastic behavior is correctly captured, the increased detail in the unsteady aerodynamic loads can influence the optimum result. In Reference [36], different unsteady aerodynamic methods were used in a multi-fidelity aerostructural optimization of a helicopter rotor, leading to variation in the final optimum blade geometry. Even for the static aeroelastic case, differences due to varying fidelity of the aerodynamics solvers have indeed been pinpointed [37]. Furthermore, the presence of structural non-linearities can additionally have a serious effect on the aeroelastic behavior of a wing structure [38]. Such non-linearities are expected to become more common in the future due to the tendency to move to ever more efficient wings of very high-aspect ratio combined with highly flexible composite materials.

It becomes evident that the obtained optimized solution corresponds to a specific level of fidelity. Depending on the case, the assumptions in modeling each discipline can be justified. However, they could also represent an oversimplification of the actual physics of the problem and yield a sub-optimal design. The optimized solution can be in doubt due to the level of fidelity with which the various disciplines are approximated regardless of how tight the optimization constraints are set and how rigorous the optimization algorithm becomes. Thus, the fidelity of each component of the optimization framework must be carefully evaluated by taking into account the outstanding properties of each problem or by confirming the validity of the solution using higher fidelity analysis.

In the following brief overview, the emphasis will be mainly upon the impact of increasing the fidelity of the aerodynamic formulation. This is a conscious choice because solution and modeling procedures for even detailed 3D Shell-3D Beam Finite Element representations of the structure have become quite efficient. However, some examples covering the impact of high-fidelity structural modeling are also mentioned below.

Gains from the inclusion of high-fidelity analysis component to the optimization of future unconventional aircraft wing concepts have been reported in literature. In References [39,40], high-fidelity Computational Fluid Dynamics (CFD) was used to provide realistic predictions of engine-pylon-strut interference and viscous drag that was fed to the optimization procedure. Significant reduction in structural mass for a strut braced wing based upon the Common Research Model (CRM) wing planform was obtained by Variyar et al. [41] by using a complete multi-fidelity optimization framework that encompassed both high-fidelity aerodynamics and structural dynamics. For a similar aircraft configuration, the accurate derivation of aerodynamic loads is stressed as a crucial parameter by Variyar et al. [41] and a multi-fidelity approach even in conceptual stage deemed advantageous. Qian and Alonso [42] have included a high-fidelity structural model in an aerostructural optimization procedure to discover gains in range and a reduction in structural mass. Smith et al. [43] mention that including high-fidelity CFD and non-linear structural effects in the aeroelastic modeling of high aspect ratio wings can lead to important differences compared to linear methods. In this particular case, linear methods were found to lead to an overly conservative result for tip displacement and aerodynamic loads.

Unsurprisingly, most recently published studies regarding high-fidelity optimization frameworks deal with innovative wing configurations that are prone to aeroelastic instabilities (high-aspect ratio wings, transonic aircraft) and may also contain structural non-linearities. Quantifying the impact of higher-fidelity optimization to current state-of-the-art wings has not been as prominent in the literature. Liem et al. [44] have shown that significant benefits from high-fidelity aerostructural optimization can also be realized for current state-of-the-art wings.

The decision on the required analysis capabilities that dictates the adequate level of fidelity has important ramifications from a computational-cost perspective. Further numerical analysis with high-fidelity tools is typically performed to assess the validity of the low-fidelity results and support the decision on the adequate level of fidelity. This approach has been used in industry and research centers [45–47]. Multiple strategies to bridge the gap in computational cost and capabilities between high and low-fidelity potential flow solutions represented by the 3D Euler Equations and simple Vortex Lattice and Doublet Lattice Method (VLM and DLM) formulations, respectively have been presented in the literature. A practical approach is to employ medium-fidelity aerodynamics. This level of fidelity is represented by advanced 3D Panel Methods. These can allow for a more detailed geometrical modeling of aeronautical structures, including non-planar aerodynamic shapes. Furthermore, by employing advanced wake modeling techniques, reasonable approximations for re-circulation effects and induced drag can be achieved at a lower cost than high-fidelity solutions. Kennedy and Martins [48] have employed a 3D

Panel Method in a parallel aerostructural optimization framework and have additionally checked the validity of the obtained aerodynamic solutions against high-fidelity ones. Mieloszyk and Goetzendorf-Grabowski [49] have used a 3D Panel method in conjunction with corrections for viscous effects to evaluate the flight dynamics stability constraints in the multidisciplinary optimization of a joined wing box configuration. Recently, a finite element beam model was coupled to 3D Panel Aerodynamics and employed in aeroelastic optimization by Conlan-Smith and Schousboe Andreasen [50].

A natural path to reduce the computational cost associated with increased fidelity is to search for an optimized solution in smaller regions of the parameter space (thus requiring a smaller number of optimization iterations). This approach requires a priori some knowledge of the region in parameter space where the optimum is likely to be detected. In our multi-fidelity optimization framework, the low-fidelity part is entrusted with obtaining an initial optimized design with a high numerical certainty. Then, the high-fidelity modules explore further gains in performance by modeling the physics that the low-fidelity modules cannot. Overall, only a fraction of the iterations using high-fidelity are performed (in the range of 5–10%). Such an approach prevents a drastic increase in computational cost. The procedure of gradually moving to increasing level of fidelity but with a smaller number of solutions is typical of a multi-fidelity optimization framework [51,52].

Still, a very high computational cost due to each high- fidelity solution is incurred, particularly so if a higher number of runs is needed. This cost is further exacerbated in the aeroelastic solutions. The cost of solving the unsteady aerodynamics with a high-fidelity method (typically a form of the Navier–Stokes equations with spatial approximation provided by the finite volume or element methods) tends to be higher than the structural solution. A combined computational cost of solving the structure and unsteady aerodynamics plus the interpolation of data between the two systems is then incurred. To mitigate this cost significantly, particularly for multiple runs around the same reference flow conditions, Order Reduction techniques can be used. Such techniques strive to approximate the physical behavior of the system on a greatly reduced (but physically meaningful) set of degrees of freedom. Most reduced order models are trained using samples from full solutions. The cost of obtaining those samples represents the main cost considering their use, since, afterward, they provide extremely rapid solution times. Several types of reduced order models (ROMs) have been used to treat aeroelasticity, including those based on (a) Mathematical decompositions, such as proper orthogonal decomposition (POD) [53,54], (b) Series approximations (for example, Volterra or Wiener) [55], and (c) Machine Learning techniques [56]. In our work, we integrate a ROM to the optimization framework that uses a Volterra series approximation for the unsteady aerodynamics coupled with mode superposition on the structural side. It serves as an intermediate step between low-fidelity aeroelastic solutions and the full coupled high-fidelity final run.

Although multiple analysis techniques of varying fidelity have been successfully used in optimization, the different computational cost that they incur combined with their setup complexity leads to a different combination of methods being adopted in each design stage of an aircraft structure. Specifically in industry, the current state-of-the-art is to include higher fidelity modules as design maturity increases (preliminary and detailed design stages), while exploiting the low computational cost of low-fidelity methods in the conceptual stage [47,57]. In the conceptual stage, efficient low-fidelity methods (panel method aerodynamics and equivalent beam or plate structural models) are commonly preferred due to the requirement to evaluate the feasibility of a vast number of design configurations.

Advances in computational capabilities combined with robust and optimized high-fidelity analysis tools has resulted in a trend to incorporate such methods already from the preliminary stage. In order to bridge the gap between capabilities and computational cost between high and low-fidelity methods, two of the most attractive methodologies consist of employing correction factors to the lower-fidelity tools [37,58], as well as constructing surrogate models. Furthermore, the construction of multi-fidelity optimization frameworks represents a highly active research topic. Such frameworks strive to combine the advantages of both worlds while mitigating the associated computational cost [47].

The primary research goal of the present study is to evaluate the gains in the structural optimization of a current state-of-the-art composite airliner wing by implementing a multi-fidelity optimization framework. Structural and aeroelastic constraints are used to formulate the optimization problem. The developed framework and the insight gained in the structural design and optimization of the composite wing model was used in the GRETEL project [59]. As a case study, a modified, planar (untwisted) version of the Common Research Model (CRM) [60] wing has been treated. The optimization framework is based on the Equivalent Plate Model (EPM) and the Mixed Integer Ant Colony Optimization (MIDACO) algorithm. At the first stage, an optimized configuration for the wing under low-fidelity aerodynamic loading is obtained. Higher fidelity aerodynamic solutions are then generated, and possible changes in the optimized solution are explored. Finally, the static aeroelastic response of the wing under consideration is also investigated. An overview of the performed actions of the developed optimization framework for the current case study is presented in Appendix B, highlighting input-output relations.

Although multiple studies in the literature assess the gains of using multi-fidelity optimization frameworks for innovative wing configurations, the gains for the current-state-of-the-art wings are less prominently covered. The present study aims to contribute to closing this gap. We demonstrate that, through the use of this optimization tool, reasonable gains in structural mass of the test case wing were realized.

2. Materials and Methods

The developed multi-fidelity structural optimization framework is comprised of several distinct modules of varying fidelity and computational cost. The flowchart for the optimization framework is presented in Figure A4. These are combined in a manner that enables efficient turnaround times, as well as an optimized solution, with high levels of certainty accounting for the most relevant physics. In the next paragraphs, each of the consisting modules of the optimization framework will be mentioned, and their contribution, advantages, and disadvantages assessed. In addition, we will then present their combination in the developed optimization framework. The path from initial design to optimized solution referring to the potential changes at each separate stage will be described.

2.1. *Low-Fidelity Modules*

The present module, consisting mainly of low-fidelity computational tools and physics approximations, initializes the design process and, hence, is responsible for narrowing down the design space, while simultaneously steering the design towards optimum solutions, as well as accelerating the design procedure. The following sub-modules are included and described in detailed fashion:

- EPM Structural Model
- Low-Fidelity Aerodynamics
- Optimization Framework

2.1.1. EPM Structural Model

Starting from the cornerstone of this work, the EPM enables a great reduction in the dimensionality of the problem since under the assumption of a relatively small thickness-to-chord ratio one can assume that an aircraft wing pertains a plate-like behavior. The problem is then reduced from a 3D representation to a 2D equivalent plate with smeared properties, reducing the associated DOFs. Plate kinematics, and specifically the First Order Shear Deformation Theory (FSDT) [61], can be utilized to describe and calculate the stress and strain state of each wing component. In particular, the wing skins, spar, and rib webs are treated as laminated plates, while, on the other hand, and as per common practice, spar and rib caps are assumed to resist axial loads only and are, thus, considered as one-dimensional rods. To obtain the governing equation of the system the principle of minimum total potential energy is applied [18]. Under the FEA method, interpolation functions are used to describe the variation of the displacement field within an element, as well as to obtain the stiffness and mass matrices, $[\boldsymbol{K}]$ and $[\boldsymbol{M}]$, respectively. Another important assumption of the EPM is that the superposition principle holds; hence, the individual stiffness and mass matrices of each wing component can be summed in order to obtain the respective wing global matrices:

$$\begin{aligned} [\boldsymbol{K}]_{wing} &= [\boldsymbol{K}]_{skin} + [\boldsymbol{K}]_{spar^{web}_{cap}} + [\boldsymbol{K}]_{rib^{web}_{cap}} \\ [\boldsymbol{M}]_{wing} &= [\boldsymbol{M}]_{skin} + [\boldsymbol{M}]_{spar^{web}_{cap}} + [\boldsymbol{M}]_{rib^{web}_{cap}} \end{aligned} \tag{1}$$

Let F be a generic function that describes the stiffness and mass distribution along the three-dimensional space for each wing component. The contribution of each EPM element to the global matrices is derived via the following integrals. Gaussian numerical integration is employed for the solution of the integrals, along with a coordinate transformation between the physical coordinates (x, y) and the generalized coordinates (ξ,η). The limits of integration in the through-the-thickness (z axis) coordinate correspond to the maximum extent of the local part of the actual wing structure.

$$\iiint_V F(x,y,z)dV = \int_{-1}^{1}\int_{-1}^{1}\left(\sum_{n=1}^{N_z}\int_{z_{in}}^{z_{jn}} F\{x(\xi,\eta),y(\xi,\eta),z\}\,|J|\,dz\right)d\xi\,d\eta, \tag{2}$$

where N_z is the number of integration points in the z-direction, z_{in} and z_{jn} the integration limits of the n-th integration point, and $|J|$ the determinant of the Jacobian of the coordinate transformation. Each wing component pertains different geometric characteristics; hence, specific treatment is required for the integration procedure:

1. Skins:

$$\iiint_V F(x,y,z)dV = \int_{-1}^{1}\int_{-1}^{1}\left(\int_{z_L-\frac{1}{2}t_L}^{z_L+\frac{1}{2}t_L} F\cdot|J|\,dz + \int_{z_U-\frac{1}{2}t_U}^{z_U+\frac{1}{2}t_U} F\cdot|J|\,dz\right)d\xi\,d\eta, \tag{3}$$

where $t_{L,U} = t_0\sqrt{1+tan^2 a_{L,U}}$, the thickness of the upper and lower skins, respectively, and $a_{L,U}$ the local airfoil angle of the lower and upper skin, as illustrated in Figure 1a. For a laminated plate, the numerical integration in the z-direction is performed for each lamina individually, along with its corresponding constitutive matrix. The rest of the components contribute in an analogous manner, albeit the integration limits are modified according to their relevant geometric characteristics, as presented in Figure 1b.

2. Spar Webs:

$$\begin{aligned} &\iiint_V F(x,y,z)dV = \int_{-1}^{1}\int_{\xi_s(n)-\frac{t_1}{c}}^{\xi_s(n)+\frac{t_1}{c}}\int_{z_L+\frac{1}{2}t_L+h_1}^{z_U-\frac{1}{2}t_U-h_1} F\{x(\xi,\eta),y(\xi,\eta),z\}\,|J|dzd\xi d\eta = \\ &\int_{-1}^{1}\int_{-1}^{1}\frac{t_1}{c}\int_{z_L+\frac{1}{2}t_L+h_1}^{z_U-\frac{1}{2}t_U-h_1} F\left\{x\left[\frac{t_1}{c}\xi+\xi_s(n),n\right],y\left[\frac{t_1}{c}\xi+\xi_s(n),n\right],z\right\}\,|J|dzd\xi d\eta \end{aligned} \tag{4}$$

3. Spar Caps:

$$
\begin{aligned}
\iiint_V F(x,y,z)dV =& \int_{-1}^{1}\int_{\xi_s(n)-\frac{l_1}{c}}^{\xi_s(n)+\frac{l_1}{c}} \left(\int_{z_L+\frac{1}{2}t_L}^{z_L+\frac{1}{2}t_L+h_1} F\{x(\xi,\eta),y(\xi,\eta),z\}\,|J|dz\right.\\
&\left.+\int_{z_U-\frac{1}{2}t_U-h_1}^{z_U-\frac{1}{2}t_U} F\{x(\xi,\eta),y(\xi,\eta),z\}\,|J|dz\right)d\xi d\eta =\\
&\int_{-1}^{1}\int_{-1}^{1}\frac{l_1}{c}\left(\int_{z_L+\frac{1}{2}t_L}^{z_L+\frac{1}{2}t_L+h_1} F\left\{x\left[\frac{l_1}{c}\xi+\xi_s(n),n\right],y\left[\frac{l_1}{c}\xi+\xi_s(n),n\right],z\right\}|J|dz\right.\\
&\left.+\int_{z_U-\frac{1}{2}t_U-h_1}^{z_U-\frac{1}{2}t_U} F\left\{x\left[\frac{l_1}{c}\xi+\xi_s(n),n\right],y\left[\frac{l_1}{c}\xi+\xi_s(n),n\right],z\right\}|J|dz\right)d\xi d\eta
\end{aligned}
\tag{5}
$$

4. Rib Webs:

$$
\begin{aligned}
\iiint_V F(x,y,z)dV =& \int_{-1}^{1}\int_{\eta_r(\xi)-\frac{t_2}{s}}^{\eta_r(\xi)+\frac{t_2}{s}}\int_{z_L+\frac{1}{2}t_L+h_2}^{z_U-\frac{1}{2}t_U-h_2} F\{x(\xi,\eta),y(\xi,\eta),z\}\,|J|dzd\xi d\eta =\\
&\int_{-1}^{1}\int_{-1}^{1}\frac{t_2}{s}\int_{z_L+\frac{1}{2}t_L+h_2}^{z_U-\frac{1}{2}t_U-h_2} F\left\{x\left[\xi,\frac{t_2}{s}\eta+\eta_r(\xi)\right],y\left[\xi,\frac{t_2}{s}\eta+n_r(\xi)\right],z\right\}\,|J|dzd\xi d\eta
\end{aligned}
\tag{6}
$$

5. Rib Caps:

$$
\begin{aligned}
\iiint_V F(x,y,z)dV =& \int_{-1}^{1}\int_{\eta_r(\xi)-\frac{l_2}{s}}^{\eta_r(\xi)+\frac{l_2}{s}} \left(\int_{z_L+\frac{1}{2}t_L}^{z_L+\frac{1}{2}t_L+h_2} F\{x(\xi,\eta),y(\xi,\eta),z\}|J|dz\right.\\
&\left.+\int_{z_U-\frac{1}{2}t_U-h_2}^{z_U-\frac{1}{2}t_U} F\{x(\xi,\eta),y(\xi,\eta),z\}|J|dz\right)d\xi d\eta =\\
&\int_{-1}^{1}\int_{-1}^{1}\frac{l_2}{s}\left(\int_{z_L+\frac{1}{2}t_L}^{z_L+\frac{1}{2}t_L+h_2} F\left\{x\left[\xi,\frac{l_2}{s}\eta+\eta_r(\xi)\right],y\left[\xi,\frac{l_2}{s}\eta+\eta_r(\xi)\right],z\right\}\,|J|dz\right.\\
&\left.+\int_{z_U-\frac{1}{2}t_U-h_2}^{z_U-\frac{1}{2}t_U} F\left\{x\left[\xi,\frac{l_2}{s}\eta+\eta_r(\xi)\right],y\left[\xi,\frac{l_2}{s}\eta+\eta_r(\xi)\right],z\right\}\,|J|dz\right)d\xi d\eta,
\end{aligned}
\tag{7}
$$

where h_1,h_2 the thickness of a spar/rib cap, l_1,l_2 the width of a spar/rib cap, and t_1,t_2 the thickness of a spar/rib web, respectively.

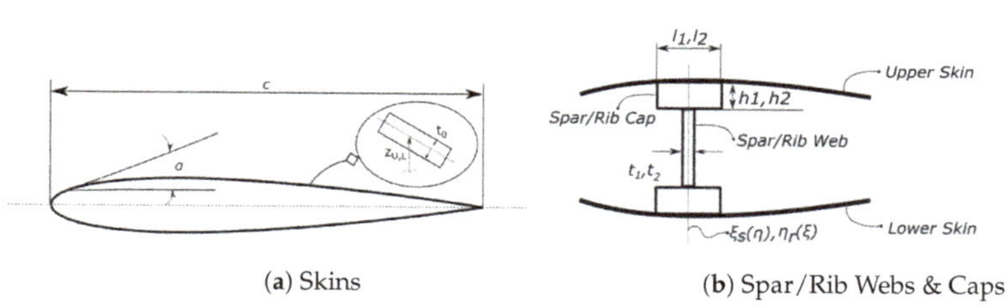

(**a**) Skins (**b**) Spar/Rib Webs & Caps

Figure 1. Wing cross-section.

2.1.2. Low-Fidelity Aerodynamics

One of the most attractive merits of the EPM is the facilitated coupling with contemporary aerodynamic panel methods, and specifically the Vortex Lattice Method (VLM) [62], since both methods are based on flat-plate representations of their respective discipline. The VLM, a well-established numerical method used in the early design stages of aircraft design, assumes an incompressible, inviscid and irrotational flow field, along with horseshoe vortex elements, in order to model an aerodynamic surface, thus neglecting the presence of thickness and viscosity in the solution. For an aerodynamic mesh consisting of N panels, at each collocation point i, the contribution of all the horseshoe vortices is added, thus generating an Aerodynamic Influence Coefficient (AIC) matrix. Introducing, also, the

Neumann boundary condition of zero normal velocity across the surface, we obtain the following set of linear equations, which are solved for the unknown vortices strength, $[\Gamma_j]$:

$$[AIC_{ij}] * [\Gamma_j] = [b_j], \tag{8}$$

where $[b_j]$ the product of the free stream velocity and the panel surface normal vectors.

Tornado VLM [63], a VLM implementation in MATLAB, has been used in our case for the generation of the critical aerodynamic loading case, corresponding to a 2.5G pull-up maneuver at sea-level conditions. Prior studies [64,65] have identified the 2.5G condition at a Mach number of 0.64 as the critical load case for the studied wing, as summarized in Table 1. Equation (8) can then be evaluated, and the aerodynamic pressure is distributed to the nodes of each element of the EPM model, since similar meshes for the structural and aerodynamic analysis have been considered. Nevertheless, more elaborate 2D interpolation schemes for load transfer purposes between the two meshes can be easily adapted, as demonstrated in subsequent modules.

Table 1. Critical aerodynamic loading summary as a function of the Maximum Take-Off Weight (MTOW).

Condition	Lift Constraint	Mach	Altitude (m)
2.5G maneuver	2.5 · MTOW	0.64	0

2.1.3. Optimization Framework

The present low-fidelity module is completed by integrating the EPM, along with the aerodynamic loading procedure, into an efficient optimization scheme for the sizing of the structural components of the wing. In a typical structural optimization scheme, the minimization of the wing structural mass is set as the objective function with stiffness, static strength, modal and dynamic aeroelastic constraints completing, and rationalizing the optimization problem. Concerning the optimization variables, the number of the 0°, (45°, −45°), 90° plies of a baseline layup, corresponding to a component of the wing, and expressed as multiples of an integer number of the material under consideration, along with the depth of the spar and rib caps, constitute the design variables vector [66]. As an example, the composite layup parametrization technique for a wing component is presented in Equation (9).

$$[0_{(n_i)}/45_{(n_{i+1})}/-45_{(n_{i+1})}/90_{(n_{i+2})}]_S \tag{9}$$

where n an integer-valued variable allowed to range between its upper and lower bounds, and i an internal variable counter, ranging from 1 to the number of variables of the optimization problem. This formulation was deemed advantageous due to a straightforward and less complex implementation than advanced lamination parameters techniques, such as References [3,4,11]. The results, expressed as the required integer number of plies to achieve an optimized percentage of different ply orientations up to the required thickness is deemed satisfactory for the purpose of this study.

Conformity to composite materials design guidelines [67], by means of generating symmetric and balanced lay-ups, is achieved by setting the number of 45° and −45° plies equal. The upper and lower bounds, along with the nature of the variables under consideration, are also presented in Table 2. The spar/rib caps depth variables are allowed to vary between their designated bounds with a 0.5 mm step. To allow for a wider design space and an increased structural design freedom, the wing is divided into 6 spanwise evenly spaced zones, as illustrated in Figure 2.

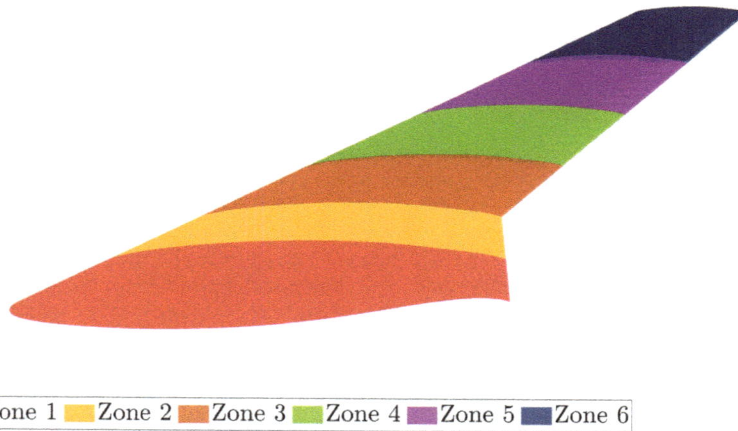

Figure 2. Optimization zones.

Table 2. Optimization variables type, lower and upper bounds.

Variable	Lower Bound	Upper Bound	Type
Ply Count, 0°	1	20	Integer
Ply Count, 90°	1	20	Integer
Ply Count, (45°,−45°)	1	20	Integer
Spar/Rib Caps Ply Count	1	10	Integer
Spar/Rib Caps Depth, mm	50	100	Integer

Regarding the constraints of the optimization problem, and although not clearly stated as design requirements per regulation, static stiffness constraints are accounted for in the majority of optimization studies of aircraft wings present. For the present study, the maximum deflection at the tip of the wing is constrained, along with a maximum twist angle induced at the tip of the wing [68]. The corresponding values are chosen to represent typical maximum deformation values reached for current state-of-the-art airliner wings [69]. Additionally, aeroelastic analysis was carried out to ensure safety against divergence and flutter. To avoid possible local structural designs during the optimization process and to ensure the required amount of modes for an accurate dynamic aeroelastic analysis, the first eigenfrequency is also constrained. The static strength of the structure is also examined via the following algorithm:

- Extraction of nodal displacements.
- Calculation of the membrane and curvature strains.
- Calculation of strains at various wing cross-sections via plate kinematic equations.
- Calculation of forces and moments per unit length.
- Given the forces and moments, the material strength values, and the maximum stress criterion, calculate the failure indices (FI) for each ply and conduct a First-Ply-Failure (FPF) analysis.

To avoid local maximum stress driven designs, while simultaneously keeping the number of constraints for the optimization problem to a minimum, the constraint aggregation technique of Kreisselmeier–Steinhauser (KS) has been employed [70,71]. For an optimization problem consisting of N_c constraints, g, having a maximum value of g_{max}, the KS functions are of the following form:

$$KS(g_j) = g_{max} + \frac{1}{\rho} ln[\sum_{j}^{N_c} exp(\rho(g_j - g_{max}))] \leq 1 \tag{10}$$

and are formed separately for the calculated FI of each of the components involved in the static strength evaluation procedure. The aggregation parameter ρ is set to 100 for all KS functions. This value has been reported in structural optimization studies of the CRM wing [72,73] and is shown to provide accuracy of the optimal solution. The aeroelastic stability by means of flutter speed is also investigated for each candidate design. A MATLAB code developed by NASA, EZ-ASE [74], is utilized for generating the unsteady aerodynamic loads calculation and conducting the flutter analysis in the frequency domain via the p-k method. EZ-ASE features flat-plate aerodynamics and particularly the VLM and DLM, thus allowing for straightforward coupling with the existing structural mesh of the EPM model; however, since, for higher reduced frequencies, a denser aerodynamic mesh is needed for accuracy purposes, a bilinear interpolation scheme is used within. The optimization objective function and constraints are summarized in the following, Table 3.

Table 3. Optimization problem setup.

Objective Function	**Minimize Structural Mass**
under the constraints	
Constraint Type	Limit Value
Maximum Deflection	$\leq$0.15 · Span
Tip Torsion Angle	$\leq$8°
First Eigenfrequency	$\leq$1 Hz
Flutter Speed	$\leq$1.2 · Dive speed
KS(FI), Upper Skin	$\leq$1
KS(FI), Lower Skin	$\leq$1
KS(FI), Spar Caps	$\leq$1
KS(FI), Spar Webs	$\leq$1

The flowchart of the resulting low-fidelity optimization framework is presented in the following, Figure 3. The MIDACO solver [75], adopting a combination of an extended Ant Colony optimization algorithm (ACO) [76], along with the Oracle Penalty Method [77], an advanced method developed for metaheuristic search algorithms for constraint handling of the solution process has been chosen for carrying out the optimization problem. A sequential approach with multiple runs has been adopted, implying that the predefined number of runs is divided into multiple runs pertaining different algorithmic parameters. Particularly, initial runs are mainly focused on extensive design space exploration and are accompanied by a relaxed constraint satisfaction tolerance. As the solution advances, the search becomes increasingly local by tweaking accordingly the internal FOCUS parameter that forces the MIDACO solver to focus mostly on the current best solution. In particular, the ACO algorithm implemented in MIDACO generates samples of iterates based on multi-kernel Gaussian probability density functions (PDF). For a generic variable k with upper and lower bounds x_u and x_l, respectively, the FOCUS parameter applies an upper bound for the standard deviation of a Gaussian PDF given by $\frac{x_u(k)-x_l(k)}{FOCUS}$ and $max(\frac{x_u(k)-x_l(k)}{FOCUS}, \frac{1}{\sqrt{FOCUS}})$ for continuous variables and integer variables, respectively. As a result, smaller values of the FOCUS parameter is recommended for the initial runs, with larger ones used for refinement purposes. In parallel, the constraint satisfaction tolerance is tightened. At each succeeding run, the previous best solution obtained serves as the starting point for the current run. As stated earlier, this procedure is repeated for a predefined number of runs, satisfying user-defined stopping criteria. The parameters of the optimization approach are summarized at the following Table 4.

Table 4. Optimization algorithm parameters.

Run	Iterations	Tolerance	FOCUS Parameter	Starting Point
1	200	0.1	0	from scratch
2	100	0.01	10	previous solution
3	50	0.001	100	previous solution
4	50	0.001	1000	previous solution

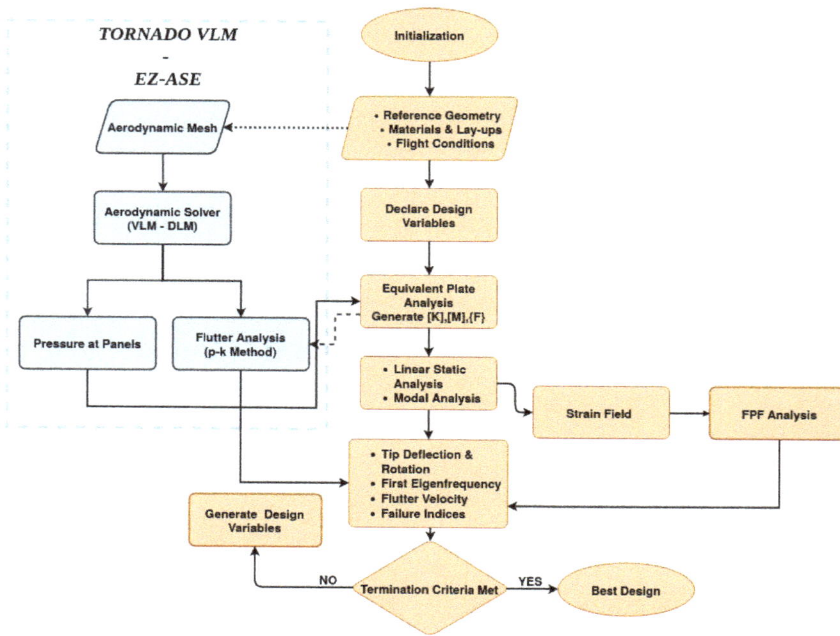

Figure 3. Low-Fidelity Module Optimization Flowchart (Figure A4).

2.2. High-Fidelity Modules

The high-fidelity analysis modules encompass aerodynamics of the undeformed body and static and dynamic aeroelasticity. The aerodynamic component is treated using the OpenFOAM framework while the high-fidelity structural system is formulated in ANSYS APDL Mechanical. The aerodynamic solution is provided by the Navier–Stokes equations discretized using the Finite Volume Method (FVM), including turbulence modeling using the Spalart-Allmaras model. Either the EPM approach described above or a 3D-shell 3D-beam FEA model is used to represent the structure. Each of those components are discussed in detail below.

2.2.1. Steady (Undeformed Body) Aerodynamic Loads Module

The main components that comprise this module are:

- High-Fidelity Finite Volume CFD Solver.
- High-Fidelity or Medium Fidelity FEA solver.
- Load Interpolation Module using General Grid Interface (GGI) method.

For the CFD solution, the 3D Navier–Stokes equations in compressible form are discretized with the finite volumes method and are solved with an implicit integration scheme. Turbulence modeling in the form of RANS methods can be included. There is voluminous bibliography on CFD with Finite Volumes Methods, to which the reader is referred for a detailed description of the respective numerical methods [78,79]. Below, only a brief overview of the CFD formulation for the studied wing geometry will be provided, with emphasis on important choices that influence the solution procedure.

The compressible Navier–Stokes equations in integral form were solved. The spatial terms of the Navier–Stokes equations are discretized on the computational finite volume mesh with appropriate discretization schemes. The mesh itself is a hybrid C-grid consisting mainly of unstructured tetrahedron volumes but using inflated prism layers near the wing's surface to achieve the first cell wall distances (Y+) necessary for capturing the boundary layer using RANS turbulence models. In particular, a Y+ of 50–70 was targeted. The CFD grid is shown in Figure 4.

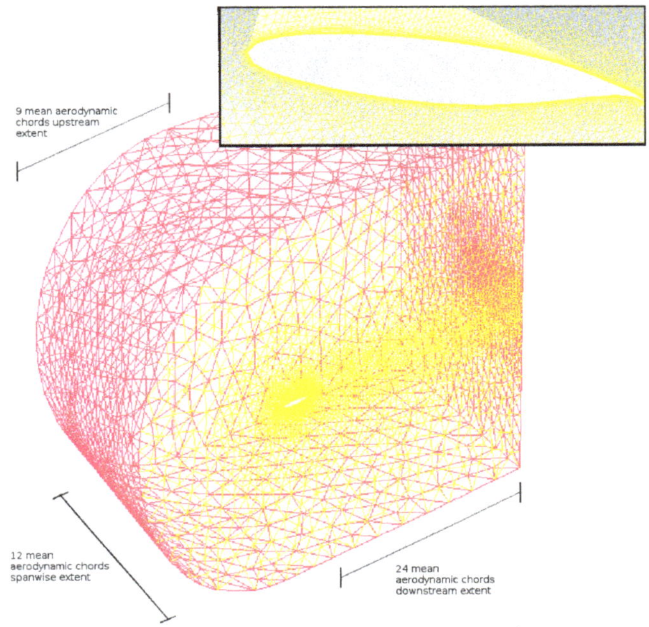

Figure 4. Finite Volume C-Grid Mesh for the coarse density (Table A4).

The unstructured internal volume mesh is advantageous for mesh deformation purposes, as will be emphasized in the description the aeroelastic module at the next stage of analysis. The Pressure Implicit Splitting of Operators (PISO) method is used to treat the problem. More details can be found in Demirdzic et al. [80].

The spatial convection terms are discretized using second order upwind schemes. These schemes offer a good balance between accuracy and mitigation of numerical oscillations. The spatial diffusion and source terms are discretized using central differences.

First order implicit time integration is used for the numerical solution. The form of the system of equations is represented in Equation (11):

$$\phi^n = \phi^{n-1} + \Delta t \cdot f(\phi^n), \tag{11}$$

where ϕ denotes any flow quantity that takes part in the Navier–Stokes equations.

The implicit formulation allows to use a wide range of timesteps with Courant numbers higher than unity. This is important for using the same mesh in the aeroelastic system as the flexibility in selecting the timestep makes the synchronization of data transfers between the structural and aerodynamic systems easier.

Turbulence was successfully treated with the Spalart-Allmaras model with wall functions providing near wall treatment. This turbulence model is integrated in most modern

CFD solvers. A first cell Y+ of 30–100 is required for this formulation. Due to the size of the wing, turbulence models that require lower values of Y+ incur a very high computational cost. The mesh convergence study at a load corresponding to a 2.5G maneuver is provided in Table A4.

For the structural component, the finite element method using either the EPM (low-fidelity) or a 3D-Shell (Shell 181 of ANSYS MEchanical APDL), 3D-Beam (Beam 188 of ANSYS MEchanical APDL) medium fidelity representation of the wing structure is used. The latter uses 4-node, 3D layered Shell Elements to discretize the wing skins, spar and rib webs, and 3D beam elements to discretize the spar and rib caps. Linear shape functions were used. The structural mesh is presented below in Figure 5.

Figure 5. Finite Element Structural Mesh.

In the current module, this component was only used to receive the steady aerodynamic loads and provide a steady state solution of the stress/strain field for optimization purposes. The interpolation of the aerodynamic loads from the CFD mesh to the structural mesh was realized using the GGI methodology [81]. This technique has seen widespread use in the transfer of loads between sliding mesh interfaces in CFD. A significant advantage compared to distance-based interpolation after projection of nodes upon a common boundary is that it takes into account the overlap between the elements. As a result, the area loads, such as pressure, are transferred consistently. This is highlighted in Figure 6 and Equation (12).

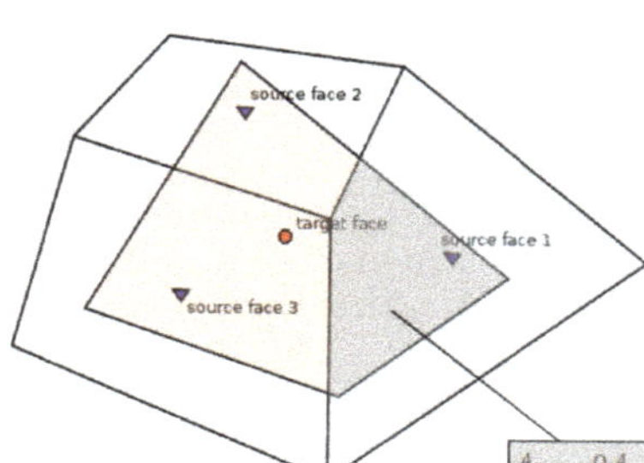

Figure 6. Transfer of pressure loads upon a common boundary with GGI.

$$P_{target\,face} = \sum_{i=1}^{n} \left(\frac{A_{target\,face} \cap A_{source\,face,i}}{A_{target\,face}} P_i \right). \tag{12}$$

A second interpolation layer is used if the loads must be transferred to the EPM. Then, after transfering them to the 3D Finite Element Mesh with the GGI approach, each node is mapped to the nearest ones of the EPM using a distance-based approach. This is less accurate, but, in any case, if maximum accuracy in interpolation is required the 3D-Shell, 3D-Beam structural formulation is preferred, and this layer not used. The complete flowchart of the High-Fidelity Steady Aerodynamic Loads Module is given in Figure 7.

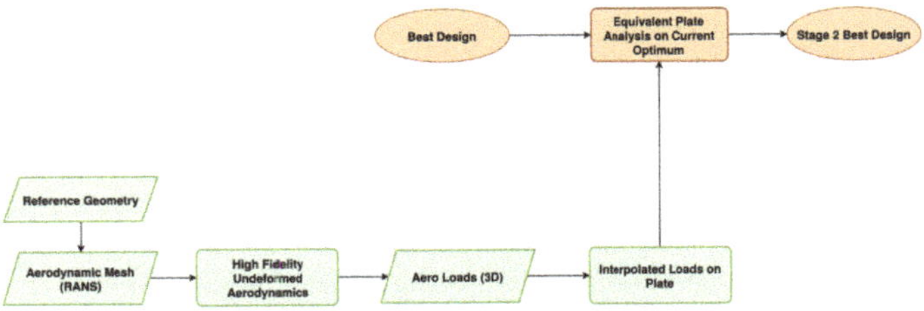

Figure 7. Steady Airloads High-Fidelity Module (Figure A4).

2.2.2. Formulation of the Aeroelastic Problem and Order Reduction

The methodology to consistently interpolate surface pressure from the CFD to the structural mesh with the GGI technique was outlined in the previous subsection. To enable a coupled aeroelastic solution, the structural displacement must additionally be passed into the wing surface boundary of the CFD mesh and the internal mesh deformed, accordingly. For passing the structural displacements to the aerodynamic mesh, distance-based interpolation is sufficient. This contrasts with surface loads (aerodynamic pressure), for which the overlap of element face areas between the structural and aerodynamic mesh is taken into account by the GGI technique. An overview of the final staggered coupling scheme is given in Figure 8.

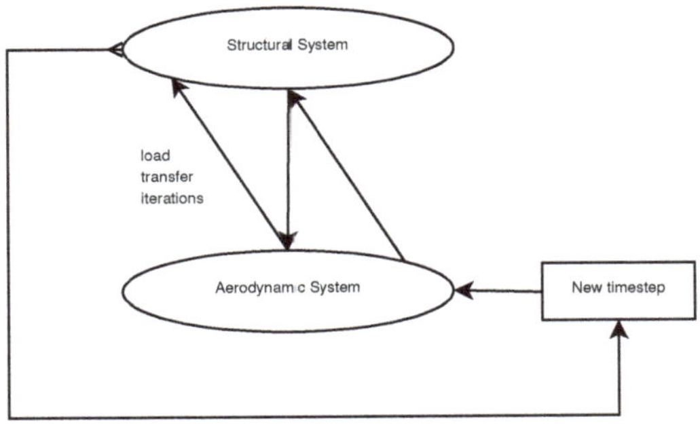

Figure 8. Staggered coupling scheme.

During deformation of the CFD mesh, the deterioration of element quality must be avoided. To that end, a Laplacian smoothing function with variable diffusivity based on distance from moving boundary is used. A detailed overview of the implementation of this mesh deformation technique, along with application examples, can be found in References [82,83]. Remeshing is only used if several cells violate the quality limits. The usage of unstructured tetrahedron internal meshes is advantageous because it is easier to maintain mesh quality upon deformation compared to structured meshes. A triangular-shaped element can retain quality upon movement of each nodes much more robustly than a rectangular one. The same applies to 3D for tetrahedrons.

Order reduction was achieved by coupling a series approximation of the unsteady aerodynamics with a modal space description of the structure. The unsteady aerodynamics component of the aeroelastic reduced order model (ROM) is based on the Volterra series theory. Volterra series ROMs for aerodynamics have been pioneered by Silva [84] and have seen implementation in complex aeroelastic problems [85].

In our work, a more basic form of a Volterra series ROM is used for unsteady aerodynamics. Linear kernels up to 50–100 timesteps in length were used to model the unsteady aerodynamic response. Although the first 2–3 non-linear kernels of the series were built, their influence was minor. This is a result of the combination of flow regime (0.64 Mach at atmospheric conditions) and the behavior of test case wing that does not generally exhibit non-linear aerodynamic phenomena at those conditions.

A Volterra series consists of linear and/or non-linear convolution integrals. The final assembled series represents an expansion of the linear convolution theorem and associates the time history of input states to the one of the outputs. For discrete systems, the convolutions are formulated as multiplications of scaled and shifted characteristic responses (or kernels of the series) with the time history of the input. For a multi-input multi-output (MIMO) system, an output state is then dependent on the input states of the system and the kernels of the series as briefly presented in Equation (13):

$$
\begin{aligned}
y_{out,i} = y_{0,i} + \sum_{j=1}^{mode\,number} \Big[\sum_{k=1}^{n} h_{ji}(n-k)\cdot x_j(k) \Big] + \\
\sum_{j=1}^{mode\,number} \Big[\sum_{k=1}^{n}\sum_{l=1}^{n} H_{ji}(n-k, n-l)\cdot x_j(k)\cdot x_j(l) \Big],
\end{aligned} \tag{13}
$$

where:
$y_{0,i}$ denotes the initial state of the ith modal coordinate,
$y_{out,i}$ is the output state of the ith modal coordinate,
x_j is the jth modal force,
h, H are the linear and non-linear kernels, respectively, and
n denotes the length of kernels in timesteps.

The Volterra series kernels were obtained by interpolating impulsive motions in each generalized coordinate to the aerodynamic system and obtaining the associated aerodynamic load response. For more information considering the construction of Volterra series models for aeroelasticity, the reader is referred to the corresponding references given in the start of this subsection.

Order reduction of the structural system was achieved by casting it in modal space and then formulating the resulting equations of motion in state-space form. Their form is provided in (14) and (15):

$$
\begin{aligned}
[\dot{x}] &= A_{ss}[x] + B_{ss}[F] \\
[y] &= C_{ss}[x] + D_{ss}[F]
\end{aligned} \tag{14}
$$

$$
\begin{aligned}
A_{ss} &= \begin{bmatrix} [0]_{nxn} & [I]_{nxn} \\ [-\omega_n^2]_{nxn} & [-2\zeta_n\omega_n]_{nxn} \end{bmatrix} \quad & B_{ss} &= \begin{bmatrix} [0]_{nxn} \\ [I]_{nxn} \end{bmatrix} \\
C_{ss} &= \big[[I]_{2nx2n}\big] \quad & D_{ss} &= \big[[0]_{2nxn}\big],
\end{aligned} \tag{15}
$$

where ω_i,ζ_i are the natural frequencies and modal damping ratios.

The assembled aeroelastic ROM, together with the associated input parameters that can be used, is presented in Figure 9 below.

The output of the ROM consists of the time history of the state variables (modal forces and modal displacements). The aeroelastic oscillation frequency at each mode is obtained by pinpointing the position of maximum for the absolute value of the Fast Fourier Transform (FFT). A window of greater length than the period of oscillation is used to scan the signal for each mode. The window moves a period of oscillation corresponding to the aeroelastic oscillation frequency at each time if it is found to be non-zero; otherwise, this is a period of oscillation corresponding to the specific structural eigenfrequency. The maximum of the absolute value of the windows is recorded. Termination is triggered when one of the following is true:

- The oscillation frequency approaches zero and the maxima of successive windows show an ascending pattern *in any mode*, indicating static divergence. This criterion exploits the fact that as divergence progresses the motion becomes non-oscillatory and the frequency approaches zero.
- The oscillation frequency approaches zero and the maxima of successive windows converge to a constant finite value within a specific predefined tolerance *in all modes*. This indicates that a static equilibrium position is reached.
- The oscillation frequency does not approach zero, but the system oscillates near a specific single frequency *in all modes*. The maxima of successive windows show an ascending pattern *in any mode*. Such behavior is indicative of flutter.

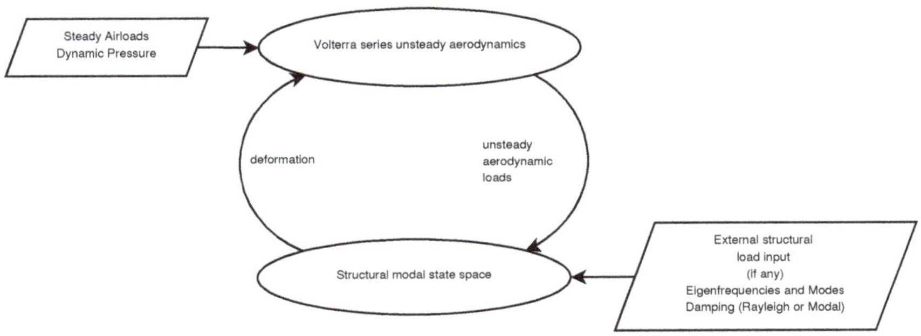

Figure 9. Aeroelastic reduced order model.

The third criterion is only used for dynamic aeroelastic analyses.

The time history of the state variables is also printed to file to allow for manual inspection. The complete module for aeroelasticity using order reduction as implemented in the interpolation framework is shown in Figure 10.

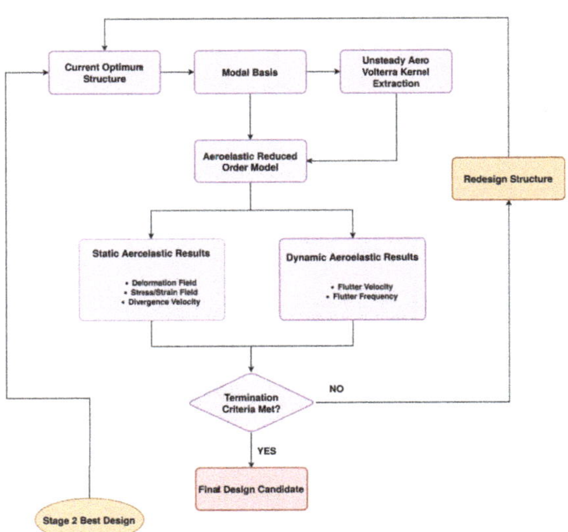

Figure 10. Implementation of aeroelasticity using order reduction to the optimization framework (Figure A4).

The final validation consists of a full order staggered (or loosely) coupled solution between high-fidelity aerodynamics and structural solvers. This full order module is run only once to validate the final optimized design candidate.

3. Results and Discussion

3.1. Benchmark Solution

A benchmark structural optimization case study is initially solved in order to enhance our confidence in the MIDACO solver. This problem refers to the optimization of the dimensions of the cross sections of a stepped cantilever beam under the application of a point load at the free end [86]. The minimization of the beam volume subject to various engineering design constraints formulate the mathematical optimization problem. In particular, the bending stress in each part of the cantilever should not exceed the maximum allowable stress, σ_{max}. The deflection at the free end is also constrained by the maximum deflection, δ_{max}. For each cross section, the height of the beam should also be twenty times less the corresponding width. The stepped cantilever beam, along with the constraints and the input data, are illustrated in the following, Figure 11.

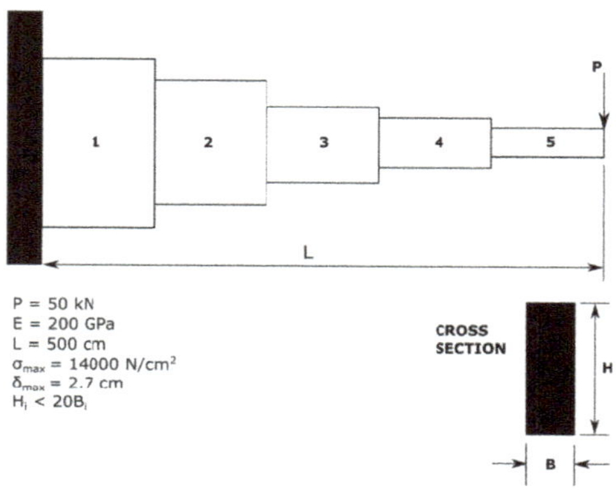

Figure 11. Stepped cantilever beam benchmark problem.

Regarding the nature of the variables of the problem, the height and width (H_i and B_i for the i-th cross-section) of the first section are integers; for the second and third section, these are chosen from a discrete set, while, for the fourth and fifth, they are continuous, resulting into a mixed-integer design problem. Several optimization techniques, including the non-linear branch and bound method, simulated annealing or genetic algorithms, and approximation methods, were investigated in Thanedar and Vanderplaats [86]. The results of the optimization study, along with those obtained by the MIDACO solver, are reported in Table 5.

Overall, the optimum height and width of each cross section, as well as the overall beam volume values, are similar to the ones obtained by the MIDACO solver. The convergence history of the objective function and of the constraints satisfaction, expressed as the L-1 norm, are illustrated in the following, Figures 12 and 13.

Table 5. Benchmark problem optimization results.

Design Variable	Continuous Optimum	Precise Discrete Optimum	Linear Approximate Discrete Optimum	Conservative Approximate Discrete Optimum	MIDACO
B1 (cm)	3.06	3	3	3	3
B2 (cm)	2.81	3.1	3.1	3.1	3.1
B3 (cm)	2.52	2.6	2.6	2.6	2.6
B4 (cm)	2.2	2.276	2.262	2.279	2.2834
B5 (cm)	1.75	1.75	1.75	1.75	1.75
H1 (cm)	61.16	60	60	60	60
H2 (cm)	56.24	55	55	55	55
H3 (cm)	50.47	50	50	50	50
H4 (cm)	44.09	45.528	45.233	45.553	45.598
H5 (cm)	35.03	34.995	34.995	35.004	34.998
Volume (cm^3)	61.110	64.537	64.403	64.558	64.562

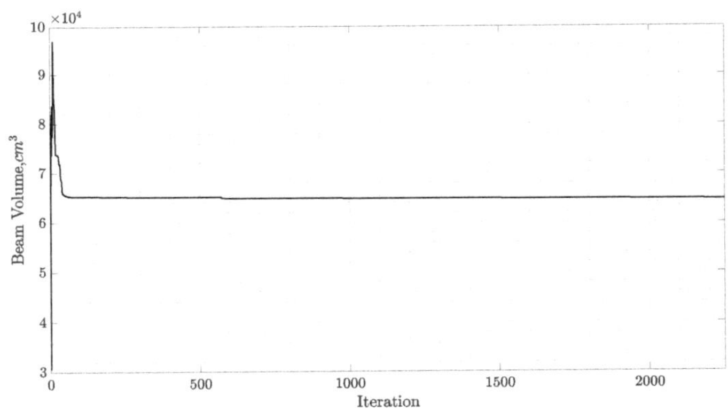

Figure 12. Stepped cantilever beam benchmark problem.

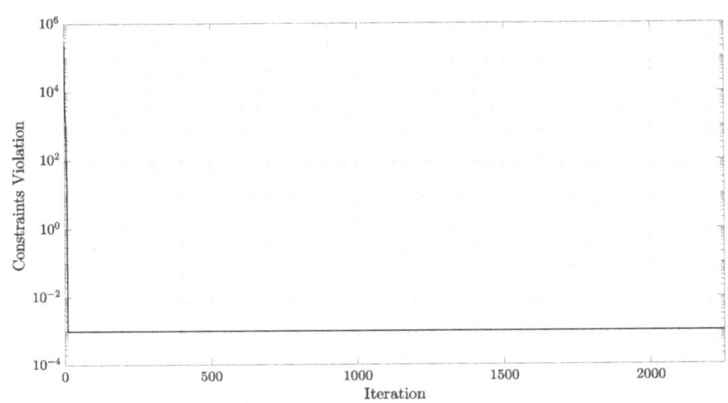

Figure 13. Stepped cantilever beam benchmark problem.

A study on the effect of tightening the constraints satisfaction tolerance value is included in Appendix C.

3.2. Main Case Study

A modified version of the original CRM wing serves as a baseline model for the analyses described earlier, with the relevant geometric parameters being presented at Table 6. The baseline CRM wing is modified by neglecting the twist distribution, the resulting geometry being planar. This modification was necessary due to the limitations of the VLM code at our disposal. Additional effort would be required for the implementation of twist angle in the low-fidelity aerodynamics software, and, since the present work constitutes a structural optimization study, it was decided to proceed with a planar representation of the original CRM wing. Furthermore, the lack of accounting for the twist distribution constitutes a limitation of the current realization of the EPM.

Table 6. CRM wing geometric data.

Wingspan	58.76 m
Root Chord	13.56 m
Tip Chord	2.73 m
Wing Gross Area	383.8 m^2
Taper Ratio	0.375
Leading Edge Sweep	35°
Yehudi Chord	7.56 m

As far as the internal configuration of the CRM wing is concerned, a three-spar configuration, with each spar laying at 10%, 30%, and 70% of the local chord, respectively, along with thirty-eight evenly spaced and aligned with the airflow ribs, similar to the one presented in Jutte et al. [87], is considered. Spar caps, along with full-depth rib caps of rectangular cross section, are also present. A view of the Outer Mold Line (OML), as well as of the internal structure of the CRM wing, can be seen in Figure 14. Concerning the material model and pertaining to all numerical models, all relevant components are assumed to be manufactured of composite materials with linear elastic behavior, and specifically the Hexcel 8552/IM7 prepreg [88] with unidirectional carbon fibers, whose properties are summarized in Table 7.

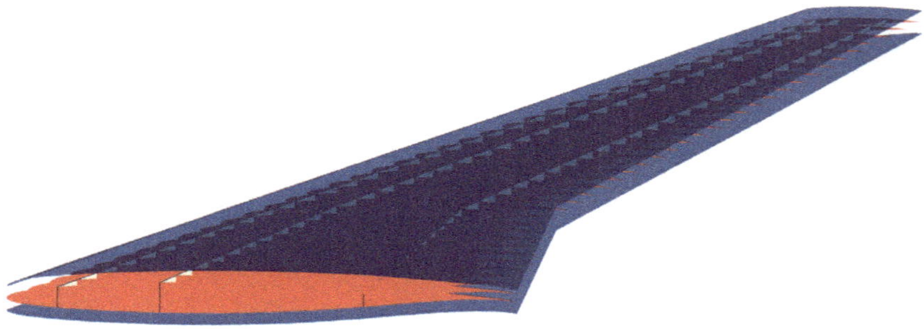

Figure 14. CRM wing OML and internal structure.

Table 7. Hexcel 8552/IM7 UD CFRP properties.

E1 [GPa]	158.5
E2 [GPa]	8.96
G12 [GPa]	4.68
G13 [GPa]	4.68
G23 [GPa]	2.63
$\nu 12$	0.356
$\nu 13$	0.356
$\nu 23$	0.5
Ply Thickness [mm]	0.183
ρ [kg/m^3]	1590

3.3. Low-Fidelity Module

The dimensionality reduction capability of the EPM in comparison with the corresponding 3D FEA model is initially investigated. As demonstrated in the following, Table 8, the EPM achieves a significant reduction of DOFs when compared with the 3D FEA model of the CRM wing, which eventually translates into faster computational time by a factor of 4.

Table 8. Numerical models size comparison.

	EPM		NASTRAN	
	Element Type	**Nr of Elements**	**Element Type**	**Nr of Elements**
Skins	QUAD-9	370	CQUAD4	14,839
Spar, Rib Webs				13,248
Spar, Rib Caps			CROD	4694
Associated DOF	7875		149,648	

The MIDACO optimizer was executed for the optimization problem presented in Table 3 with the parameters demonstrated in Table 4. In brief, the KS stress constraints represent the aggregate value of the Failure Indices of each wing component considered, with the limit value of 1 indicating failure under the maximum stress criterion. A strategy of tightening the constraints tolerance value was employed with the final value at each stage serving as an improved initial prediction for the next, as presented in Table 4. The convergence of the objective function, as well as of the constraint satisfaction, expressed via the L-1 norm, being demonstrated in the following, Figures 15 and 16.

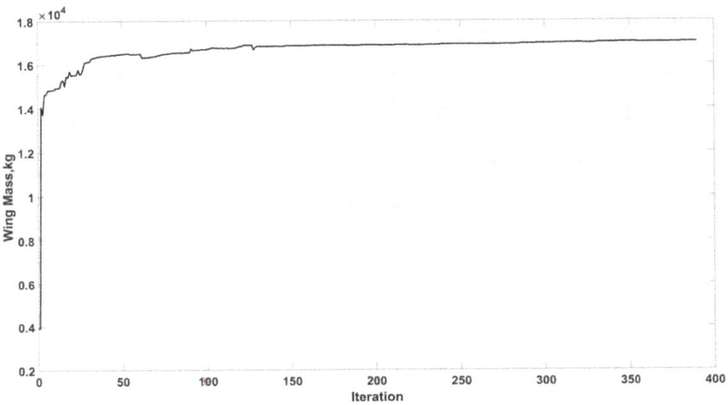

Figure 15. Wing mass convergence.

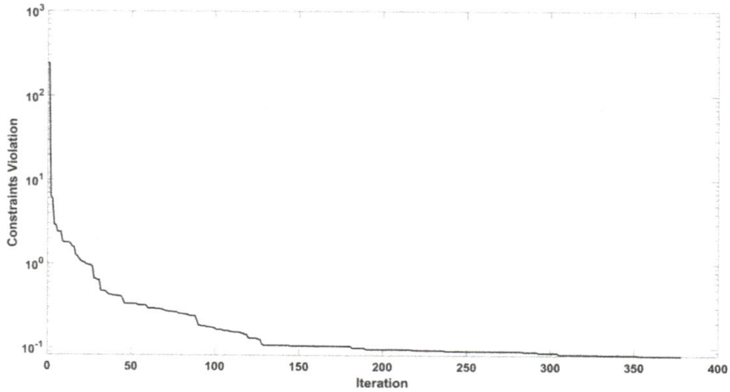

Figure 16. Constraint satisfaction, expressed via the L-1 norm.

Convergence to a minimum mass, along with non-violating design constraints, has been achieved. The corresponding mass, as well as design constraints, at the optimized solution are presented in Table 9. The values of the constraints, and especially the static stiffness and strength related ones, approach their limit values, indicating a well approximated true optimum solution.

Table 9. Low-Fidelity module optimization results summary.

Mass, kg	1.7005×10^4
Tip Deflection, m	4.1131
Tip Torsion, °	7.6
First Eigenfrequency, Hz	2.1010
Flutter Speed, m/s	509.2607
KS, Upper Skin	0.9831
KS, Lower Skin	0.9755
KS, Spar Caps	0.9766
KS, Spar Webs	0.9369

As far as the thickness distribution for the relevant wing components is concerned, a general spanwise decrease in thickness for the relevant components is demonstrated in Figures 17 and 18. Of particular interest is the fact that the aforementioned decrease appears to start at the outboard section of the wing, indicating a highly stressed region near the yehudi break, namely the coalescing region between the inboard and outboard sections of the wing. On the front of the individual thickness of each component, the thickness values associated with the wing skins are higher in comparison to the other components mainly due to the fact that they contribute more to the secondary moment of inertia of the wing. The thickness values of the spars are close to the corresponding skin thicknesses, albeit lower, being closer to the neutral axis of the wing section. Nevertheless, they take torsional loads; thus, relatively high thickness values are necessary. On the other hand, the wing ribs are associated with relatively low thickness values, as expected, since they are not the main load carrying components of an aircraft wing; they, however, aid at the relief of a portion of the developing shear stresses. The spanwise ply percentages for each component of the wing are illustrated in Figure 19.

Figure 17. CRM wing components optimal thickness distribution.

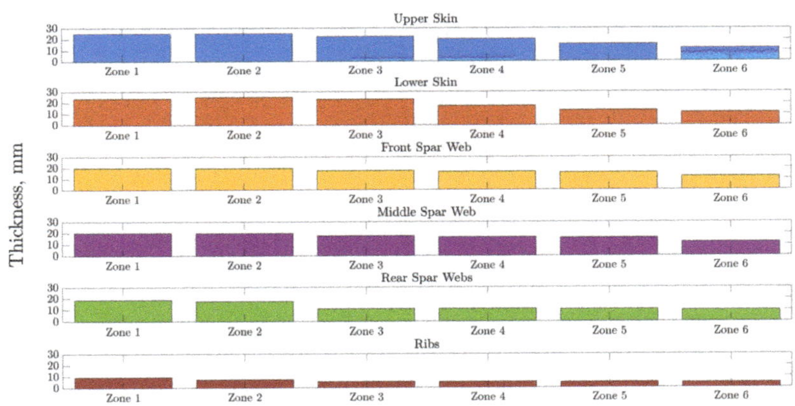

Figure 18. CRM wing components optimal spanwise thickness distribution.

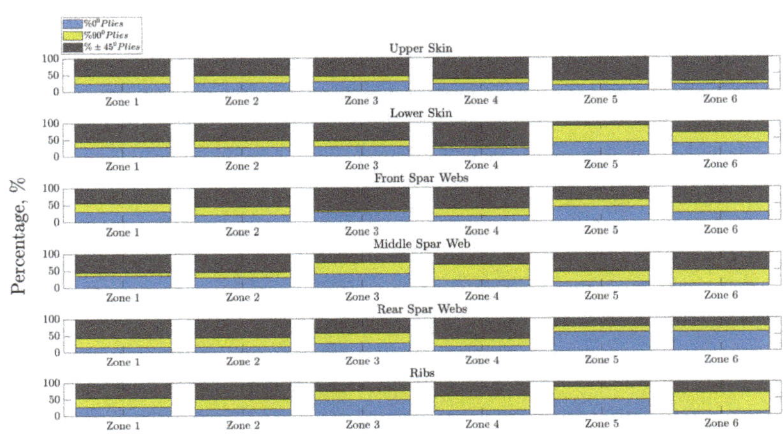

Figure 19. CRM wing components spanwise layups percentages.

3.4. High-Fidelity Module

High-fidelity aerodynamics using the RANS formulation described in Section 2.2.1 were incorporated in structural optimization. The optimization was performed for a specified lift load (2.5G maneuver condition) rather than a constant angle of attack.

A single optimization pass that includes multiple high-fidelity solutions to pinpoint the new optimized solution is conducted. The main goal is to pinpoint a good balance between loss in lift and reduced structural mass. The change to the required load for a 2.5G maneuver due to the change in wing structural mass was accounted for. For the termination of the high-fidelity aeroelastic optimization, an attained lift within 5% of the rigid wing was demanded. A tighter termination criterion for the attainable lift could have been implemented; nevertheless, due to limited computational resources, it was deemed satisfactory for the purpose of this study.

Significant changes in the required angle of attack to attain the targeted lift load were obtained by the inclusion of high-fidelity aerodynamics and are summarized in Table 10.

Table 10. Change in required angle of attack to attain a 2.5G Lift Load.

	VLM	RANS Static	RANS Static Aeroelastic
AoA	1.6	3.3	4.4

The initial optimization pass uses static structural solutions for the wing with the loads provided by high-fidelity rigid body aerodynamics. Specifically, the loads were interpolated onto the EPM mesh, providing a more realistic loadset than VLM for the static solution. Following convergence to a new optimized design, a final optimization pass that treated the coupled problem by using a high-fidelity aeroelastic formulation of the wing was performed. In this stage, the loads are transferred directly onto the wetted surface of the 3D-wing model. Both stages fail to satisfy the minimum lift that is required for the 2.5G maneuver scenario at the angle of attack of the base geometry. The inclusion of viscous effects reduces the maximum lift compared to the VLM method, the main culprit being pressure recirculation effects near the wingtip and trailing edge.

Using high-fidelity rigid body aerodynamics, an increase of the angle of attack to 3.3 degrees was deemed necessary to attain the same C_L. The static deformation of the new optimized wing geometry after the inclusion of the change in angle of attack is presented in Figure 20, for both the EPM method and a full 3D model of the structure.

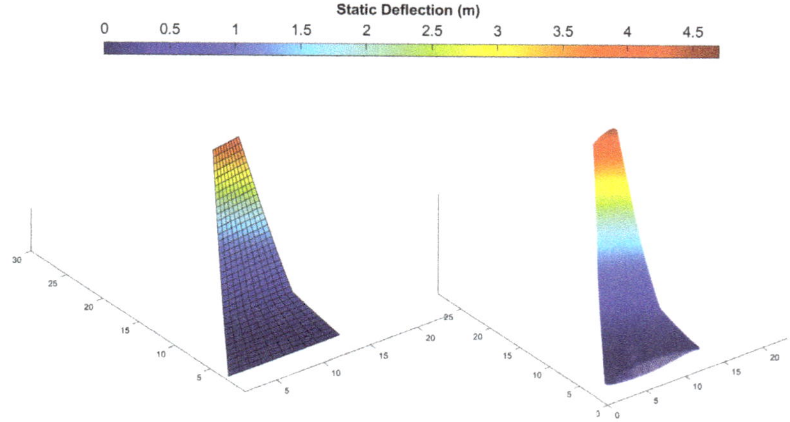

Figure 20. Deflection to static aerodynamic RANS load. EPM (**left**); 3D Shell Model (**right**).

The reasonable agreement between the two approaches points to the success of the EPM structural representation and a proper analogy between the loads on the 3D model and the ones on the plate. The structural mass remained largely unchanged after the inclusion of the new loads. Possible mass gains were deemed too small (in the region of 1–2%) to conduct further optimization iterations at this stage. From an aerodynamic perspective, however, the change in angle of attack is not negligible.

The high-fidelity aeroelastic ROM pinpointed both a further loss in lift compared to the high-fidelity static structural solution and a lower aeroelastic tip deflection. The wing candidate is characterized by strong bending-torsion coupling at structural level, and, as a result, the aeroelastic twist tends to cause negative twist angle near the wingtip upon aeroelastic deformation. The loss in lift is mainly a result of this bend-torsion coupling of the wing combined with the planar (untwisted) geometry at the undeformed state; upon aeroelastic deformation, negative twist angle is attained near the wingtip resulting in loss of lift. Near the root the aeroelastic twist angle changes are smaller; thus, a portion of the lift continues to be generated. In order to compensate for the loss of lift, a higher angle of attack for the entire wing is selected. An additional factor that leads to a reduction in lift (although minor in this specific case study due to the requirement for a reasonable bending deflection) is the curvature of the outer part of the wing due to aeroelastic bending deflection. The generated load at each section of the wing, being normal to its surface, becomes slightly tilted with respect to the vertical axis upon aeroelastic deflection. A loss of a part of the total lift ensues due to the lift load becoming tilted with respect to the vertical axis. This reduction due to bending is minor compared to the one due to change in twist angle for the current case. However, it could be significant for wings that exhibit a high magnitude of tip deflection (for example, in high aspect ratio configurations). Consequently, a further increase in AoA compared to the rigid case is required to attain the target lift load, as presented in Table 10.

Additionally, the difference between negative twist angles at the tip and much lower aeroelastic twist values towards the root also provides an opportunity for a reduction in structural mass. At the new increased angle of attack, the negative twist at the tip is increased due to the higher aeroelastic deflection magnitude and a restoring bend-torsion coupling tendency. At the same time, a much smaller change in aeroelastic twist results near the root. Thus, an even greater portion of the of the lift is generated near the root, and the bending moment is reduced significantly as a result. This reduction in bending moment at a similar lift load causes a smaller maximum deflection for the coupled problem, as presented in Figure 21, in which the static (right) and static aeroelastic (left) equilibrium deformation for the final design candidate of each stage are compared. The reduction of bending moment due to aeroelastic twist cascades to a reduction in structural mass as captured in Table 11.

Table 11. Change in structural mass.

Property	EPM *	3D Static Aeroelastic Solution *
Mass, kg	17,005	15,498

* high-fidelity aerodynamics.

In a non-planar baseline backswept wing with an optimized twist distribution, lower gains are expected. Since negative twist angles are already used near the wingtip together with a higher incidence angle at the root a lower bending moment even for the static case would be obtained compared to the planar wing results. It needs to be, therefore, stressed that the non-optimized aerodynamic shape of the considered wing (due to lack of proper twist distribution) exacerbates the reductions in structural mass that were attained in the present study.

A structural mass reduction in the order of 9% was realized using a static aeroelastic description of the wing in optimization as opposed to the static high-fidelity one. A loss of approximately 4% resulted. The final mass is compared against the static case in Table 11.

The wing was additionally checked against divergence and flutter using high-fidelity aeroelasticity. Divergence was pinpointed as the dominant aeroelastic instability, but the dynamic pressure of occurrence is more than three times higher than the operational one for a constant Mach number of 0.64. The modified wing is, therefore, Flutter- and Divergence-free at the considered point of the flight envelope, confirming the results of the low-fidelity aeroelastic analysis.

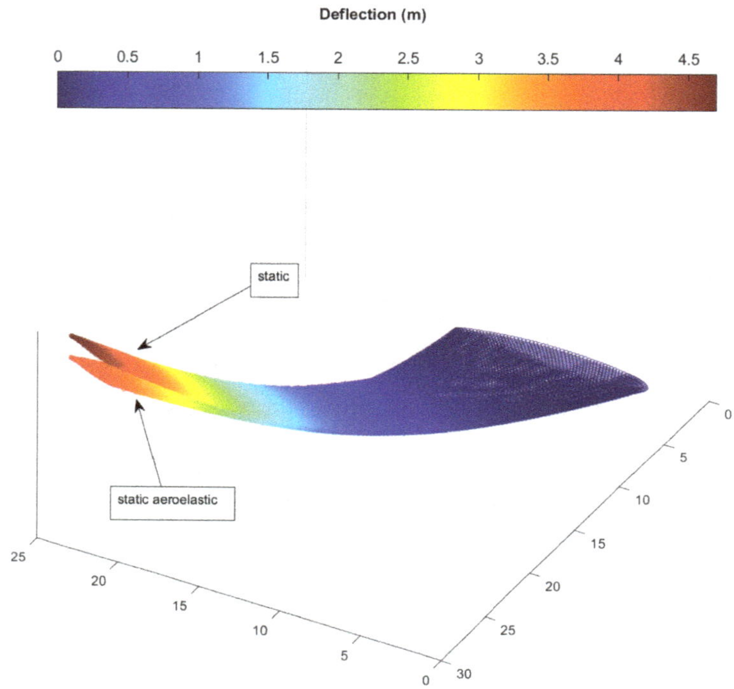

Figure 21. Comparison of static deflection distribution versus aeroelastic solution (z-axis scale not actual). Note that a higher angle of attack was required for the aeroelastic solution to attain the same coefficient of lift. Only deformation is shown in the figure; angle of attack is not plotted.

4. Conclusions

A multi-fidelity optimization framework for current state-of-the-art composite aircraft wings, based on the Mixed Integer Distributed Ant Colony Optimization (MIDACO) has been presented. The main case study consisted of a modified CRM wing with a planar geometry. The structure was represented using the Equivalent Plate Method (EPM) or a 3D shell and rod model, while aerodynamics were treated with a Vortex Lattice Method (VLM) or Reynolds Averaged Navier–Stokes (RANS) solutions, depending on level of fidelity. The aeroelastic solutions were provided by Reduced Order Models (ROMs). Upon completion of the optimization procedure, optimized layups for the various wing components were obtained, as illustrated in Figures 18 and 19. As a general trend, a spanwise decrease in thickness was observed, as expected. A region near the yehudi break was accompanied with an increase in the relevant thicknesses, indicating that this region is in fact a highly stressed one due to the slope change in geometry of the wing.

Comparison between the different fidelity modules was performed with the aim of highlighting the merits and contribution of each to the optimization process. The EPM enabled a great reduction of the associated DOFs of the problem, as discussed in Table 8 and in Section 3.3, while a good comparison with the high-fidelity structural model was demonstrated in Figure 20. Significant changes to the angle of attack to attain the target lift load resulted from the inclusion of high-fidelity aerodynamics and aeroelasticity. The differences have been attributed to negative aeroelastic twist angles near the wingtip and resulted in a lift degradation in the order of 4% and reduced bending moment which cascaded to a structural mass reduction in the order of 9%. It should be stressed that the reduction in mass is exacerbated by the planar baseline geometry of the wing since, for an optimized twist distribution, a smaller bending moment is exhibited, even in the undeformed state, as discussed in Section 3.4. The inability of the current framework to account for a realistic twist angle distribution at all levels of fidelity is acknowledged as an important shortcoming.

Overall, the present methodology presents a successful blend of current state-of-the-art multi-fidelity structural, aerodynamic, and fluid-structure interaction analysis for structural optimization. It is shown that employing such frameworks can result in performance gains even for current state-of-the-art-wings. Further possible enhancements to the current optimization framework could include the modification of the VLM and EPM code to account for twist angle distributions, implementation of buckling constraints in the high-fidelity structural model, and incorporation of aeroelastic tailoring capabilities. Another major addition would be the development aerodynamic shape optimization procedures.

Author Contributions: Conceptualization, A.K. (Angelos Kafkas) and S.K.; methodology, A.K. (Angelos Kafkas) and S.K.; software, A.K. (Angelos Kafkas) and S.K.; validation, A.K. (Angelos Kafkas) and S.K.; formal analysis, S.K., A.K. (Angelos Kafkas), A.K. (Athanasios Kotzakolios), V.K. and G.L.; investigation, S.K. and A.K. (Angelos Kafkas); resources, S.K., A.K. (Angelos Kafkas), A.K. (Athanasios Kotzakolios), G.L. and V.K.; data curation, S.K. and A.K. (Angelos Kafkas); writing—original draft preparation, S.K., A.K. (Angelos Kafkas), A.K. (Athanasios Kotzakolios), V.K. and G.L.; writing—review and editing, S.K., A.K. (Angelos Kafkas), A.K. (Athanasios Kotzakolios), G.L. and V.K.; visualization, S.K., A.K. (Angelos Kafkas), A.K. (Athanasios Kotzakolios), V.K. and G.L.; supervision, A.K. (Athanasios Kotzakolios), V.K. and G.L.; project administration, A.K. (Athanasios Kotzakolios), V.K. and G.L.; funding acquisition, S.K., A.K. (Athanasios Kotzakolios), V.K. and G.L. All authors have read and agreed to the published version of the manuscript.

Funding: This research was financially supported by the State Scholarships Foundation of Greece (IKY), which granted a scholarship to Spyridon Kilimtzidis. The scholarship is co-financed by Greece and the European Union (European Social Fund—ESF) through the Operational Programme << Human Resources Development, Education and Lifelong Learning >> in the context of the project "Strengthening Human Resources Research Potential via Doctorate Research" (MIS-5000432), implemented by the State Scholarships Foundation (IKY).

The authors also gratefully acknowledge the financial support of EU, H2020 CS2 project under the acronym GRETEL (Grant agreement ID: 737671).

Institutional Review Board Statement: Not applicable.

Informed Consent Statement: Not applicable.

Data Availability Statement: Not applicable.

Conflicts of Interest: The authors declare no conflict of interest.

Abbreviations

The following abbreviations are used in this manuscript:

AoA	Angle of Attack
CFD	Computational Fluid Dynamics
CRM	Common Research Model
EPM	Equivalent Plate Method
FEA	Finite Element Analysis
ROM	Reduced Order Model

Appendix A. EPM Validation Test Case

In this section, the accuracy and efficiency of the EPM is thoroughly investigated via a test case concerning a built-up wing composed of composite materials skins, spar/ribs webs and caps. A typical internal geometry, consisting of two spars placed at 20% and 70% of each chord and 12 evenly spaced and aligned with the airflow ribs is considered. In addition, spar and rib caps of rectangular cross-section are also present. The relative dimensions of the various components were selected such that no local modes, which the EPM is incapable of capturing, will be present to the results as much as possible. Additionally, two numerical models are developed and compared, namely a 3D FEA model in NASTRAN and its corresponding EPM model. Regarding the latter, the FEA is employed for the solution of the numerical integrals present in the Materials and Methods section, with the solution procedure being carried out entirely in MATLAB. Initially, and based on the external geometry of the corresponding wing, the geometry of the equivalent plate is generated, upon which a FEA mesh is built. To avoid shear locking phenomena and reduce the spurious mechanisms present, isoparametric nine-noded Lagrange plate elements with selective integration for the membrane, bending and shear terms have been chosen for the analysis. Eigenfrequencies and eigenmodes extracted from a free vibration analysis constitute the quantities of interest for the comparison between the two methods, since both the stiffness, as well as the mass matrix, are involved in this type of analysis. Concerning the 3D FEA mesh, the wing upper and lower skins, along with the spar and rib webs, are modeled via shell elements, (CQUAD4), while the spar and rib caps are assumed to resist axial loads only and are, thus, modeled via rod elements (CROD). For the EPM, QUAD-9 elements have been chosen to model the equivalent plate. The particulars of the two FEA meshes along are presented in Table A1, while the resulting FEA mesh for both of the methods is presented in the following Figure A1a,b. Each component pertains a [45/0/−45/90]s quasi-isotropic, balanced and symmetric layup, with similar material properties to the one presented in the Results section. Regarding the boundary conditions, the wing is considered clamped at its root for both of the numerical models, fixing all 6 DOFs.

Table A1. Numerical models comparison.

	EPM		NASTRAN	
	Element Type	**Nr of Elements**	**Element Type**	**Nr of Elements**
Skins	QUAD-9	675	CQUAD4	16,932
Spar, Rib Webs				5448
Spar, Rib Caps			CROD	1888
Associated DOF′	14,105		136,168	

The efficiency of the EPM is further emphasized for this test case, since the addition of more components (spars/ribs webs and caps) is accompanied by an increase in the DOF's for the 3D FEA, which is not the case for the EPM, where the associated DOFs remain unaltered. The natural frequencies, mode shapes (as illustrated in Figures A2 and A3) and the mass of the wing for the two methods, presented in Table A2, are also examined.

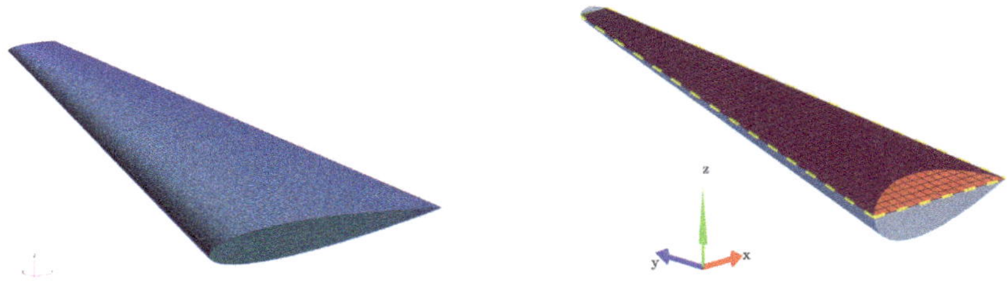

(a) 3D-FEA Mesh (b) EPM Mesh

Figure A1. FEA & EPM mesh.

Table A2. NASTRAN-EPM natural frequencies and mass.

Mode Nr	NASTRAN Value, Hz		Error, %
1	16.12	16.521	−2.492
2	61.009	61.603	−0.973
3	98.908	94.546	4.409
4	139.348	140.791	−1.036
5	160.549	166.408	−3.649
6	245.620	250.112	−1.828
7	311.941	311.121	0.262
8	325.066	326.562	−0.460
9	373.481	384.695	−3.003
10	468.341	483.222	−3.177
Total Mass, kg	250.765	244.4	2.539

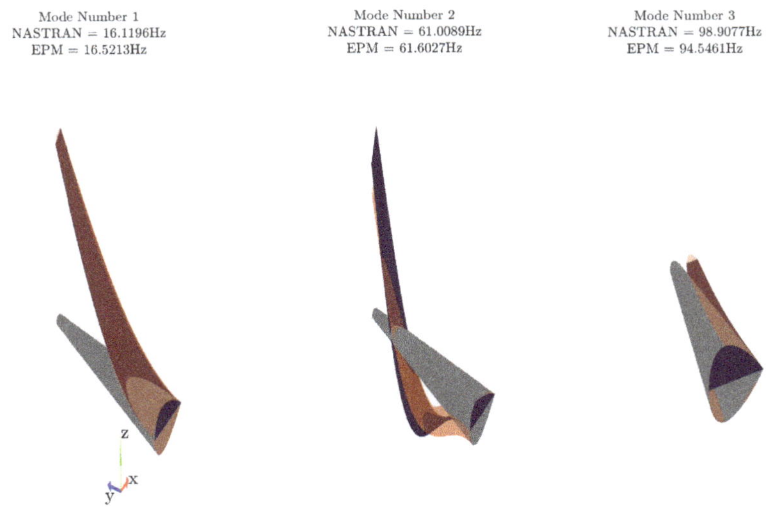

Figure A2. Eigenmodes comparison—Modes 1–3.

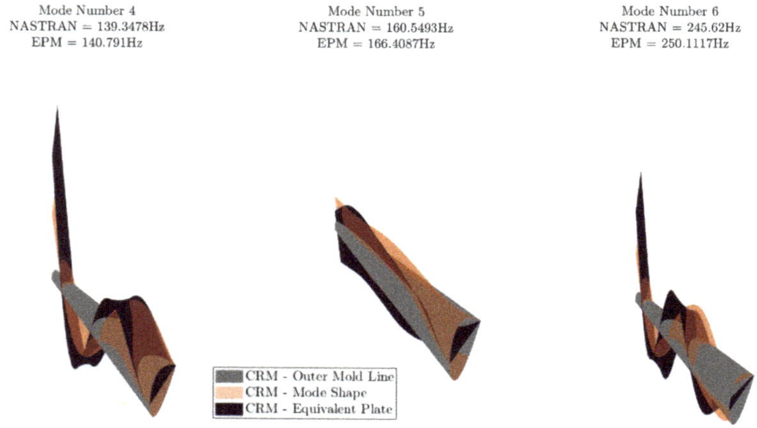

Figure A3. Eigenmodes comparison—Modes 4–6.

A close agreement between the shapes of the eigenmodes is observed in general, with the EPM being capable of predicting the trends of the 3D-FEA model.

Appendix B. Flowchart of the Developed Optimization Framework

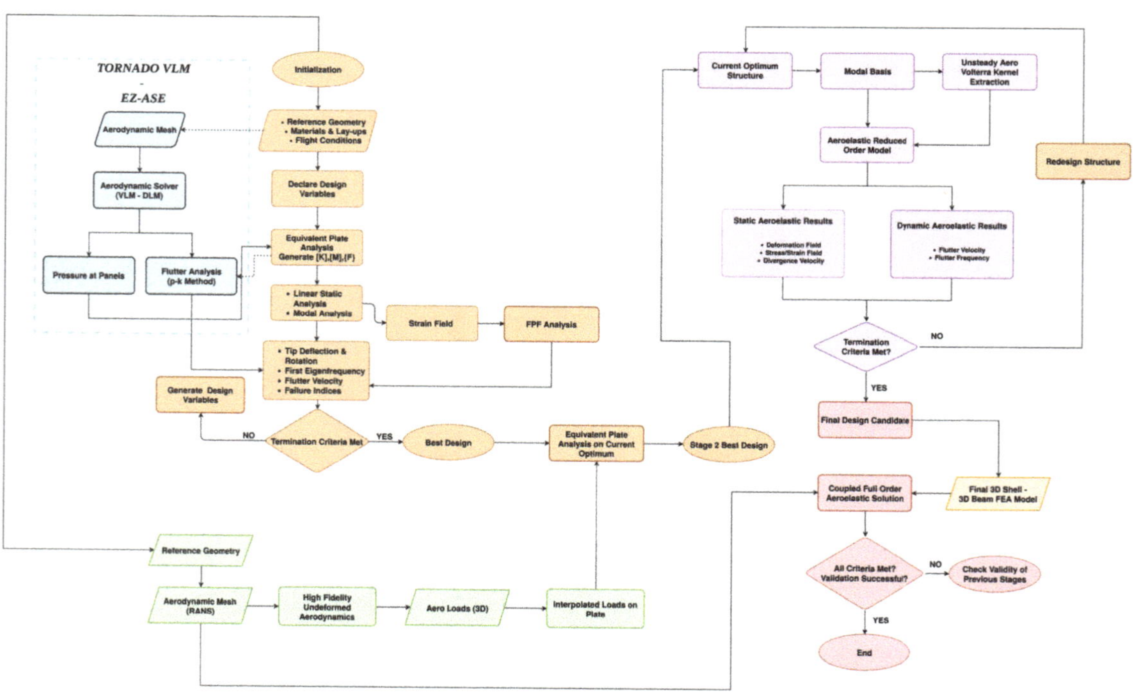

Figure A4. Flowchart of the optimization framework.

Action ID	Action Type	Participants	Triggered by	Receives Input from Action ID
1	VLM Loads	TORNADO VLM EPM	-	Reference Geometry Flight Conditions
2	MIDACO Optimization (Loose Tolerance) Obtain Initial Starting Point Prediction	MIDACO EPM EZASE	Optimization Framework Start	1
3-5	MIDACO Optimization Receive Current Best Solution Tighten Tolerance	MIDACO EPM EZASE	Initial Starting Point	1,2
6	Manual Inspection of Results			
7	High-Fidelity Structure Generation	ANSYS APDL	MIDACO Termination	5
8	High-Fidelity Loads Generation Correct AoA	OpenFoam	-	Reference Geometry Initial AoA from 1
9	Static Solution-Loads Correlation	ANSYS APDL EPM	-	5,6,7
10	MIDACO Optimization	MIDACO EPM	End of Loads Correlation Action 8	5,7
11	MIDACO High-Fidelity Optimization	MIDACO APDL	End of MIDACO Optimization Action 9	5,6,7,9
12	Manual Inspection of Results			
13-18	Build ROM MIDACO High-Fidelity Aeroelastic Optimization Pass Correct ROM ... Evaluate Divergence & Flutter Limits	MIDACO ROM	MIDACO High-Fidelity Optimization	10
19	Final Candidate Evaluation			

Figure A5. Action list of the developed optimization framework for the CRM case.

Appendix C. Benchmark Solution Tolerance Value Study

Table A3. Effect of tolerance on the optimized solution of the benchmark case study.

Constraint Tolerance Value	1×10^{-1}	1×10^{-2}	1×10^{-3}	1×10^{-4}	1×10^{-5}	1×10^{-6}	1×10^{-7}
Optimized Volume (cm^3)	63,880	64,419	64,562	64,577	64,578	64,578	64,578

Appendix D. High-Fidelity CFD Grid Convergence Study

Table A4. CFD grid convergence study to a 2.5G lift load.

	Lift (N)	Drag (N)	Grid Convergence Indexes	
Mesh 1 (1.4×10^6 Cells)	3.003×10^6	1.306×10^5	GCI (lift)	GCI (drag)
Mesh 2 (2.6×10^6 Cells)	3.027×10^6	1.295×10^5	8.68×10^{-1}	5.98×10^{-1}
Mesh 3 (5.4×10^6 Cells)	3.016×10^6	1.264×10^5	4.07×10^{-1}	1.69×10^0

References

1. Kundu, A.K. *Aircraft Design*; Cambridge University Press: Cambridge, UK, 2009. [CrossRef]
2. Shirk, M.H.; Hertz, T.J.; Weisshaar, T.A. Aeroelastic tailoring—Theory, practice, and promise. *J. Aircr.* **1986**, *23*, 6–18. [CrossRef]
3. Miki, M.; Sugiyama, Y. Optimum Design of Laminated Composite Plates Using Lamination Parameters. *AIAA J.* **1993**, *31*, 921–922. [CrossRef]
4. Fukunaga, H.; Sekine, H.; Sato, M. Optimal Design of Symmetric Laminated Plates for Fundamental Frequency. *J. Sound Vib.* **1994**, *171*, 219–229. [CrossRef]
5. Liu, B.; Haftka, R.; Trompette, P. Maximization of buckling loads of composite panels using flexural lamination parameters. *Struct. Multidiscip. Optim.* **2004**, *26*, 28–36. [CrossRef]
6. Setoodeh, S.; Abdalla, M.M.; Gürdal, Z. Design of variable–stiffness laminates using lamination parameters. *Compos. Part B Eng.* **2006**, *37*, 301–309. [CrossRef]
7. Thuwis, G.A.A.; Breuker, R.D.; Abdalla, M.M.; Gürdal, Z. Aeroelastic tailoring using lamination parameters. *Struct. Multidiscip. Optim.* **2009**, *41*, 637–646. [CrossRef]
8. IJsselmuiden, S.T.; Abdalla, M.M.; Gurdal, Z. Optimization of Variable-Stiffness Panels for Maximum Buckling Load Using Lamination Parameters. *AIAA J.* **2010**, *48*, 134–143. [CrossRef]

9. Dillinger, J.K.S.; Klimmek, T.; Abdalla, M.M.; Gürdal, Z. Stiffness Optimization of Composite Wings with Aeroelastic Constraints. *J. Aircr.* **2013**, *50*, 1159–1168. [CrossRef]
10. Macquart, T.; Maes, V.; Bordogna, M.T.; Pirrera, A.; Weaver, P. Optimisation of composite structures—Enforcing the feasibility of lamination parameter constraints with computationally-efficient maps. *Compos. Struct.* **2018**, *192*, 605–615. [CrossRef]
11. Bordogna, M.T.; Lancelot, P.; Bettebghor, D.; Breuker, R.D. Static and dynamic aeroelastic tailoring with composite blending and manoeuvre load alleviation. *Struct. Multidiscip. Optim.* **2020**, *61*, 2193–2216. [CrossRef]
12. Giles, G.L. Equivalent plate analysis of aircraft wing box structures with general planform geometry. *J. Aircr.* **1986**, *23*, 859–864. [CrossRef]
13. Giles, G.L. Further generalization of an equivalent plate representation for aircraft structural analysis. *J. Aircr.* **1989**, *26*, 67–74. [CrossRef]
14. Giles, G.L. Equivalent plate modeling for conceptual design of aircraft wing structures. In Proceedings of the Aircraft Engineering, Technology, and Operations Congress, Los Angeles, CA, USA, 19–21 September 1995; American Institute of Aeronautics and Astronautics: Reston, VA, USA, 1995. [CrossRef]
15. Livne, E.; Schmit, L.A.; Friedmann, P.P. Towards integrated multidisciplinary synthesis of actively controlled fiber composite wings. *J. Aircr.* **1990**, *27*, 979–992. [CrossRef]
16. Livne, E. Equivalent plate structural modeling for wing shape optimization including transverse shear. *AIAA J.* **1994**, *32*, 1278–1288. [CrossRef]
17. Livne, E.; Navarro, I. Nonlinear Equivalent Plate Modeling of Wing-Box Structures. *J. Aircr.* **1999**, *36*, 851–865. [CrossRef]
18. Kapania, R.K.; Liu, Y. Static and Vibration Analyses of General Wing Structures Using Equivalent-Plate Models. *AIAA J.* **2000**, *38*, 1269–1277. [CrossRef]
19. Krishnamurthy, T.; Eldred, L. Frequency Response of an Aircraft Wing with Discrete Source Damage Using Equivalent Plate Analysis. In Proceedings of the 48th AIAA/ASME/ASCE/AHS/ASC Structures, Structural Dynamics, and Materials Conference, Honolulu, HI, USA, 23–26 April 2007; American Institute of Aeronautics and Astronautics: Reston, VA, USA, 2007. [CrossRef]
20. Krishnamurthy, T.; Tsai, F. Static and Dynamic Structural Response of an Aircraft Wing with Damage Using Equivalent Plate Analysis. In Proceedings of the 49th AIAA/ASME/ASCE/AHS/ASC Structures, Structural Dynamics, and Materials Conference, Schaumburg, IL, USA, 7–10 April 2008; American Institute of Aeronautics and Astronautics: Reston, VA, USA, 2008. [CrossRef]
21. Krishnamurthy, T. Frequencies and Flutter Speed Estimation for Damaged Aircraft Wing Using Scaled Equivalent Plate Analysis. In Proceedings of the 51st AIAA/ASME/ASCE/AHS/ASC Structures, Structural Dynamics, and Materials Conference, Orlando, FL, USA, 12–15 April 2010; American Institute of Aeronautics and Astronautics: Reston, VA, USA, 2010. [CrossRef]
22. Na, Y.; Shin, S. Equivalent-Plate Analysis for a Composite Wing with a Control Surface. *J. Aircr.* **2013**, *50*, 853–862. [CrossRef]
23. Henson, M.C.; Wang, B. Efficient Methods for Design and Analysis of Tow Steered Wing Structures. In Proceedings of the 58th AIAA/ASCE/AHS/ASC Structures, Structural Dynamics, and Materials Conference, Grapevine, TX, USA, 9–13 January 2017; American Institute of Aeronautics and Astronautics: Reston, VA, USA, 2017. [CrossRef]
24. Triplett, W.E. Aeroelastic Tailoring Studies in Fighter Aircraft Design. *J. Aircr.* **1980**, *17*, 508–513. [CrossRef]
25. Love, M.; Bohlman, J. *Aeroelastic Tailoring Studies in Fighter Aircraft Design*; Technical Report; NASA Langley Research Center: Hampton, VA, USA, 1989.
26. Haftka, R.T. *Automated Procedure for Design of Wing Structures to Satisfy Strength and Flutter Requirements*; Technical Report; NASA Langley Research Center: Hampton, VA, USA, 1973.
27. Haftka, R.T. Optimization of flexible wing structures subject to strength and induced drag constraints. *AIAA J.* **1977**, *15*, 1101–1106. [CrossRef]
28. Grossman, B.; Gurdal, Z.; Strauch, G.J.; Eppard, W.M.; Haftka, R.T. Integrated aerodynamic/structural design of a sailplane wing. *J. Aircr.* **1988**, *25*, 855–860. [CrossRef]
29. Grossman, B.; Haftka, R.T.; Kao, P.J.; Polen, D.M.; Rais-Rohani, M.; Sobieszczanski-Sobieski, J. Integrated aerodynamic-structural design of a transport wing. *J. Aircr.* **1990**, *27*, 1050–1056. [CrossRef]
30. Dababneh, O.; Kipouros, T.; Whidborne, J. Application of an Efficient Gradient-Based Optimization Strategy for Aircraft Wing Structures. *Aerospace* **2018**, *5*, 3. [CrossRef]
31. Raveh, D.E. Computational-fluid-dynamics-based aeroelastic analysis and structural design optimization—A researcher's perspective. *Comput. Methods Appl. Mech. Eng.* **2005**, *194*, 3453–3471. [CrossRef]
32. Raveh, D.E.; Levy, Y.; Karpel, M. Structural Optimization Using Computational Aerodynamics. *AIAA J.* **2000**, *38*, 1974–1982. [CrossRef]
33. Cavagna, L.; Quaranta, G.; Mantegazza, P. Application of Navier–Stokes simulations for aeroelastic stability assessment in transonic regime. *Comput. Struct.* **2007**, *85*, 818–832. [CrossRef]
34. McDaniel, D.R.; Cummings, R.M.; Bergeron, K.; Morton, S.A.; Dean, J.P. Comparisons of computational fluid dynamics solutions of static and manoeuvring fighter aircraft with flight test data. *Proc. Inst. Mech. Eng. Part G J. Aerosp. Eng.* **2009**, *223*, 323–340. [CrossRef]
35. Guruswamy, G.P.; Obayashi, S. Study on the Use of High-Fidelity Methods in Aeroelastic Optimization. *J. Aircr.* **2004**, *41*, 616–619. [CrossRef]
36. Wilke, G. Variable Fidelity Optimization of Required Power of Rotor Blades: Investigation of Aerodynamic Models and their Application. In Proceedings of the 38th European Rotorcraft Forum, Amsterdam, The Netherlands, 4–7 September 2012.

37. Crovato, A.; Almeida, H.S.; Vio, G.; Silva, G.H.; Prado, A.P.; Breviglieri, C.; Guner, H.; Cabral, P.H.; Boman, R.; Terrapon, V.E.; et al. Effect of Levels of Fidelity on Steady Aerodynamic and Static Aeroelastic Computations. *Aerospace* **2020**, *7*, 42. [CrossRef]
38. Afonso, F.; Vale, J.; Oliveira, É.; Lau, F.; Suleman, A. A review on non-linear aeroelasticity of high aspect-ratio wings. *Prog. Aerosp. Sci.* **2017**, *89*, 40–57. [CrossRef]
39. Grasmeyer, J.; Naghshineh-Pour, A.; Tetrault, P.A.; Grossman, B.; Haftka, R.; Kapania, R.; Mason, W.; Schetz, J. *Multidisciplinary Design Optimization of a Strut-Braced Wing Aircraft with Tip-Mounted Engines*; Multidisciplinary Analysis and Design Center for Advanced Vehicles: Blacksburg, VA, USA, 1998.
40. Gern, F.; Gundlach, J.; Ko, A.; Naghshineh-Pour, A.; Sulaeman, E.; Tetrault, P.A.; Grossman, B.; Kapania, R.; Mason, W.; Schetz, J.; et al. *Multidisciplinary Design Optimization of a Transonic Commercial Transport with a Strut-Braced Wing*; American Institute of Aeronautics and Astronautics: Reston, VA, USA, 1999. [CrossRef]
41. Variyar, A.; Economon, T.D.; Alonso, J.J. Multifidelity Conceptual Design and Optimization of Strut-Braced Wing Aircraft using Physics Based Methods. In Proceedings of the 54th AIAA Aerospace Sciences Meeting, San Diego, CA, USA, 4–8 January 2016. [CrossRef]
42. Qian, J.; Alonso, J.J. High-Fidelity Structural Design and Optimization of Blended-Wing-Body Transports. In Proceedings of the 2018 Multidisciplinary Analysis and Optimization Conference, Atlanta, GA, USA, 25–29 June 2018. [CrossRef]
43. Smith, M.; Patil, M.; Hodges, D. CFD-based analysis of nonlinear aeroelastic behavior of high-aspect ratio wings. In Proceedings of the 19th AIAA Applied Aerodynamics Conference, Anaheim, CA, USA, 11–14 June 2001. [CrossRef]
44. Liem, R.P.; Kenway, G.K.W.; Martins, J.R.R.A. Multimission Aircraft Fuel-Burn Minimization via Multipoint Aerostructural Optimization. *AIAA J.* **2015**, *53*, 104–122. [CrossRef]
45. Haar, D.; Brezillon, J. Engine integration based on multi-disciplinary optimisation technique. *CEAS Aeronaut. J.* **2012**, *3*, 17–24. [CrossRef]
46. Ronzheimer, A.; Natterer, F.J.; Brezillon, J. Aircraft Wing Optimization Using High Fidelity Closely Coupled CFD and CSM Methods. In Proceedings of the 13th AIAA/ISSMO Multidisciplinary Analysis Optimization Conference, Fort Wort, TX, USA, 13–15 September 2010. [CrossRef]
47. Garrigues, E. A Review of Industrial Aeroelasticity Practices at Dassault Aviation for Military Aircraft and Business Jets. *Aerosp. Lab.* **2018**, 1–34. [CrossRef]
48. Kennedy, G.; Martins, J. A parallel aerostructural optimization framework for aircraft design studies. *Struct. Multidiscip. Optim.* **2014**, *50*, 1079–1101. [CrossRef]
49. Mieloszyk, J.; Goetzendorf-Grabowski, T. Introduction of full flight dynamic stability constraints in aircraft multidisciplinary optimization. *Aerosp. Sci. Technol.* **2017**, *68*, 252–260. [CrossRef]
50. Conlan-Smith, C.; Schousboe Andreasen, C. Aeroelastic Optimization of Aircraft Wings Using a Coupled Three-Dimensional Panel-Beam Model. *AIAA J.* **2021**, *59*, 1374–1386. [CrossRef]
51. Mitrotta, F.M.A.; Rajpal, D.; Sodja, J.; Breuker, R.D. Multi-Fidelity Design of an Aeroelastically Tailored Composite Wing for Dynamic Wind-Tunnel Testing. In Proceedings of the AIAA Scitech 2020 Forum, Orlando, FL, USA, 6–10 January 2020. [CrossRef]
52. Choi, S.; Alonso, J.J.; Kroo, I.M. Two-Level Multifidelity Design Optimization Studies for Supersonic Jets. *J. Aircr.* **2009**, *46*, 776–790. [CrossRef]
53. Ghoreyshi, M.; Jirasek, A.; Cummings, R. Reduced order unsteady aerodynamic modeling for stability and control analysis using computational fluid dynamics. *Prog. Aerosp. Sci.* **2014**, *71*, 167–217. [CrossRef]
54. Walton, S.; Hassan, O.; Morgan, K. Reduced order modelling for unsteady fluid flow using proper orthogonal decomposition and radial basis functions. *Appl. Math. Model.* **2013**, *37*, 8930–8945. [CrossRef]
55. Silva, W.A. AEROM: NASA's Unsteady Aerodynamic and Aeroelastic Reduced-Order Modeling Software. *Aerospace* **2018**, *5*, 41. [CrossRef] [PubMed]
56. Mannarino, A.; Mantegazza, P. Nonlinear aeroelastic reduced order modeling by recurrent neural networks. *J. Fluids Struct.* **2014**, *48*. [CrossRef]
57. Piperni, P.; DeBlois, A.; Henderson, R. Development of a Multilevel Multidisciplinary-Optimization Capability for an Industrial Environment. *AIAA J.* **2013**, *51*, 2335–2352. [CrossRef]
58. Dillinger, J.K.S.; Abdalla, M.M.; Meddaikar, Y.M.; Klimmek, T. Static aeroelastic stiffness optimization of a forward swept composite wing with CFD-corrected aero loads. *CEAS Aeronaut. J.* **2019**, *10*, 1015–1032. [CrossRef]
59. H2020 Clean Sky 2 GRETEL Project. Available online: https://www.cleansky2gretel.eu (accessed on 28 November 2021).
60. Vassberg, J.; Dehaan, M.; Rivers, M.; Wahls, R. Development of a Common Research Model for Applied CFD Validation Studies. In Proceedings of the 26th AIAA Applied Aerodynamics Conference, Honolulu, HI, USA, 18–21 August 2008; American Institute of Aeronautics and Astronautics: Reston, VA, USA, 2008. [CrossRef]
61. Jones, R.M. *Mechanics of Composite Materials*; CRC Press: Boca Raton, FL, USA, 2018. [CrossRef]
62. Katz, J.; Plotkin, A. *Low-Speed Aerodynamics*; Cambridge University Press: Cambridge, UK, 2001. [CrossRef]
63. Melin, T. A Vortex Lattice MATLAB Implementation for Linear Aerodynamic Wing Applications. Master's Thesis, Royal Institute of Technology (KTH), Stockholm, Sweden, 2000.
64. Brooks, T.R.; Kenway, G.K.; Martins, J.R.R.A. Undeflected Common Research Model (uCRM): An Aerostructural Model for the Study of High Aspect Ratio Transport Aircraft Wings. In Proceedings of the 35th AIAA Applied Aerodynamics Conference, Denver, CO, USA, 5–9 June 2017; American Institute of Aeronautics and Astronautics: Reston, VA, USA, 2017. [CrossRef]

65. Martins, J.R.R.A.; Kenway, G.K.W.; Burdette, D.; Jonsson, E.; Stanford, B.K.; Kennedy, G.J. *High-Fidelity Multidisciplinary Design Optimization of Aircraft Configurations*; Technical Report; NASA: Washington, DC, USA, 2017.
66. Kilimtzidis, S.; Kotzakolios, A.; Kostopoulos, V. Efficient Structural Optimisation of Composite Materials Aircraft Wings. *Compos. Struct.* **2021**, under review.
67. Kassapoglou, C. *Design and Analysis of Composite Structures*; John Wiley & Sons Ltd.: Hoboken, NJ, USA, 2013. [CrossRef]
68. Liu, Q.; Mulani, S.B.; Kapania, R.K. Global/Local Multidisciplinary Design Optimization of Subsonic Wing. In Proceedings of the10th AIAA Multidisciplinary Design Optimization Conference, National Harbor, MD, USA, 13–17 January 2014; American Institute of Aeronautics and Astronautics: Reston, VA, USA, 2014. [CrossRef]
69. Starnes, J.H.; Haftka, R.T. Preliminary Design of Composite Wings for Buckling, Strength, and Displacement Constraints. *J. Aircr.* **1979**, *16*, 564–570. [CrossRef]
70. Kreisselmeier, G.; Steinhauser, R. Systematic Control Design by Optimizing a Vector Performance Index. *IFAC Proc. Vol.* **1979**, *12*, 113–117. [CrossRef]
71. Poon, N.M.K.; Martins, J.R.R.A. An adaptive approach to constraint aggregation using adjoint sensitivity analysis. *Struct. Multidiscip. Optim.* **2006**, *34*, 61–73. [CrossRef]
72. Lambe, A.B.; Kennedy, G.J.; Martins, J.R.R.A. An evaluation of constraint aggregation strategies for wing box mass minimization. *Struct. Multidiscip. Optim.* **2016**, *55*, 257–277. [CrossRef]
73. Lambe, A.B.; Martins, J.R.R.A. Matrix-free aerostructural optimization of aircraft wings. *Struct. Multidiscip. Optim.* **2015**, *53*, 589–603. [CrossRef]
74. Suh, P.M. *EZASE Easy Aeroelasticity: A Tool to Simulate Aircraft Wing Geometry*; NASA: Washington, DC, USA, 2011.
75. Schlüter, M.; Erb, S.O.; Gerdts, M.; Kemble, S.; Rückmann, J.J. MIDACO on MINLP space applications. *Adv. Space Res.* **2013**, *51*, 1116–1131. [CrossRef]
76. Schlüter, M.; Egea, J.A.; Banga, J.R. Extended ant colony optimization for non-convex mixed integer nonlinear programming. *Comput. Oper. Res.* **2009**, *36*, 2217–2229. [CrossRef]
77. Schlüter, M.; Gerdts, M. The oracle penalty method. *J. Glob. Optim.* **2009**, *47*, 293–325. [CrossRef]
78. Joel, H.; Ferziger, M.P. *Computational Methods for Fluid Dynamics*; Springer: Berlin/Heidelberg, Gemany, 2002. [CrossRef]
79. Fletcher, C. *Computational Techniques for Fluid Dynamics 2*; Springer: Berlin/Heidelberg, Gemany, 1991. [CrossRef]
80. Demirdzic, I.; Lilek, Z.; Peric, M. A collocated finite volume method for predicting flows at all speeds. *Int. J. Numer. Methods Fluids* **1993**, *16*, 1029–1050. [CrossRef]
81. Beaudoin, M.; Jasak, H. Development of a Generalized Grid Interface for Turbomachinery simulations with OpenFOAM. In Proceedings of the Open Source CFD International Conference 2008, Berlin, Germany, 4–5 December 2008.
82. Jasak, H. Dynamic Mesh Handling in OpenFOAM. In Proceedings of the 47th AIAA Aerospace Sciences Meeting including The New Horizons Forum and Aerospace Exposition, Orlando, FL, USA, 5–8 January 2009. [CrossRef]
83. Kassiotis, C. *Which Strategy to Move the Mesh in the Computational Fluid Dynamic Code OpenFOAM (2008)*; Chalmers TH Technical Report for OpenFOAM: Gothenburg, Sweden, 2008.
84. Silva, W. Identification of Nonlinear Aeroelastic Systems Based on the Volterra Theory: Progress and Opportunities. *Nonlinear Dyn.* **2005**, *39*, 25–62. [CrossRef]
85. Jirasek, A.; Cummings, R. Application of Volterra Functions to X-31 Aircraft Model Motion. In Proceedings of the 27th AIAA Applied Aerodynamics Conference, San Antonio, TX, USA, 22–25 June 2009. [CrossRef]
86. Thanedar, P.B.; Vanderplaats, G.N. Survey of Discrete Variable Optimization for Structural Design. *J. Struct. Eng.* **1995**, *121*, 301–306. [CrossRef]
87. Jutte, C.V.; Stanford, B.K.; Wieseman, C.D. *Internal Structural Design of the Common Research Model Wing Box for Aeroelastic Tailoring*; Technical Report; NASA: Washington, DC, USA, 2015.
88. Marlett, K. *HEXCEL 8552 IM7 Unidirectional Prepreg 190 gsm 35% RC Qualification Statistical Analysis Report*; Rept. NCP-RP-2009-028 Rev B; Technical Report; National Institute for Aviation Research: Wichita, KS, USA, 2011.

aerospace

MDPI

Article

Compressive Behaviour of 3D-Printed PETG Composites

Sara Valvez [1], Abílio P. Silva [1] and Paulo N. B. Reis [2,*]

[1] Department of Electromechanical Engineering, C-MAST, University of Beira Interior, 6201-001 Covilhã, Portugal; sara.valvez@ubi.pt (S.V.); abilio@ubi.pt (A.P.S.)
[2] Department of Mechanical Engineering, CEMMPRE, University of Coimbra, 3030-194 Coimbra, Portugal
* Correspondence: paulo.reis@dem.uc.pt

Abstract: It is known that 3D-printed PETG composites reinforced with carbon or Kevlar fibres are materials that can be suitable for specific applications in the aeronautical and/or automotive sector. However, for this purpose, it is necessary to understand their mechanical behaviour, which is not yet fully understood in terms of compression. Therefore, this study intends to increase the knowledge in this domain, especially in terms of static behaviour, as well as with regard to creep and stress relaxation due to the inherent viscoelasticity of the matrix. In this context, static, stress relaxation and creep tests were carried out, in compressive mode, using neat PETG and PETG composites reinforced with carbon and Kevlar fibres. From the static tests, it was found that the yield compressive strength decreased in both composites compared to the neat polymer. Values around 9.9% and 68.7% lower were found, respectively, when carbon and Kevlar fibres were added to the PETG. Similar behaviour was observed for compressive displacement, where a reduction of 20.4% and 46.3% was found, respectively. On the other hand, the compressive modulus increased by 12.4% when carbon fibres were added to the PETG matrix and decreased by 39.6% for Kevlar fibres. Finally, the stress relaxation behaviour revealed a decrease in compressive stresses over time for neat PETG, while the creep response promoted greater compressive displacement. In both situations, the response was very dependent on the displacement/stress level used at the beginning of the test. However, when the fibres were added to the polymer, higher stress relaxations and compressive displacements were observed.

Keywords: additive manufacturing; fused filament fabrication (FFF); PETG; composites; compressive properties; creep and stress relaxation behaviour; mechanical testing

Citation: Valvez, S.; Silva, A.P.; Reis, P.N.B. Compressive Behaviour of 3D-Printed PETG Composites. *Aerospace* **2022**, *9*, 124. https://doi.org/10.3390/aerospace9030124

Academic Editors: Spiros Pantelakis, Andreas Strohmayer and Liberata Guadagno

Received: 15 December 2021
Accepted: 25 February 2022
Published: 28 February 2022

1. Introduction

According to the technical committee of the American Society for Testing and Materials, additive manufacturing (AM) is described as the process of material joining for the fabrication of three-dimensional (3D) parts. In this context, fused filament fabrication (FFF) is a 3D printing process based on thermoplastic polymers that uses a continuous filament to produce complex three-dimensional parts. This technique is also often confused with Fused Deposition Modelling (FDM), which is a trademark created by Stratasys co-founder Scott Crump in 1988 [1].

In terms of materials, poly(ethylene terephthalate)-glycol (PETG) is one of the most used materials in 3D printing technology due to the chemical alkali resistance, transparency, gloss, low haze and good printability, among other benefits. Furthermore, with the correct print settings, it is possible to obtain excellent layer adhesion and very low shrinkage properties. At the same time, it is extremely strong, which allows one to print objects that can function at high temperatures or in food-safe applications and possess exceptional impact performance. All these advantages make this material suitable for both the food and medical industries [2]. In the last case, for example, its rigid structure allows it to survive harsh sterilisation processes, making it a perfect material to be used in medical implants, as well as pharmaceutical packaging and medical devices [3,4]. However, adding carbon fibres reduces the risk of warpage even further, making the material more resistant and resilient, making it an excellent choice for automotive and other industrial applications [5]. The addition of carbon fibres extends its field of application to the automotive and aeronautical sector, and others (such as prosthetics or adaptable parts of wheelchairs), because the material becomes more resistant and resilient, in addition to further decreasing the risk of warping [6]. On the other hand, when reinforced with aramid fibres, the applications can be extended to the aeronautical and aerospace industries, where high resistance to friction and impact is expected [7].

Kasmi et al. [6], for example, developed studies involving PETG and PETG reinforced with carbon fibres and found, for both materials, that the best mechanical properties were obtained with raster orientations of 0°. Furthermore, a decrease in Young's modulus and ultimate tensile strength was observed with increasing raster orientation. Therefore, it was possible to conclude that the mechanical properties of PETG with layers printed unidirectionally were significantly influenced by the raster orientation and shell presence. Similar conclusions were observed for PETG reinforced with continuous carbon fibres (CCF). For unidirectional layups at 0°, the Young's modulus and ultimate tensile strength were significantly improved by the presence of the CCF composite, with values from 1.47 GPa and 30 MPa to 25 GPa (17 times higher) and 268 MPa (9 times higher), respectively. On the other hand, the elongation at break was around 1.9% for PETG-CCF specimens against 2.8% for the PETG ones, evidencing a more brittle behaviour. Similar studies were developed by Jiang et al. [8] with commercially available carbon-fibre-filled (CFF) polymer composite filaments, and the results were compared to unfilled polymers from the same filament suppliers. Fibre length was less than 100 μm. In this case, the tensile modulus increased by 313.2% when CFF was added, for 0° printing orientation, while the tensile strength increased by around 48.2%. Kumar et al. [9] used a Taguchi L9 approach to optimise the printing parameters in order to maximise the mechanical performance of carbon-fibre-reinforced PETG thermoplastics. The optimal tensile and flexural strength, as well as the hardness, were around 34.6 MPa, 36.5 MPa and 68.7 BHN, respectively, and the values obtained with a printing speed of 60 mm/s, infill density 80% and 200 μm layer height. The authors also noted that, while the print speed and infill density govern the tensile strength and hardness, flexural strength is influenced by layer height and infill density. In another study, Kumar et al. [10] studied the infill density (25%, 50%, 75% and 100%) and the annealing treatment of PETG and PETG with 20 wt.% of carbon fibres (PETG+CF). It was observed that the best mechanical properties were obtained with annealed samples and 100% infill density, but compared to the annealed PETG, they were higher for annealed PETG+CF. For example, improvements of 21%, 25%, 23% and 18% were observed in terms of hardness, tensile, impact and flexural strength. Bhandari et al. [11] studied the interlayer mechanical properties of PETG and PETG reinforced with short carbon fibres (SCF) and observed that the composites showed lower interlayer tensile strength than the neat PETG. This decrease was explained by the increase in the melt viscosity and consequent slower interlayer diffusion bonding. However, this decrease can be recovered by annealing when the temperature is higher than the glass transition temperature of the amorphous polymer (PETG). The post-processing treatment, for example, increased the interlayer Young's modulus (1.65 times compared to unannealed composite and 1.32 times compared to neat annealed PETG) and ductility of the printed composites. Ferreira et al. [12] evaluated the benefits of adding short carbon fibres (20 wt.%) to PETG polymer and found benefits of 191.4% and 5.1% in terms of flexural modulus and strength, respectively, while tensile modulus improvements were only approximately 70.1%. However, a decrease of around 28.1% was observed in terms of tensile strength. Mansour et al. [13], using similar materials (PETG and PETG with 20 wt.% of short carbon fibres), studied the compression properties and found a decrease in the compressive strain of 66% when the short carbon fibres (SCF) were added, while the modulus and hardness increased by around 30% and 27%, respectively. In terms of the loss factor and damping, those obtained from the cyclic compression and models tests dropped from 17.3% to 15.4% and 13.8% to 12.3%, respectively. Kromoser and Pachner [14] developed a study on the load-bearing behaviour of knots made from different kinds of polymer-based printing materials, including PETG, and the results showed that PACF (a carbon-fibre-infused polyamide-based material) had the best compressive mechanical properties and a brittle failure behaviour. A similar behaviour was observed for PETG but for lower compressive values (2–3 times smaller). Sörensen et al. [15] used the FFF process to manufacture a tooling to obtain a small series of sheet metal parts in combination with the rubber pad forming process for the automotive sector and, for this purpose, a variety of common AM polymeric materials (PETG, PLA and ABS) were compared in compression tests. Both PETG and PLA were able to fulfil the proposed performance (loading capacity), but in this study, PLA was selected due to its bio-based structure and its industrial composting possibilities. Hsueh et al. [16] compared the compression properties of PLA and PETG specimens at different printing temperatures and speeds. They found that the values of the compression properties increased with the increasing printing temperature due to the decrease in the contact stress at the line and line contacts. It was also observed that there was a minimal effect on the compression properties of PETG with respect to the printing speed. However, the compression properties of PLA were higher than those of PETG, similar to their tensile properties. Mercado-Colmenero et al. [17] developed a numerical and experimental study on PETG produced by the FDM technology in order to obtain its mechanical characterisation under uniaxial compression loads. They found that the specimens manufactured in the X, Y and Z directions had a completely linear behaviour until reaching the elastic limit, the point from which the plasticisation process begins until the final fracture. However, different structural failures were observed. For the specimens produced along the Z direction, fracture was brittle due to the delamination of adjacent layers and breakage of the filaments, while, along the X and Y directions, they presented permanent plastic deformations due to the lower strength compared to the specimens manufactured in Z. Finally, Patterson et al. [18] studied the IZOD impact properties of different materials and observed, for example, that polylactic acid (PLA), PLA + wood fibre (WPLA), PLA + chopped carbon fibre (CFPLA) and PLA + aluminium powder (AlPLA) showed little or no local plastic deformation, while some plastic deformation on the fracture surface (mainly localised and not too deep into the part) was noted for acrylonitrile butadiene styrene (ABS), high-temperature PLA (HTPLA) and high-impact polystyrene (HIPS). Finally, polycarbonate (PC), nylon (a synthetic polyamide) and polyethylene terephthalateþglycol (PETG) showed significant plastic deformation, including large and deep scars on the part, stretching and layer pullout.

Therefore, from the literature [2–7], it is evident that PETG and PETG-based composites have applications in many industries due to their resistance to heat, impact and solvents. In terms of compressive properties, the focus of the present study, the literature is not abundant on this topic and what exists is not enough to establish a consolidated knowledge base [13–17], because most of the available studies focus essentially on the tensile mode. On the other hand, as consequence of the inherent viscoelasticity of the matrix phase, polymer composites are prone to creep and stress relaxation, making it a great challenge when used in long-term applications. However, in terms of creep and stress relaxation behaviour, these subjects are totally absent in the open literature for PETG and PETG-based composites, limiting their use in many applications due to lack of knowledge. In this context, the novelty of this study is related to the effect of fibre type on the compression properties of PETG-based composites, to consolidate knowledge, but essentially at the level of creep and stress relaxation behaviour due to the absence of studies in this domain. For this purpose, an experimental study was developed to characterise the compressive properties of PETG and PETG reinforced with carbon and aramid fibres, as well as the stress relaxation and creep behaviour of these materials.

2. Materials and Methods

The B2X300 printer was used to extrude the initial 1.75-mm-diameter filament through a 0.6-mm-diameter nozzle. The extruded filament was placed on a platform heated by a print head in a user-defined pattern to achieve the desired flat shape. Once a particular layer was finished, the print head was raised and continued to deposit the next layer. Using this equipment and methodology (see Figure 1), specimens with dimensions of 12.7 × 12.7 × 25.4 mm^3 were printed, values that are recommended by the ASTM D695-15 standard [19].

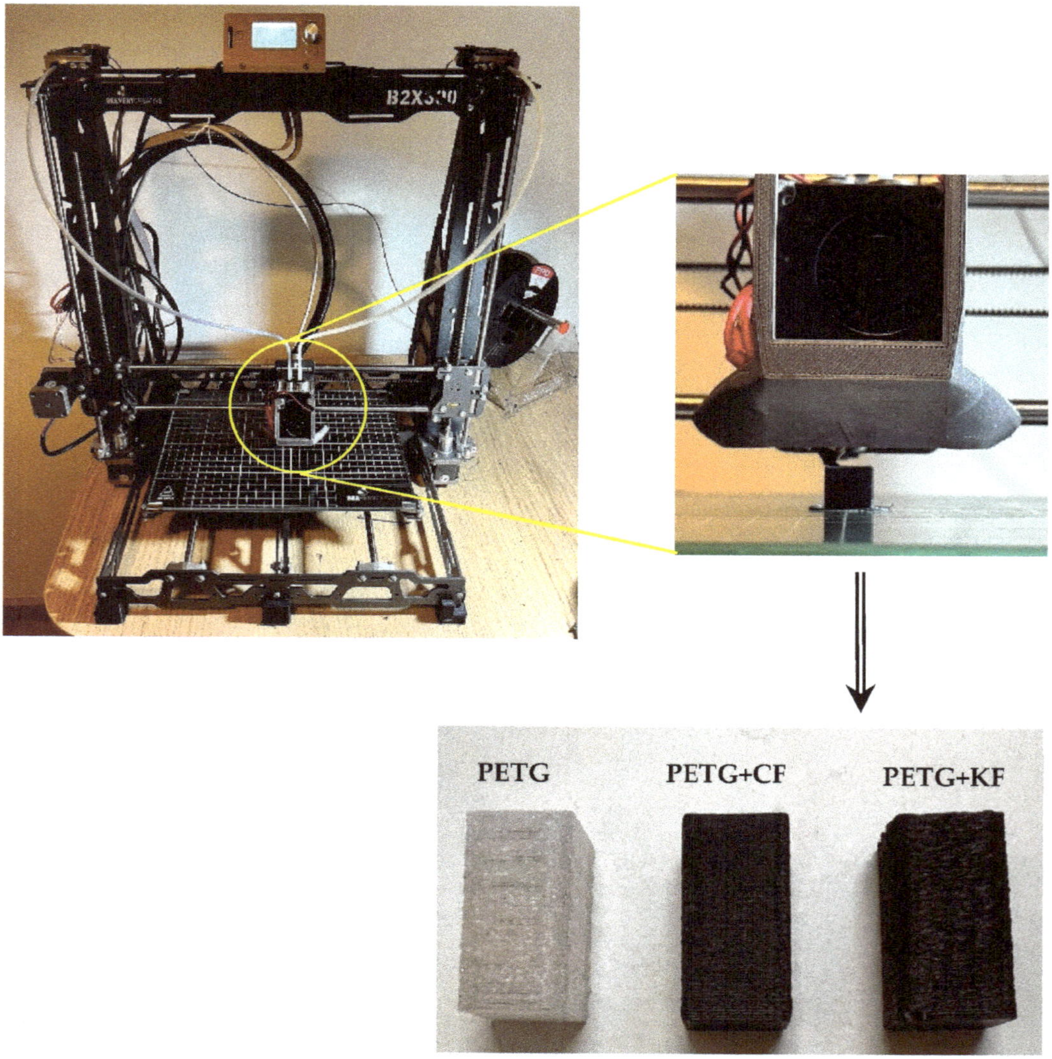

Figure 1. Details of the 3D printer producing a specimen in an environmentally controlled room.

In terms of materials, poly(ethylene terephthalate)-glycol (PETG) is one of the most used materials in 3D printing technology due to its chemical alkali resistance, transparency, gloss, low haze and good printability, among other benefits. However, the addition of carbon fibres (CF) further expands its field of application because the composite becomes more resistant and resilient, as well as significantly reducing the risk of warping. In this case, it becomes an excellent choice for automotive, aeronautical and aerospace applications, besides other industrial applications. On the other hand, when reinforced with aramid fibres (KF), applications can be extended to components where high resistance to friction and impact is expected. Therefore, given the benefits reported for these materials, PETG, PETG + CF and PETG + KF were the filaments selected to obtain the specimens used in this study. While PETG composites containing fibres were supplied by Nanovia (Louargat, France), neat PETG was supplied by FilTech (Baesweiler, Germany). Different printing parameters were used, and Table 1 shows those that maximised the mechanical properties of these materials. For this purpose, a study was previously carried out. The criterion of maximum mechanical performance was adopted, because, in any structural application, we expect, in general, to maximise the mechanical properties.

Table 1. Printing parameters for each material.

Material	Extrusion Temperature [°C]	Speed [mm/s]	Layer Height [mm]	Infill [%]
PETG	265	20	0.40	100
PETG + CF	195	60	0.52	100
PETG + KF	265	20	0.35	100

The deposition angle direction was [0/45/−45/90/90/−45/45/0], which means that the first layer was printed in a unidirectional pattern at an angle of 0° from the horizontal. The second layer was rotated 45° away from the horizontal, the third rotated −45° and the fourth rotated 90°, and this sequence continued as a mirror until the code was completed. This is a quasi-isotropic sequence and was selected because the extensional stiffness of the laminate was practically the same in each in-plane direction.

Finally, compressive tests were carried out in accordance with the ASTM D695-15 standard [19] in a Shimadzu universal testing machine (Duisburg, Germany), model Autograph AGS-X, equipped with a 10 kN load cell (Figure 2). For each condition, 5 specimens were tested at room temperature with a displacement rate of 1.3 mm/min. Compressive yield strength was calculated by dividing the yield compressive load carried by the cross-sectional area of the specimen. The compressive modulus was calculated by drawing a tangent to the initial linear portion of the load deformation curve.

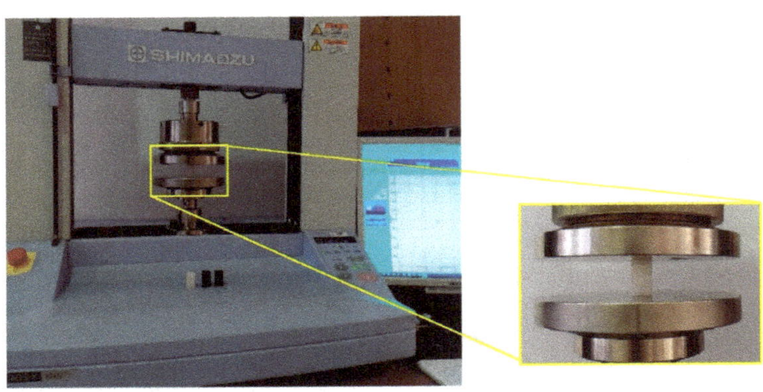

Figure 2. Compression test setup.

Stress relaxation and creep tests were also carried out in the same machine (Shimadzu, model Autograph AG-X), at room temperature, and using specimens with geometry similar to those used on the compressive tests. For the first tests, a fixed strain was applied (correspondent to 75%, 50% and 25% of the yield compressive stress for all configurations) and the stress recorded during the loading time. In terms of creep tests, a fixed stress was applied (correspondent to 75%, 50% and 25% of the yield compressive stress for all configurations) and the displacement recorded during the loading time. These values were selected to guarantee that all tests were performed in the elastic regime of all configurations analysed.

3. Results and Discussion

Experimental tests were performed to obtain the compressive properties. In this context, Figure 3 shows the load–displacement curves obtained for each condition, which are representative of the others. Figure 3b shows the repeatability of the curves for PETG with carbon fibres, evidencing the low dispersion represented in Table 2. The low dispersion observed extends to the other materials analysed.

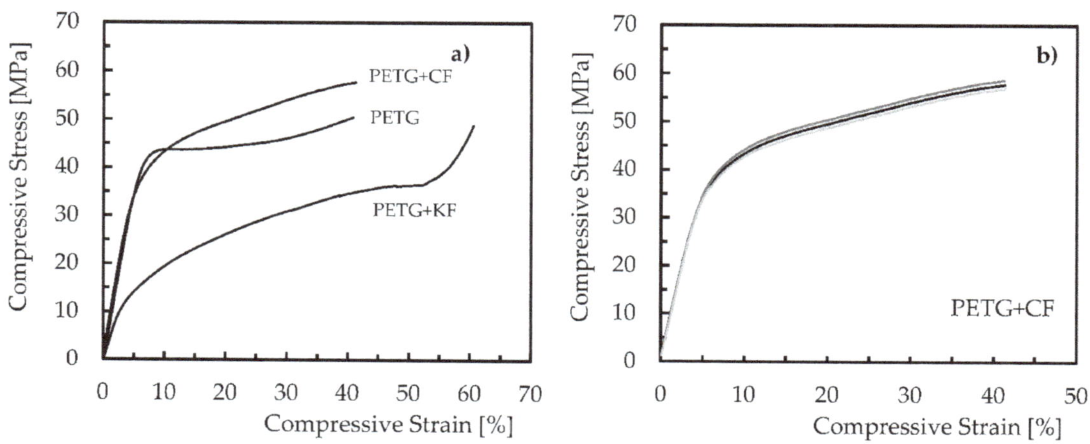

Figure 3. (**a**) Typical compressive curves of all materials studied; (**b**) Repeatability of the curves for PETG with carbon fibres.

Table 2. Compressive properties.

Material	Yield Compressive Strength [MPa]	Compressive Modulus [MPa]	Compressive Displacement [mm]
PETG	35.02 ± 2.35	680.8 ± 45.97	5.4 ± 2.35
PETG + CF	31.55 ± 0.83	765.2 ± 6.77	4.3 ± 0.07
PETG + KF	10.97 ± 0.79	411.1 ± 8.37	2.9 ± 0.12

For all materials, a linear region is observed, in which the material deforms elastically and returns to its original length when the stress is removed. This region terminates when the yield point is reached, and the material starts to have a plastic behaviour. Under these conditions, it no longer regains its original length when the load is removed. However, depending on the material, the linear region presents different extensions, denoting that there are materials with longer elastic regimes than others. Therefore, different compressive properties are expected, which are summarised in terms of mean values and respective standard deviations in Table 2.

Regarding the yield compressive strength, the highest value was obtained for the neat PETG, and when the carbon and Kevlar fibres were added, a decrease of around 9.9% and 68.7% was found, respectively. A similar behaviour is also observed for the compressive displacement, where an average value of 5.4 mm is obtained for pure PETG, but, in this case, the decrease is around 20.4% and 46.3%, respectively. Concerning the compressive modulus, and when compared to the neat PETG (680.8 MPa), there is an increase of 12.4% when carbon fibres are added to the PETG matrix and a decrease of 39.6% for Kevlar fibres. These results agree with the studies developed by Mansour et al. [13], where, for similar materials (PETG and PETG with 20 wt.% of short carbon fibres), they found a decrease in the compressive strain and an increase in modulus when carbon fibres were added. In terms of compressive strength, the decrease that is observed is in line with the studies developed by Mahesh et al. [20], where the addition of SCF to PETG composites (5 and 10 wt.% of SCF) promoted decreases between 3% and 12%. According to the authors, most of the load is supported by the matrix, which means poor matrix–fibre adhesion and load transfer at these points. Furthermore, these authors also mention that excessive shear stresses acting on the layer interface and fibre–matrix interface, as well as the air gaps between the adjacent printed layers, contribute to this decrease [20]. However, the literature reports good fibre–matrix adhesion for glass and carbon fibres, while Kevlar fibres are recognised for establishing weak fibre–matrix adhesion [21]. In this context, the poor results obtained with Kevlar fibres can be explained by the weak interfaces between the Kevlar fibres and the PETG matrix, while, for the PETG composite reinforced with carbon fibres, it can also be explained by the damage mechanisms observed in Figure 4. This figure shows details of the specimens during the compressive tests for PETG and the PETG composite reinforced with carbon fibres (PETG+CF). In terms of PETG+CF, for example, in the linear region, the specimen geometry maintains its shape (Figure 4a), and the compressive load closes the intra- and inter-layer porosities generally exhibited within the print beads and adjacent layers of the print [22]. Subsequently, this phenomenon promotes internal instability that leads to the minor buckling of the samples (buckling initiated at the arrowhead in Figure 4b), which, when increasing, allows the adjacent fibre–matrix and matrix–matrix interfaces buckled within the specimen to bend in a similar way (see Figure 4c) and, consequently, shear stresses/strains arise [22]. Continuing to apply the load, a significant shear deformation is observed with the consequent sudden occurrence of a dominant crack between layers (see Figure 4d). In terms of neat PETG, Figure 4e,f report a similar behaviour, but instead of a dominant crack, there are several small cracks between layers.

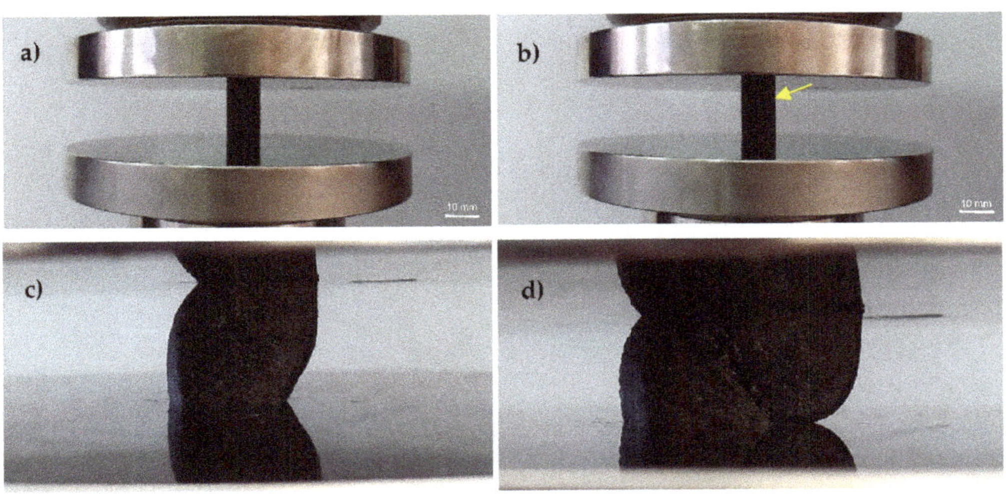

Figure 4. *Cont.*

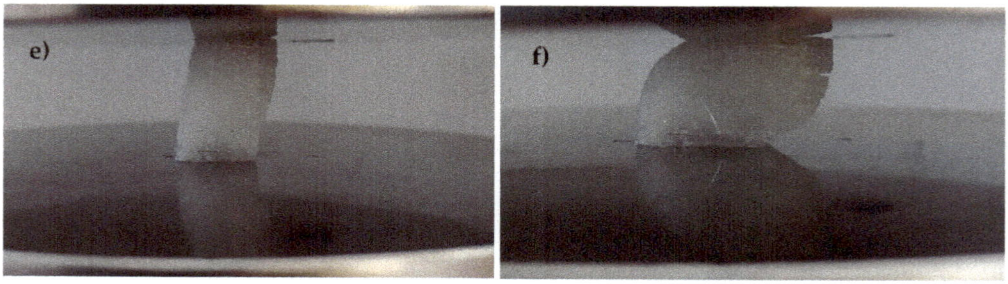

Figure 4. Details of the specimens during the compressive tests for (**a**) PETG + CF in the linear region; (**b**) PETG + CF at the beginning of the nonlinear region; (**c**) PETG + CF in the nonlinear region; (**d**) PETG + CF before the collapse; (**e**) PETG in the nonlinear region; (**f**) PETG before the collapse.

Regarding the creep and stress relaxation behaviour, the results obtained from these tests are shown in Figure 5 for neat PETG. Three curves obtained from the experimental tests and the respective average curve (blue colour) are shown, which, despite representing the observed behaviour for 50% of the yield compressive stress, are also representative of the tests performed for the other stress levels.

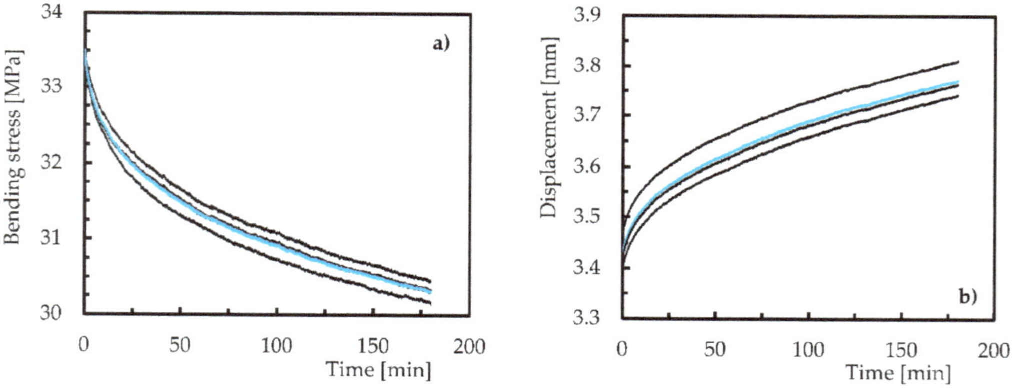

Figure 5. Repeatability of the curves for PETG for 50% of the maximum bending stress: (**a**) stress relaxation; (**b**) creep.

It is possible to notice a consistent behaviour and without significant dispersion, which justifies the low scattering presented in the figures below and the respective discussion. Therefore, Figure 6 plots the average compressive stress versus time obtained for a fixed strain correspondent to 75%, 50% and 25% of the yield compressive stress, where σ is the compressive stress at any given moment of the test and σ_0 is the initial compressive stress. The final bands represent the maximum and minimum values obtained for each condition analysed. In this case, the compressive stress decreases with time for all analysed strains, but this tendency becomes more pronounced with the increase in the strain for which the experiment is carried out. For example, for a constant displacement corresponding to 25% of the yield compressive stress, the compressive stress continuously decreases over the analysed time, reaching, for example, values around 10.5% lower after 10,800 s (3 h). However, this difference increases to 13.8% and 16.4% for compressive displacements correspondent to 50% and 75% of the yield compressive stress, respectively. In fact, after some time, it would be expected that this reduction would tend towards a constant value, but, in the present work, this was not achieved because short-term tests were considered.

According to the open literature, these tests are an easy, fast and reliable method to predict long-term behaviour [23–25]. On the other hand, it is also reported that an initial stage occurs in which the stress decreases considerably in relation to the remaining time [26–28], and this evidence is also observed in the present study for all compressive strains analysed.

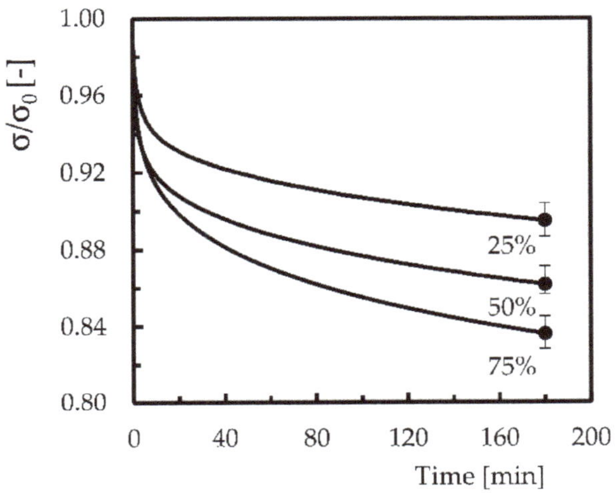

Figure 6. Stress relaxation curves for PETG.

The bibliography justifies the typical decreases observed in plastics during stress relaxation tests due to physical and/or chemical phenomena. In the first case, molecular rearrangements that require little formation or rupture of primary bonds explain the observed decrease, while, from a chemical point of view, it is justified by the chain scission, crosslink scission or crosslink formation [29–31], both expected in this study. Furthermore, any decohesion that may occur between the printed interface layers or the air gaps between adjacent printed layers can also contribute to the observed decrease. Regarding the displacement effect (correspondent to 25%, 50% and 75% of the yield compressive stress), it is possible to observe that its increase promotes higher stress relaxation due to a change in the mechanisms mentioned above. For example, at lower displacement levels, stress relaxation is mainly induced by the molecular rearrangements and/or chain scission, while, for higher displacement levels, it is due to damage and its propagation. This is line with the studies developed by Mirzae et al. [32], where these authors also observed that increasing the strain level promotes more stress relaxation.

The creep behaviour observed for the different compressive stress levels is shown in Figure 7, where the displacement is the value obtained at any instant of the test (D) divided by its initial value (D_0). Once again, the final bands represent the maximum and minimum values obtained for each condition analysed. It can be noted that all curves present an instantaneous displacement, which is dependent on the compressible stress level, followed by the primary and secondary stage. This is a typical behaviour of creep curves, where the third stage is only expected for higher stress values or longer times. In quantitative terms, for example, after 180 min, the compressive strain increases by 5.5%, 7.4% and 16.4% for compressive stresses of 25%, 50% and 75% of the yield compressive stress. These results prove that the creep phenomenon in polymers occurs even at room temperature and for stresses below the ultimate strength, which is explained by the molecular motion in the backbone polymer arrangement [33–36].

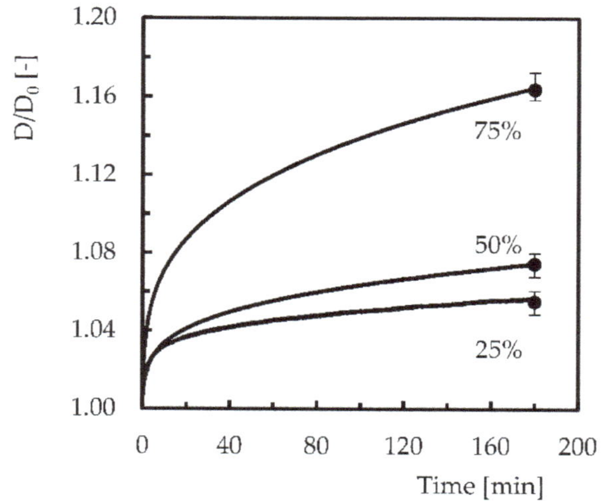

Figure 7. Creep curves for PETG and different compressive stress levels.

It is also possible to notice that the stress levels significantly influence the creep displacement of the material. This behaviour agrees with what has been described for stress relaxation and is also justified by the change in the previously reported mechanisms. In this case, at lower stress levels, creep is mainly induced by molecular movements, while at higher stress levels, creep is dominated by damage and its propagation. For the highest stress levels, the load transfer on the interface regions is significantly affected and, inclusively, some debonding is expected [22]. However, as displacement is held constant in stress relaxation, any mechanism reported above will have lower severity than in creep, because the deflection is always increasing [22].

In terms of PETG-based composites, stress relaxation and creep behaviour are compared in Figure 8 for 50% of the yield compressive stress. In this figure, the type of fibres used is also compared.

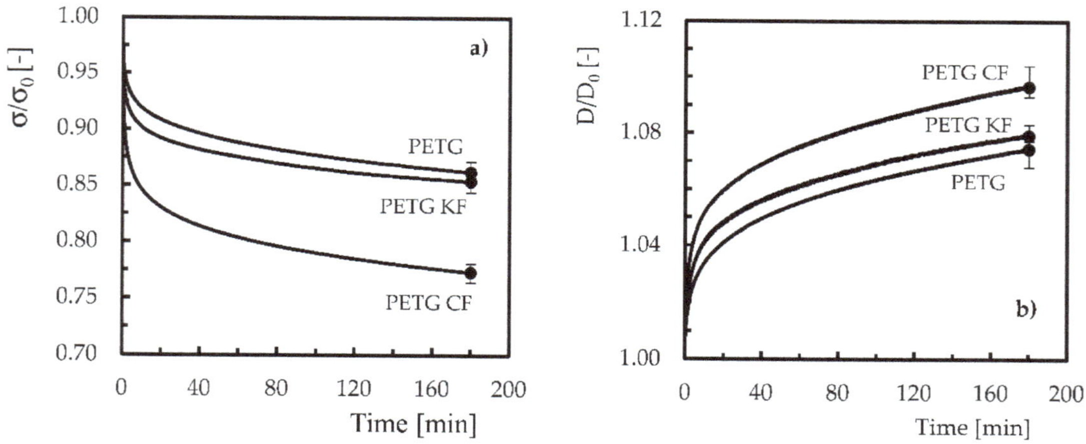

Figure 8. Comparison for 50% of the yield compressive stress for PETG composites in terms of (**a**) stress relaxation curves; (**b**) creep curves.

Concerning the stress relaxation behaviour (Figure 8a), the neat polymer is, of all materials, the one with the lowest relaxation. For example, compared to the initial compressive stress, after 180 min, the neat PETG showed a decrease of around 13.8%, while, for PETG + KF and PETG + CF, this value was around 14.7% and 22.7% less than that obtained for the neat polymer, respectively. On the other hand, the creep behaviour (Figure 8b) showed a similar trend, where the compressive displacement after 180 min increased by around 7.4% compared to the initial value for neat PETG. However, when Kevlar and carbon fibres were added to the polymer, the compressive displacements after 180 min were, respectively, 7.9% and 9.7% higher than the value obtained for the neat polymer. In terms of polymers, and as reported above, while stress relaxation occurs due to physical and/or chemical phenomena, molecular motion in the backbone polymer arrangement explains the creep behaviour. However, when fibres are used to reinforce polymers, they hinder the molecular flow in the matrix and, consequently, delay the relaxation process [37]. In this context, the interface properties play a relevant role because relaxation is due to the breaking of bonds and their propagation. In terms of creep behaviour, the fibre–matrix interface is also very important because the bonds' breakage and their propagation control the creep displacement [34,38]. Therefore, the creep process is delayed because both elastic deformation and viscous flow are retarded by the presence of fibres. Nevertheless, reducing the size of the reinforcements and an increase in their weight content promotes aggregation and the consequent loss of mechanical properties [39]. In addition, Mahesh et al. [20], when using a similar filament, found that most of the load was supported by the matrix, revealing, in this case, poor matrix–fibre adhesion. In such contexts, these authors reported that the excessive shear stresses acting on the layer interface and fibre–matrix interface, accompanied by the air gaps between printed layers, justified the lower compressive strength of the composites compared to neat PETG. In fact, the inter-layer adhesion proves to be determinant regarding thermal and mechanical properties, including tensile, fatigue, toughness and creep [40–42]. Voids promote low inter-layer strength and are sites of crack growth due to reduced inter-layer adhesion [42]. Furthermore, voids created during printing processes can reduce the material density and cause lower thermal, mechanical and conductive properties [42,43]. Therefore, all these facts reported in the open literature explain the higher compressive displacement, stress relaxation and creep behaviour observed in composites compared to neat PETG.

4. Conclusions

Poly(ethylene terephthalate)-glycol (PETG) is a thermoplastic widely used in 3D printing due to its excellent layer adhesion, very low shrinkage, good mechanical strength and high service temperature. These advantages make it suitable for several applications, but when reinforced with fibres, it becomes even more attractive. In previous studies, the characterisation of neat PETG and PETG-based composites essentially focuses on the tensile mode; however, the compression behaviour should be considered early in the design phase because it is also a very common loading mode. In this context, the compressive properties of PETG and PETG reinforced with carbon and aramid fibres were analysed in terms of static properties as well as creep and stress relaxation behaviour.

From the static characterisation, it was possible to conclude that, in relation to the neat PETG, the yield compressive strength decreased for both composites, with the largest drop observed for the PETG reinforced with Kevlar fibres (around 68.7% lower). Similar behaviour was observed for the compressive strain for PETG reinforced with carbon and Kevlar, where there was a decrease of 20.4% and 46.3%, respectively. Nevertheless, the highest compressive modulus was observed for PETG reinforced with carbon fibres (increase of 12.4%), while the lowest modulus was observed for PETG with Kevlar fibres (decrease of 39.6%).

In terms of stress relaxation behaviour, it was possible to observe, for the neat PETG, as expected, a decrease in compressive stresses over time, but more expressive for higher strain levels. Compared to the initial compressive stress, after 180 min, the neat PETG showed a decrease of 13.8%, while, for PETG reinforced with Kevlar and carbon fibres, decreases of around 14.7% and 22.7%, respectively, were observed. On the other hand, the creep response promoted higher compressive displacements, which were shown to be dependent on the level of compressive stresses. Thus, the compressive displacements after 180 min were, for PETG reinforced with Kevlar and carbon fibres, respectively, 7.9% and 9.7% higher, than the value obtained for the neat PETG. However, in the design of components subject to the compression mode, contrary to what is typical for the tensile mode, when fibres are added to PETG, higher values of stress relaxation and compressive displacement are observed.

Author Contributions: S.V. produced the materials, performed the tests and analysed the results; A.P.S. helped to analyse the results; P.N.B.R. wrote the manuscript. All authors have read and agreed to the published version of the manuscript.

Funding: This research received external funding from the SAICT project n° 31296, "COMP4UAVs", supported by POCI in its FEDER component and by FCT-IP.

Institutional Review Board Statement: Not applicable.

Informed Consent Statement: Not applicable.

Data Availability Statement: Not applicable.

Acknowledgments: This research is sponsored by national funds through FCT—Fundação para a Ciência e a Tecnologia, under the projects UID/EMS/00151/2020 and UIDB/00285/2020.

Conflicts of Interest: The authors declare no conflict of interest.

References

1. Sood, A.K.; Ohdar, R.K.; Mahapatra, S.S. Parametric Appraisal of Mechanical Property of Fused Deposition Modelling Processed Parts. *Mater. Des.* **2010**, *31*, 287–295. [CrossRef]
2. Panneerselvam, T.; Raghuraman, S.; Vamsi Krishnan, N. Investigating Mechanical Properties of 3D-Printed Polyethylene Terephthalate Glycol Material Under Fused Deposition Modeling. *J. Inst. Eng. Ser. C* **2021**, *102*, 375–387. [CrossRef]
3. Yuan, L.; Ding, S.; Wen, C. Additive Manufacturing Technology for Porous Metal Implant Applications and Triple Minimal Surface Structures: A Review. *Bioact. Mater.* **2019**, *4*, 56–70. [CrossRef] [PubMed]
4. Gantenbein, S.; Masania, K.; Woigk, W.; Sesseg, J.P.W.; Tervoort, T.A.; Studart, A.R. Three-Dimensional Printing of Hierarchical Liquid-Crystal-Polymer Structures. *Nature* **2018**, *561*, 226–230. [CrossRef]
5. Melenka, G.W.; Cheung, B.K.O.; Schofield, J.S.; Dawson, M.R.; Carey, J.P. Evaluation and Prediction of the Tensile Properties of Continuous Fiber-Reinforced 3D Printed Structures. *Compos. Struct.* **2016**, *153*, 866–875. [CrossRef]
6. Kasmi, S.; Ginoux, G.; Allaoui, S.; Alix, S. Investigation of 3D Printing Strategy on the Mechanical Performance of Coextruded Continuous Carbon Fiber Reinforced PETG. *J. Appl. Polym. Sci.* **2021**, *138*. [CrossRef]
7. Dong, G.; Tang, Y.; Li, D.; Zhao, Y.F. Mechanical Properties of Continuous Kevlar Fiber Reinforced Composites Fabricated by Fused Deposition Modeling Process. *Procedia Manuf.* **2018**, *26*, 774–781. [CrossRef]
8. Jiang, D.; Smith, D.E. Anisotropic Mechanical Properties of Oriented Carbon Fiber Filled Polymer Composites Produced with Fused Filament Fabrication. *Addit. Manuf.* **2017**, *18*, 84–94. [CrossRef]
9. Kumar, M.A.; Khan, M.S.; Mishra, S.B. Effect of Machine Parameters on Strength and Hardness of FDM Printed Carbon Fiber Reinforced PETG Thermoplastics. *Mater. Today Proc.* **2020**, *27*, 975–983. [CrossRef]
10. Sathish Kumar, K.; Soundararajan, R.; Shanthosh, G.; Saravanakumar, P.; Ratteesh, M. Augmenting Effect of Infill Density and Annealing on Mechanical Properties of PETG and CFPETG Composites Fabricated by FDM. *Mater. Today Proc.* **2021**, *45*, 2186–2191. [CrossRef]
11. Bhandari, S.; Lopez-Anido, R.A.; Gardner, D.J. Enhancing the Interlayer Tensile Strength of 3D Printed Short Carbon Fiber Reinforced PETG and PLA Composites via Annealing. *Addit. Manuf.* **2019**, *30*, 100922. [CrossRef]
12. Ferreira, I.; Vale, D.; Machado, M.; Lino, J. Additive Manufacturing of Polyethylene Terephthalate Glycol /Carbon Fiber Composites: An Experimental Study from Filament to Printed Parts. *Proc. Inst. Mech. Eng. Part L J. Mater. Des. Appl.* **2019**, *233*, 1866–1878. [CrossRef]
13. Mansour, M.; Tsongas, K.; Tzetzis, D.; Antoniadis, A. Mechanical and Dynamic Behavior of Fused Filament Fabrication 3D Printed Polyethylene Terephthalate Glycol Reinforced with Carbon Fibers. *Polym. Plast. Technol. Eng.* **2018**, *57*, 1715–1725. [CrossRef]

14. Kromoser, B.; Pachner, T. Optiknot 3D-Free-Formed Frameworks out of Wood with Mass Customized Knots Produced by FFF Additive Manufactured Polymers: Experimental Investigations, Design Approach and Construction of a Prototype. *Polymers* **2020**, *12*, 965. [CrossRef] [PubMed]
15. Frohn-Sörensen, P.; Geueke, M.; Tuli, T.B.; Kuhnhen, C.; Manns, M.; Engel, B. 3D Printed Prototyping Tools for Flexible Sheet Metal Drawing. *Int. J. Adv. Manuf. Technol.* **2021**, *115*, 2623–2637. [CrossRef]
16. Hsueh, M.H.; Lai, C.J.; Wang, S.H.; Zeng, Y.S.; Hsieh, C.H.; Pan, C.Y.; Huang, W.C. Effect of Printing Parameters on the Thermal and Mechanical Properties of 3d-Printed Pla and Petg, Using Fused Deposition Modeling. *Polymers* **2021**, *13*, 1758. [CrossRef]
17. Mercado-Colmenero, J.M.; La Rubia, M.D.; Mata-Garcia, E.; Rodriguez-Santiago, M.; Martin-Doñate, C. Experimental and Numerical Analysis for the Mechanical Characterization of PETG Polymers Manufactured with FDM Technology under Pure Uniaxial Compression Stress States for Architectural Applications. *Polymers* **2020**, *12*, 2202. [CrossRef]
18. Patterson, A.E.; Pereira, T.R.; Allison, J.T.; Messimer, S.L. IZOD Impact Properties of Full-Density Fused Deposition Modeling Polymer Materials with Respect to Raster Angle and Print Orientation. *Proc. Inst. Mech. Eng. Part C J. Mech. Eng. Sci.* **2021**, *235*, 1891–1908. [CrossRef]
19. *ASTM D695−15*; Standard Test Method for Compressive Properties of Rigid Plastics. American Society for Testing and Materials: Philadelphia, PA, USA, 2015; Volume 08.01, pp. 1–8.
20. Mahesh, V.; Joseph, A.S.; Mahesh, V.; Harursampath, D.; Chethan, V.N. Investigation on the Mechanical Properties of Additively Manufactured PETG Composites Reinforced with OMMT Nanoclay and Carbon Fibers. *Polym. Compos.* **2021**, *42*, 2380–2395. [CrossRef]
21. Monjon, A.; Santos, P.; Valvez, S.; Reis, P.N.B. Hybridization Effects on Bending and Interlaminar Shear Strength of Composite Laminates. *Materials* **2022**, *15*, 1302. [CrossRef]
22. Adeniran, O.; Cong, W.; Bediako, E.; Aladesanmi, V. Additive Manufacturing of Carbon Fiber Reinforced Plastic Composites: The Effect of Fiber Content on Compressive Properties. *J. Compos. Sci.* **2021**, *5*, 325. [CrossRef]
23. Lim, S.D.; Rhee, J.M.; Nah, C.; Lee, S.-H.; Lyu, M.-Y. Predicting the Long-Term Creep Behavior of Plastics Using the Short-Term Creep Test. *Int. Polym. Process.* **2004**, *19*, 313–319. [CrossRef]
24. Reis, P.N.B.; Silva, M.P.; Santos, P.; Parente, J.M.; Bezazi, A. Viscoelastic Behaviour of Composites with Epoxy Matrix Filled by Cork Powder. *Compos. Struct.* **2020**, *234*, 111669. [CrossRef]
25. Reis, P.N.B.; Silva, M.P.; Santos, P.; Parente, J.M.; Valvez, S.; Bezazi, A. Mechanical Performance of an Optimized Cork Agglomerate Core-Glass Fibre Sandwich Panel. *Compos. Struct.* **2020**, *245*, 112375. [CrossRef]
26. Ferreira, J.A.; Costa, J.D..; Reis, P.N. Static and Fatigue Behaviour of Glass-Fibre-Reinforced Polypropylene Composites. *Theor. Appl. Fract. Mech.* **1999**, *31*, 67–74. [CrossRef]
27. Reis, P.N.B.; Gorbatikh, L.; Ivens, J.; Lomov, S.V. Strain-Rate Sensitivity and Stress Relaxation of Hybrid Self-Reinforced Polypropylene Composites under Bending Loads. *Compos. Struct.* **2019**, *209*, 802–810. [CrossRef]
28. Reis, P.N.B.; Silva, M.P.; Santos, P. Stress Relaxation in Delaminated Carbon/Epoxy Composites. *Fibers Polym.* **2019**, *20*, 1284–1289. [CrossRef]
29. Varghese, S.; Kuriakose, B.; Thomas, S. Stress Relaxation in Short Sisal-Fiber-Reinforced Natural Rubber Composites. *J. Appl. Polym. Sci.* **1994**, *53*, 1051–1060. [CrossRef]
30. George, J.; Sreekala, M.S.; Thomas, S.; Bhagawan, S.S.; Neelakantan, N.R. Stress Relaxation Behavior of Short Pineapple Fiber Reinforced Polyethylene Composites. *J. Reinf. Plast. Compos.* **1998**, *17*, 651–672. [CrossRef]
31. Sreekala, M.; Kumaran, M.; Joseph, R.; Thomas, S. Stress-Relaxation Behaviour in Composites Based on Short Oil-Palm Fibres and Phenol Formaldehyde Resin. *Compos. Sci. Technol.* **2001**, *61*, 1175–1188. [CrossRef]
32. Mirzaei, B.; Tajvidi, M.; Falk, R.H.; Felton, C. Stress-Relaxation Behavior of Lignocellulosic High-Density Polyethylene Composites. *J. Reinf. Plast. Compos.* **2011**, *30*, 875–881. [CrossRef]
33. Wang, W.-H.; Huang, H.-B.; Du, H.-H.; Wang, H. Effects of Fiber Size on Short-Term Creep Behavior of Wood Fiber/HDPE Composites. *Polym. Eng. Sci.* **2015**, *55*, 693–700. [CrossRef]
34. Park, B.-D.; Balatinecz, J.J. Short Term Flexural Creep Behavior of Wood-Fiber/Polypropylene Composites. *Polym. Compos.* **1998**, *19*, 377–382. [CrossRef]
35. Houshyar, S.; Shanks, R.A.; Hodzic, A. Tensile Creep Behaviour of Polypropylene Fibre Reinforced Polypropylene Composites. *Polym. Test.* **2005**, *24*, 257–264. [CrossRef]
36. Bouafif, H.; Koubaa, A.; Perré, P.; Cloutier, A. Creep Behaviour of HDPE/Wood Particle Composites. *Int. J. Microstruct. Mater. Prop.* **2013**, *8*, 225–238. [CrossRef]
37. Obaid, N.; Kortschot, M.; Sain, M. Understanding the Stress Relaxation Behavior of Polymers Reinforced with Short Elastic Fibers. *Materials* **2017**, *10*, 472. [CrossRef]
38. Georgiopoulos, P.; Kontou, E.; Christopoulos, A. Short-Term Creep Behavior of a Biodegradable Polymer Reinforced with Wood-Fibers. *Compos. Part B Eng.* **2015**, *80*, 134–144. [CrossRef]
39. Hassanzadeh-Aghdam, M.K.; Ansari, R.; Mahmoodi, M.J.; Darvizeh, A. Effect of Nanoparticle Aggregation on the Creep Behavior of Polymer Nanocomposites. *Compos. Sci. Technol.* **2018**, *162*, 93–100. [CrossRef]
40. Yao, B.; Imani, F.; Sakpal, A.S.; Reutzel, E.W.; Yang, H. Multifractal Analysis of Image Profiles for the Characterization and Detection of Defects in Additive Manufacturing. *J. Manuf. Sci. Eng.* **2018**, *140*, 031014. [CrossRef]

41. Braga, R.A.; Magalhaes, P.A.A. Analysis of the Mechanical and Thermal Properties of Jute and Glass Fiber as Reinforcement Epoxy Hybrid Composites. *Mater. Sci. Eng. C* **2015**, *56*, 269–273. [CrossRef]
42. Blok, L.G.; Longana, M.L.; Yu, H.; Woods, B.K.S. An Investigation into 3D Printing of Fibre Reinforced Thermoplastic Composites. *Addit. Manuf.* **2018**, *22*, 176–186. [CrossRef]
43. Mohammadizadeh, M.; Imeri, A.; Fidan, I.; Elkelany, M. 3D Printed Fiber Reinforced Polymer Composites—Structural Analysis. *Compos. Part B Eng.* **2019**, *175*, 107112. [CrossRef]

aerospace

MDPI

Article

Electro-Thermal Parameters of Graphene Nano-Platelets Films for De-Icing Applications

Khitem Lahbacha [1,2], Sarah Sibilia [1], Gianmarco Trezza [1], Gaspare Giovinco [3], Francesco Bertocchi [4], Sergio Chiodini [4], Francesco Cristiano [4] and Antonio Maffucci [1,*]

1 Department of Electrical and Information Engineering, University of Cassino and Southern Lazio, 03043 Cassino, Italy; khitem.lahbacha@unicas.it (K.L.); sarah.sibilia@unicas.it (S.S.); gianmarco.trezza@unicas.it (G.T.)

2 Laboratory of Electronics and Microelectronics, University of Monastir, Monastir 5060, Tunisia

3 Department of Civil and Mechanical Engineering, University of Cassino and Southern Lazio, 03043 Cassino, Italy; giovinco@unicas.it

4 Nanesa S.r.l., 52100 Arezzo, Italy; francesco.bertocchi@nanesa.com (F.B.); rd@nsys-engineering.it (S.C.); francesco.cristiano@nanesa.com (F.C.)

* Correspondence: maffucci@unicas.it

Abstract: This paper provides a study of some relevant electro-thermal properties of commercial films made by pressed graphene nano-platelets (GNPs), in view of their use as heating elements in innovative de-icing systems for aerospace applications. The equivalent electrical resistivity and thermal emissivity were studied, by means of models and experimental characterization. Macroscopic strips with a length on the order of tens of centimeters were analyzed, either made by pure GNPs or by composite mixtures of GNPs and a small percentage of polymeric binders. Analytical models are derived and experimentally validated. The thermal response of these graphene films when acting as a heating element is studied and discussed.

Keywords: de-icing; electro-thermal models; electrical conductivity; graphene nano-platelets; thermal emissivity

Citation: Lahbacha, K.; Sibilia, S.; Trezza, G.; Giovinco, G.; Bertocchi, F.; Chiodini, S.; Cristiano, F.; Maffucci, A. Electro-Thermal Parameters of Graphene Nano-Platelets Films for De-Icing Applications. *Aerospace* **2022**, 9, 107. https://doi.org/10.3390/aerospace9020107

Academic Editors: Spiros Pantelakis, Andreas Strohmayer and Liberata Guadagno

Received: 15 January 2022
Accepted: 14 February 2022
Published: 16 February 2022

1. Introduction

Developing efficient and cost-effective de-icing systems is a major challenge in several industrial applications, not only related to aerospace [1], but also, for instance, in fields such as energy production [2] or energy distribution [3]. To this purpose, many techniques have been proposed and implemented over the years, based on three main approaches, involving mechanical, chemical and thermal methods and implementing either active or passive mitigation techniques [1,4]. By comparing these approaches, it turns out that de-icing based on Joule heating in electrical conductors conjugates a higher energy efficiency with easier system controllability, compared to the other existing techniques [5,6]. However, a major critical point in view of implementing such an electro-thermal technique is the proper choice of the heating element. Indeed, the use of heaters made by conventional conductors, such as copper and aluminum, holds some disadvantages, such as excessive weight or limited capability of handling the electrical current density required for efficient local heating.

To overcome these limits, recently, attention has been paid to the possibility of realizing the heaters by means of lightweight composite materials, such as those based on graphene-related materials [6,7]. Indeed, such materials are demonstrated to provide outstanding thermal and electrical behavior, as well as superior mechanical features, such as lightness [8,9]. However, the materials that can be strictly denoted as graphene or graphene-related are of a limited interest for industrial applications, given the well-known issues related to their fabrication and integration costs. Therefore, recently, attention has been paid to the so-called commercial graphene materials, which, strictly speaking, are more correctly classified as graphitic rather than graphene materials. Although their performance is not so outstanding as the real graphene, they can, however, realize a satisfactory compromise between the need for improving the performance and the need to use an industrially appealing material [10]. Materials based on graphene nanoplatelets (GNPs) are among the most promising candidates to meet the above trade-off. Indeed, GNPs are irregular flakes of few-layers graphene [11,12] that can be produced in several ways, including high-yield industrially scalable techniques such as microwave irradiation (e.g., [10,13]). Starting from GNPs, industrial materials suitable for applications such as those discussed in this paper are proposed in the form of composites.

Industrial nanocomposites made by macroscopic polymeric matrices embedding nanofillers such as GNPs, as well as carbon nanotubes, have been investigated over the last few years [14–16], highlighting their potential performance, but also their current limits. Indeed, the electrical, thermal and mechanical properties of nanocomposites with a small percentage of carbon nanofillers have been extensively studied [17–19]. Their electrical conductivity is high enough to provide good electromagnetic properties, such as shielding effectiveness [20,21], but is too low for them to be efficiently used as heaters for de-icing applications.

Recently, industrial composites with a high percentage of GNPs (up to 100%) have been investigated by the authors [22,23], in view of their possible use as heating elements of a de-icing system. To this purpose, a thorough study of the behavior of the main electro-thermal parameters is needed, since, for this class of composites, few works are so far available; a study is carried out, for instance, in [24], but it is only related to electrical conductivity.

The materials analyzed in this paper are macroscopic strips, industrially fabricated by assembling GNPs produced from expandable graphite. Strictly speaking, this material should be classified as graphite micro-platelets, but, for the sake of simplicity, we will, hereafter, refer to them as industrial graphene strips or films, with the meaning so far discussed. The strips analyzed are designated to become heating elements embedded in a glass-fibre reinforced material, in the typical multilayer structure of an airwing. In view of these applications, it is essential to investigate their electrical and thermal properties.

In this paper, we investigate the electrical resistivity and the thermal emissivity. The first parameter is, of course, essential to correctly manage the Joule effect, responsible for de-icing and anti-icing capability. The second one is fundamental to properly include radiative exchange with the surrounding environment.

This paper follows previous works done by the authors (see [22,23]). Compared to such references, here, a more accurate analysis of the electrical resistivity behavior is provided, including a new material, assessing a sensitivity analysis on the effect of strip lengths, and proposing a physically meaningful method to explain the obtained negative derivative of the resistivity with respect to temperature.

Furthermore, a new experimental setup is presented that is specifically intended to discuss and assess the Joule heating capability with respect to the power density targets for de-icing and anti-icing applications. This new setup provides much more controllable and stable conditions, so that a transient thermal analysis can be performed.

Finally, another novel contribution is a detailed description of a methodology proposed for deriving the thermal emissivity, by exploiting the feature of the set-up realized for the Joule heating assessment.

The paper is organized as follows: in Section 2, the fabrication of the graphene films is discussed, as well as the set-ups and methods proposed both for the characterization of the electrical resistivity and for the heating element. In Section 3, the results are presented and discussed, and an interpretation is given to the behavior of the resistivity vs. temperature.

2. Materials and Methods

2.1. Material Fabrication and Text-Fixture for Electrical Resistivity Characterization

The heating element studied in this paper is made from graphene strips connected to external electrodes, through which a controlled flow of current is injected. In view of characterizing its electrical resistivity, these films are put into the text-fixture reported in Figure 1. The graphene film is on a dielectric substrate and on copper electrodes at its end, which are connected to the amperometric circuit. For the purpose of characterization, the text fixture is made by two parts that are connected; the top frame in Figure 1 is placed on the bottom part, so that the film ends are sandwiched between two conducting plates (electrodes). Such a frame would not be strictly needed when realizing the heater.

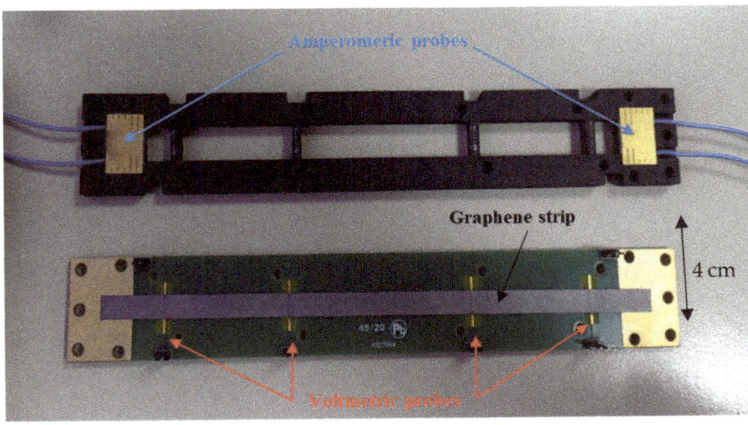

Figure 1. The text-fixture realized for the electro-thermal characterization, with a 2- or 4-probe measurement of the electrical resistance.

The films of industrial graphene analysed in this paper are strips of a lateral dimension of 1 cm, lengths between 10 and 18 cm, and thicknesses ranging from 55 to 75 μm. These films are realized by Nanesa [25], through the following industrial process:

(i) Graphene nano platelets (GNPs) are synthesized from a low-cost graphitic precursor (intercalated expandable graphite), through a process of thermal expansion followed by liquid exfoliation.

(ii) A mixture is obtained by dispersion of GNPs in a solvent (acetone) or aqueous solution, with magnetic stirring and a final sonication phase. In case of inclusion of any polymeric binder, this is added during sonication phase. From a mechanical point of view, the binders that are found to be suitable to the purposes of this work are polyurethane (used here) or epoxy.

(iii) the strips are then obtained by spraying the mixture at a controlled pressure, by using a semiautomatic 3-axes pantograph (Computer Numeric Control plotter EXTREMA, model Basic);

(iv) a final step of calendering (optionally joined to annealing) is applied, to compact the strip and provide optimized thickness/alignment ratio.

It is known that size and thickness of the GNPs exert significant influence on the physical properties of the final composite material. The effect of GNP thickness has been, for instance, discussed in [26], showing that thinner GNPs are recommended for improving global mechanical and thermal performance. The impact of size/thickness aspect ratio on the electrical conduction has been, instead, analysed in [27], showing that the percolation threshold in the composite increased with increasing values of aspect ratio. Finally, in [28] the electrical conductivity of the composite was demonstrated to improve as the GNP size and surface area increased. Therefore, in order to ensure stability to the behaviour of the GNP films, it is important to assess a fabrication process that is able to control the GNP dimensions. The GNPs produced by the above-mentioned process exhibit an average particle size of 30 μm, with a standard deviation of 5 μm (Figure 2), and an average thickness around 12 nm, with a standard deviation of 3 nm.

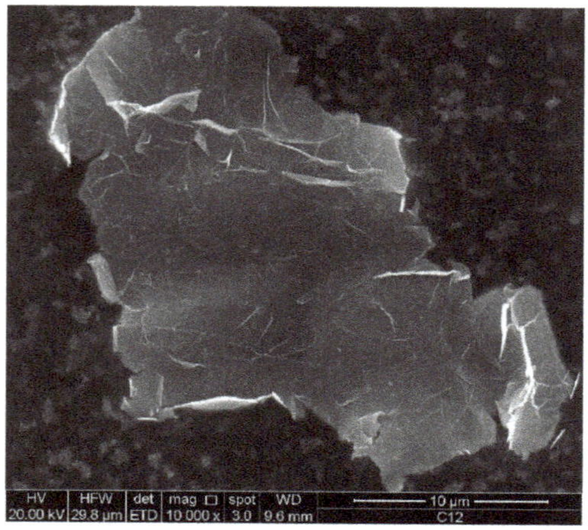

Figure 2. A single GNP in a Scanning Electron Microscope picture.

Table 1 summarizes the characteristics of the industrial graphene samples analysed in this paper.

Table 1. Characteristics of the graphene strips analyzed in this paper.

Material	%GNPs	Binder	Thickness (μm)	Length (cm)	Width (mm)
G-paper	100	none	75	7–18	7–11
G-PREG (95/5)	95	Polyurethane 5%	75	7–18	7–11
G-PREG (80/20)	80	Polyurethane 20%	75	7–18	7–11

2.2. Electrical Resistivity Characterization and Modelling

A model of the electrical resistivity versus temperature was here derived, by using the results of experimental characterizations and numerical simulations. The first ones consisted in measuring the DC electrical resistance at various values of temperature, while the text-fixture was undergoing a controlled thermal cycle inside a climatic chamber. Specifically, the temperature range (40 °C, +60 °C) was investigated. As shown in Figure 1, the text-fixture was designed so that the resistance could be measured by using a 2-probe technique (when the voltage probes were placed on the amperometric electrodes) or a 4-probe one (when using the voltage electrodes). Note that the voltage probe pair can be placed at different lengths, in order to estimate and remove the effect of contact resistance, which can be an issue when dealing with carbon interconnects/conventional electrodes interfaces [29]. The experimental characterization carried out here was performed with current levels up to 10 mA, which has been verified to be too low for inducing a significant Joule heating in the considered strips and, thus, temperature can be only varied externally, by controlling the climatic chamber environmental conditions.

For each temperature value T, the measured value of the electrical resistance $R_m(T)$ was used to provide a first estimation of the equivalent electrical resistivity, as follows:

$$\rho_m(T) = \frac{R_m(T) \cdot w \cdot t}{L_v} \tag{1}$$

being L_v the distance between the voltage probe pair, and w and t the nominal width and thickness of the graphene strip. However, a more accurate estimation of such a parameter is obtained by using the numerical result of a full 3D numerical electro-thermal model implemented in COMSOL Multiphysics [30], whose CAD is reported in Figure 3.

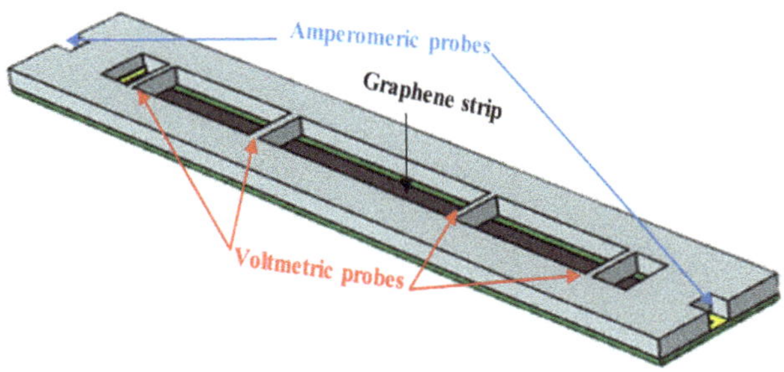

Figure 3. CAD model of the text-fixture adopted for the electro-thermal numerical simulation.

The equivalent electrical resistivity of the strip, $\rho_{eq}(T_i)$, is among the inputs of the numerical model; hence, its evaluation follows from the solution of an inverse problem. Indeed, starting from the value (1), which is assumed as the initial guest, the numerical model is used to calculate the values of resistance obtained by letting the film resistivity vary in a suitable neighborhood of the initial guess. By indicating with $R_s\left(\rho_{eq,k}, T_i\right)$ the simulated resistance corresponding to the k-th value of such a range, the equivalent electrical resistivity is assigned the value minimizing the mean square error between R_m and R_s:

$$MSE\left(\rho_{eq,n}\right) = \sqrt{\sum_{k=1}^{n}\left(R_m(T_i) - R_s\left(\rho_{eq,k}, T_i\right)\right)^2} \tag{2}$$

The equivalent resistivity is then recast in analytical form, by using the popular linear model that can be used in many cases also for nanoscale materials such as those derived from conventional conductors [31], and those based on graphene conductors [13,32]:

$$\rho_{eq}(T) = \rho_0(1 + \alpha(T - T_0)) \tag{3}$$

being T_0 a reference temperature, $\rho_0 = \rho_{eq}$ at T_0, and α the so-called temperature coefficient of the resistance (TCR).

The experiments have been carried out by using the climatic chamber ACS DY110, the DC power supply QJE QJ-3005 to impose the current, and two $6\frac{1}{2}$ digits multimeters to read the current (Agilent 34401A, Santa Clara, CA, USA) and the voltage values (Keithley 2700, Cleveland, OH, USA).

2.3. Thermal Emissivity Characterization and Modelling, Joule Heating Set-Up

The thermal emissivity parameter was here estimated starting from the experimental results obtained with the set-up to be used for demonstrating the capability of the graphene strip as a heating element. To this end, a new text-fixture has been fabricated, reported in Figure 4a, made by a plastic support to which the graphene film was vertically connected at two terminal electrodes. The support with the film was then placed inside a hollow black body (cavity), closed by a wall with a thermal InfraRed (IR) camera; see Figure 4b. This allowed us to insulate the region of the experiment from the outside environment, thus, obtaining a controlled cavity for the experiment. The temperature of the inner walls of the cavity and of the industrial graphene strip was measured not only by the IR camera, but also by two additional thermocouples, one of which was placed on the strip (contact probe) and the other on the inner walls of the cavity. In this way, it is possible to verify the temperature uniformity.

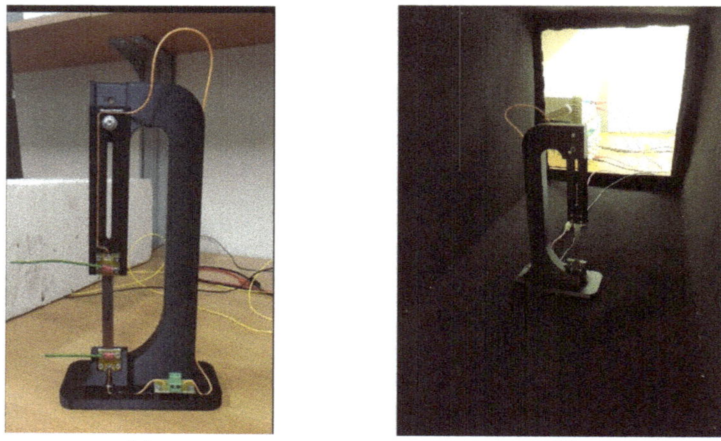

(**a**) (**b**)

Figure 4. Set-up for the Joule heating test and the emissivity characterization: (**a**) the heating element is fixed vertically to a plastic support, and connected to two electrodes; (**b**) the support is placed inside a hollow black body, closed by a wall with an IR camera (not shown here). Two additional thermocouples are placed inside, to control the temperature of the box and on the graphene strip.

In this set-up, the graphene strip was working as a heating element, with the heat self-produced by the Joule effect. Preliminary tests have shown that the current level must be increased to at least 0.7 A in order to have a significant effect: with such a current, the steady state temperature has been found to be at least +10 °C with respect to the baseline temperature recorded at the beginning. In the following experiment, we have used a current level of 1 A.

As already pointed out, this set-up can be used to estimate the thermal emissivity: to this end, the following model is adopted for estimating such a parameter, see [33]:

$$\varepsilon_m = \varepsilon_{IR} \cdot \frac{T_{IR}^4 - T_r^4}{T_G^4 - T_r^4} \tag{4}$$

where T_r, T_{IR}, and T_G are, respectively, the temperature values measured inside the cavity, by the IR camera and by the contact thermocouple on the graphene strip, whereas ε_{IR} is the reference emissivity value set on the camera, and, finally, ε_m is the measured emissivity. Note that Equation (4) came from a radiative heat transfer balance on the strip surface. An additional equation can be written by equating the heat flow between the graphene surface and the IR camera with the one between the graphene and the cavity walls:

$$\sigma.\varepsilon_m \cdot A_G \cdot \left(T_G^4 - T_r^4\right) = \frac{\sigma\left(T_G^4 - T_{cav}^4\right)}{\frac{1-\varepsilon_G}{\varepsilon_G \cdot A_G} + \frac{1}{F_{G-cav} \cdot A_G} + \frac{1-\varepsilon_{cav}}{\varepsilon_{cav} \cdot A_{cav}}} \tag{5}$$

where $\sigma = 5.67 \cdot 10^{-8}$ W/m^2 K^4 is the Stefan–Boltzmann constant, T_{cav} is the temperature of the cavity walls measured by the second thermocouple, ε_{cav} is the emissivity of the material adopted for the cavity walls (in our case, $\varepsilon_{cav} = 0.98$), A_G and A_{cav} are the surface areas of the graphene strip and the cavity inner wall facing the strip, and, finally, F_{G-cav} is the view factor between the graphene strip surface and the cavity, which can be here assumed to be 1. The solution of the equation system (4) and (5) provide the values of T_r and of the emissivity ε_m.

From the practical point of view, the method proceeds as follows: the strip is fed by a controlled current of 1A, which starts producing heat by the Joule effect, until the steady-state condition is reached. At that time, all the temperature values in (4) are measured and a thermal image of the strip is taken by the IR camera (see, for instance, Section 3). The emissivity is identified through an iterative process of minimization of the difference between the temperature values T_{IR}, and T_G. Indeed, T_{IR} can be changed by changing the reference value ε_{IR}, that in turns controls the estimated radiative heat flow coming from the strip surface.

The same set-up is used to study the Joule heating effect and so the characteristics of the graphene heater. To this end, currents of the same order of magnitude are used (about 1 A) and the thermocamera is used to provide the meaurement of the spatial distribution of the remperature over the whole text-fixture, both during the thermal transient and at the finale steady-state condition. This fixed current level of 1 A corresponds to a variable value of the electrical power, as the equivalent resistance of the strip varies as an effect of temperature rise. In the experiment carried out here and described in the next Section, the baseline temperature was T = 26.8 °C (see Section 3.2): the electrical power adsorbed at such a temperature was 1.05 W.

In these experiments, a Fluke Ti480 PRO thermocamera has been used, together with 3 calibrated K-type thermocouples. The current is provided by the DC power supply QJE QJ-3005, and monitored by a multimeter Agilent 34401A.

3. Results and Discussion

3.1. Electrical Resistivity Evaluation

The set-up and methods described in Section 2.2 have been used to retrieve the equivalent electrical resistivity of the three kinds of graphene strips under investigation (as listed in Table 1).

A preliminary characterization has been carried out to demonstrate the linearity of the V-I characteristic in the considered range of current values (up to 1 A). The parasitic effects of the setup, including any contact resistance, have been successfully removed by following the same as done in [22], namely by means of a proper preliminary calibration and by using the 4-probe technique.

For the G-paper strip, Figure 5 shows the equivalent electrical resistivity vs. temperature, obtained by using the methods described in Section 2.2. Here, the red dots are the values identified by the optimization of the numerical solution, whereas the blue line is the result of a linear fitting of such values, according to (3). The fitting coefficients are reported in Table 2, along with an estimation of the uncertainty.

Indeed, as already announced, a linear fitting matches the behavior of the resistivity of such materials very well. This result is in good agreement with that found in [22,23], despite the different dimensions of the strips (especially the reduced variability of the thickness), which demonstrates the robustness of the result. A similar behavior has been found for the other two materials, whose results are listed in Table 2. For the composite mixtures G-Preg, the results are slightly different from those reported in [22,23], since a different annealing has been used here.

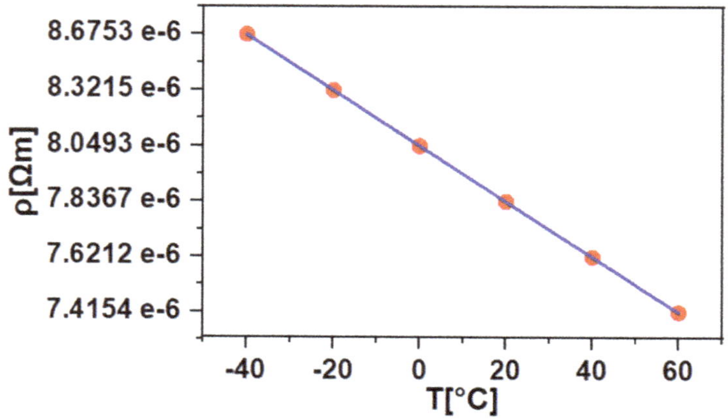

Figure 5. Equivalent resistivity vs. temperature for the pure graphene strip (G-paper): values identified from the experimental data (red dots), and linear fitting according to (3), whose coefficients are given in Table 2 (blue line).

Table 2. Fitting coefficients for model (3), for the three graphene materials films and for copper.

Material	ρ_0 (μΩm)	α (1/°C)	u (ρ)/ρ (%)
Cu	1.68×10^{-2}	3.90×10^{-3}	-
G-paper	7.86	-1.55×10^{-3}	2.1
G-PREG (95/5)	19.32	-1.31×10^{-3}	2.0
G-PREG (80/20)	41.57	-0.84×10^{-3}	2.0

An assessment of the result has been carried out here by checking its sensitivity to the strip length, and specifically to L_v, the distance between the voltage probe pair (see Figure 1). The result of such an assessment is reported in Table 3, with reference to a strip of G-preg 95/5 of surface dimensions 11 × 180 mm, at T = 20 °C. As it can been seen from the results, the proposed method provides a resistivity value that is independent from the strip length, with an error that is about 2% at maximum, for shorter lengths (for which it is expected that the contact resistance may start having a role).

Table 3. Assessment of the results with respect to the strip length.

L_v (mm)	R_m (Ω)	R_s (Ω)	ρ_{eq}(μΩm)	Err%
140	3.363	3.325	19.82	1.14
100	2.390	2.372	19.72	0.76
60	1.437	1.415	19.77	1.57
40	0.958	0.938	19.76	2.10

The results of the electro-thermal characterization of the electrical resistivity confirm what has been already found in the previous papers of [22,23]: these graphene films behave as negative temperature coefficient (NTC) materials. This nice and desirable behavior can be extremely useful in easing the control of the final heater system, since it helps stability, as, for instance, pointed out in applications such thermoelectrical sensors [34].

Note that the electrical performance of the industrial graphene is far from those exhibited by conventional conductors such as copper, see Table 2. However, although the resistivity is much higher, one of the main reasons for proposing graphene instead of copper is the possibility of reducing the weight, which is a crucial constraint in aeronautical applications. Indeed, copper density is about 9 g/cm^3, whereas that for the strips analysed here is in the range 0.0020–0.045 g/cm^3. Therefore, given the same dimensions, a copper strip weight will be about 2100 or 2500 times larger than the graphene strips.

To gain more insight into this phenomenon, we propose here a simple model that can be represented as in Figure 6.

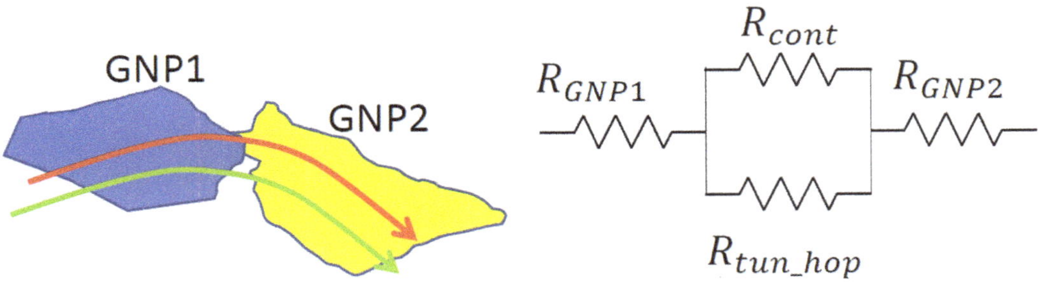

(a) (b)

Figure 6. Interpretation of the NTC behavior of the graphene strips: (**a**) two adjacent GNPs in the strip, with two possible conduction mechanisms: contact (red arrow) or tunneling and hopping (green arrow); (**b**) equivalent circuital description.

The electrical conduction between two adjacent GNPs in the strip may be modelled as the series of the conduction phenomena in the single GNPs (corresponding to the resistances R_{GNP} in Figure 6b, and the phenomena at interface, which, in turn, can be associated to the contact and to tunneling and hopping mechanisms. The first is schematically represented as the red path in Figure 6a, corresponding to the resistor R_{cont} in Figure 6b, whereas the second is described by the green path in Figure 6a, corresponding to R_{cont} in Figure 6b.

The equivalent resistance R_{GNP}, that can be approximately modeled as the resistance of a graphene ribbon [29,35]:

$$R_{GNP}(T) = \frac{r_0}{l_{mfp}(T)M(T)} len_{GNP} \tag{6}$$

where r_0 is a constant value related to the quantum resistance, $l_{mfp}(T)$ is the mean free path, $M(T)$ is the equivalent number of conducting channels [36], and len_{GNP} is the average length of the GNP. As pointed out in [36,37], these parameters display different behavior with respect to the temperature: l_{mfp} decreases as T^{-2}, whereas M increases linearly with T. The interface resistances can be modelled as [37]

$$R_{cont}(T) = \frac{r_0 + R_P(T)}{M(T)}, \tag{7}$$

with $R_P(T)$ linearly increasing with T, whereas the tunneling and hopping term can be demonstrated to depend on T as it follows, see [38]:

$$R_{tun\_hop}(T) = \frac{K}{T^{-4}}, \tag{8}$$

where K is a quantum term not depending on T.

By analyzing the expressions (6)–(8) and the dependence of their parameters from temperature, it is easy to realize that there can be operating conditions for which the counteracting behavior of these parameters may lead the equivalent resistance of the circuit in Figure 6b to become a decreasing function of the temperature.

3.2. Thermal Emissivity Evaluation and Joule Heating Capability

Starting from the set-up and methods reported in Section 2.3, the graphene strips have been used as heating element. A G-paper film has been used, with dimensions of 11 × 75 mm, with an imposed current of 1.05 A. The steady-state distribution of the temperature is given in Figure 7 (1 pixel corresponds to about 1.5 mm).

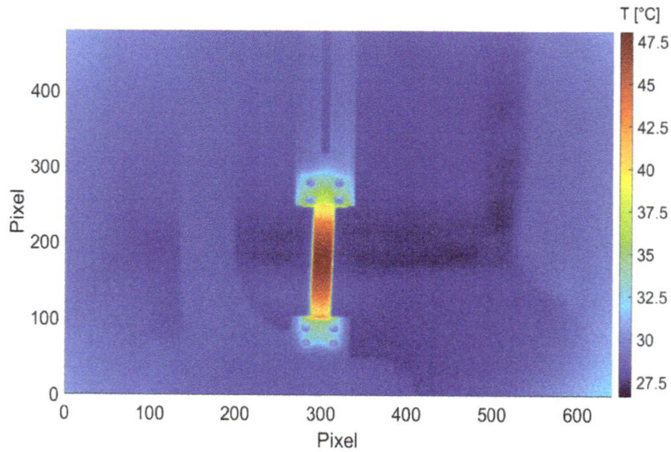

Figure 7. Temperature distribution over the graphene strip at steady-state (G-paper film fed by a current of 1.05 A). 1 pixel corresponds to about 1.5 mm.

These experiments allow us to estimate the thermal emissivity, by means of the method previously described, and by using (4). The results are summarized in Table 4, and are coherent with those reported in [14]. Note that the emissivity of the binder (polyurethane) can be assumed equal to 0.90, hence by increasing the percentage of binder, the emissivity is expected to increase, as coherently reported in Table 4.

Table 4. Thermal emissivity computed for the three graphene films (combined standard uncertainty is 0.01).

	G-Paper	G-Preg (95/5)	G-Preg (80/20)
Emissivity	0.50	0.56	0.81

In Figure 8 the temperature distribution calculated along the middle line of the strip is plotted for each time instant. The thermal transient duration is about 2 min, hence the graphene heater is characterized by a fast responsitivity, if compared to conventional conductors.

Let us now discuss the Joule heating capability in view of the use of such strips as heating elements. Standard requirements for the heat flux needed for de-icing and anti-icing are falling in the range of tens of kW/m^2, depending on the specific systems and conditions [39]. For our purposes, the project targets are 50 W/m^2 and 25 W/m^2 for de-icing and anti-icing, respectively. As mentioned, the power dissipated by the strip is about 1.05 W, that provide an average value of the power density of 95.8 kW/m^2 (assuming a uniform distribution of the temperature along the whole strip). This value would be enough both for de-icing and anti-icing applications. However, it can be observed from Figure 6 that the graphene heater is characterized by an almost flat distribution of the temperature on almost the entire strip, except for two regions of few mm in touch with the electrodes. Considering this behavior, if we refer to this inner zone, the power density can be estimated as equal to 213 kW/m^2, thus, providing an excellent performance.

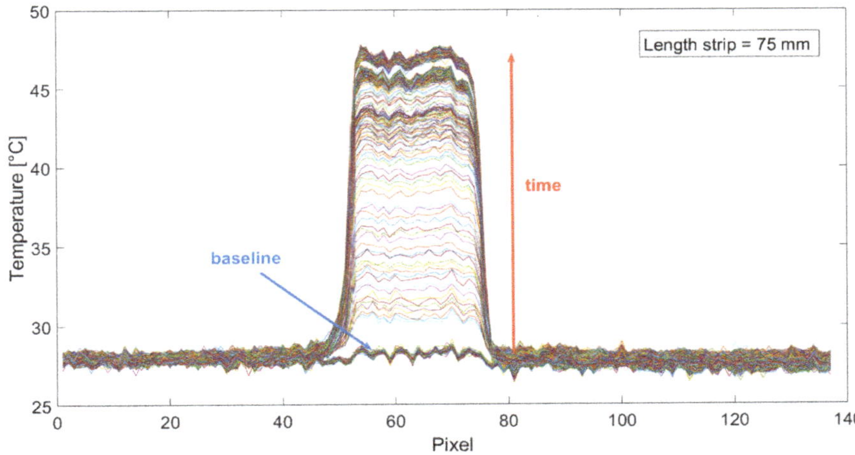

Figure 8. Temperature distribution calculated along the middle line of the strip, for each time instant from the beginning (baseline), to the steady state, reached after about 2 m (1 pixel corresponds to about 1.5 mm).

Note that in real de-icing and anti-icing applications, these values of power intensity must be imposed in continuous mode with no failure for hot-spots over the whole heating surface. Therefore, the uniform distribution of temperature in the inner part of the strip and the absence of hot spots (Figure 7) is a highly desirable feature for a heater. Note that this behavior is observed not only at steady state but also at each time instant during the transient (see the flat areas in Figure 8). Similar behavior is observed for the other two graphene films.

4. Conclusions

This paper has studied commercial graphene-based macroscopic strips, in view of their potential use as heating elements in de-icing systems. Set-ups and methods have been proposed to identify two parameters of major interest for such an application: the electrical resistivity and the thermal emissivity. The electrical resistivity is found to fall in the range (8–40) μΩm at room temperature, with a negative coefficient of temperature, whereas the thermal emissivity values ranged from 0.5 to 0.8, at room temperature. The thermal response of these graphene films is characterized by short transients, compared to conventional conductors, and by a highly desirable flatness and uniformity of the temperature distribution over the strip, not only at steady-state but also during the thermal transient.

As for the Joule heating capability, the analyzed strips were demonstrated to provide under a current of 1A an average power density of 95.8 kW/m^2, matching the targets requested for de-icing and anti-icing applications.

Future works will be devoted to complete the analysis of parameters by estimating also the thermal conductivity, and carrying on a systematic study of the characteristics of the thermal response of the heating elements.

In addition, work is in progress to assess the mechanical performance of the strips studied in this paper, when considering them in a realistic assembly, where they will be enveloped in a glass-fibre reinforced material. In order to enhance the mechanical performance of such multi-layered structures, different methodologies are under examination, and will be discussed in future works.

Author Contributions: Conceptualization, A.M.; methodology, A.M., G.G., F.B., F.C., S.C.; investigation, K.L., S.S., G.T.; writing, K.L. and A.M. All authors have read and agreed to the published version of the manuscript.

Funding: This research was partially funded by EU Commission program H2020, Graphene Flagship Graphene, Core3/H2020-SGA-FET-GRAPHENE-2019, Project SH10 GICE "Graphene-Based, Thermoelectric Ice Protection System," grant # 881603, and by Regione Lazio, Italy, under the program POR FESR LAZIO 2014–2020, Project AIRETEC, grant #A0320-2019-28139.

Institutional Review Board Statement: Not applicable.

Informed Consent Statement: Not applicable.

Data Availability Statement: Not applicable.

Conflicts of Interest: The authors declare no conflict of interest.

References

1. Yamazaki, M.; Jemcov, A.; Sakaue, H. A Review on the Current Status of Icing Physics and Mitigation in Aviation. *Aerospace* **2021**, *8*, 188. [CrossRef]
2. Wang, Z. Recent Progress on Ultrasonic De-Icing Technique Used for Wind Power Generation, High-Voltage Transmission Line and Aircraft. *Energy Build.* **2017**, *140*, 42–49. [CrossRef]
3. Fakorede, O.; Feger, Z.; Ibrahim, H.; Ilinca, A.; Perron, J.; Masson, C. Ice Protection Systems for Wind Turbines in Cold Climate: Characteristics, Comparisons and Analysis. *Renew. Sustain. Energy Rev.* **2016**, *65*, 662–675. [CrossRef]
4. Thomas, S.K.; Cassoni, R.P.; Macarthur, C.D. Aircraft anti-icing and de-icing techniques and modeling. *J. Aircr.* **1996**, *33*, 841–854. [CrossRef]
5. Botura, G.; Sweet, D.; Flosdorf, D. Development and Demonstration of Low Power Electrothermal De-icing System. In Proceedings of the 43rd AIAA Aerospace Sciences Meeting and Exhibit, Reno, NV, USA, 10–13 January 2005.
6. Karim, N.; Zhang, M.; Afroj, S.; Koncherry, V.; Potluri, P.; Novoselov, K.S. Graphene-based surface heater for de-icing applications. *RSC Adv.* **2018**, *8*, 16815–16823. [CrossRef]
7. Redondo, O.; Prolongo, S.G.; Campo, M.; Sbarufatti, C.; Giglio, M. Anti-icing and de-icing coatings based Joule's heating of graphene nanoplatelets. *Compos. Sci. Technol.* **2018**, *164*, 65–73. [CrossRef]
8. Sahu, D.; Sutar, H.; Senapati, P.; Murmu, R.; Roy, D. Graphene, Graphene-Derivatives and Composites: Fundamentals, Synthesis Approaches to Applications. *J. Compos. Sci.* **2021**, *5*, 181. [CrossRef]
9. Jia, X.; Campos-Delgado, J.; Terrones, M.; Meunier, V.; Dresselhaus, M.S. Graphene edges: A review of their fabrication and characterization. *Nanoscale* **2011**, *3*, 86–95. [CrossRef]
10. Kovtun, A.; Treossi, E.; Mirotta, N.; Scidà, A.; Liscio, A.; Christian, M.; Valorosi, F.; Boschi, A.; Young, R.J.; Galiotis, C.; et al. Benchmarking of graphene-based materials: Real commercial products versus ideal graphene. *2D Mater.* **2019**, *6*, 025006. [CrossRef]
11. Cataldi, P.; Athanassiou, A.; Bayer, I.S. Graphene Nanoplatelets-Based Advanced Materials and Recent Progress in Sustainable Applications. *Appl. Sci.* **2018**, *8*, 1438. [CrossRef]
12. Jiménez-Suárez, A.; Prolongo, S.G. Graphene Nanoplatelets. *Appl. Sci.* **2020**, *10*, 1753. [CrossRef]
13. Maffucci, A.; Micciulla, F.; Cataldo, A.; Miano, G.; Bellucci, S. Bottom-up Realization and Electrical Characterization of a Graphene-Based Device. *Nanotechnology* **2016**, *27*, 095204. [CrossRef] [PubMed]
14. Vertuccio, L.; De Santis, F.; Pantani, R.; Lafdi, K.; Guadagno, L. Effective de-icing skin using graphene-based flexible heater. *Compos. Part B Eng.* **2019**, *162*, 600–610. [CrossRef]
15. Vertuccio, L.; Foglia, F.; Pantani, R.; Romero-Sánchez, M.; Calderón, B.; Guadagno, L. Carbon nanotubes and expanded graphite based bulk nanocomposites for de-icing applications. *Compos. Part B Eng.* **2021**, *207*, 108583. [CrossRef]
16. Prolongo, S.G.; Moriche, R.; Del Rosario, G.; Jiménez-Suárez, A.; Prolongo, M.G.; Ureña, A. Joule Effect Self-Heating of Epoxy Composites Reinforced with Graphitic Nanofillers. *J. Polym. Res.* **2016**, *23*, 189. [CrossRef]
17. Mohiuddin, M.; Hoa, S.V. Temperature-dependent electrical conductivity of CNT–PEEK composites. *Compos. Sci. Technol.* **2011**, *72*, 21–27. [CrossRef]
18. Han, S.; Chand, A.R.; Araby, S.; Cai, R.; Chen, S.; Kang, H.; Cheng, R.; Meng, Q. Thermally and electrically conductive multifunctional sensor based on epoxy/graphene composite. *Nanotechnology* **2019**, *31*, 075702. [CrossRef] [PubMed]
19. Kargar, F.; Barani, Z.; Balinskiy, M.; Magana, A.S.; Lewis, J.S.; Balandin, A.A. Dual-Functional Graphene Composites for Electromagnetic Shielding and Thermal Management. *Adv. Electron. Mater.* **2019**, *5*, 1800558. [CrossRef]
20. Ayub, S.; Guan, B.H.; Ahmad, F.; Oluwatobi, Y.A.; Nisa, Z.U.; Javed, M.F.; Mosavi, A. Graphene and iron reinforced polymer composite electromagnetic shielding applications: A review. *Polymers* **2021**, *13*, 2580. [CrossRef] [PubMed]
21. Jen, Y.M.; Huang, J.C. Synergistic Effect on the Thermomechanical and Electrical Properties of Epoxy Composites with the Enhancement of Carbon Nanotubes and Graphene Nano Platelets. *Materials* **2019**, *12*, 255. [CrossRef]

22. Sibilia, S.; Bertocchi, F.; Chiodini, S.; Cristiano, F.; Ferrigno, L.; Giovinco, G.; Maffucci, A. Temperature-dependent electrical resistivity of macroscopic graphene nanoplatelet strips. *Nanotechnology* **2021**, *32*, 275701. [CrossRef] [PubMed]
23. Maffucci, A.; Bertocchi, F.; Chiodini, S.; Cristiano, F.; Ferrigno, L.; Giovinco, G.; Trezza, G. Electrothermal Parameters of Graphene Nanoplatelets Films. In Proceedings of the IEEE 21st International Conference on Nanotechnology (IEEE NANO), Montreal, QC, Canada, 28–30 July 2021; pp. 323–326.
24. Wu, H.; Drzal, L.T. Graphene nanoplatelet paper as a light-weight composite with excellent electrical and thermal conductivity and good gas barrier properties. *Carbon* **2012**, *50*, 1135–1145. [CrossRef]
25. NANESA. Available online: http://www.nanesa.com/en-US/Graphene (accessed on 10 January 2022).
26. Sutar, H.; Mishra, B.; Senapati, P.; Murmu, R.; Sahu, D. Mechanical, Thermal, and Morphological Properties of Graphene Nanoplatelet-Reinforced Polypropylene Nanocomposites: Effects of Nanofiller Thickness. *J. Compos. Sci.* **2021**, *5*, 24. [CrossRef]
27. Fang, C.; Zhang, J.; Chen, X.; Weng, G.J. Calculating the Electrical Conductivity of Graphene Nanoplatelet Polymer Composites by a Monte Carlo Method. *Nanomaterials* **2020**, *10*, 1129. [CrossRef]
28. Ravindran, A.R.; Feng, C.; Huang, S.; Wang, Y.; Zhao, Z.; Yang, J. Effects of Graphene Nanoplatelet Size and Surface Area on the AC Electrical Conductivity and Dielectric Constant of Epoxy Nanocomposites. *Polymers* **2018**, *10*, 477. [CrossRef]
29. Todri-Sanial, A.; Dijon, J.; Maffucci, A. *Carbon Nanotubes for Interconnects: Process, Design and Applications*; Springer: Dordrecht, The Netherlands, 2016.
30. COMSOL. Available online: https://www.comsol.com/comsol-multiphysics (accessed on 12 December 2021).
31. Steinhögl, W.; Schindler, G.; Steinlesberger, G.; Traving, M.; Engelhardt, M. Comprehensive study of the resistivity of copper wires with lateral dimensions of 100 nm and smaller. *J. Appl. Phys.* **2005**, *97*, 023706. [CrossRef]
32. Chiariello, A.G.; Maffucci, A.; Miano, G. Size and Temperature Effects on the Resistance of Copper and Carbon Nanotubes Nano-interconnects. In Proceedings of the 19th IEEE Topical Meeting EPEPS 2010, Austin, TX, USA, 25–27 October 2010; pp. 97–100.
33. Arpino, F.; Buonanno, G.; Giovinco, G. Thermal conductance measurement of windows: An innovative radiative method. *Experim. Therm. Fluid Sci.* **2008**, *32*, 1731–1739. [CrossRef]
34. Koskinen, T.; Juntunen, T.; Tittonen, I. Large-Area Thermal Distribution Sensor Based on Multilayer Graphene Ink. *Sensors* **2020**, *20*, 5188. [CrossRef] [PubMed]
35. Forestiere, C.; Maffucci, A.; Miano, G. Hydrodynamic model for the signal propagation along carbon nanotubes. *J. Nanophotonics* **2010**, *4*, 041695.
36. Maffucci, A.; Miano, G. Number of Conducting Channels for Armchair and Zig-Zag Graphene Nanoribbon Interconnects. *IEEE Trans. Nanotechnol.* **2013**, *12*, 817–823. [CrossRef]
37. Maffucci, A.; Micciulla, F.; Cataldo, A.; Miano, G.; Bellucci, S. Modeling, Fabrication, and Characterization of Large Carbon Nanotube Interconnects with Negative Temperature Coefficient of the Resistance. *IEEE Trans. Compon. Packag. Manuf.* **2017**, *7*, 485–493. [CrossRef]
38. Zhao, S.; Lou, D.; Zhan, P.; Li, G.; Dai, K.; Guo, J.; Zheng, G.; Liu, C.; Shena, C.; Guo, Z. Heating-induced negative temperature coefficient effect in conductive graphene/polymer ternary nanocomposites with a segregated and double-percolated structure. *J. Mater. Chem. C* **2017**, *32*, 8233–8242. [CrossRef]
39. Hann, R.; Enache, A.; Nielsen, M.C.; Stovner, B.N.; van Beeck, J.; Johansen, T.A.; Borup, K.T. Experimental Heat Loads for Electrothermal Anti-Icing and De-Icing on UAVs. *Aerospace* **2021**, *8*, 83. [CrossRef]

aerospace

MDPI

Article

Investigation of the Impact of a Particle Foam Insulation on Airflow, Temperature Distribution, Pressure Profile and Frost Buildup on the Aircraft Structure

Victor Norrefeldt * and Gerhard Riedl

Fraunhofer Institute for Building Physics IBP, Fraunhoferstr. 10, D-83626 Valley, Germany; gerhard.riedl@ibp.fraunhofer.de
* Correspondence: victor.norrefeldt@ibp.fraunhofer.de

Abstract: Aircraft insulation separates the thermally comfortable cabin interior environment from the extremely cold outside conditions. However, the fabrication and installation of the insulation in the aircraft is a labor-intensive task. Tailored, rigid particle foam parts could be a solution to speed up the installation process. The presented study investigates the feasibility of such a concept from a hygrothermal point of view. Due to the temperature difference between the cold air trapped between aircraft skin and insulation on one side and the warm cabin air on the other side, a buoyancy-induced pressure difference forms. This effect drives the warmer air through leakages in the insulation system towards the cold skin. Here, moisture contained in the air condenses on the cold surfaces, increasing the risk for uncontrolled dripping ("rain in the plane") when it melts. Therefore, this study compares the condensate build-up of different installations of a rigid particle foam frame insulation with the classical glass fiber capstrip. Tests are hosted in the Fraunhofer Lining and Insulation Test Environment chamber. It is shown that careful installation of the particle foam frame insulation provides similar level of moisture protection as the current state of the art insulation, and that the condensate amount does not depend on the amount of airflow directly behind the sidewall.

Keywords: aircraft insulation; leakage; moisture; ice; frost

Citation: Norrefeldt, V.; Riedl, G. Investigation of the Impact of a Particle Foam Insulation on Airflow, Temperature Distribution, Pressure Profile and Frost Buildup on the Aircraft Structure. *Aerospace* **2021**, *8*, 359. https://doi.org/10.3390/aerospace8120359

Academic Editors: Spiros Pantelakis, Andreas Strohmayer and Liberata Guadagno

Received: 21 October 2021
Accepted: 18 November 2021
Published: 23 November 2021

Publisher's Note: MDPI stays neutral with regard to jurisdictional claims in published maps and institutional affiliations.

1. Introduction

The typical aircraft cabin operative temperature is in the range of 18.3 to 26.7 °C (65 to 80 °F) [1], whereas the external temperature is around −56.5 °C [2]. This temperature difference can be maintained due to the presence of thermal insulation in the aircraft, leading to a sufficiently warm sidewall surface. In [1], it is specified that the sidewall surface temperature shall not differ more than 5.6 K (10 °F) from the cabin temperature to maintain passenger thermal comfort.

The main material for insulating an aircraft is glass fiber [3] wrapped by a cover film (e.g., [4]). Besides the thermal protection, the insulation system provides acoustic dampening of the aircraft noise, and most importantly for certification aspects, it protects passengers in the case of fire and therefore undergoes thorough testing [5]. These tests cover the flame propagation, heat and smoke release and burn through resistance. An insulation material may only be installed onboard the aircraft if all four tests are successfully passed.

The insulation has the form of blankets and covers the frames and fields (Figure 1). These blankets are flexible and compressible and thus can be wrapped around the aircraft structure The insulation concept used within this publication is deduced from [6] and extended where needed (Figure 1). Within this publication, the "cavity" denotes the air gap between skin and the outboard side of the primary insulation blanket, and "gap" denotes the air gap between the primary insulation blanket and the sidewall and secondary insulation. Both air volumes are of undefined shape as they depend on the blanket deformations, structural obstacles, etc.

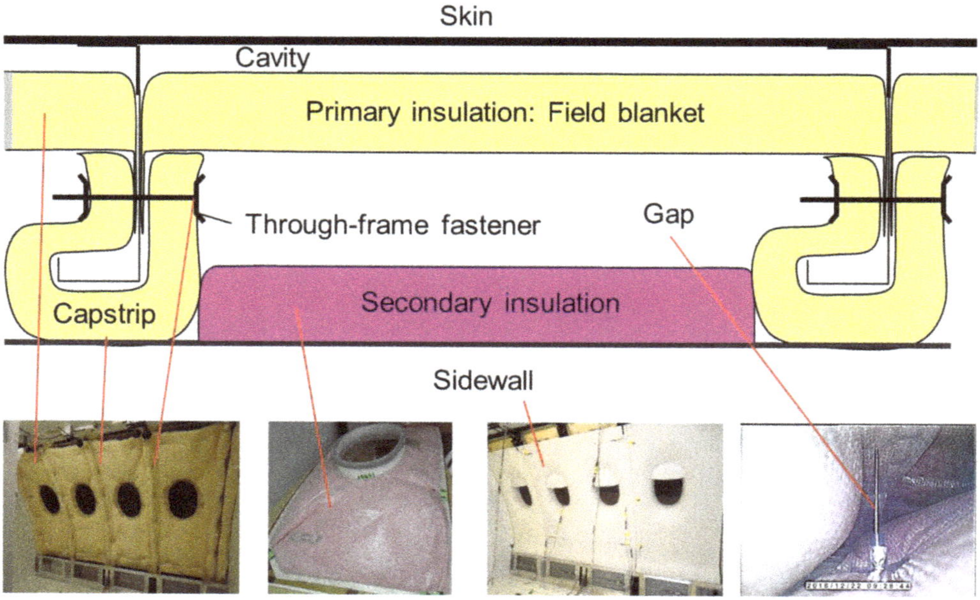

Figure 1. Pieces of the aircraft insulation.

Whereas field blankets are comparably easy to install, the capstrip covering the frame requires more manual work. [6] demands that the capstrip should be fixed either with so called through frame fasteners or with brackets at least every 14" (35.5 cm, Figure 2). With this installation, the flexible capstrip blanket is adequately pushed against the field blanket to provide sufficient burnthrough resistance without leaving large, unprotected air gaps. Thus, a high amount of manual work is required to install this piece of insulation on the frame. With the use of a rigid particle foam capstrip, this installation effort is reduced because it clamps on the frame and does not require the fasteners.

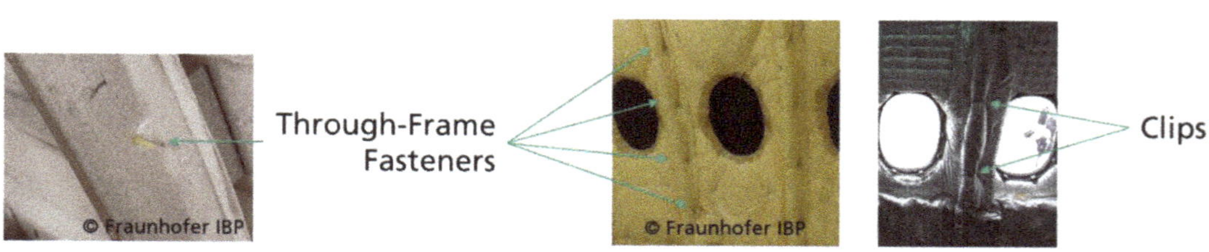

Figure 2. Comparison of different capstrip fixation principles, right picture from Olivier Cleynen, CC BY-SA 3.0, via Wikimedia Commons.

In addition to the thermal protection, the insulation has a function of airflow and moisture barrier in the aircraft. The major source of moisture in the aircraft cabin is the water vapor emitted by passengers. In [7], a literature review was conducted on cabin air quality measurements and it was concluded that the average cabin relative humidity level amounts to 16% with a minimum of 0.9%. In a subject study, [8] found that the perception of dryness significantly increases after 90 min at low humidity levels of 10%. In [9], measurements were performed on domestic short-haul flights and found relative humidity levels in the cabin between 17.9% and 27%, which are typically reached after 15 min of flight. Based on the CO_2 balance, the resulting fresh airflow rate was back-computed and a parametric study was conducted showing that a reduction in the fresh airflow rate would contribute to a higher and more comfortable cabin moisture content. [10] conducted a subject study with variable fresh airflow rates between 1.1 and 5.2 L/s per passenger in a cabin mock-up and found resulting relative humidity levels in the cabin between 13% and 33%.

Most of the emitted moisture vents out by the cabin ventilation system. Nevertheless, some moisture can accumulate by two major paths: through molecular diffusion into the blanket, driven by the water vapor pressure gradient between the warm inbound and the cold outbound blanket sides, and by convective transport of air towards cold structures. In order to prevent moisture from diffusing into the insulation blanket and being trapped, the cover film should have a sufficient moisture resistance. Reference [11] measured a 420 kg weight increase in the insulation blankets in an A310 during a D-check. The dependence of water uptake from the cover film's moisture diffusion resistance is experimentally proven in [12], where a 75% reduction in accumulated water in the blanket could be achieved by using a cover film with the moisture diffusion equivalent length s_d = 5.3 m instead of s_d = 0.66 m. An s_d-value of 1 m corresponds to the moisture diffusion resistance of a 1 m thick layer of stagnant air.

Due to the temperature difference between the warm cabin and crown section and the cold exterior skin, a stack pressure forms [13,14], resulting in a pressure difference and consequently an airflow towards the skin cavity in the upper section of the aircraft envelope. Reference [13] estimates the magnitude of the driving stack pressure to around 4 Pa and shows the correlation between of the sizes of leakages in the insulation layer and the resulting air ingress. Reference [15] developed a simplified test setup representing a field section of the cabin sidewall and investigated the frost buildup with and without insulation blanket installed. It was found that the installation of an insulation blanket reduces the frost by a factor of 40, compared with guiding the airflow across the uncovered skin. Hence, a careful installation of the insulation blankets helps reducing the airflow to cold structure. Even though passengers often perceive the air as dry, its dew point in cruise is at about −10 °C and thus above typical fuselage temperatures in cruise. As a result, frost will form on the cold structure outboard the insulation. On ground, most of this ice drains after melting by a suited installation of the insulation blankets similar to roof shingles. However, some water may flow in an uncontrolled way and ultimately drip into the cabin [16] ("rain in the plane"). Reference [11] reviews data from a survey conducted by [16] on the B757 fleet and concludes an average daily amount of water condensation of 91 g per field (area between two frames).

Hence, the careful installation of the insulation blankets is crucial to reduce moisture-related issues in the aircraft. As this installation requires a high amount of manual workforce in the production line, new concepts with at least equal functional performance are of interest. The research presented in this paper conducts an experimental investigation into what extent a rigid particle foam insulation on the frame provides the same level of protection against air ingress and thus frost formation as the conventional capstrip.

2. Materials and Methods

The tests of the particle foam insulation are carried out in the Lining and Insulation Test Environment (LITE) chamber of the Fraunhofer IBP in Holzkirchen, Germany. The chamber is constructed for the experimental investigation of aircraft insulation systems. It consists of an aluminum fuselage with five frames thus resulting in four fields. The floor area is 2.89 × 1.74 m, and it is 1.86 m high (Figure 3). The chamber walls are insulated to limit thermal losses. In an airliner, the chamber section would accommodate approximately nine economy passengers.

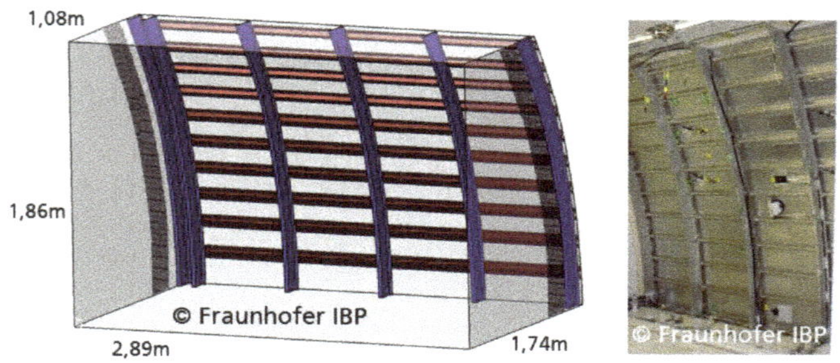

Figure 3. LITE chamber.

2.1. Airflow Patttern

Figure 4 depicts the ventilation system of the LITE chamber. In order to condition the exterior fuselage, a cocoon is built leaving an approximately 5 cm wide air gap. This gap is flushed with conditioned air in order to represent cold exterior conditions or hot day on ground. An air conditioning system providing heating, cooling and humidity control ventilates the chamber to represent the cabin climatic conditions. In the real aircraft, the cabin extraction would flow through the dado panel resulting in a pressure drop between the cabin and the gap behind the sidewall. As a result, some air may overflow behind the sidewall through leakages, for example behind the stowage bins. This air flows downwards through the gap between primary and secondary insulation and some air may leak into the cavity behind the primary insulation. This leakage is actively generated using an exhaust fan in the LITE chamber. An opening on the suction side of the HVAC system ducting compensates the excess of air removed by the extraction fan behind the dado panel.

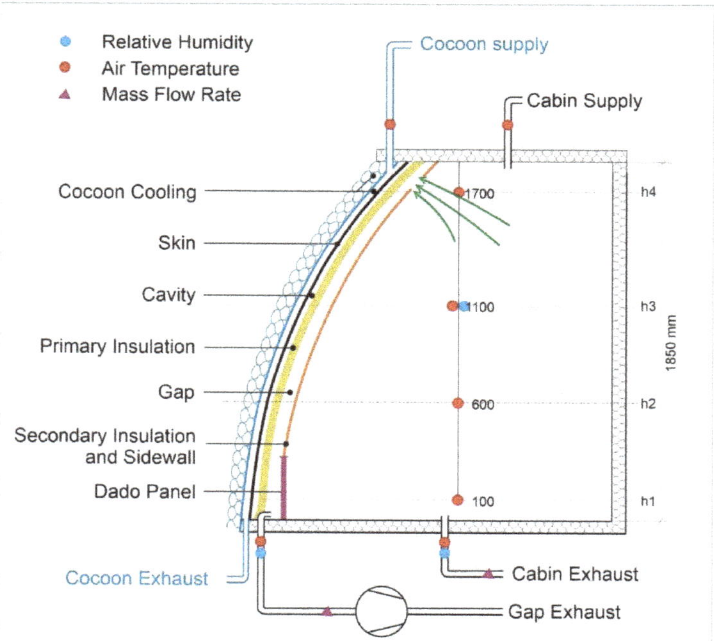

Figure 4. Ventilation system of the LITE chamber.

2.2. Sensor Distribution

Figure 5 shows the skin sensor distribution for the test conduct. Surface temperatures of the fuselage are measured in four heights in all four fields. Two frames are equipped with surface temperature sensors and air temperature sensors to determine the air temperature in the cavity between frame and insulation. To measure temperatures, four-wire PT100 sensors are used with an accuracy of 0.1 K @ 20 °C according to DIN EN 60751 class A [17]. The pressure difference between cavity and cabin is measured in four positions using Fischer DE23 sensors (FISCHER Mess- und Regeltechnik GmbH, 32107 Bad Salzuflen, Germany) with an accuracy of 1% in the range ±25 Pa.

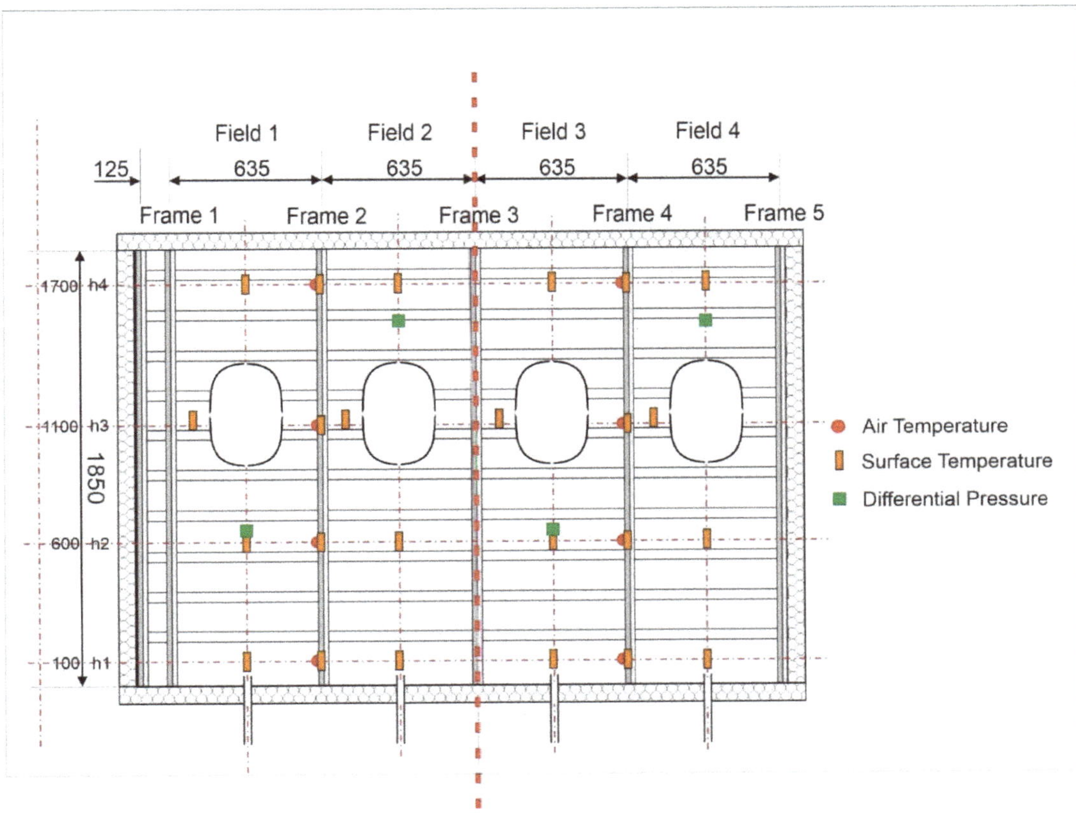

Figure 5. Sensor distribution on the skin.

Figure 6 shows the sensor distribution on the primary insulation. The temperature and humidity (Rotronic HygroClip HC2-C05 sensor with an accuracy of ±1.5% RH, rotronic messgeräte GmbH, 76275 Ettlingen, Germany) in the gap are measured, as well as the pressure difference between cabin and gap. The left pressure difference at 600 mm height is used as reference pressure for the control of the extraction fan behind the dado panel.

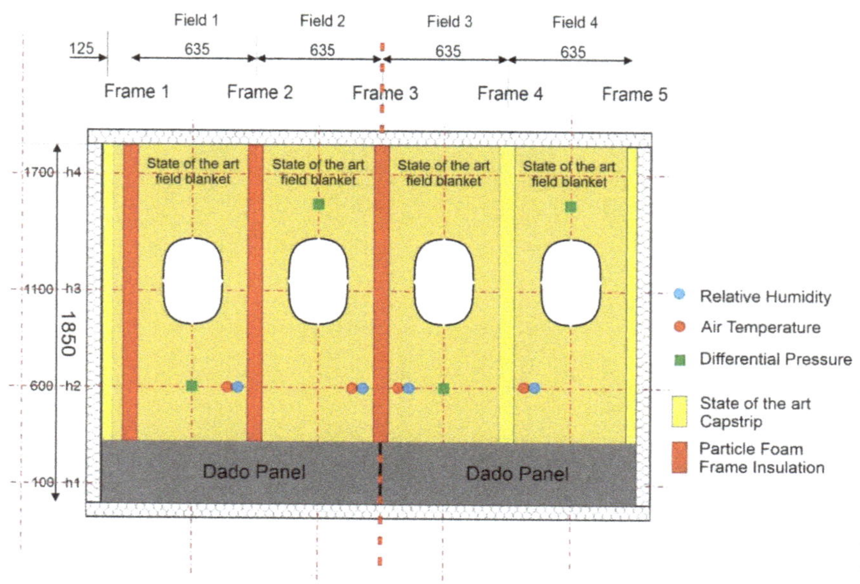

Figure 6. Sensor distribution on primary field blanket insulation.

The surface temperature of the sidewall is monitored using infrared cameras taking one picture every minute (type Optris PI 640, post processing with software Optris PI connect, Optris GmbH, 13127 Berlin, Germany).

2.3. Specimen Installation

For the experimental testing program, the fuselage section is split into two subsections. On the left section, the frames are covered by the particle foam insulation, whereas on the right side, the conventional capstrip with through-frame fasteners is used. The primary field insulation blankets are of the same type on all sections. Through this installation, three major types of leakages emerge (Figure 7):

- Between the particle foam frame insulation and the field blanket.
- Between the capstrip and the field blanket.
- On the upper and lower edges of the field blankets.

Gaps in the middle frame connecting the second and third field are thoroughly sealed to avoid lateral flows behind the primary insulation.

The particle foam insulation is manufactured from EPERAN PP MH24 [18], an expanded polypropylene material. The density of the particle foam insulation is 60 kg/m^3 and its thermal conductivity is 0.043 $W/(m{\cdot}K)$. All other insulation blankets are made of classical glass fiber insulation wrapped in foil, as used in the aircraft, with a thermal conductivity of 0.03 to 0.04 $W/(m{\cdot}K)$ and a density between 6.7 to 24 kg/m^3 [3]. Thus, the thermal conductivities of particle foam and state of the art insulation are comparable, whereas densities noticeably differ.

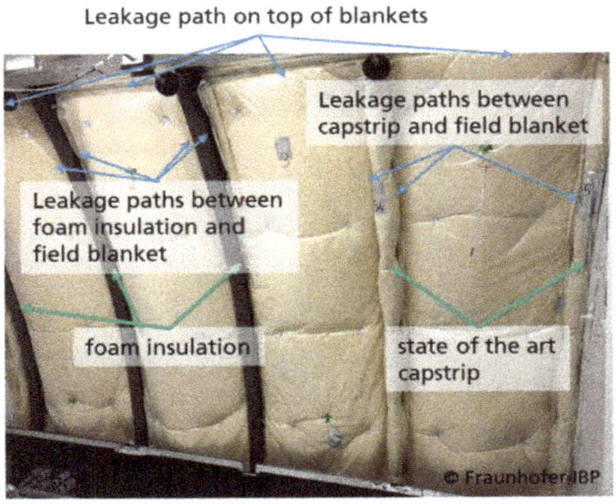

Figure 7. Particle foam insulation and capstrip installed in the LITE test chamber.

The primary insulation layer is covered with sidewall lining panels (Figure 8). On the rear side, state of the art secondary insulation is fixed. This insulation consists of glass fiber blankets wrapped in foil. The window openings in the sidewall are plugged with Styrofoam bluff bodies to align with the field blanket and seal the opening against air ingress.

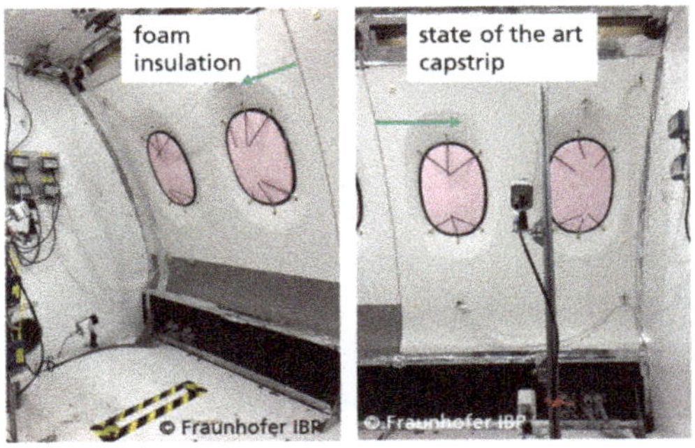

Figure 8. Fixation of sidewall in front of the primary insulation.

The sidewall is supported on its lower edge by a U-bar mounted on top of the reconstructed dado panel (black lower panel in Figure 8). On the upper edge, it is prolonged with a self-constructed structure and pressed against the frame using stamps attached to threaded bars that are fixed on the ceiling. The top structure leaves an overflow path between cabin air and the gap between primary and secondary insulation. In the aircraft, such a gap is not visible to the passenger but is present behind the stowage bin.

2.4. Test Matrix

The testing program is designed to answer two major questions from a hygrothermal point of view:

- Does the installation sequence of the insulation parts influence the condensate?
- Can the material thickness be reduced to save weight?

The first question addresses the sensitivity of condensate buildup towards optimizations in the installation process. In order to address this, the installation sequence of the insulation pieces is altered (Figure 9). In a first test, the frame insulation is first installed and then the field blankets are put between the frames. Thus, a somewhat undefined gap prevails between frame insulation and blanket. As this procedure requires less careful manual work, it is considered the low quality, uncareful installation. In the second test, the field blankets are first installed, and it is made sure the blankets' edges are clamped by the particle foam insulation. This is considered to be a careful and high-quality installation. Both tests are conducted using a frame insulation of 20 mm thickness.

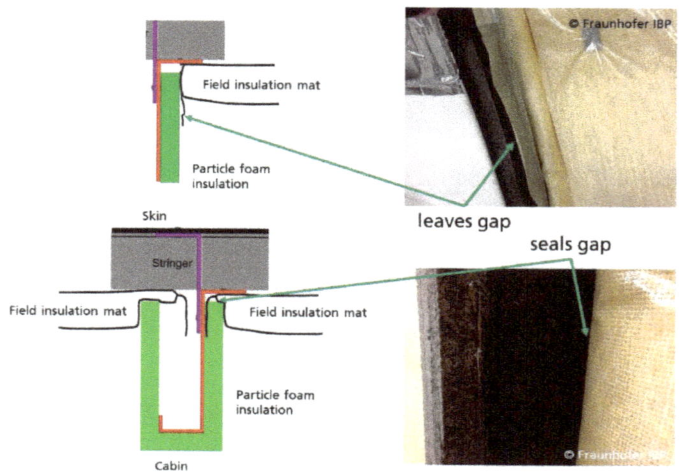

Figure 9. Comparison of the low-quality installation (**top**) with the high-quality installation (**bottom**).

In order to investigate the potential to reduce material, a third test is performed using 10 mm thick frame insulation carefully installed to clamp the field insulation.

2.5. Test Conduct

The boundary conditions summarized in Table 1 are applied. Cabin humidity is exaggerated in order to receive a clearer frost buildup distinction between the test cases.

Table 1. Test boundary conditions.

Boundary Condition	Value
Exterior fuselage temperature	−30 °C
Cabin ventilation flow rate	368 kg/h
Cabin temperature	24 °C
Cabin humidity	30%
Pressure difference across sidewall	4 Pa
Exposure duration	7 h

After the exposure, the sidewall and the insulation are removed. The condensate buildup is photographed, and the quantity is assessed by weight difference after scratching frost into a bag and wiping melted water from the structure with a sponge (Figure 10). The inherent uncertainty of this method is that not all moisture can be collected because some water may be unreachable in corners, and drips off when melting or evaporates. In order to reduce the impact of such uncertainty, the exposure time was selected as several hours, and the cabin moisture was increased to obtain a noticeable amount of condensate. The uncertainty of the weight difference between the dry and wet sponges and bags is considered negligible because a sufficiently accurate scale was used. Overall, the weighted water mass thus presents a lower limit measurement, but covers the major amount.

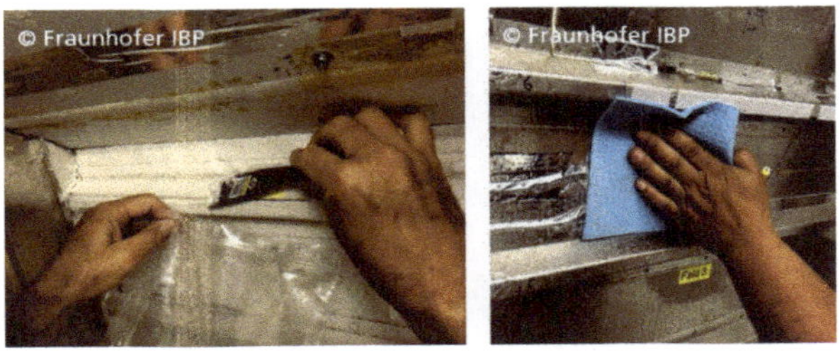

Figure 10. Collection of accumulated frost from the structure.

3. Results

Tests are started at 9:30 in the morning. 1 to 1.5 h later (11:00), the skin surface temperature, cabin temperature and flow rates, differential pressures, etc., show stable conditions. For the following analysis, temperatures are averaged between 13:30 and 14:30 in order to ease comparisons between the tests. The exposure finishes at 16:30 with switching off the ventilation and cooling system, dismounting the sidewall and collecting frost and water from the skin.

3.1. Cabin Boundary Conditions

For all three tests, the boundary conditions in the cabin are similar within 1.5 K and lead to comparable results as depicted in Table 2. Thermal stratification remains below 0.5 K and is thus negligible.

Table 2. Cabin boundary conditions.

Cabin Boundary Conditions	20 mm, Uncareful	20 mm, Careful	10 mm, Careful
Temperature in °C	22.2	23.0	23.7
Absolute humidity in g/m^3	6.3	6.4	6.7

3.2. Sidewall Surface Temperature

Figure 11 shows the IR signature of the frame insulations on the lower sidewall surface. It can be seen that the section on top of the particle foam frame insulation is warmer than the section on top of the state-of-the-art capstrip. The reason is that the capstrip is compressed by the sidewall, whereas the particle foam insulation remains rigid. The thicker (20 mm) particle foam insulation results in higher surface temperature than the thinner one (10 mm). Even through the effect is visible, its magnitude is not considered to have an impact on the passenger comfort due to the relatively local confinement.

3.3. Skin Temperatures

For all tests, skin temperatures between −21.5 and −30.1 °C are measured with an average value of −27.2 °C. This corresponds to a saturation water vapor fraction of 0.5 to 1.0 g/m^3. Thus, the difference between the cabin humidity and this residual humidity contained in air after contact with the cold skin determines the frost buildup potential.

Figure 12 shows the measured inboard skin surface temperatures. The pulldown after activation of the cocoon cooling is performed within approximately 1 to 1.5 h. For the remaining 5.5 to 6 h of the test, the temperatures remain rather constant. A real aircraft in operation probably would have faster pulldown because the external heat transfer coefficient during flight is higher due to the elevated cruising speed of the aircraft. Furthermore, typical air temperatures at flight altitudes are lower than the cocoon cooling provides. The duration of the temperature stabilized phase, and thus a constant frost buildup rate would depend on the length of the flight, ranging from only few minutes for short-haul flights to several hours for long-haul flights.

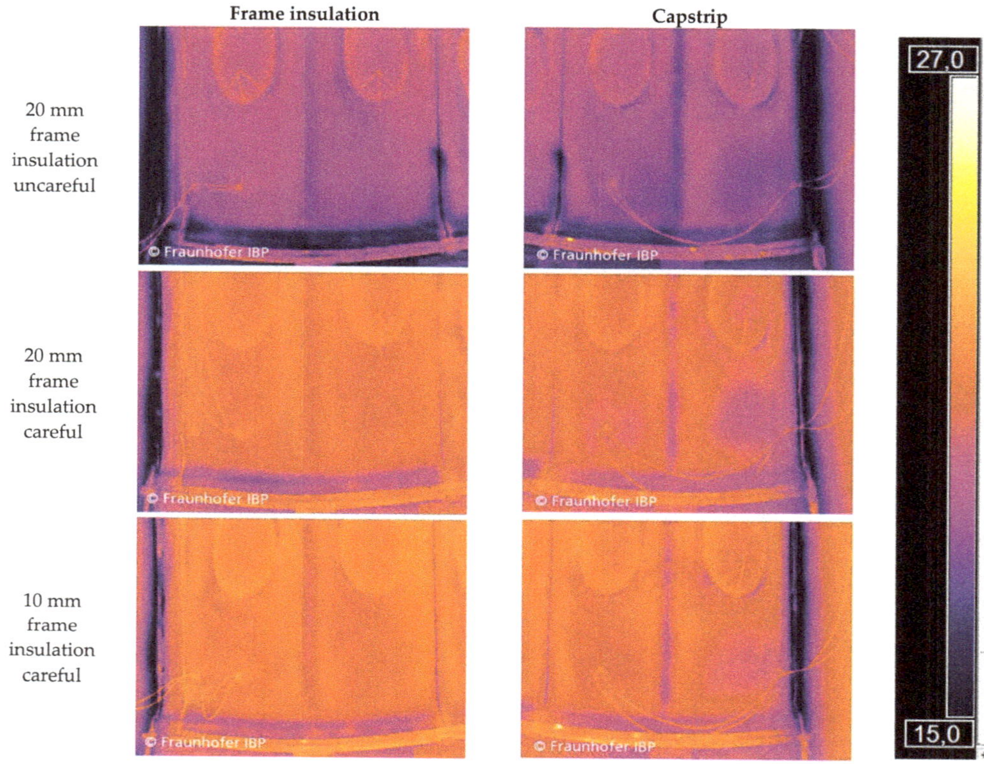

Figure 11. IR image of the lower sidewall.

3.4. Frame Temperatures

Two frame surface temperature profiles are measured:

- Frame 2 with particle foam insulation.
- Frame 4 with state-of-the-art capstrip.

Figure 13 shows the temperature profiles for both frames. The following main observations are noted, and a possible explanation is provided:

- Seemingly, some effect affects the lowest positions' (0.1 m) temperature on Frame 4. Possibly, there is a heat bridge warming the lower frame section or air exchange is inhibited in the lower section, leading to dominant heat flow through the capstrip.
- For the uncareful installation (blue line), Frame 2 is approximately 8 K warmer in the upper half than Frame 4. At 0.6 m height, both show similar measurements, in the lower section, Frame 2 is considerably colder. Due to the uncareful installation, more air ingresses from gap to cavity by buoyancy in the upper part, leading to a higher frame temperature. This air falls downwards in the cavity along the cold structure and thus leads to convective cooling of the lower frame part.
- For the careful installation with 20 mm thickness, Frame 2 shows a similar temperature as Frame 4. Further downwards, Frame 2 shows approximately 5 K higher temperatures than Frame 4.
- Using a 10 mm thick frame insulation leads to a higher frame temperature compared with the other concepts and compared with the capstrip. The thinner insulation results in a higher frame temperature.

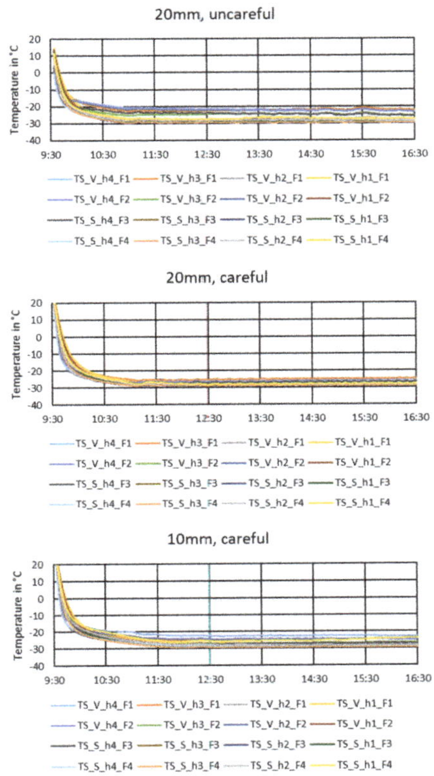

Figure 12. Skin surface temperature and pulldown behavior.

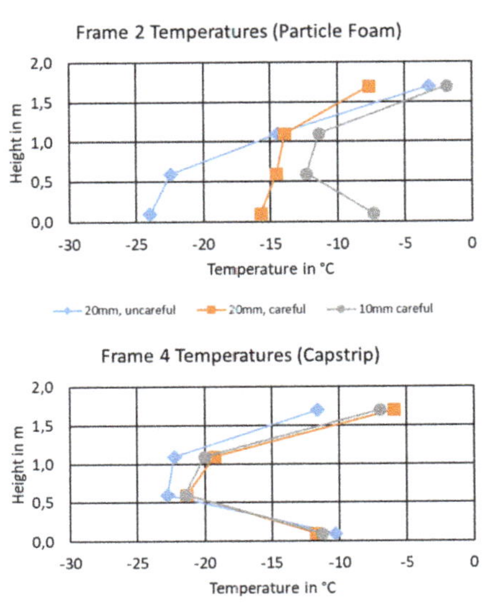

Figure 13. Surface temperatures on Frame 2 and Frame 4 (dots mark the measurement data).

3.5. Gap pressure and Flow Rate

Figure 14 compares the measured extraction flow rate of the gap extraction (between primary insulation and sidewall), and the measured pressure differences between cabin and gap. A negative pressure reflects that pressure is lower in the gap than in the cabin.

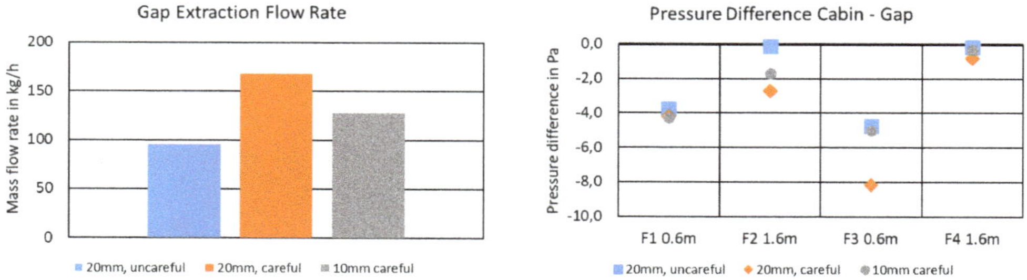

Figure 14. Extraction flow rate and pressure difference between cabin and gap.

The air extraction system behind the sidewall accurately maintains a 4 Pa pressure difference at the reference sensor (0.6 m height in Field 1).

The pressure difference in Field 2 at 1.6 m height shows a lower magnitude as the measurement is closer to the overflow opening above the sidewall. For example, the window structure obstacle has not yet been encountered at this point. The differential pressure magnitude increases with increasing extraction flow rate.

In Field 3 the same position as Field 1 is investigated but for the capstrip arrangement. For the highest extraction flow rate, the differential pressure magnitude noticeably increases to more than 8 Pa.

Field 4 again shows relatively low-pressure magnitude as it is close to the overflow opening.

Measurement reveals that the thickness and installation quality of the frame and field insulation have an impact on the air overflowing from cabin to the gap between primary and secondary insulation. This behavior is explainable by considering the shape of this gap.

The uncareful installation reflects that first the frame insulation and then the field blanket are installed. Thus, the blanket is not subjected to any specific compression and can buckle behind the sidewall (cf. Figure 9). As a result, the available space for air to overflow becomes smaller, leading to a higher flow resistance and thus lower flow rate when applying the same differential pressure.

The careful installation reflects that first, the field blanket is installed, and then the frame insulation clamps and compresses the edges. As a result, a gap between sidewall and primary insulation emerges that provides a lower flow resistance and thus higher flow rate when applying the same differential pressure.

The same is valid for the 10 mm frame insulation, but the gap becomes narrower than for the 20 mm insulation thickness, and thus flow resistance increases and flow rate decreases.

The gap flow rate amounts to 26 to 46% of the chamber supply airflow rate (368 kg/h). This leakage is a shortcut, and thus negatively affects the cabin ventilation effectiveness. As a result, the demand for the energetically expensive bleed air would increase to meet regulatory requirements [1]. Therefore, for a future integration of such insulation system, the use of horizontal flow blockers between sidewall and primary insulation as, e.g., suggested by [14] should be foreseen. Furthermore, [14] suggests using dry air injection for envelope overpressurization with regard to the cabin. This would inhibit the frost buildup from air leakages to the cold structure and reduce airflow shortcuts, leading to higher cabin ventilation effectiveness.

Even though not tested here, it is considered that the gap flow in aircrafts is lower because the compressibility of the insulation blankets and the capstrip would lead to a smaller and less defined gap than the rigid particle foam insulation does.

3.6. Gap Air Temperatures

Figure 15 shows the measured air temperatures in the gap between field blanket and sidewall. Overall, it can be concluded that the gap air temperature is close to the cabin air temperature. The reason is that the larger fraction of air flowing behind the sidewall originates from the cabin. The only exception is in the "10 mm, careful" measurement in Field 1. The exact reason for this deviation is not known; it might be a local leakage providing cold air from the cavity close to the sensor.

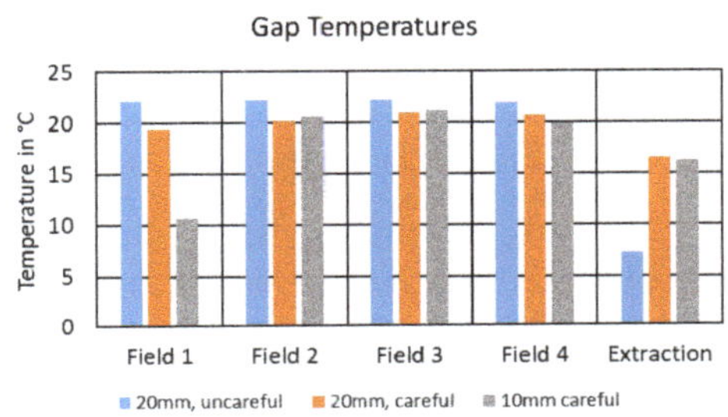

Figure 15. Measured gap air temperatures.

The extraction air is shown to be colder than the gap air. This air consists of a mixture of the gap air and air that passed the cavity, and thus became noticeably colder.

3.7. Differential Pressure between Gap and Skin Cavity and Condensate on the Skin

Figure 16 shows the differential pressure measured between the gap and the skin cavity. This pressure is mainly buoyancy induced and results in an airflow between the two air volumes. In this plot, a positive pressure depicts a gradient from skin cavity to the gap behind the sidewall, whereas a negative pressure results in an ingress of air to the cold structure. Again, pressure is measured in the four fields alternating a low (0.6 m) and high (1.6 m) measurement position. Figure 17 shows a picture of the frost formation pattern directly after removing the insulation on top of Field 1.

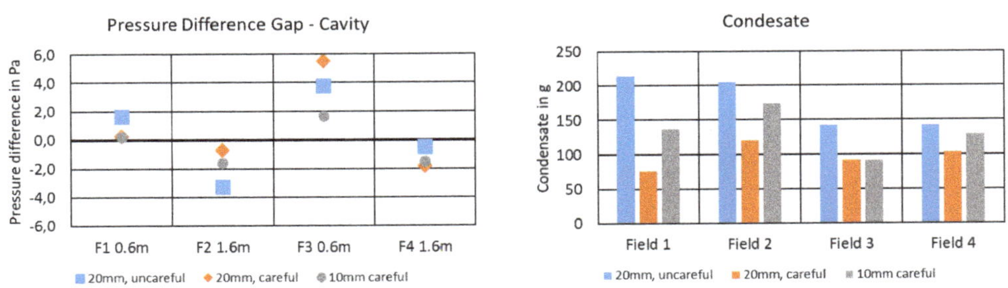

Figure 16. Driving buoyant pressure between gap and skin cavity and measured amount of condensate on the skin.

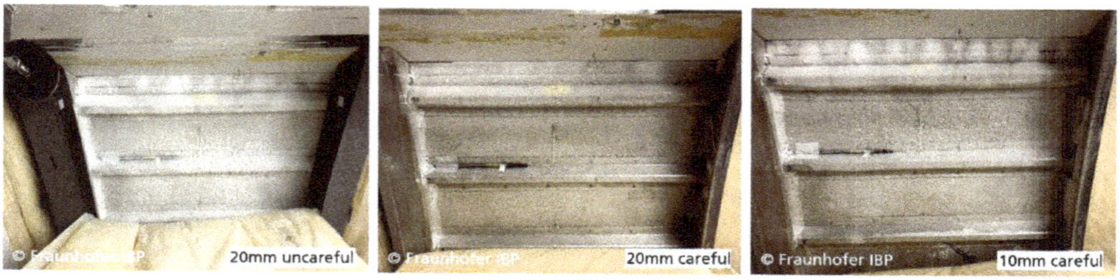

(**a**) (**b**) (**c**)

Figure 17. Frost formation on Field 1 for the 20 mm thick frame insulation uncarefully installed (**a**), 20 mm thick frame insulation carefully installed (**b**) and 10 mm thick frame insulation carefully installed (**c**).

A clear correlation between higher pressure differences and the amount of condensate is not obvious. The reason could be that larger leakage paths result in higher flow rate at lower pressure, whereas well-sealed insulations are able to sustain a higher differential pressure.

On the other hand, the quality of installation is reflected by the amount of condensate. The uncareful installation leads to a larger leakage path between the frame insulation and the field blanket, resulting in a close to doubled amount of condensate compared with the clamped installation, where the major leakage paths is expected to be on the top and bottom edge of the field blanket (cf. Figure 7). Handling shows that the 10 mm thick frame insulation is less rigid and tends to show higher torsion. Hence, its sealing potential is seemingly lower.

On frame 3, located between Field 2 and Field 3, the particle foam frame insulation is applied. As a result, the Field 3 measurement is a mixture of a particle foam frame insulation on the left and the conventional capstrip on the right. Hence, some influence of the specimen is found here, too.

Comparing the amount of condensate with the leakage rate into the gap (Figure 14 vs. Figure 16) shows that a higher flow rate in the gap inboard the primary insulation does not necessarily result in a higher amount of condensate on the structure. In order to analyze this effect, the flow rate exchanged between the gap and the skin cavity are estimated. For this, the amount of frost is divided by the potential for frost formation (difference between cabin moisture content and skin saturation moisture content) and the exposure duration of 7 h.

$$\dot{V}_{estimated} = \rho_{air} \cdot \frac{m_{condensate}}{(x_{cabin} - x_{sat,skin}) \cdot 7h} \quad (1)$$

This estimation leads to the flow rate estimation per field shown in Figure 18. In sum, the total airflow to the cavities varies between 10.6 to 19.9 kg/h. This is an order of magnitude smaller than the airflow through the gap behind the sidewall. Hence, the independence of frost buildup from the airflow in the gap becomes comprehensive as only a small fraction of air contributes to condensate formation.

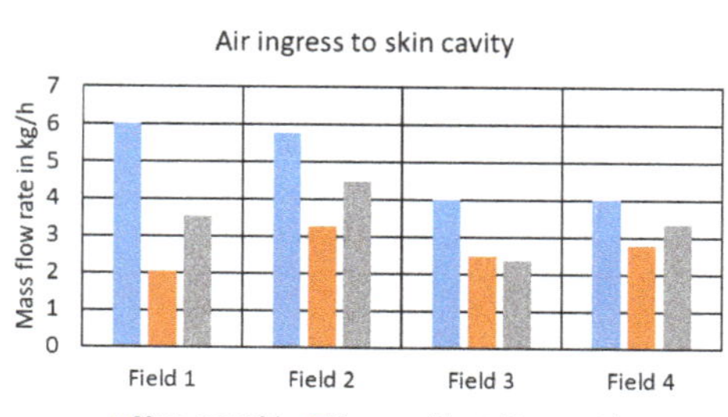

Figure 18. Estimated air ingress to skin cavity.

4. Discussion

This research investigates the effect of a particle foam insulation and its installation procedure on the hygrothermal conditions behind the cabin sidewall. The experimental study is conducted in a representative test chamber. Nevertheless, some simplifications compared with the real flight are made. First, the typical cabin cruise pressure is 750 hPa whereas the lab ambient air pressure is approximately 940 hPa (the lab is at 694 m elevation). In a comparison performed by [13], pressure differences are estimated approximately 20% lower in cruise than in ground pressure conditions. On the other hand, typical cruise exterior conditions are −56.5 °C [2], whereas the LITE chamber achieves around −30 °C. Furthermore, the chamber considers a rather undisturbed section of the fuselage, without the presence of, e.g., doors or riser ducts. Additionally, the crown section is not present, potentially reducing the buoyancy effect due to lower elevation. Nevertheless, the chamber tests provide a thorough insight into the driving effects of the airflows behind the sidewall.

Comparing the measured amount of condensate on the frames proves to be in the order of magnitude of the value of 91 g postulated by [11,16]. Measured stack pressure is merely within the estimated 0 to 4.5 Pa of [13].

Therefore, the experiments are considered providing a valuable step to the understanding of the major effects driving the hygrothermal behavior of the sidewall section, and persistent conclusions can be drawn from it. The experiments prove that novel insulation systems can be assessed in this type of chamber.

5. Conclusions

This paper compares three tests with different installation procedures and different material thicknesses of a particle foam frame insulation replacing the state-of-the-art capstrip. These tests are a step forward, and with them, the Lining and Insulation Test Environment has proven very useful in addressing the problem of condensate. The conducted experiments give a valuable insight into the hygrothermal process behind the cabin sidewall and allow the following major conclusions:

- The quality of the insulation installation is the key parameter to reduce condensate on the cold exterior skin. It is proven that clamping the field blanket below the particle foam frame insulation results in only half the amount of condensate than a loose joint of both insulation pieces.
- With a 20 mm thick particle foam frame insulation, the same amount of protection against condensate on the structure is achieved as with the conventional capstrip, but less parts are needed for installation.
- The particle foam insulation leads to slightly higher local sidewall surface temperatures than the capstrip. However, due to the confinement of this area, no comfort relevant effect is expected from this.
- The leakage of air into the gap between sidewall and primary insulation is one order of magnitude larger than the leakage of air towards the skin cavity. Therefore, the amount of condensate formation is independent from this leakage.
- Even through the gap behind the sidewall is of undefined shape, measures affecting its size (e.g., thicker frame insulation) influence the amount of air leakage into the gap. This can have an impact on cabin ventilation, as this presents a shortcut of cabin air and thus requires higher bleed air offtake to meet ventilation requirements [1]. A flow blocker located between the sidewall and primary insulation could reduce this shortcut.

Figure 19 summarizes these findings in brief.

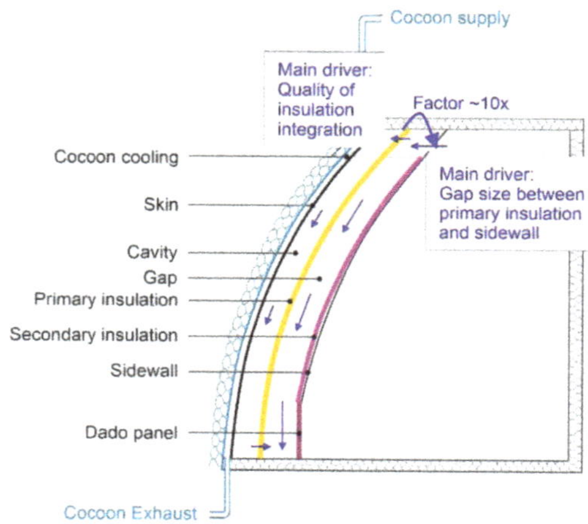

Figure 19. Main conclusion on hygrothermal situation behind sidewall.

Author Contributions: Conceptualization, V.N. and G.R.; methodology, V.N.; formal analysis, V.N.; investigation, G.R.; resources, V.N.; data curation, V.N.; writing—original draft preparation, V.N.; writing—review and editing, G.R.; visualization, V.N.; supervision, V.N.; project administration, V.N.; funding acquisition, V.N. All authors have read and agreed to the published version of the manuscript.

Funding: This research is conducted with financial support from the German Federal Ministry for Economic Affairs and Energy (FKZ: 20X1715B). The authors are responsible for the content of this publication.

Gefördert durch:

Bundesministerium
für Wirtschaft
und Energie

aufgrund eines Beschlusses
des Deutschen Bundestages

Data Availability Statement: Some data presented in this study are available upon request from the corresponding author.

Acknowledgments: We thank Diehl Aviation for their support in the SOPHIA project and Neue Materialien Bayreuth for the provision of the rigid foam capstrip material.

Conflicts of Interest: The authors declare no conflict of interest. The funders had no role in the design of the study; in the collection, analyses, or interpretation of data; in the writing of the manuscript; or in the decision to publish the results.

References

1. ANSI/AHSHRAE. *ASHRAE Standarad 161–2007: Air Quality within Commercial Aircraft*; American Society of Heating, Refrigerating and Air-Conditioning Engineers: Atlanta, GA, USA, 2007.
2. ICAO. *Manual of the ICAO Standard Athmosphere Extended to 80 km*, 3rd ed.; Internation Civil Aviation Organization: Montréal, QC, Canada, 1993.
3. Johns Manville. *Microlite(R) AA Blankets: Aircraft Thermal and Acoustical Insulation Data Sheet*; Johns Manville: Denver, CO, USA, 2020.
4. Lamart Corporation. Orcotek Strip Blankets. Available online: http://www.orcon.com/orcotek_strip_blankets (accessed on 13 September 2021).
5. U.S. Deportment of Transportation Federal Aviation Administration. *Thermal/Acoustical Insulation Flame Propagation Test Method Details*; U.S. Department of Transportation: Washington, DC, USA, 2005.

6. U.S. Deportment of Transportation Federal Aviation Administration. *Installation of Thermal/Acoustic Insulation for Burnthrough Protection*; U.S. Department of Transportation: Washington, DC, USA, 2008.
7. Chen, R.; Fang, L.; Liu, J.; Herbig, B.; Norrefeldt, V.; Mayer, F.; Fox, R.; Wargocki, P. Cabin air quality on non-smoking commercial flights: A review of published data on airborne pollutants. *Indoor Air* **2021**, *31*, 926–957. [CrossRef]
8. Grün, G.; Trimmel, M.; Holm, A. Low humidity in the aircraft cabin environment and its impact on well-being–Results from a laboratory study. *Build. Environ.* **2012**, *47*, 23–31. [CrossRef]
9. Giaconia, C.; Orioli, A.; Di Gangi, A. Air quality and relative humidity in commercial aircrafts: An experimental investigation on short-haul domestic flights. *Build. Environ.* **2013**, *67*, 69–81. [CrossRef]
10. Norrefeldt, V.; Mayer, F.; Herbig, B.; Ströhlein, R.; Wargocki, P.; Lei, F. Effect of Increased Cabin Recirculation Airflow Fraction on Relative Humidity, CO_2 and TVOC. *Aerospace* **2021**, *8*, 15. [CrossRef]
11. Wörner, M. Wärme- und Stofftransport in einer Flugzeugkabine unter Besonderer Berücksichtigung des Feuchtetransports. Ph.D. Thesis, University of Hamburg, Harburg, Germany, 2006.
12. Riedl, G.; Rösler, D.; Stratbücker, S.; Grün, G. *Interdisziplinäre Kabinenarchitekturen (INDIKAR)*; Schlussbericht; Fraunhofer-Institute for Building Physics: Valley, Germany, 2006.
13. Walkinshaw, D.; Preston, K. Controlling Cabin and Envelope Air Flows and Pressure Differentials to Prevent Envelope Condensation, Enable Cabin Humidification, Improve Fire Safety, and Decrease Fuel Use. *SAE Int. J. Aerosp.* **2011**, *4*, 1243–1253. [CrossRef]
14. Walkinshaw, D.S.; Horstman, R.H. Stack Pressure-Created Airflows in Insulation Envelopes, Part 2: Passenger Aircraft. *ASHRAE J.* **2020**, *5*, 32–46.
15. Westhoff, A.; Wagner, C. Experimental study of moist air flow in the gap between the aircraft's fuselage and its cabin wall. *CEAS Aeronaut J.* **2020**, *11*, 591–607. [CrossRef]
16. Huber, P.; Schuster, K.; Townsend, R. Controlling Nuisance Moisture in Commercial Airplanes. Aero 05 1999. Available online: https://www.boeing.com/commercial/aeromagazine/aero_05/m/m01/index.html (accessed on 13 September 2021).
17. DIN. *Industrial Platinum Resistance Thermometer Sensors*; German Version EN 60751:2008; Beuth Verlag: Berlin, Germany, 2009.
18. Kaneka Belgium NV. *EPERAN PP MH24/Expanded Propylene*, 1st ed.; FR-EPEM-022-E/2, Issue Date: 4 March; Kaneka Belgium NV: Westerlo-Oevel, Belgium, 2016.

aerospace

MDPI

Article

Development of a Novel Hybrid Thermoplastic Material and Holistic Assessment of Its Application Potential

Christos V. Katsiropoulos [1,*], Spiros G. Pantelakis [1], Marco Barile [2] and Leonardo Lecce [2]

[1] Laboratory of Technology & Strength of Materials, Department of Mechanical Engineering & Aeronautics, University of Patras, Panepistimioupolis Rion, 26500 Patras, Greece; pantelak@mech.upatras.gr
[2] NOVOTECH Aerospace Advanced Technology S.r.l., Via G. Pascoli, 7, 80026 Casoria, NA, Italy; marco.barile@novotech.it (M.B.); leonardo.lecce@novotech.it (L.L.)
* Correspondence: xkatsiro@mech.upatras.gr; Tel.: +30-261-0969498

Abstract: The development of a novel hybrid thermoplastic prepreg material enabling the fabrication of next-generation recyclable composite aerostructures produced by affordable, automated technologies is presented in the present work. The new hybrid material is produced using automated equipment designed and developed for this reason. A preliminary assessment of the application of the new material is made to obtain material properties related to its processability as well as to its strength. A typical aeronautical flat skin panel has been identified and produced using an autoclave-based process in order to assess the potential of the new material for producing aircraft structural parts. Moreover, a newly developed holistic index is implemented to enable a more holistic comparison of the suitability of the materials used for the panel production. The aspects considered for the material comparison are the quality, the environmental footprint, and the cost. The results of the study pointed out that the hybrid thermoplastic material that has been developed represents a viable manufacturing option from an industrial point of view and that its implementation in structural component manufacturing leads to clear cost and environmental advantages.

Keywords: thermoplastics; holistic assessment; PEEK/PEI; life cycle analysis; life cycle costing

Citation: Katsiropoulos, C.V.; Pantelakis, S.G.; Barile, M.; Lecce, L. Development of a Novel Hybrid Thermoplastic Material and Holistic Assessment of Its Application Potential. *Aerospace* **2021**, *8*, 351. https://doi.org/10.3390/aerospace8110351

Academic Editor: Giuliano Allegri

Received: 6 October 2021
Accepted: 12 November 2021
Published: 18 November 2021

1. Introduction

Over the past decades, carbon-epoxy thermosets have proved to be the most used composite material types for aviation applications. The use of these materials is related to long curing cycles and, thus, relatively reduced volume production. Furthermore, emerging environmental issues related to their waste handling are also among the issues linked to their wide use. These shortcomings and the imposition of related, rigorous environmental policies have attracted the attention of the aviation industry in evaluating thermoplastic composites as a promising alternative, as mentioned, e.g., by Oliveux et al. [1]. In terms of material properties, it is well known that the most significant opportunity of thermoplastic resins is related to their processing capabilities. Thermoplastics can be processed by single-stage fabrication technologies, and the assembling of components can be performed by welding, both allowing the elimination of adhesives or fasteners. Furthermore, thermoplastic composite recycling is much simpler than the recycling of thermosetting composites, since thermoplastic resins are recyclable and can be re-melted through heating. Currently, one of the vital barriers to the wide-ranging adoption of thermoplastic composites for producing primary aircraft structures [2,3] lies in the limitations of the existing production processes, which make the manufacturing of such structures unaffordable.

Indeed, for manufacturing components with dimensions not suitable for hot press-based processes, the autoclave is the most straightforward alternative for reaching adequate crystallinity levels. However, the related consolidation cycles at high temperature, with high energy consumption and costs as well as the usage of very expensive auxiliary materials, are not negligible. Additionally, the reduced mechanical properties of the recycled materials limit the benefit of their recyclability and pose a further burden for fully exploiting the potential of this important class of materials. It is worth noting that aeronautical structural materials require not only a relatively high glass transition temperature, T_g, but also a high glass transition temperature of the moisture-saturated material, T_g-wet, both no lower than 110 °C. Therefore, a number of thermoplastic resins which offer satisfactory structural properties (e.g., polycarbonate) cannot be used for such applications, due to their low T_g. Thermoplastic composites which are currently available in the market and suitable for potential use in aeronautical structures involve as matrices either amorphous resins with very high T_g (e.g., polyetherimide—PEI, polyethersulfone—PES) or semi-crystalline resins with high melting point temperature, T_m, (e.g., polyphenylsulfide—PPS, polyetheretherketone—PEEK, polyetherketoneketone—PEKK) [4,5]. On the basis of the above engineering considerations, PEEK and PEKK are currently the most suitable thermoplastic resins for structural applications, while PPS is preferred for interior applications [6]. However, the usage of PEEK and PEKK matrix-based composites remains up to now limited, due to cost and environmental considerations. In addition to the high cost of the raw material, the necessary prepreg fabrication process is expensive, due to its processing conditions (e.g., melting temperature above 350 °C). Furthermore, the time/cost advantages of working with a material already consolidated, and, hence, not exhibiting long curing cycles, are compensated by the cost limitations related to the high working temperature range as well as to the specific cooling rate necessary to achieve the desired crystallinity level in the produced components [7]. Finally, the long and complex cycles at high temperatures required are associated with high energy consumption and, hence, with an increased environmental footprint.

Therefore, to facilitate the exploitation of this desirable class of materials, several research actions have been undertaken aiming to develop faster and flexible out-of-autoclave processes; in parallel, efforts for developing novel structural materials which can be processed at lower temperatures are also in progress [8–11]. As for the processes, they should be "faster" in terms of their capability to allow the rapid cooling of the material down to room temperature. The potential use of in situ consolidation processes is also an asset related to the need of fast processes. In addition, "flexible" processes, in terms of reducing limitations related to the geometry and size of the parts, are under development.

A straightforward approach to achieving cost and manufacturing time reductions is the exploitation of manufacturing processes dedicated to thermoplastics, such as, for instance, thermoforming in hot press. However, hot press processes cannot be used in the case of large structures. A number of further promising processes have been proposed, such as the automated continuous compression molding (CCM), which has the capacity to offer fast processing of high-performance thermoplastic profiles of theoretically unlimited length; however, it is limited to low-complexity shaped profiles or flat panels [12,13]. Like CCM, the diaphragm forming (DF) process, which is also referred to as superplastic forming (SPF), was first introduced in [14] and is an out-of-autoclave process developed for processing thermoplastic structural composites. DF is considered for the production of structural components of good quality by appreciably reducing processing time and energy consumption [15]. Nevertheless, the industrial exploitation of this process remains limited to date, mainly due to some limitations on the shapes and dimensions of the components which can be produced; the low degree of process automation so far is also a constraint.

A very promising alternative, which targets a simultaneous consolidation of the laid material during the automated fiber placement (AFP) process directly after the layup step, is known as the "in situ consolidation (ISC)-AFP process" [4,16]. This technology aims to exclude the necessary autoclave post-consolidation step; the consolidation temperature must be higher than T_g for amorphous polymers or higher than T_m for semi-crystalline polymers. Considering the criticality of controlling the cooling rate during the laydown phase, a sufficient crystallization level could be achieved by involving the ISC-AFP process to produce thermoplastic composite parts.

In parallel to the efforts to deploy fast and flexible out-of-autoclave techniques, the development of alternative material solutions that enable the reduction of processing temperatures has been presented in the literature; these works target the processing of semi-crystalline thermoplastic prepreg structures. The apparent solution has been to interchange superior but high-cost semi-crystalline thermoplastics, such as PEEK, with other cheaper/lower-performance semi-crystalline thermoplastics [10,17,18] or other amorphous thermoplastics, as studied by Botelho et al. [19]. However, this approach leads to shortening the range of possible structural aviation applications substantially. Towards this direction, particularly in cutting back the cost of the component by lowering the thermoplastic resin process temperature, a cheaper but still high-performance material is the obvious solution; this has not been developed so far. Moreover, an alternative option lies behind the current research on evolving thermoplastic blends, as presented in Mitschang et al. [20,21]. The latter technology, however, is still not considered mature for advanced applications [22].

Alternatively, the addition of a different amorphous polymer layer, the T_g of which is lower than the T_m of the semi-crystalline one, offers the advantage of forming a stack of layers by operating at a lower melting temperature with respect to the one needed to consolidate the semi-crystalline prepreg [23]. Consequently, it is possible to weld the parts by avoiding the deconsolidation of material that is not involved in the welding. An additional advantage of the use of the above amorphous bonding concept is the resulting lower processing time, as the resin is already consolidated and does not need to be further processed.

Today, regardless of whether working with thermosets or thermoplastics, the cost and environmental impacts of manufacturing an aeronautic component are of vital interest. In addition, the necessity of keeping the weight as low as possible whilst maintaining sufficient quality remains permanent. Therefore, it needs to be underlined that by introducing a novel material, the assessment of cost and environmental footprint aspects is mandatory. For this reason, life cycle costing (LCC) and life cycle assessment (LCA) models are used prior to manufacturing as tools for the selection of the most appropriate process [24,25]. These analyses are performed in addition to the need to meet the non-negotiable requirement of satisfactory quality.

However, in most cases, a quality improvement is associated with an increase in cost and, in several cases, with an increase in the environmental footprint as well and vice versa. Therefore, quality criteria for the component as well as the overall environmental footprint and cost of the product, including the component manufacturing process and end-of-lifecycle, need to be considered as component optimization inter-dependent objective functions at the component design phase. Recently, a holistic design index was presented in [16]. The index was introduced as a tool for the selection of the suitable material or manufacturing process among several materials or techniques, respectively. The criteria involved in the index are quality, cost, and environmental footprint functions, which are considered inter-dependent.

Therefore, the scope of this work is the development of a new material and its preliminary assessment regarding its application potential and use in an aeronautic structural component. In this framework, a novel hybrid thermoplastic prepreg material is introduced, answering the needs for reduced weight and, consequently, reduced fuel consumption and emissions of an aircraft as well as reduced manufacturing costs and increased recyclability. The innovative hybrid thermoplastic prepreg material will provide an advantage in terms of process simplification, in particular, improved cycle times and lower energy consumption, since it does not require the use of an autoclave phase. A first evaluation of the new material was made by involving differential scanning calorimeter (DSC) tests to obtain material properties related to its processability as well as short-beam shear tests related to its interlaminar shear strength (ILSS). To assess the potential of the new material for producing aircraft structural parts, a typical aeronautical flat skin panel has been identified and produced using an autoclave-based process. A newly developed holistic index was implemented to enable a more holistic comparison of the suitability of the materials used for the panel production; in this framework, life cycle assessment (LCA) and cost analysis were carried out. The flow diagram of the methodology used is shown in Figure 1.

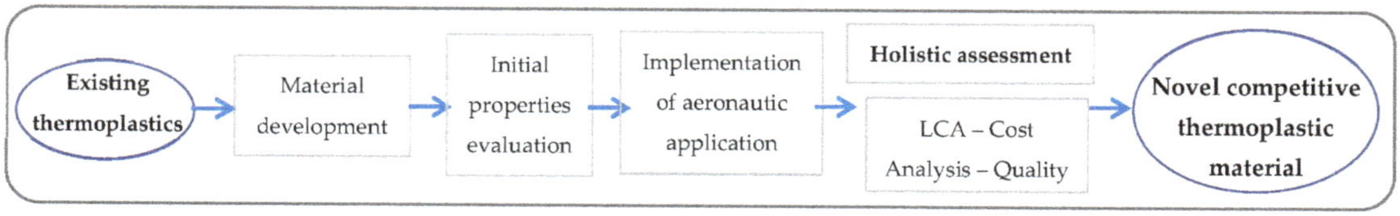

Figure 1. Flow diagram of methodology used.

2. Hybrid Thermoplastic Prepreg Material Development and Production

To face the challenges mentioned above, a novel thermoplastic material was developed in the framework of the New Hybrid Thermoplastic Composite Aerostructures Manufactured by Out-of-Autoclave Continuous Automated Technologies (NHYTE) project [11]. It is a hybrid thermoplastic material which is based on the amorphous bonding concept (Figure 2). More specifically, a prepreg composite ply is produced by PEEK–carbon fiber composite layers with the addition of external amorphous PEI layers. In this way, two sets of plies (or two plies) can be joined at a temperature above the PEI T_g and below the PEEK T_m. Recall that during the production of multilayer structures, only the melting of PEI (T_m up to 250 °C) is necessary, since the PEEK is already consolidated during the hybrid material production process. On the other hand, when only PEEK–PEEK is implemented, a T_m of at least 370 °C is necessary for the consolidation. The adhesion between PEEK and PEI in the hybrid material is very good, owing to their intrinsic compatibility [11]. A similar statement has also been made in the literature for the same blend in several studies [26,27]. The typical thickness of the hybrid material is about 0.25 mm, comprised of two PEI amorphous plies of 0.05 mm (each) and a UD APC-2/AS4 PEEK ply of 0.137 mm thickness. Overall, for each ply of APC-2/PEEK, there are two plies of PEI (Figure 2). The obtained fiber volume fraction, V_F, is about 50% [11] and is only estimated by the amount of the constituent materials used; its value lies below the standard value of composite materials used in industrial applications (e.g., $V_F \approx 67$) [3], because, at the early stage of the material development, the main aim was, on the one hand, to demonstrate the feasibility of the material production process and, on the other hand, to achieve the affordability of the fabrication of composite structures by using out-of-autoclave and automated technologies.

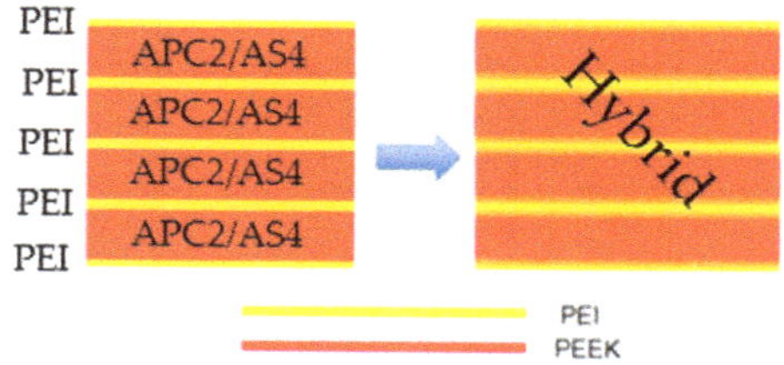

Figure 2. Amorphous bonding concept developed in [9].

This material concept was initially introduced under the name "Thermabond" and was implemented solely for bonding applications [28]; however, it was patented as an individual structural material by Leonardo [23]. It is sometimes mentioned in the literature as "interleafing". On the one hand, it provides advantages from a structural point of view, as it offers good impact damage performance, while on the other hand, it results in processing simplification, including reduced cycle times and lower energy consumption. Furthermore, the above material concept is based on tapes already consolidated, and, thus, it does not need processing temperatures much higher than the ones used for the already certified aerospace grade thermosetting resins. As a result, it can be processed by exploiting the ISC-AFP process or by involving autoclaves currently used for thermosetting materials. Moreover, the necessary crystallinity level (>23%) is obtained.

As mentioned above, the fabrication process of the hybrid prepreg starts with the heating of both the PEEK–carbon fiber prepreg and the PEI films above the PEEK's melting temperature (343 °C), thus having both resin types in a fluid state. Afterwards, the obtained prepreg is consolidated under pressure and cooled down under precise cooling conditions. Regarding the system used for the integration and combination of the PEEK–carbon fiber prepreg material that should be sandwiched between the amorphous PEI films, a specifically designed automated device was used; this equipment enabled the material feeding, alignment, trailing, and winding using a setup with different spools. For the manufacturing process, a hot press system with plates heated at different temperatures and pressures was involved. The hybrid prepreg motion was achieved by opening the normal press, transferring the assembled tape in the direction of the length of the press, and closing it again. The press process steps as well as the respective thermal conditions are sketched in Figure 3a–d. Through the devised concept, the different prepreg areas adapted their temperature progressively; the temperature was kept constant in the different press segments. This automated process facilitates energy and cost savings, because the sequential shifting of the material in the hot press during the entire working cycle eliminates the need for repeated heating and cooling of the press plates.

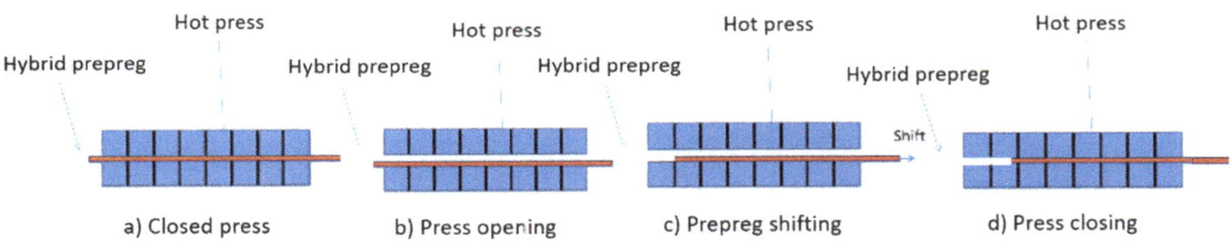

Figure 3. Hybrid Prepreg Motion under the Press.

The qualitative temperature profile of the material moving through the press exhibiting different cooling rates, indicated as the respective gradients of the curves, is shown in Figure 4. The typical cooling rate was 10 °C/min, with the maximum and minimum cooling rate being 320 °C/min and 1 °C/min, respectively.

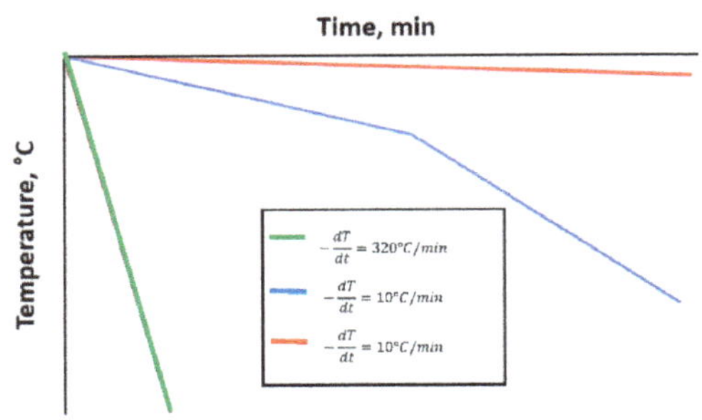

Figure 4. Qualitative temperature profile for the processed prepreg moving through the press and for different cooling rates applied.

Figure 5a shows the automated prototype equipment developed at NOVOTECH (Italy) in the framework of [11] for the production of the novel hybrid material, and in Figure 5b, a flat skin produced using the developed hybrid thermoplastic material is exhibited. The dimensions of the panel were 1.5 m × 1.0 m.

a 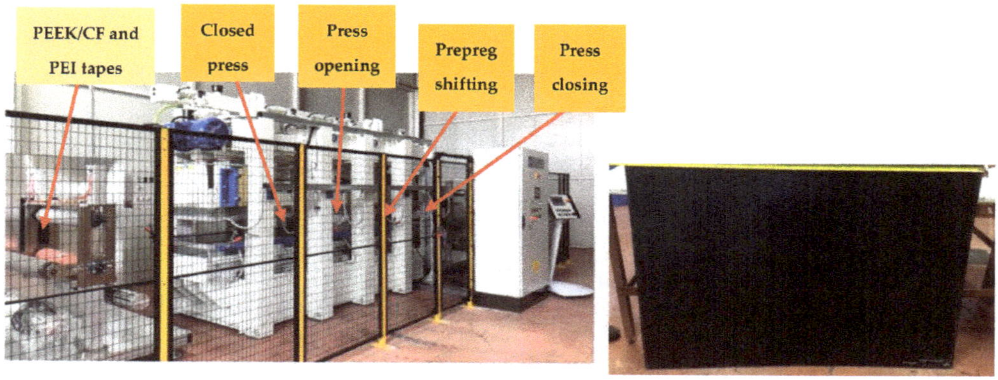 b

Figure 5. Hybrid material manufacturing equipment at NOVOTECH (Italy) (**a**) and a flat skin panel produced using the developed hybrid thermoplastic material (**b**).

A point that needs to be verified during the production of the material is the soundness of the amorphous resin molecular structure when processed at a temperature far above its T_g. It is known that the crystallization of PEEK is affected by the cooling rate [29]; a high crystallinity level of about 40% was obtained by cooling the melt at very low rates (1 °C/min). However, when the cooling rate ranges between 10 and 300 °C/min, a constant level of crystallinity of about 30% is obtained [29] (Figure 6). Quenching the melt polymer at higher cooling rates, a lower crystallinity content is obtained, leading finally to amorphous material [30]. The latter is an unstable status, and when the quenched material is heated above T_g, a crystallization occurs. However, the produced crystals are no longer arranged in a spherulitic geometry (cold crystallization).

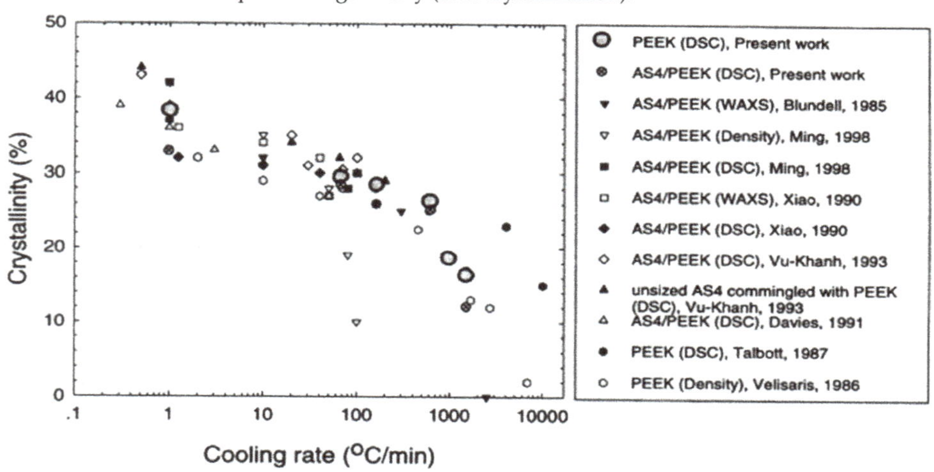

Figure 6. Crystallinity versus cooling rate for PEEK resin and PEEK–carbon fiber composites. Measurement methods are shown in parentheses [29].

The above results were exploited to optimize the temperature distribution within the plates of the developed tool (Figure 5a). To allow the consolidation of the layers and suitable bonding between PEEK and PEI, the temperature in the first zone must lie above the melting temperature of the PEEK crystalline phase (343 °C); both resin types should be liquid to allow for the blending creation. This zone is characterized by the gradual transition from 100% PEEK to 100% PEI. When PEEK and PEI blends are obtained, a crystalline phase like the standard PEEK material is obtained. The percentage of this phase depends on both the PEEK amount and the remaining blended PEEK/PEI amorphous phase. The T_g of the mentioned crystalline phase lies between the T_g of the involved materials. As a next step, a crystalline zone is created by cooling down to the crystalline temperature. The precise cooling rate is a prerequisite in this phase; a rate between 1 and 320 °C/min is necessary (Figure 4) to obtain the desired crystallinity level. For aeronautic structural applications, the crystallinity level should exceed 25%. It should be noted that, although the cooling rate window is very wide, the gap between the plates could result in a rapid cooling that might cause the temperature difference exhibited in the zone of T_m (Figure 3). It should also be noted that in the present study, the temperature of the hybrid profile was considered to be equal to the temperature of the press plate. This assumption is justified when the thickness of the material is small, as was the case in the present work.

3. Evaluation of the Implementation Potential of the New Material

3.1. Derivation of Critical Features

For the preliminary assessment of several significant technological features of the novel hybrid material produced by involving the developed continuous consolidation production process, several samples were extracted from different zones along the tapes. Said samples were subjected to the following tests, as part of the activities carried out in [11]:

- micrometer mapping to check the material thickness uniformity;
- cross section microscopy;
- differential scanning calorimeter (DSC) with double scanning both on pristine material (to verify it was provided in amorphous state) and on the new hybrid material by heating at 10 °C/min, cooling at 10 °C/min, and then heating again at 10 °C/min;
- areal weight evaluation (where possible, fiber weight after acid digestion as well).

The DSC tests revealed two distinct transitions during both runs (Figure 7):

- T_g: 207 °C–215 °C (close to the T_g of the PEI film).
- T_m: 340 °C–343 °C (no cold crystallization of PEEK).

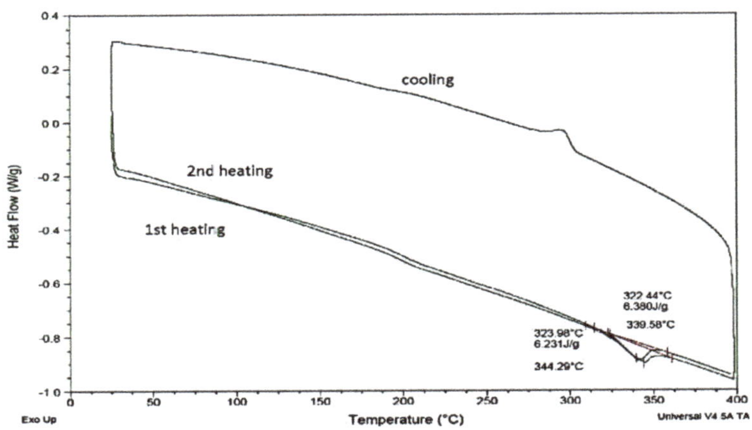

Figure 7. Two distinct transitions were revealed during the DSC test [11].

The crystallinity level was measured to be >23% and thus considered acceptable for industrial applications. Furthermore, the slight (≤3%) variability of thickness of the tested specimens that was found is also considered acceptable for industrial applications.

In addition, for validating the reliability of the "amorphous bonding concept", a preliminary mechanical test campaign was conducted by performing short-beam shear tests on specimens of the produced hybrid material. A set of six coupons produced according to ASTM D2344 were tested [11]. The obtained ILSS results are reported in Table 1.

Table 1. ILSS results for both material versions tested.

Property	Specimen Description	Average Value [MPa]	Standard Deviation [MPa]
Interlaminar Shear Strength	APC-2/AS4 PEEK + 2 PEI LAYERS (2 MIL)	81.47	±2.95
	APC-2/AS4 PEEK + 2 PEI LAYERS (1 MIL)	71.78	±2.65

The outcomes of the short-beam shear testing campaign, carried out at the coupon level on the two different versions of the new hybrid thermoplastic produced, are shown in Table 1. Specifically, the derived ILSS experimental value was found to vary between 71.78 ± 2.65 MPa (version 1) and 81.47 ± 2.95 MPa (version 2).

The achieved values are in line and comparable with data reported in the literature [29], as shown in Figure 8. Specifically, the ILSS experimental value is expected to be 66.1 ± 5 MPa in the case of carbon fiber–PEI composite material, while in case of carbon–PEEK composites, it is in the range of 85–95 MPa. The above results confirm the appropriate choice of the manufacturing process parameters.

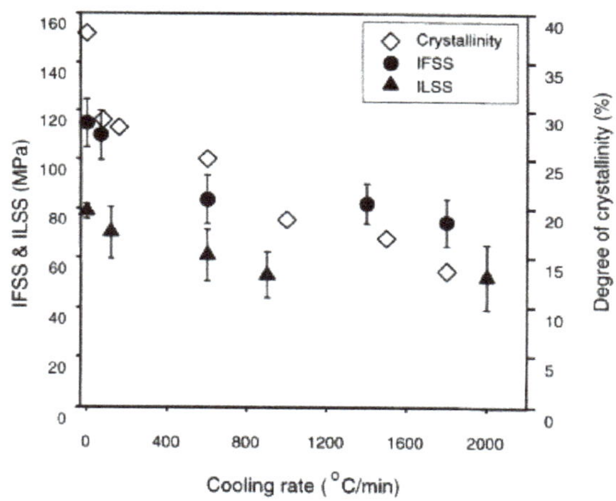

Figure 8. Comparison of IFSS, ILSS, and crystallinity level versus cooling rate [29].

3.2. Holistic Assessment of the Material's Application Potential

For assessing the application potential for aircraft structural applications, the developed hybrid material was compared to the aircraft structural thermoplastic composite material PEEK reinforced with carbon fibers, under the commercial name APC2/AS4. However, a comparison between two materials is not a straightforward issue, as it involves a variety of aspects, such as the mechanical properties, the processability, the cost, the impact on the environment, etc. To facilitate a more holistic comparison, a holistic index presented in [16] will be implemented. According to the mentioned index, quality is understood as compliance with certain critical mechanical properties of the component. In the present investigation, the ILSS property was considered a main critical mechanical property, since two different TP resins were combined, namely PEEK and PEI, and, therefore, their interlaminar behavior is significant. An additional critical point that is considered extensively in the literature is the connection between this property and the crystallinity [30,31]; the latter is also assessed in the present study. In particular, a chemical modification of the thermoplastic APC-2/AS4 (PEEK) material is obtained by the addition of PEI films. Thus, the crystallinity considerably influences the interphase adhesion of the material, which is strongly related to the interlaminar behavior. This supports the claim that the ILSS is the most sensitive mechanical property. To simplify the quality assessment process, in the present work ILSS was considered as the sole critical mechanical property. Moreover, cradle-to-cradle life cycle analysis (LCA) and life cycle costing (LCC) models were developed and put into effect for the case of the skin panel production using the two materials under comparison. In particular, for the total cost calculation as well as for identifying the main factors in the total cost, a life cycle costing model using the activity-based costing (ABC) method was created. LCA was obtained by implementing the ISO 14040:2006 [32] and involving the ReCiPe method [33]. According to the latter method, a critical evaluation of the environmental potentials for the produced fuselage panel's entire life (including production and recycling) was carried out. For the convenience of the reader, the index will be presented in brief below; more details can be found in [16].

Apart from the features involved in the index, the processability of the novel material as well as its comparison to the processability of the classical APC2/AS4 thermoplastic composite should be also taken into account. For this reason, the comparison was made through the production of flat skin panels, representative of typical aerostructures for a short-haul regional aircraft such as the ATR aircraft. The flat skin panel used in the present study may be seen in Figure 5b.

4. Material Comparison and Discussion

For the quality analysis, the derived quality function (QF), which relates the considered mechanical property with the main autoclave process parameters, as well as the fitting constants f_1, f_2, and f_3 involved in this equation are given in Table 2. T refers to temperature, t_{dwell} to the dwell time, and P to the pressure.

Table 2. The QF derived for the autoclave process.

Quality Property	Quality Function—QF [MPa]	Fitting Constants
Interlaminar shear strength ILSS	$ILSS = f_1 \cdot T + f_2 \cdot t_{dwell} + f_3 \cdot P - 322$	$f_1 = 1.45$ $f_2 = 0.06$ $f_3 = 1.04$

The results from the cost analysis (Figure 9a) showed that the main contributor to the total cost is the labor cost, associated with the hand layup involved in the component production and using both materials analyzed. However, by using the developed hybrid material, both the energy cost and the cost of the infrastructure process materials (bags, sealant tapes, valves, etc.) are reduced as compared to the respective necessary costs for the conventional APC2/AS4. In addition, for the latter case, elevated heating temperatures up to 400 °C during the autoclave process require expensive secondary materials which should be resistant to high temperatures. For the case of the novel hybrid material, the total time of the autoclave process is reduced, as the presence of amorphous PEI does not demand any specific cooling rate control; for the case of standard PEEK/AS4, the cooling rate plays a crucial role.

For the present LCA, the categories analyzed are: climate change, human toxicity, water acidification potential, and terrestrial eutrophication. Obviously, the carbon fiber production occupies more than 50% of the total environmental footprint. It further confirms the need for developing recycling processes maintaining the carbon fiber quality and keeping the involved environmental footprint as low as possible. The environmental categories analyzed are displayed in Figure 9b; their values were calculated using the Open LCA commercial software [34].

a

b

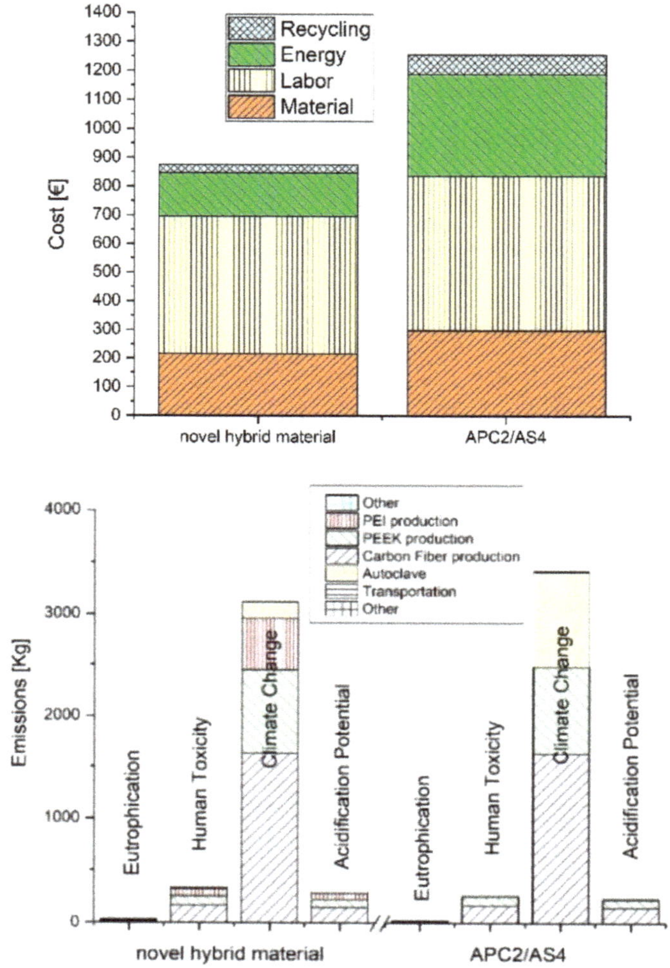

Figure 9. Allocated cost (**a**) and main environmental impact factors for both materials considered (**b**).

According to the holistic index methodology implemented, the weight factors were determined using an analytic hierarchy process (AHP) analysis [35]. The maximum priority has been assigned to the quality, as it is related to safety. The weight factors were computed to be:

Quality: $K_c = 0.08$;
Environmental Footprint: $K_E = 0.49$;
Cost: $K_Q = 0.43$.

Using the above weight factors, the index P takes the form:

$$P = 0.43 \cdot Q - 0.49 \cdot C - 0.08 \cdot E \quad (1)$$

The computed index P as well as the calculated quality, cost, and environmental footprint factors for both materials investigated are summarized in Figure 10. However, it should be underlined that this value is not an absolute quantity nor a material property. It depends on the choice of the significance, assigned by the engineer, of the aspect's quality, cost, and environmental footprint for assessing the application potential of a material. The index P should be understood as a tool exploited by the engineer to quantify his/her qualitative judgment about the significance of those controversial but interrelated aspects by a single quantity and thus meet a justified decision. Obviously, a comparison between different materials is reliable, as it is made by assigning the same significance to quality, cost, and environment, independently of the material that will be used.

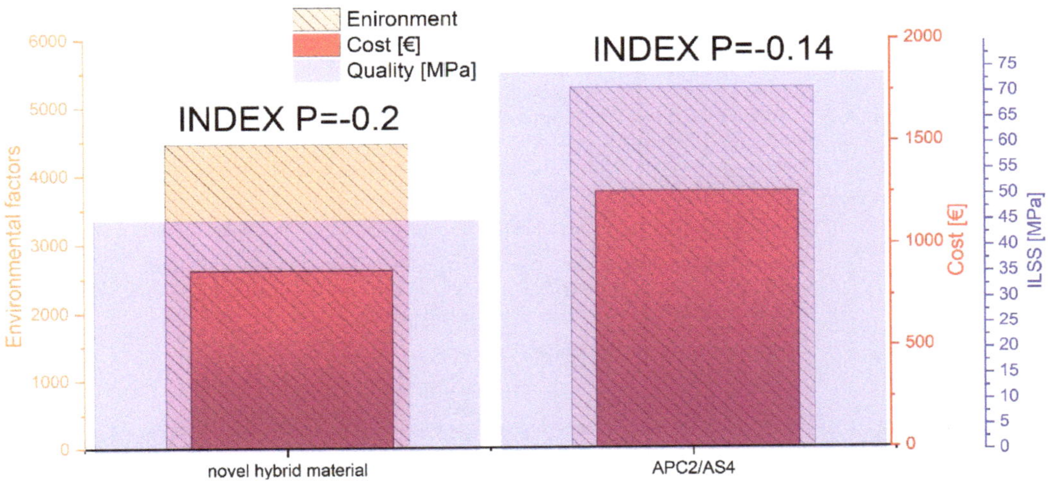

Figure 10. Overall comparison of the materials investigated.

To assess the suitability of the new material for producing parts like the one under consideration in this study, the suitability of the classical structural material APC2/AS4 was taken as the reference. Therefore, in formulating the involved index P for the APC2/AS4 material, all terms in Equation (1) are taking by definition the value 1. In this case, the index P is the sum of the weight factors. For the hybrid as well as the conventional material, the index values computed were −0.2 and −0.14, respectively, as shown in Figure 10.

From Figure 10, it is shown that the novel material (P = −0.2) is a material capable of replacing APC2/AS4 in several applications. It seems to demonstrate numerous cost and environmental advantages as compared to APC2/AS4. These advantages are quantified by means of the features *C* and *E*, respectively, in the index. Furthermore, it is made evident that once the quality of the parts produced using the new material are refined, the conventional material, which now exhibits an appreciably higher ILSS value, will be outpointed and potentially replaced.

5. Conclusions

The development of a novel hybrid thermoplastic material is presented in this study. The development of the material is described in detail, exhibiting the theoretical background involved as well as the equipment developed and utilized. The derivation of selected critical features for the preliminary evaluation of the novel material is presented as well. Moreover, the potential of the developed material for being exploited as structural aircraft material was assessed by involving a holistic index. For the assessment of the processability, which is a significant parameter for the choice of a material, the newly developed material was assessed with regard to its suitability for producing a certain component using the autoclave process. The classical material APC2/AS4 was considered as an alternative for producing the same component by involving the same manufacturing process and compared against the hybrid material developed. The results of the study pointed out that the amorphous bonding concept represents a viable manufacturing option from an industrial point of view, owing to the advantages offered by the hybrid material. The holistic index that was implemented in the component manufacturing as a decision-making tool showed the high potential of the new material, as the determined value (−0.2) was very close to the well-established material APC2/AS4. The results definitely indicate that for the case where the quality requirements of the parts produced are modest, the developed material is already competitive and advantageous from the economical and environment viewpoints. If high quality is the issue, a quality improvement of the hybrid material is necessary and would step it up to a material choice that would be even superior to the typical APC2/AS4.

Author Contributions: Conceptualization, C.V.K. and M.B.; methodology, C.V.K., S.G.P. and M.B.; validation, S.G.P. and L.L.; investigation, C.V.K. and M.B.; resources, M.B. and L.L.; writing—original draft preparation, C.V.K. and M.B.; writing—review and editing, S.G.P. and L.L.; supervision, S.G.P. and L.L.; project administration, M.B.; funding acquisition, S.G.P. and L.L. All authors have read and agreed to the published version of the manuscript.

Funding: This research was funded by European Union, grant number 723309.

Acknowledgments: The present work is a preliminary study performed within the NHYTE project (NHYTE, 2017–2020). The NHYTE project has received funding from the European Union's Horizon 2020 research and innovation program under grant agreement No 723309.

Conflicts of Interest: The authors declare no conflict of interest.

References

1. Oliveux, G.; Dandy, L.O.; Leeke, G. Current status of recycling of fibre reinforced polymers: Review of technologies, reuse and resulting properties. *Prog. Mater. Sci.* **2015**, *72*, 61–99. [CrossRef]
2. Vaidya, U.; Chawla, K. Processing of fibre reinforced thermoplastic composites. *Int. Mater. Rev.* **2008**, *53*, 185–218. [CrossRef]
3. Dijkstra, H.S. Materials for Future Aircraft Structures. Master's Thesis, Delft University, Delft, The Netherlands, 2008.
4. Barile, M.; Lecce, L.; Iannone, M.; Pappadà, S.; Roberti, P. Thermoplastic Composites for Aerospace Applications. In *Revolutionizing Aircraft Materials and Processes*; Pantelakis, S., Tserpes, K., Eds.; Springer: Berlin/Heidelberg, Germany, 2020.
5. Saenz-Castillo, D.; Martín, M.I.; Calvo, S.; Rodriguez-Lence, F.; Güemes, A. Effect of processing parameters and void content on mechanical properties and NDI of thermoplastic composites. *Compos. Part A Appl. Sci. Manuf.* **2019**, *121*, 308–320. [CrossRef]
6. Pérez-Martín, H.; Mackenzie, P.; Baidak, A.; Brádaigh, C.Ó.; Ray, D. Crystallinity studies of PEKK and carbon fibre/PEKK composites: A review. *Compos. Part B Eng.* **2021**, *223*, 109127. [CrossRef]
7. Croft, K.; Lessard, L.; Pasini, D.; Hojjati, M.; Chen, J.H.; Yousefpour, A. Experimental study of the effect of automated fiber placement induced defects on performance of composite laminates. *Compos. Part A Appl. Sci. Manuf.* **2011**, *42*, 484–491. [CrossRef]
8. Centea, T.; Grunenfelder, L.K.; Nutt, S.R. A review of out-of-autoclave prepregs—Material properties, process phenomena, and manufacturing considerations. *Compos. Part A Appl. Sci. Manuf.* **2015**, *70*, 132–154. [CrossRef]
9. Lee, J.; Ni, X.; Daso, F.; Xiao, X.; King, D.; Gómez, J.S.; Varela, T.B.; Kessler, S.S.; Wardle, B.L. Advanced carbon fiber composite out-of-autoclave laminate manufacture via nanostructured out-of-oven conductive curing. *Compos. Sci. Technol.* **2018**, *166*, 150–159. [CrossRef]
10. European Aeronautic Defence and Space Company (EADS). *Development and Innovation for Advanced ManufacturIng of Thermoplastics (DINAMIT): FP6 Dinamit Final Publishable Activity Report*; AST3-CT-2003-502831 FP6 Dinamit, EU FP6 Project: AST3—CT—2003 —5022831, CEC Final Report; European Aeronautic Defence and Space Company (EADS): Paris, France, 2007.
11. New Hybrid Thermoplastic Composite Aerostructures manufactured by Out of Autoclave Continuous Automated Technologies (NHYTE): EU Project, 2017–2020. Available online: https://cordis.europa.eu/project/id/723309/results (accessed on 10 February 2021).
12. Christmann, M.; Medina, L.; Mitschang, P. Effect of inhomogeneous temperature distribution on the impregnation process of the continuous compression molding technology. *J. Thermoplast. Compos. Mater.* **2017**, *30*, 1285–1302. [CrossRef]
13. Pantelakis, S.; Baxevani, E.; Spelz, U. An automated technique for manufacturing thermoplastic stringers in continuous length. *Compos. Struct.* **1993**, *26*, 115–121. [CrossRef]
14. Cattanach, J.B.; Barnes, A.J. Forming Fiber-Plastic Composite. U.S. Patent 4657717A, 14 April 1987.
15. Pantelakis, S.G.; Baxevani, E.A. Optimization of the diaphragm forming process with regard to product quality and cost. *Compos. Part A.* **2002**, *33*, 459–470. [CrossRef]
16. Katsiropoulos, C.V.; Pantelakis, S.G. A Novel Holistic Index for the Optimization of Composite Components and Manufacturing Processes with Regard to Quality, Life Cycle Costs and environmental Performance. *Aerospace* **2020**, *7*, 157. [CrossRef]
17. Pantelakis, S.G.; Katsiropoulos, C.V.; Lefebure, P. Effect of thermal treatment on the tensile and in-plane shear behavior of Carbon fiber reinforced PPS composite specimens. *J. Appl. Polym. Sci.* **2008**, *107*, 3190–3199. [CrossRef]
18. Spruiell, J.E.; Janke, C. *A Review of The Measurement and Development of Crystallinity and Its Relation to Properties in Neat Poly (Phenylene Sulfide) and Its Fiber Reinforced Composites*; Technical report; OSTI: Washington, DC, USA, 2005.

19. Botelho, E.C.; Figiela, L.; Rezende, M.C.; Lauke, B. Mechanical behavior of carbon fiber reinforced polyamide composites. *Compos. Sci. Technol.* **2003**, *63*, 1843–1855. [CrossRef]
20. Mitschang, P.; Blinzler, M.; Woginger, A. Processing Technologies for continuous fiber reinforced thermoplastics with novel polymer blends. *Compos. Sci. Technol.* **2003**, *63*, 2099–2110. [CrossRef]
21. Wang, J.; Liu, R.; Jian, X. Introduction to Epoxy/Thermoplastic Blends. In *Handbook of Epoxy Blends*; Parameswaranpillai, J., Hameed, N., Pionteck, J., Woo, E., Eds.; Springer: Cham, Switzerland, 2017.
22. Iannone, M.; Esposito, F. Thermoplastic matrix composites for aeronautical applications—The amorphous/semi-crystalline blends option. *AIP Conf. Proc.* **2014**, *66*, 1599.
23. Iannone, M. Preimpregnated materials with semi-crystalline matrix and amorphous surface layers. European Patent 2 109 532 B1, 2 March 2011.
24. Loukopoulos, A.; Katsiropoulos, C.V.; Pantelakis, S.G. Carbon and financial evaluation of an aeronautic component production using different manufacturing processes. *Int. J. Struct. Integr.* **2018**, *10*, 425–435. [CrossRef]
25. Timmis, A.J.; Hodzic, A.; Koh, L.; Bonner, M.; Soutis, C.; Schäfer, A.W.; Dray, L. Environmental impact assessment of aviation emission reduction through the implementation of composite materials. *Int. J. Life Cycle Assess* **2015**, *20*, 233–243. [CrossRef]
26. Ramani, R.; Alam, S. Composition optimization of PEEK/PEI blend using model-free kinetics analysis. *Thermochim. Acta* **2010**, *511*, 179–188. [CrossRef]
27. Kenny, J.; D'Amore, A.; Nicolais, L.; Iannone, M.; Scatteia, B. Processing of Amorphous PEEK and Amorphous PEEK Based Composites. *SAMPE J.* **1989**, *25*.
28. Smiley, A.J.; Halbritter, A.; Cogswell, N.; Meakin, P.J. Dual Polymer Bonding of Thermoplastic Composite Structures. *Polym. Eng. Sci.* **1991**, *31*, 526–532. [CrossRef]
29. Shang-Lin, G.; Jang-Kyo, K. Cooling rate influences in carbon fibre/PEEK composites: Part 1: Crystallinity and interface adhesion. *Compos. Part A Appl. Sci. Manuf.* **2000**, *31*, 517–530.
30. Comer, A.J.; Ray, D.; Obande, W.O.; Jones, D.; Lyons, J.; Rosca, I.; O'Higgins, R.M.; McCarthy, M.A. Mechanical characterisation of carbon fibre–PEEK manufactured by laser-assisted automated tape placement and autoclave. *Compos. Part A Appl. Sci. Manuf.* **2015**, *69*, 10–20. [CrossRef]
31. Lee, I.G.; Kim, D.H.; Jung, K.H.; Kim, H.J.; Kim, H.S. Effect of the cooling rate on the mechanical properties of glass fiber reinforced thermoplastic composites. *Compos. Struct.* **2017**, *177*, 28–37. [CrossRef]
32. The International Standards Organisation (ISO). *ISO 14040:2006: The International Standards Organisation: Environmental Management—Life Cycle Assessment—Principles and Framework*; ISO: Geneva, Switzerland, 2006.
33. Huijbregts, M.A.; Steinmann, Z.J.; Elshout, P.M.; Stam, G.; Verones, F.; Vieira, M.; Zijp, M.; Hollander, A.; Van Zelm, R. ReCiPe2016: A harmonised life cycle impact assessment method at midpoint and endpoint level. *Int. J. Life Cycle Assess.* **2017**, *22*, 138–147. [CrossRef]
34. Ciroth, A. ICT for environment in life cycle applications openLCA—A new open source software for life cycle assessment. *Int. J. Life Cycle Assess.* **2007**, *12*, 209–210. [CrossRef]
35. Velasquez, M.; Hester, P. An Analysis of Multi-Criteria Decision Making Methods. *Int. J. Oper. Res.* **2013**, *10*, 56–66.

aerospace

MDPI

Article

Pre-Flight Test Verification of Automatic Stabilization System Using Aircraft Trimming Surfaces

Albert Zajdel *, Mariusz Krawczyk and Cezary Szczepański

Łukasiewicz Research Network—Institute of Aviation, Al. Krakowska 110/114, 02-256 Warszawa, Poland; mariusz.krawczyk@ilot.lukasiewicz.gov.pl (M.K.); cezary.szczepanski@ilot.lukasiewicz.gov.pl (C.S.)
* Correspondence: albert.zajdel@ilot.lukasiewicz.gov.pl

Abstract: The new requirement of installing the flight stabilization system onboard the airplanes for performing the single-pilot flights in IFR rules was issued lately. It caused the necessity of developing such a system for small aircraft. The proposed system was developed using Model-Based Design then tuned and tested in Model, Pilot and Hardware in the Loop Simulations. The paper presents the next advanced stage of testing—verification in simulation and ground tests on the PZL-130 Orlik airplane. The implementation of this system does not modify the pilot's primary manual controls. The newly introduced electrical trim is used for automatic stabilization but can be used at manual trimming as it was previously, depending on the chosen operation mode. The ground tests were planned according to civil aviation authority and aviation law requirements. Chosen results from simulated flights were analyzed and presented, confirming the effectiveness of the proposed system. The dedicated application allowing the test engineer to change stabilization system parameters during the flight on a touchscreen tablet was developed and described. The outcome of the stabilization system test campaign was a verification of its performance before the flight tests. The comparison of simulated and real flight data will allow identifying model deficiencies and flight stabilization system efficiency, which makes possible improvements implementation. Additionally, it appeared to be the cost-effective and less electrical energy-consuming automatic flight stabilization system for small aircraft. Such features benefit initiatives like Future Sky, More Electric Aircraft and aircraft stabilization system retrofit.

Keywords: flight stabilization; flight control; flight trimming system; small airplane avionics

Citation: Zajdel, A.; Krawczyk, M.; Szczepański, C. Pre-Flight Test Verification of Automatic Stabilization System Using Aircraft Trimming Surfaces. *Aerospace* **2022**, *9*, 111. https://doi.org/10.3390/aerospace9020111

Academic Editors: Spiros Pantelakis, Andreas Strohmayer and Liberata Guadagno

Received: 14 January 2022
Accepted: 15 February 2022
Published: 19 February 2022

1. Introduction

Historically, the slipstream effect became evident during World War II after introducing piston engines and again in the 1970s when propeller-powered civil purpose airplanes were equipped with turbine engines. The same case was with the PZL-130TC-II Orlik, a Polish trainer airplane shown in Figure 1, that underwent modernization and got the new Pratt & Whitney Canada PT6A 560kW engine. However, the disposable power was quite significant and resulted in apparent difficulties in piloting and the pilot's discomfort, affecting flight safety. As a result, the PZL-130 TC-II Orlik airplane, earlier classified as Good according to Cooper–Harper rating scale [1], after the engine replacement was degraded to Fair. As that airplane is used as a trainer at the Polish Air Force Aviation Academy for initial and medium level pilot training, such degradation of piloting characteristics was not acceptable. Therefore, the team of engineers from the Institute of Aviation in Warsaw commenced the development of an automatic rudder trimming system [2] aiming at yaw compensation induced by slipstream effects. As a result, it was developed, built on the airplane and substantially improved its piloting characteristics.

Figure 1. PZL-130TC II Orlik trainer aircraft. Arrows indicate trimmers location (By Maciej Szopa/Aviation International).

In recent years, a new requirement related to piloting the airplanes under the IFR rules in controlled aerospace by a single pilot has appeared that requires a two-pilot crew on board when the aircraft does not have an autopilot or flight stabilization system. For airplanes, like PZL-130 TC-II Orlik, it became a limitation in the flights performed from the airfield to the training zones through civil controlled aerospace. Several problems of many kinds appeared because of that requirement. Then, a natural move was to build such a system onboard. After extensive analyses of similar solutions, the decision on developing a flight stabilization system for that airplane was taken. Considering the high efficiency of the yaw stabilization system previously introduced onboard, a similar stabilization method was decided for the two other control channels, pitch and roll-flight altitude and course accordingly. Adding a flight stabilization function or even automatic flight control, a feature specific to autopilots should be feasible by expanding the trimming system capabilities [3,4].

Moreover, this development path for flight control automatization seemed to have another advantage. The trimming system is far less complicated than the autopilot. The difference consists in the design of autopilot's actuators that must be integrated with manual controls and equipped with anti-lock mechanisms of the control system [5]. By contrast, trimming surfaces actuators are coupled with the manual controls through the electrical control system. Their deflections create hinge moments changing the positions of the main control surfaces, which change the forces felt by the pilot on the joystick and pedals. Thus, the trimming system is mechanically less complex, and it does not require any switching mechanisms or overload clutch engaged in the case of actuator lockage. As a result, an automatic trimming system may constitute an interesting alternative for autopilots, especially for aircraft where:

- An autopilot is not of a key system, e.g., as in a trainer airplane that arrives quickly to the exercise zone and does not travel long distances;
- An autopilot is too expensive, which is the case with economic ultralight aircraft and small general aviation aircraft;
- IFR single-pilot operations in controlled airspace are planned.

The cost-benefit of a trimming system extended with an automatic flight control function over an autopilot is obvious and stems from the following facts:

- It does not require any substantial changes and extensions of the onboard elements or systems;
- It has a simple mechanical design;
- It provides automatic trimming at the same time;

- Certification costs for trimming system (classified as Level C according to the DO-178 recommendation: Software resulting in a major failure condition for the system) are lower than for autopilot (Level B DO-178 recommendation: Software resulting in a hazardous or severe-major failure condition for the system) [6].

The flight stabilization system designing process was initiated with the extensive analyses of requirements for the system functionalities and specifications of the airplane's operational capabilities, onboard systems and civil and military standards related to such systems. As a result, the specific requirements and expectations of the PZL-130 TC-II Orlik airplane's user were:

- Keeping unchanged the way of the airplane's flight controlling, among this it's trimming,
- Certifying the developed system according to the EASA Form 1 requirements,
- Keeping the onboard power supply system and other installations, like hydraulic and pneumatic, unchanged.

Automatic trimming systems for pitch and roll channels should be included in the existing classical trimming system, operated manually with the fly-by-wire type installation. It should also fit the existing frame spaces and electric installation. The new trimmers controlling hardware and software needed to be developed, designed, built and tested (on the ground and during the flight).

The hardware modules of that system were the following: trimmers' control panels for two pilot cabins, servomechanisms (one for each control channel), computer controlling automatic stabilization flight mode and cabling. The key module for the successful development of the system was the automatic flight control computer and, in fact, the development of the control laws allowing for that.

The proposed automatic stabilization and control system is a state machine performing transitions between three operating modes:

- Manual control;
- Manual control with automatic rudder trimmer control;
- Fully automatic flight control with automatic trimming of rudder, elevator, and ailerons.

While executing the third mode, the automatic trimming in all three control channels runs in the background. What is noticeable from the pilot's point of view is that when the crew takes back control, the aircraft is properly trimmed, allowing for an easy transition.

2. Initial Verification of the Assumed Hypothesis

At the first stage of an automatic trimming system design, it had to be determined if it was feasible to control the aircraft manually using trimmers. A positive result of such a test would allow for maintaining its non-hazardous system category.

The flight tests were planned to verify the possibility of maintaining the manual control using trimmers. During the tests, the pilot could use the classic set of controls: the control stick and pedals or move the trimmers only with the additional manipulator(s). Trimmers were controlled with two joysticks: one right hand operated to move the elevator and aileron trimmers, and the other left hand operated to move the rudder trimmer.

The test pilot, with many hours of experience in the PZL-130 Orlik aircraft, was instructed to begin with the simplest maneuvers, i.e., stabilizing the straight flight manually with the use of trimmers only. Then, he was supposed to move on to more difficult tasks: change of a course, change of flight altitude, and simultaneous course and altitude changes. In the pilot's opinion, the complex maneuvers were not especially demanding. Nevertheless, the aircraft behaviour was rated at least as *Fair*. The set of test flights was concluded with a complex aerobatic figure, a barrel roll, presented in Figure 2. Currently, there are works on automating such complicated maneuvers [7], which proves the feasibility of their performance by automatic flight control systems.

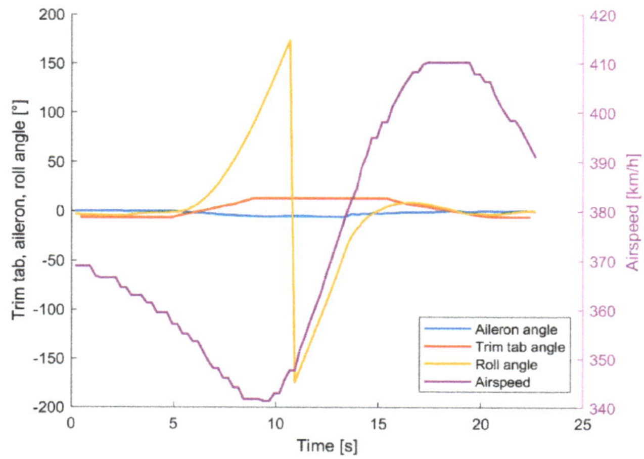

Figure 2. PZL-130TC II Orlik flight parameters during the barrel roll maneuver.

The next test phase aimed to confirm if the trimming system could be certified as a Level C system, meaning a non-hazardous one. It had to be proven that safe flight might be continued in the worst case of system failure when trimmers stay locked in an extreme position. Flight tests covered the following sequence: the pilot sets trimmers to the extreme position. After a few seconds, he tries to control the aircraft manually, compensating the significant trimmers' adverse effects.

Figure 3 shows the simulated elevator trimmer failure results when it moved and stayed locked in its extreme position at 3°, causing a sudden increase of pitch angle. The trim deflection limits of the Orlik aircraft control surfaces are as follows: −20–12° for aileron, −12–25° for rudder and −20–3° for the elevator. Asymmetries result from the nonzero neutral angle position of the trimming surface. From the flight safety point of view, the longitudinal control channel was chosen as the most difficult and dangerous one. At the beginning of the maneuver, the pilot tried to maintain a steady flight for 3 s. During this time, the airspeed dropped 12 km/h, and the pitch angle was about 6–8°. After the extreme elevator trim tab deflection in 8th s, the aircraft goes into a nosedive with a pitch angle below −40° and airspeed increase from 178 km/h to 227 km/h. Next, he tried to compensate the pitch angle changes to achieve the flat flight. He succeeded in doing it despite the significant increase of the forces on his joystick. What is important is that the pilot was still able to perform manual recovery by changing the elevator position in 11th s. After the pitch angle achieved −10°, the pilot trimmed the aircraft manually appropriately to a newly levelled flight airspeed.

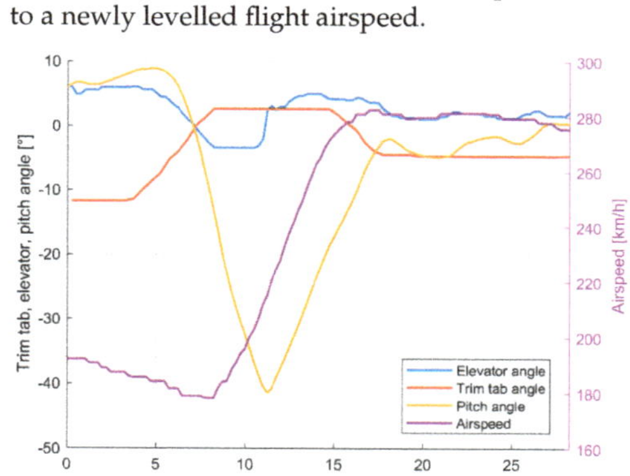

Figure 3. Manual recovery from a nosedive caused by elevator trim tab failure in a pitch control channel.

Except for the two presented maneuvers, the flight test campaign included many others required by European and American aircraft design regulations, such as:

- Nose-down recovery with an elevator trim tab at its limit, initial airspeeds from 190 to 375 km/h range,
- Nose-up recovery with an elevator trim tab at its limit, initial airspeeds from 190 to 375 km/h range,
- Right barrel roll, initial airspeeds 300 km/h and 375 km/h,
- Left barrel roll, initial airspeeds 300 km/h and 375 km/h,
- Right and left turns,
- Reaction to right and left limit rudder trim tab setting,
- Climbs and descents,
- Stick-free cruise flight controlled only by the use of trim tabs.

The flight parameters were recorded during the described flight tests to perform a post-flight analysis. Part of the obtained results is presented in Figure 4, showing how the elevator deflects due to its trim tab deflection. In this test, a pilot let go of the stick and controlled only the elevator trim table until 5 s; the flight was level at 185 km/h. Then, due to the aerodynamic coupling, the trim tab caused the elevator to deflect, and the aircraft began its descent.

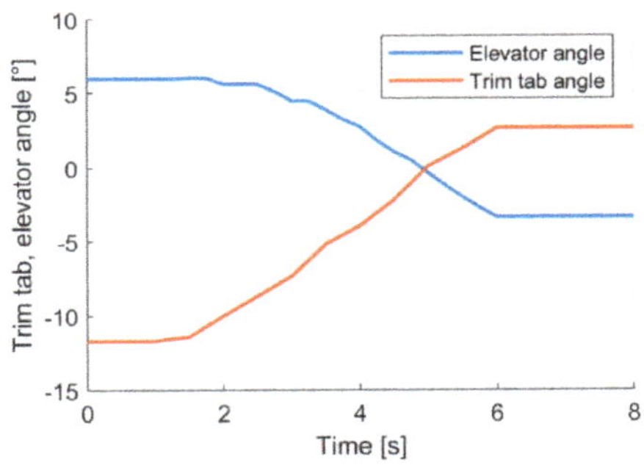

Figure 4. Example flight test results of elevator deflection caused by its trim tab deflection.

The analysis provided evidence to confirm the research hypothesis and delivered data to determine the transfer functions (1) that show the dependence between trimmer and corresponding control surface deflection.

Analyzing the Formula (1), it is evident that it is not feasible to induce the control surface movement from the upper to the lower limit by deflecting the trimmer. Thus, the designed system is inefficient at low speed when the minimum and maximum control surface deflections are necessary. Due to this limitation, the automatic flight stabilization system using trimmers installed in aircraft already in service may not be adequate for automatic take-off and landing. To make it feasible for these flight phases, changes in the design of trimmers and control surfaces may be required.

$$\delta_H = \frac{-0.6}{0.25s+1}\delta_{TH}, \delta_A = \frac{-0.42}{0.25s+1}\delta_{TA}, \delta_V = \frac{-0.75}{0.3s+1}\delta_{TV} \tag{1}$$

where: s—Laplace operator, δ_H—elevator deflection, δ_{TH}—elevator trim tab deflection, δ_A—aileron deflection, δ_{TA}—aileron trim tab deflection, δ_V—rudder deflection, and δ_{TV}—rudder trim tab deflection

3. Model-Based Design

The preliminary flight tests validated proof of concept of the trim tab stabilization system. The next stage was the system's design based on its mathematical models. At first, an airplane model was created, and its parameters were identified from flight and CFD data.

Then, before hardware manufacturing, models in the loop simulations were performed. Due to the future certification plans, a classical cascade controller was designed with outer and inner loops, including PI and PID regulators in the altitude and heading channels. Using newer, more complex methods [8,9] can significantly slow down or even prevent certification. Finally, control channels were separated—the altitude channel actuates only the elevator trim tab, the heading channel—only the aileron trim table. The third channel actuates the rudder trim tab based on a predefined lookup table based on flight test results (Figure 5) with a partial correction from the PI regulator (Figure 6). The lookup table is three-dimensional in the form of a surface called a map. Analyzing Figure 5, we conclude that the equivalent deflections of the rudder trimmer are a function of the current flight speed and the power of the propulsion system.

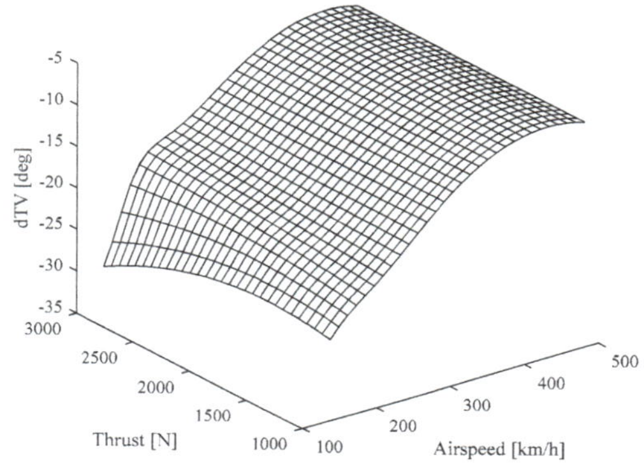

Figure 5. Three-dimensional lookup table defining rudder trim position.

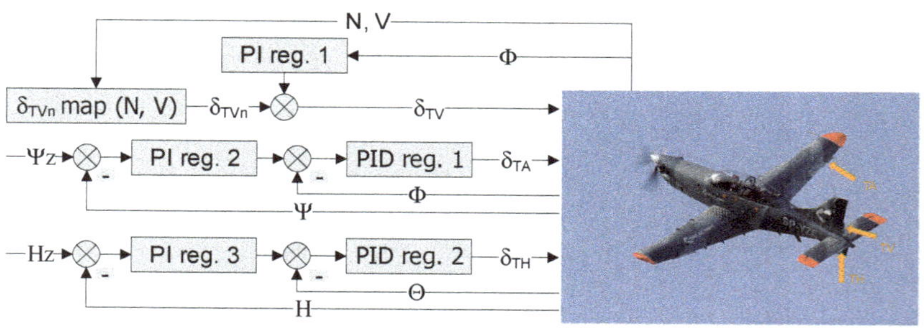

Figure 6. Stabilization system controller structure. Where: *N*—engine shaft torque, *V*—airspeed, Φ—roll angle, Θ—pitch angle, Ψ—yaw angle, H_Z—reference altitude, Ψ_Z—reference yaw angle, δ_{TVn}—rudder trim tab position from the map, δ_{TV}—rudder trim tab position output, δ_{TA}—aileron trim tab position output, and δ_{TH}—elevator trim tab position output.

Using the Bode plot forming method, the controllers were tuned with linearized models around average level cruise flight velocity, altitude operating points and airplane weight and balance typical for a training mission. First, the operating points were computed using Matlab Model Trimming tools. Then, the model was also automatically linearized at that point in the Model Linearizer tool generating the state-space form of the models. Next, controllers' coefficients tuning was done graphically using Bode plot tools. Then, those tuned coefficients values were returned with a nonlinear airplane model in the real-time simulations. The developed models were verified after the hardware parts were manufactured: stabilization system computer and trim tab servomechanisms (Figure 7). For that purpose, the hardware in the loop (HIL) simulations [10,11] were performed on a stand designed for this kind of research (Figure 8). Before installation on aircraft, a pilot in the loop (PIL) simulations performed at that stand with mounted stabilization system's hardware allowed to fine-tune the system performance, using the launch aircraft pilot's opinions.

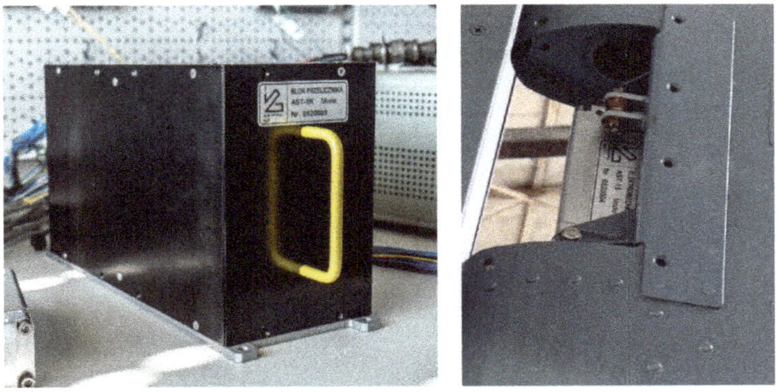

Figure 7. Stabilization system computer (**left**) and elevator trim tab servomechanism installed on the aircraft (**right**).

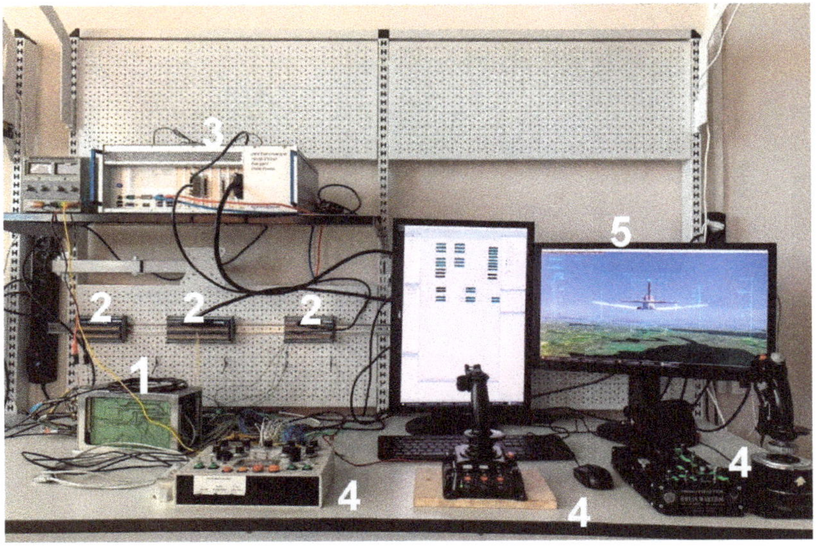

Figure 8. Hardware in the loop simulation test stand: 1—Device under test, 2—interface terminal boards, 3—real-time target computer, 4—input devices, and 5—visualization.

4. Pre-Flight Simulation Results

In pre-flight simulations, the two main goals were:

- To verify the feasibility of fulfilling the requirements by the developed system,
- To define the best values of flight stabilization system coefficients for as broad a range of the launch airplane flight envelope as possible.

Pre-flight simulation testing covered many scenarios defined in the test plan of the flight stabilization system. Finally, some chosen simulation results were presented to show and discuss the achieved results.

Example results show stabilization performance in case of atmospheric disturbance that causes pitch and roll rate 15°/s increase for 2 s (Figures 9 and 10). In this scenario, both altitude and heading channels are affected simultaneously. In the beginning, the airplane is in a stabilized cruise flight, and the disturbance starts at the 5th second of the simulation. The shape of the responses is also influenced by the Dryden turbulence used in the environment model. Initially, the disturbance causes an altitude increase of 14 m and a heading decrease of -5°. As a result, the pitch angle rises to 4.9°, and the roll angle decreases to -17°. The system stabilizes initial disturbance after about 25 s in altitude with undershoot reaching 7.5 m at 19 s of simulation. Pitch angle stabilizes after 15 s from disturbance start (Figure 9). Heading and roll angles stabilize after 15 s from the disturbance occurrence with a small 1° overshoot (Figure 10). Figure 11 shows elevator, aileron and rudder trim tab angles commanded by the stabilization system controller and corresponding primary control surfaces deflections. The graphs are limited to 20 s of simulation because the deflections were stable after that time. Those example simulation results prove the achieved effectiveness of the flight stabilization system using trimming tabs.

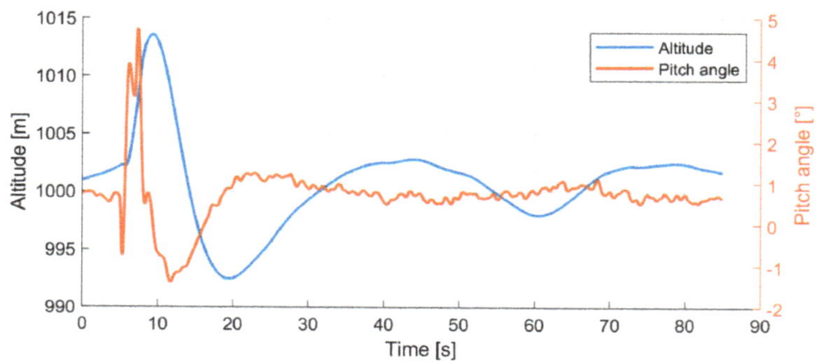

Figure 9. Altitude and pitch angle stabilization after disturbance.

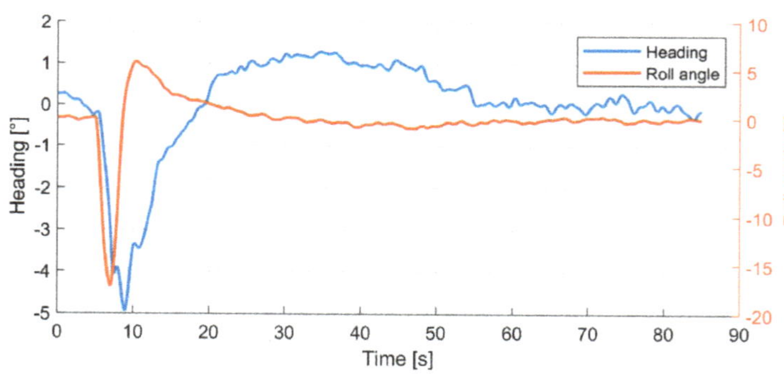

Figure 10. Heading and roll angle stabilization after disturbance.

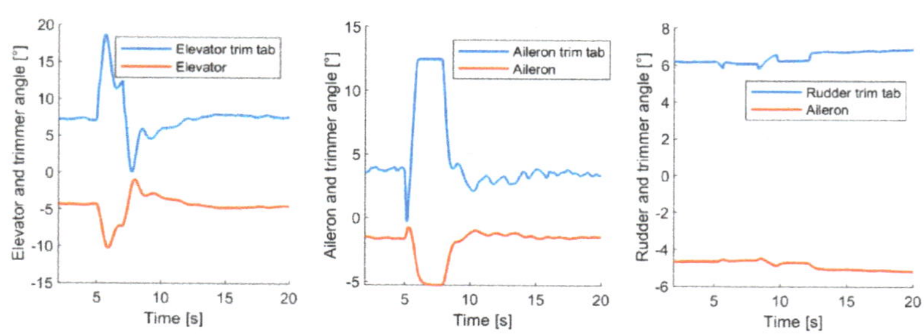

Figure 11. Elevator, aileron, rudder, and their trim tabs deflection angles during stabilization.

The possibility of applying the developed system to stabilize the launch airplane flight was tested and confirmed within the flight velocities from the landing approach velocity up to the maximum operational velocity.

5. Installation on Aircraft and Ground Tests

The set of tests preceded the installation of the stabilization system's hardware modules on the airplane, according to the DO-160G environmental standard requirements. After passing laboratory tests [12,13], the system was installed on PZL 130 TC-II Orlik airplane (Figure 12) and underwent ground tests performed under the supervision of the Polish Civil Aviation Authority. Ground tests covered typical checks: electrical wiring, trim tab calibration and polarity, interface, cabin controls, emergency cut-off, failure detection, diagnostics. In addition, a special application installed on the dedicated ruggedized laptop with a touch screen interface (Figure 13) was developed to handle diagnostics and calibration of the system at the aircraft.

Figure 12. PZL 130 TC-II Orlik during a ground test (By Juliusz Sabak/Defence24.pl).

Figure 13. In-flight controller tuning application installed on the dedicated laptop.

Flight tests are the costliest stage of an avionics project. Therefore, another application was developed to minimize the expensive test flights number. It was dedicated to the flight engineer to allow him to change the stabilization system controller parameters during the flight (Figure 13). The experienced flight test engineer can tune controller parameters and see the result immediately during one flight, eliminating the need for landing, on-ground tuning, and verifying introduced changes in the next test flight. Moreover, several parameter sets can be tested in one flight instead of one set, as in the former method. During the ground tests, the flight engineer positively assessed the application and its graphical interface.

All the performed research and tests show the capabilities and effectiveness of the proposed solution. Extensive ground test results confirmed their passing by the Polish Civil Aviation Authority. The whole developed flight stabilization system using the trimming surfaces was ready for the flight tests, which will be the final stage of its validation.

6. Conclusions and Final Remarks

The various automatic control methods for designing the stabilization flight were checked. Many of them appeared to be effective and could be applied. However, the final choice of a control method was caused by the feasibility of its certification according to the rules and requirements of the EASA Form 1 certification. Therefore, the classical type of control methods was applied to that project.

The stabilization system development described in the paper was at the final stage, which could be performed on the ground. Before initiating the flight tests, extensive research and tests were performed. They were based on the substantial use of PIL, HIL simulations and environmental testing laboratory tests. The final ground tests stage was supervised by the Polish Civil Aviation Authority, according to certification standards.

The developed airplane flight stabilization system using the trimming surfaces showed its safety and efficiency during all ground phases of development. The SIL, HIL and PIL simulations proved its capability of flight parameters stabilization in the full cruising flight velocities range. That stabilization is effective in the three channels in parallel, i.e., the effective three-dimensional automatic flight control system was developed.

It can stabilize the flight in the typical operating conditions and act as autopilot in the range of the cruising flight velocities. However, it would require connecting it with the onboard navigation system and interface, allowing the pilots for introducing the given flight parameters, like altitude and velocity.

The developed system is planned to be certified after completing the flight tests. That certification with Form 1 certificate will allow installing that system onboard the GA airplanes not equipped with the autopilot. The launch airplane will be the PZL-130 TC-II Orlik turboprop training airplane.

Such a certified stabilization system will fulfil the appropriate requirements and specifications and allow for performing single-pilot flights in the controlled airspace in the IFR conditions.

Thanks to supporting the pilots in their piloting duties, the presented stabilization system will substantially increase the safety of the GA airplanes performing the flights according to the Free Sky conditions. It will take away from the pilots the simple piloting activities and allow them to perform the navigation and communication with ground control duties. It is particularly important for the not experienced pilots exploiting the GA aircraft.

The proposed system goes into the more electric and future electric aircraft direction. It will allow replacing the heavy hydraulic autopilots' effectors onboard systems with lighter electrical solutions. In small aircraft, such a system can act as a simplified autopilot stabilizing the current flight parameters and will not require a substantial increase of the electric power installed onboard those airplanes.

Author Contributions: Conceptualization, A.Z., M.K. and C.S.; methodology, M.K. and C.S.; software, A.Z.; validation, A.Z.; formal analysis, A.Z.; investigation, A.Z.; resources, A.Z.; data curation, A.Z.; writing—original draft preparation, A.Z., M.K. and C.S.; writing—review and editing, A.Z., M.K. and C.S.; visualization, A.Z.; supervision, M.K. and C.S.; project management, C.S.; funding acquisition, C.S. and M.K. All authors have read and agreed to the published version of the manuscript.

Funding: Research described in that paper has been funded under the EU co-financed project number POIR.04.01.02-00-0006/17-00, titled "Innovative system of flight stabilization with use of trimmers"—ISSLOT.

Data Availability Statement: The data relating to that paper are stored and managed in the Lukasiewicz Research Network—Institute of Aviation. They are not publicly available as they are related to the IPR of the airplane manufacturer Airbus Poland S.A.

Conflicts of Interest: The authors declare no conflict of interest.

References

1. Cooper, G.E.; Harper, R.P. *The Use of Pilot Rating in the Evaluation of Aircraft Handling Qualities*; National Aeronautics and Space Administration: Washington, DC, USA, 1969.
2. Krawczyk, M.; Graffstein, J. Propozycja systemu eliminującego szkodliwe odziaływanie strumienia zaśmigłowego w samolotach turbośmigłowych. *Sci. Lett. Rzesz. Univ. Technol. Mech.* **2013**, *30*, 287–295. [CrossRef]
3. Levy, D. Design of a full time wing leveler system using tab driven aileron controls. In Proceedings of the Guidance, Navigation and Control Conference, Hilton Head Island, SC, USA, 10–12 August 1992. [CrossRef]
4. Jenks, G.E.; Henry, H.F.; Roskam, J. *Flight Test Results for a Separate Surface Stability Augmented Beech Model 99*; Technical Report; National Aeronautics and Space Administration: Edwards, CA, USA, April 1977.
5. *Vehicle Management Systems-Flight Control Function, Design, Installation and Test of Piloted Military Aircraft, General Specification for AS94900A*; SAE International: Warrendale, PA, USA, 2018.
6. Gohil, K. *Software Considerations in Airborne Systems and Equipment Certification*; Radio Technical Commission for Aeronautics: Washington, DC, USA, 1992.
7. Rogalski, T.; Rzucidło, P.; Prusik, J. Unmanned aircraft automatic flight control algorithm in a spin maneuver. *Aircr. Eng. Aerosp. Technol.* **2020**, *92*, 1215–1224. [CrossRef]
8. Asa, E.; Yamamoto, Y. Aircraft Flight Stabilizer System by CDM Designed Servo State-Feedback Controller. *Aerospace* **2021**, *8*, 45. [CrossRef]
9. Efremov, A.V.; Mbikayi, Z.; Efremov, E.V. Comparative Study of Different Algorithms for a Flight Control System Design and the Potentiality of Their Integration with a Sidestick. *Aerospace* **2021**, *8*, 290. [CrossRef]
10. Dolega, B.; Rogalski, T. The new conception of the laboratory testing of the FBW control system for small aircraft. *Aircr. Eng. Aerosp. Technol.* **2004**, *76*, 293–298. [CrossRef]
11. Kiesbye, J.; Messmann, D.; Preisinger, M.; Reina, G.; Nagy, D.; Schummer, F.; Mostad, M.; Kale, T.; Langer, M. Hardware-In-The-Loop and Software-In-The-Loop Testing of the MOVE-II CubeSat. *Aerospace* **2019**, *6*, 130. [CrossRef]
12. Krawczyk, M.; Zajdel, A.; Szczepański, C. Simulation and Testing of Flight Stabilisation System Using Trimmers. In *Conference on Automation*; Springer: Cham, Switzerland, 2021; Volume 1390, pp. 185–196. [CrossRef]
13. Szczepanski, C.; Krawczyk, M.; Zajdel, A. The airplane trim system–new functionalities. *Aircr. Eng. Aerosp. Technol.* **2020**, *92*, 1401–1406. [CrossRef]

aerospace

MDPI

Article

The Impact of Sensor Errors on Flight Stability

Michal Welcer, Cezary Szczepański and Mariusz Krawczyk *

Lukasiewicz Research Network—Institute of Aviation, al. Krakowska 110/114, 02-256 Warsaw, Poland; michal.welcer@Ilot.Lukasiewicz.Gov.Pl (M.W.); cezary.szczepanski@ilot.lukasiewicz.gov.pl (C.S.)
* Correspondence: mariusz.krawczyk@ilot.lukasiewicz.gov.pl

Abstract: Sensors play a significant role in flight control systems. The accuracy of the measurements of state variables affects the quality and effectiveness of flight stabilization. When designing closed-loop systems, it is desirable to use sensors of the highest class and reliability, the signals of which will be as error-free as possible. False indications lead to malfunctioning of the stabilization system, and its operation does not meet the requirements set for it. There are many types of errors—bias, white noise, hysteresis, or bias drift—which affect the measurement signals from the sensors. One of the significant problems is assessing what maximum level of sensor errors stabilization system will still operate as required. In this paper, the impact of different sensor errors on flight stabilization was presented. The research was carried out using the example of an automatic flight stabilization system using aircraft trimming surfaces in a longitudinal control channel in Hardware-in-the-Loop simulations. The model simulates various types of sensor errors during flight, while the stabilization system is implemented in hardware interfaced with a real-time computer. The results of the simulations are presented and analyzed. Their comparison indicated which sensor errors affects the flight stability the most and how the effectiveness of the stabilization system changes as error increases. The presented results show changes in flight parameters due to added sensor errors. Depending on the accuracy class of the IMU, the errors more or less disrupt the operation of the system.

Keywords: sensor errors; flight stability; trimmer; control system

Citation: Welcer, M.; Szczepański, C.; Krawczyk, M. The Impact of Sensor Errors on Flight Stability. *Aerospace* **2022**, *9*, 169. https://doi.org/10.3390/aerospace9030169

Academic Editors: Andreas Strohmayer, Spiros Pantelakis and Liberata Guadagno

Received: 14 January 2022
Accepted: 15 March 2022
Published: 19 March 2022

1. Introduction

Towards the end of designing a control system, its tolerance to errors of sensors used in the system can be checked. The outcomes verify reliability and provide information on how it affects the accuracy of the designed system. It is very important to investigate how and to what extent sensor errors change the behavior of the system.

This article focuses on the impact of errors on a flight stabilization system using trimmers designed for the single engine, turboprop trainer aircraft [1–3]. This system relieves pilots during flight, so there is no need to trim the aircraft every time when changing flight parameters. The Trimmers Stabilization System (TSS) has two modes—the first one is the STAB mode, which is used to maintain the altitude and heading of the aircraft; for this purpose, the control system calculates the deflection of elevator and aileron trimmers. The second one, the ATSK mode, is responsible for elimination of the side-slip effect caused by the propeller; in this case, only the rudder trimmer is used [4]. It was necessary to investigate how the accuracy and robustness of sensors can influence the behavior of the designed system.

For the operation of control systems, flight parameters such as airplane heading, altitude, bank angle, and pitch are essential. It is necessary to constantly check for errors between the set and measured controlled parameters. The Inertial Measurement Unit (IMU) is responsible for providing data on the acceleration acting on an object and the angular velocities to determine its orientation and position (so-called dead reckoning). One of the disadvantages of this type of navigation is sensitivity to the accuracy of the sensors used and their errors. They can lead to malfunctioning of the system. There are many IMUs available on the market; they differ in accuracy, range of measurements and, above all, price. From an economic point of view, the best solution would be to use the cheapest IMU with the highest measurement accuracy, which would ensure correct measurements throughout the flight. Therefore, a compromise between these factors is an important task. This is possible by checking which sensor error values negatively affect the operation of the system.

The aim of this work is to investigate which type of sensor errors have the most significant impact on the quality of the system and what are the critical values of the error parameters. This information is important for the validation of the designed control system.

There are many articles describing this problem. Luu describes the errors which mostly occur in Mems IMUs [5]. Tripathi presents the sensitivity analysis of various types of sensor errors, such as bias, misalignment, and scale factor and their impact on tactical flight navigation [6]. Ayre-Sampaio describes the outcomes of comparing three IMUs for strapdown airborne gravimetry [7].

This work takes into account the influence of the most common sensor errors occurring in such systems. The present article is distinguished by the fact that different IMUs with different accuracy classes are compared. The error parameters were collected from their datasheets and implemented in the model.

The performed tests showed that the IMU of the best and middle class of accuracy did not significantly affect the control system; the efficiency of the system was still acceptable. Critical values of error parameters were presented and the operation of the control system was examined. Differences in the use of various IMUs with miscellaneous accuracy classes are shown [8].

The structure of this paper is as follows: Section 2 describes the tools used to conduct the test. Then the test plan is introduced, and the problems are described. Section 3 presents the obtained research results. Section 4 summarizes the results obtained and discusses the consequences, along with a description of subsequent research plans.

2. Materials and Methods

This research was conducted at the Avionics Department of Institute of Aviation. MATLAB/Simulink r2019b was used to model and design the aircraft and the flight control system.

A test was planned to check the behavior of STAB and ATSK modes with occurrence of the chosen sensors' errors. The test plan is presented in the points below:

1. Altitude hold test with occurrence of white noise from gyroscopes during a steady flight;
2. Altitude hold test with occurrence of white noise from gyroscopes during a flight disturbed by the appearance of altitude error;
3. Heading hold test with occurrence of white noise from gyroscopes during a stable flight;
4. Heading hold test with occurrence of white noise from gyroscopes during a flight disturbed by the appearance of heading error;
5. ATSK mode test with occurrence of drift error of Torque Sensor on the Shaft (TSoS) during a stable flight;
6. ATSK mode test with occurrence of white noise from Torque Sensor on the Shaft (TSoS) during a stable flight.

The simulation logs were imported from each simulation; then, the simulation data were presented and the specific variables analysed using graphs.

The general structure of the TSS system is shown in Figure 1. The trimmer controller calculates the deflections of trimmers to ensure stable flight of the airplane. The algorithm implemented inside is based on the PID controller, which is responsible for maintaining the altitude and heading of the aircraft. A simple scheme of the controller is shown in Figure 2. The signal then goes to the trimmers causing the rotation of the aircraft; signals from sensors distorted by all kinds of errors go to the trimmer controller.

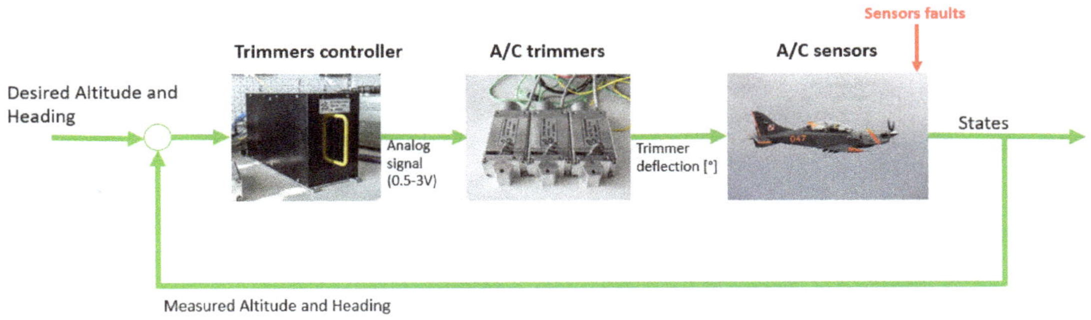

Figure 1. General structure of the TSS system.

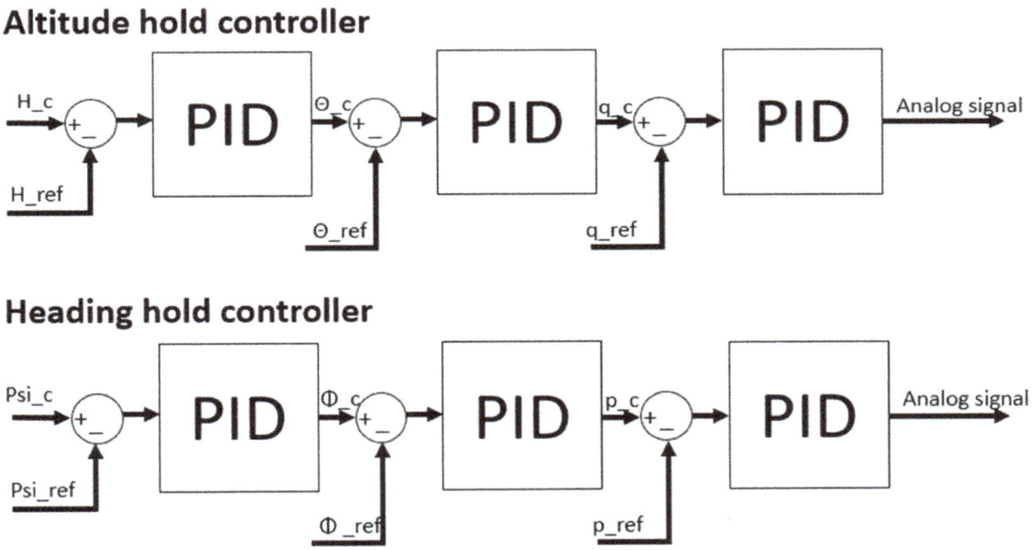

Figure 2. Simplified scheme of the STAB mode algorithm.

During the tests, white noise, bias error, and sensor drift were applied to the modelled sensors [9]. The white noise was represented by a random signal with equal strength at different frequencies. The sensors were disturbed with a constant value of power spectral density during the tests. The bias error is a systematic error that usually occurs when the sensor is incorrectly calibrated, characterized by a constant deviation of the sensor indications from the real values. Sensor drift causes an increasing error over time. The tests were planned to check the impact of each error on the system operation. Figure 3 shows the types of applied errors [10].

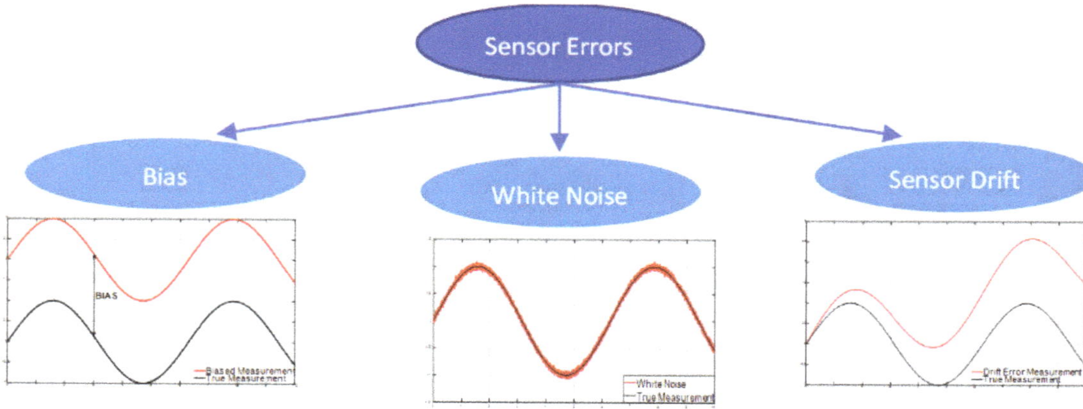

Figure 3. Types of sensor errors simulated in the tests.

The white noise error was parametrized by the constant value of PSD (Power Spectral Density) [11]:

$$PSD(f) = \lim_{T \to \infty} \frac{1}{T} |\widehat{x_T}(f)|^2 \tag{1}$$

where $\hat{x_T}$ is the Fourier transform of the signal x_T, f is a frequency, and T is the period of signal. In the test, three different values of PSD were taken into consideration; the values are presented in the table below.

The first value is taken from KVH P-1775 IMU which was tested in our laboratory. The value of PSD was calculated from data obtained from those measurements [12]. Other values of PSD correspond to IMUs with worse parameters than the KVH P-1775 IMU [13,14]. The presented IMU consists of three fibre optic gyroscopes, which provide high accuracy class measurements. In addition, the 3-axis magnetometer is included to compensate the magnetic field and high performance 3-axis accelerometer with measurements up to 30 g. This IMU can be implemented in any type of aircraft or vehicle. The outcomes from sensors are used to calculate the position and orientation of the object. To detect the described errors, such as drift error, it was necessary to install the device on a rotary table, where it was possible to check the accuracy of the measurements. The white noise of the gyroscopes was tested with an IMU installed on a stable table and the value of PSD (Power Spectral Density) was calculated; therefore, it was possible to simulate the performance of sensors during the tests. Then, two gyroscopes with a worse class of performance were selected; a comparison of the signals is presented in Figure 4. The signal colored with yellow represents the signal from KVH P-1775 IMU.

Sensor errors are modelled and added to sensors models, as shown in Figure 5. This is a simplified diagram simulating sensors with the discussed errors. The signals of white noise and bias error come from square blocks colored with dark blue; the drift error is shown as the result of temperature change during the operation. Error signals go to the selector, which chooses between errors selected for the specific test. Then, the signal goes to the sum block, where the specific error is added to the sensor model, presented as dark blue rectangle, which represents the states from the six degrees of freedom equations implemented by the Simulink library block. The output represents measurements with specified errors.

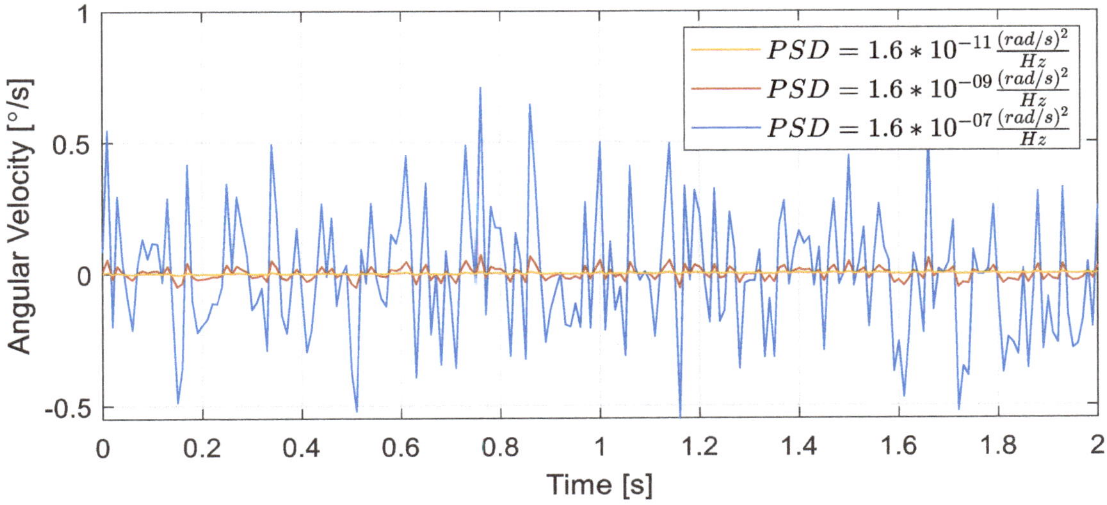

Figure 4. Measurements from gyroscopes with white noise.

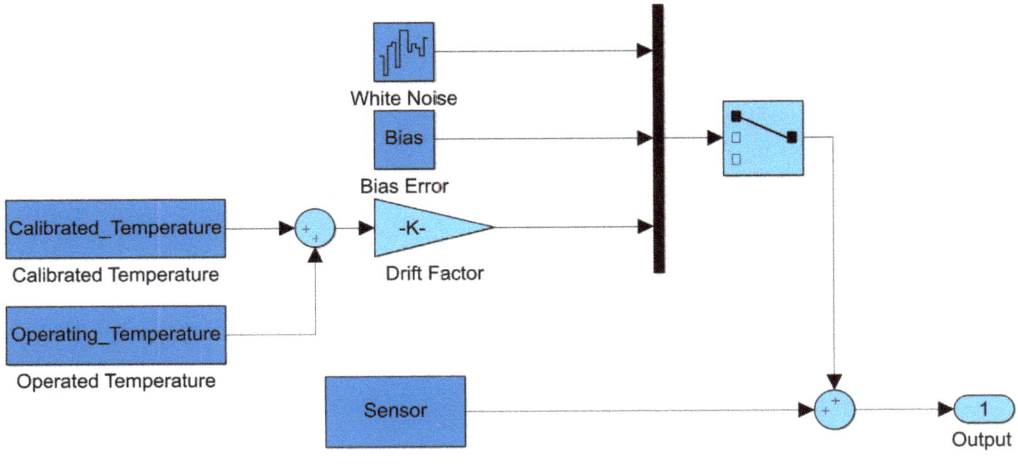

Figure 5. Modelling the sensor errors.

To maintain the attitude of the controlled aircraft, the measurements from gyroscopes were used to calculate the orientation of the object. To simplify the simulation, the angular velocities signals were directly integrated into the pitch, roll, and yaw angles. Simulations were carried out in Simulink with initial velocity and altitude of the aircraft in order to be able to ensure a stable flight during the tests.

While testing the STAB mode, five scenarios were established: first, altitude maintenance was tested. During the stable flight, the measurements of pitch angular velocity were interrupted by white noise. The aim of TSS was to continue the stable flight of the aircraft. Then, the TSS was tested to cope with the occurrence of an abrupt altitude error with the same sensor errors as in the previous test. The next two tests were carried out analogously to the previous one. This time, the task of TSS was to maintain constant heading during the flight. The last test of the STAB mode was carried out with bias error of the pitch angular velocity sensor during the stable flight.

The ATSK mode was then tested. In Figure 6, the structure of the ATSK mode is shown. Based on values of airspeed and torque on the engine shaft, the TSS calculates the deflection of the rudder trimmer. During the test, white noise and sensors' drift were applied to the torque sensor on the shaft.

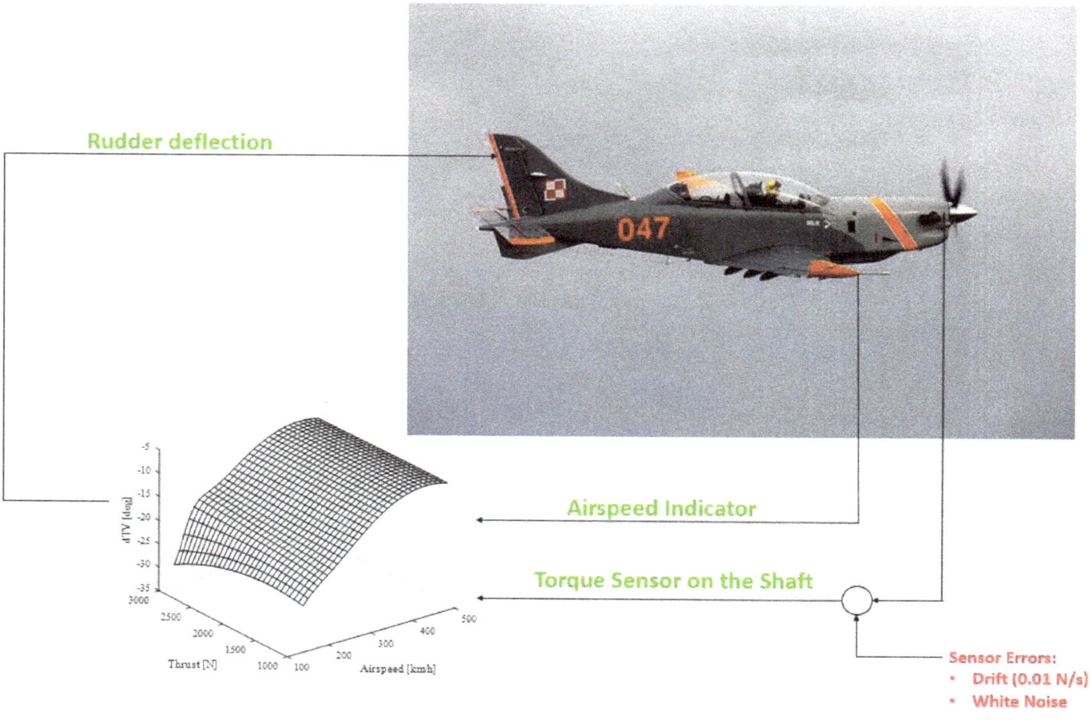

Figure 6. Structure of the ATSK mode.

For the system to meet its requirements, the error must not exceed 1 m for the altitude hold mode and 0.1 degree for the heading hold mode.

3. Results

In this section, the results of the tests are presented. First, the influence of white noise on three axis gyroscopes in STAB mode is shown. In Figure 7, the outputs of simulation are presented; in Figure 7a, the change in altitude is shown. There are three signals, each corresponding to IMUs with a different class of accuracy, which is determined by the value of the PSD parameter. The blue one represents the IMU tested in our laboratory, the yellow signal represents the IMUs where the PSD of white noise causes the TSS to malfunction, resulting in altitude error that exceeds the assumed limits equal to 1 m. In the zoomed area, the signals of the two other simulations are shown. The values of both signals are acceptable; the error is less than1 m and does not cause incorrect operation of the TSS. The amplitudes of these signals do not exceed ± 0.2 degrees. In Figure 7b, the pitch angle error during the simulation is shown. This graph presents the data calculated from IMU with gyroscopes' measurements distorted with white noise with PSD value equal to 1.6×10^{-11} [rad^2/s]; a non-distorted signal has been added for comparison. The difference between these signals is negligible. Figure 7c shows the deflection of the elevator trimmer set by the trimmer controller, calculated from the distorted and non-distorted IMUs gyroscopes. It should be noted that these signals come from the trimmer regulator and deflections of trimmers vary from these. The amplitude of noise presented in the graph is small enough to not cause movement of the trimmer.

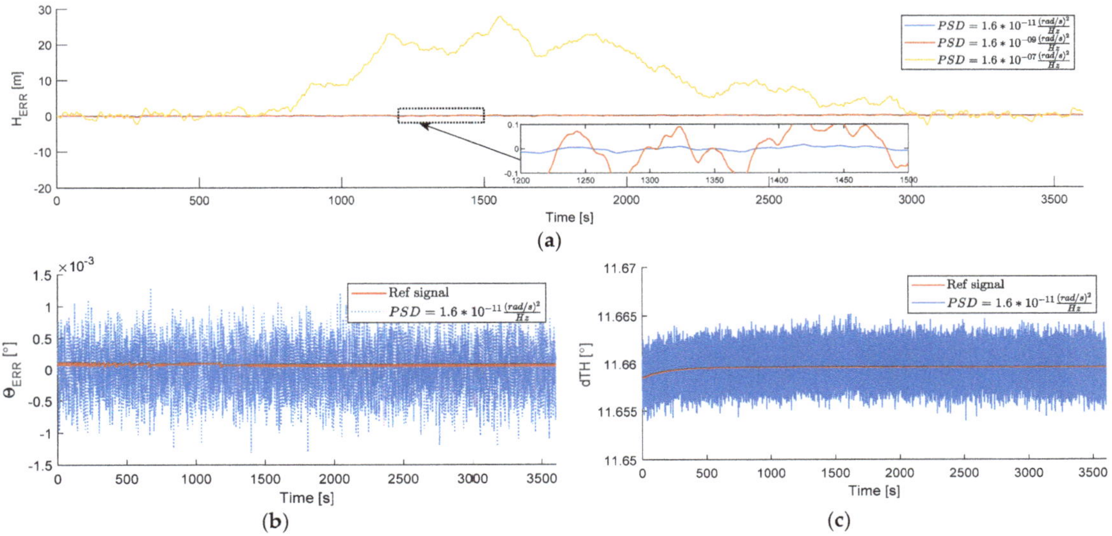

Figure 7. Influence of white noise on a stable flight: (**a**) Comparison of altitude error; (**b**) Comparison of pitch angle error; (**c**) Comparison of elevator trimmer deflection.

Figure 8 shows three graphs with data collected from simulation where a sudden altitude error occurred. In Figure 8a, the change in altitude is shown. The three signals, each one from different simulations, are compared. The yellow signal represents the data from IMU with the lowest class of accuracy. From Figure 8a, the fastest settling time is 300 s and this value meets the stated requirements. The settling time of the system using the IMU with the lowest class of accuracy is about 2200 s and is higher than the others. Like in the previous figure, Figure 8b,c show the pitch angle and deflection of the elevator trimmer. The differences between signals presented in Figure 8b are negligible. Figure 8c, similar to the previous chart, shows small differences between two signals—the elevator trimmer deflection signal of the distorted and not-distorted system.

Figure 9 shows the outcomes from simulations in which the heading hold was tested in the STAB mode. The outcomes contain the data calculated from three different IMUs. In Figure 9a, the heading errors are compared. During the steady flight, the TSS was supposed to maintain a constant heading equal to 0. The heading error of yellow signal deviates less than 0.8 degrees from 0 during the flight; other signals oscillate around 0 degree with amplitude equal to 0.2 degrees. Then, Figure 9b,c show heading error and calculated aileron trimmer deflection obtained from the simulation, in which the highest class of accuracy IMU was used. These signals are compared to outcomes from the simulation where no sensor errors were implemented. The difference between them does not exceed ± 0.015 for the phi error and ± 0.02 degrees for the aileron trimmer deflection.

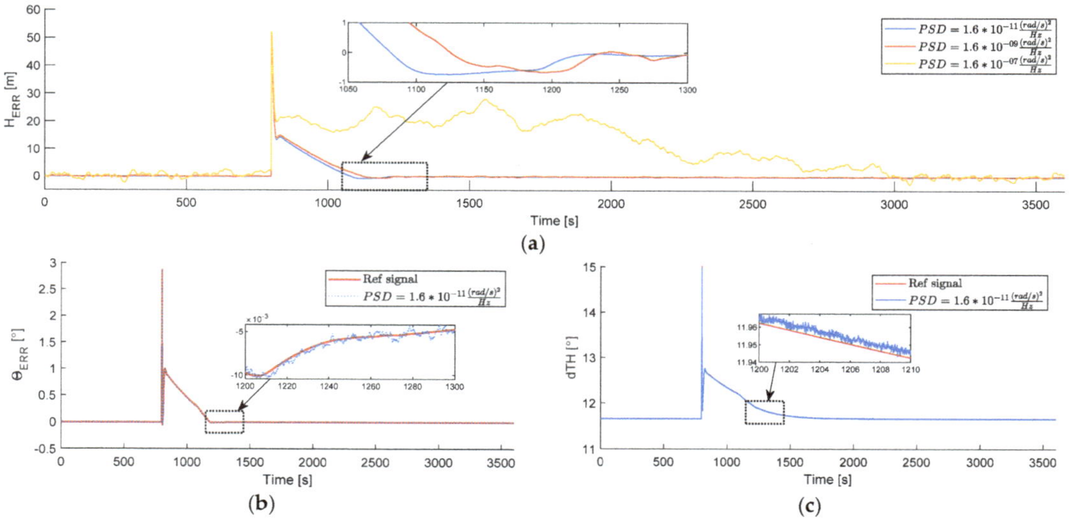

Figure 8. Influence of white noise on elimination of altitude error: (**a**) Comparison of altitude error; (**b**) Comparison of pitch angle error; (**c**) Comparison of elevator trimmer deflection.

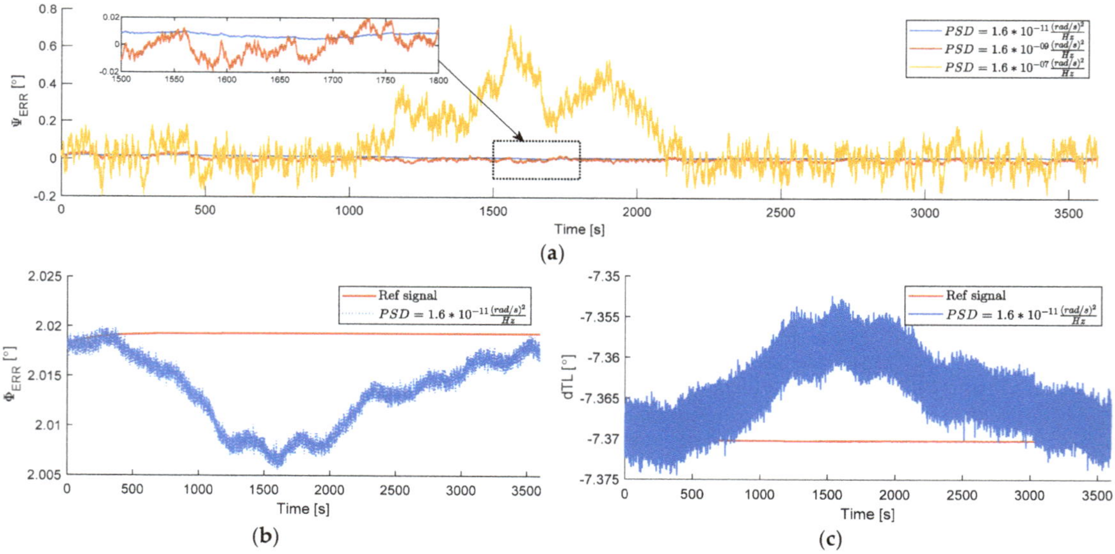

Figure 9. Influence of white noise on a stable flight: (**a**) Comparison of heading error; (**b**) Comparison of roll angle error; (**c**) Comparison of aileron trimmer deflection.

Figure 10 contains the data from simulated flights where TSS was tested with heading error scenario. Same as before, three simulations with different IMUs were carried out. To compare the performance of each of the IMU, signals of heading error are shown in Figure 10a. The results of all the simulations hover around 0 degrees; the error ranges from −0.2 to 0.6 degrees. Figure 10b,c contain data of phi error and aileron deflections for simulation where the IMU with the highest class of accuracy was used. The signal differences do not exceed ±0.1 degrees for the phi error and ±0.02 degrees for the elevator trimmer deflection. It can be assumed that the differences in both cases are negligible.

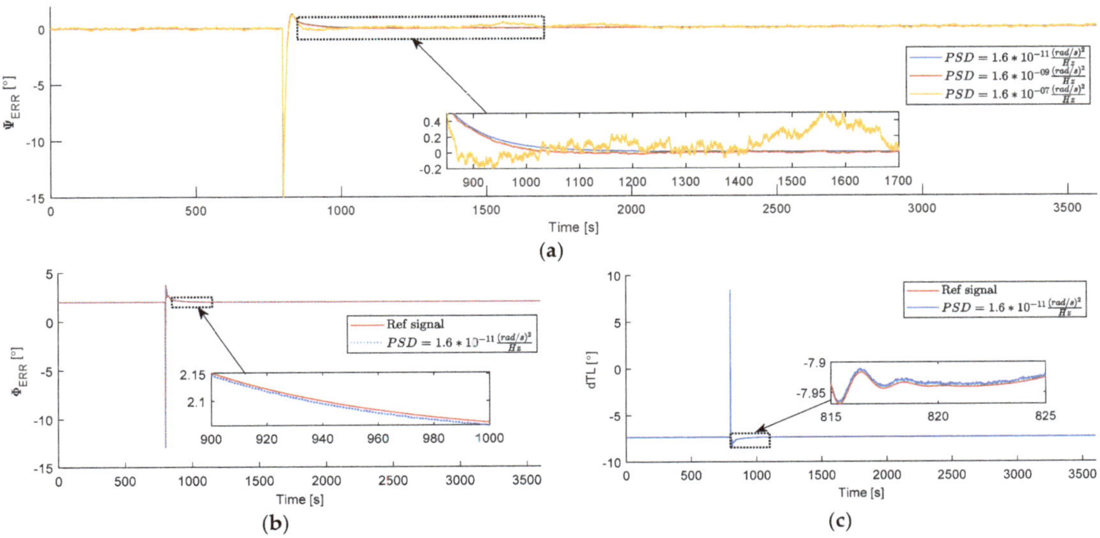

Figure 10. Influence of white noise on the elimination of sudden heading error: (**a**) Comparison of heading error; (**b**) Comparison of roll angle error; (**c**) Comparison of aileron trimmer deflection.

Figure 11 shows the results of the steady flight. The red signal represents the outcomes from simulation with not-distorted sensors, the blue one represents the outcomes from the simulation where the gyroscopes were distorted by error bias equals to 1.7435×10^{-5} m. Figure 11a shows that altitude error occurred during the flight. Up to approximately 500 s, the error keeps the value constant; then the error increases with time. The constant value of altitude error in the first 500 s is the result of the controller counteracting the angular velocity bias error until the regulator integral is saturated; then the altitude error increases with time. Figure 11b shows the pitch angle error comparison; the difference between the two values is negligible. Figure 11c represents the deflection of the elevator trimmer. There is a slight difference between two signals—less than 0.1 degrees.

Figure 12 contains the graphs which show the differences between the simulations with distorted and non-distorted torque sensors on the shaft. Here, the drift with 0.001 N/s error is presented. Figure 12a shows the differences in heading error. Similar to the previous outcomes, the difference remains constant and then increases with time. The value of heading error remains constant until the integrator reaches its saturation; then the error increases over time. The same behavior can be noticed in the angle of sideslip, which is shown in Figure 12b, and in the deflection of the rudder trimmer, which is shown in Figure 12c.

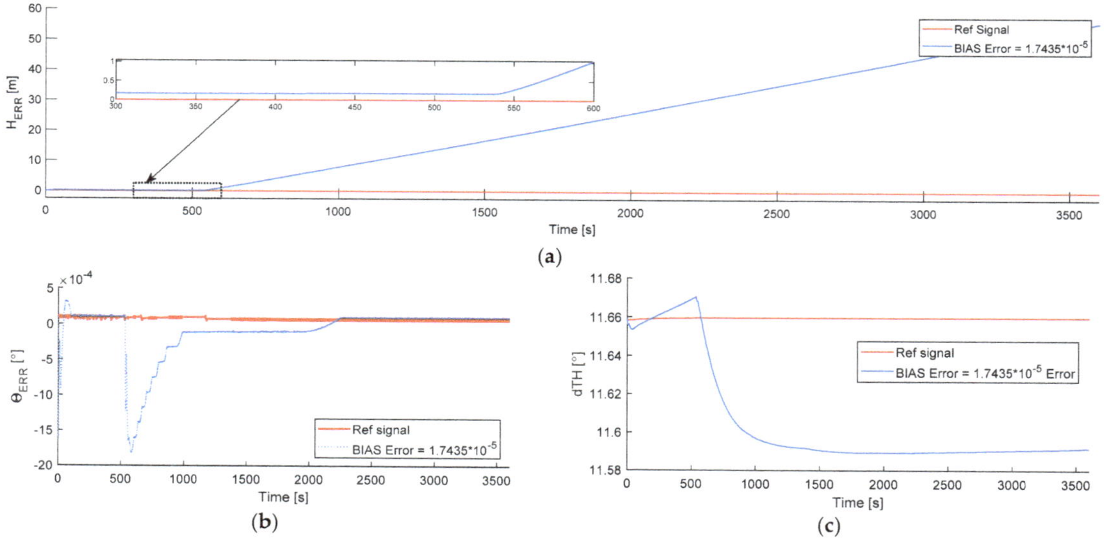

Figure 11. Influence of bias error of gyroscopes on a stable flight: (**a**) Comparison of altitude error; (**b**) Comparison of pitch angle error; (**c**) Comparison of elevator trimmer deflection.

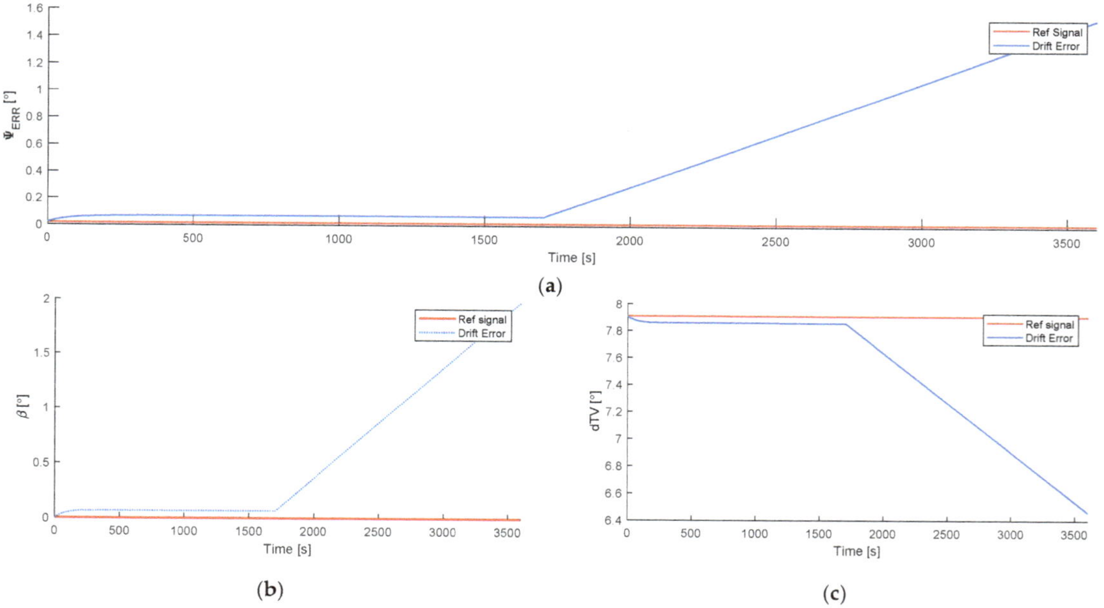

Figure 12. Influence of drift of TSoS on a stable flight: (**a**) Comparison of heading error; (**b**) Comparison of angle of sideslip; (**c**) Comparison of rudder trimmer deflection.

The last set of graphs, presented in Figure 13, shows the comparison of two simulations with distorted and non-distorted sensor of torque on the shaft. The sensor is disturbed by white noise with PSD equal to 0.001 [rad^2/s]. The change in heading error is shown in Figure 13a. Figure 13b shows the angle of sideslip during the flight and Figure 13c shows the commanded deflection of the rudder trimmer. In each of the graphs, the parameters are distorted by white noise from the torque sensor on the shaft.

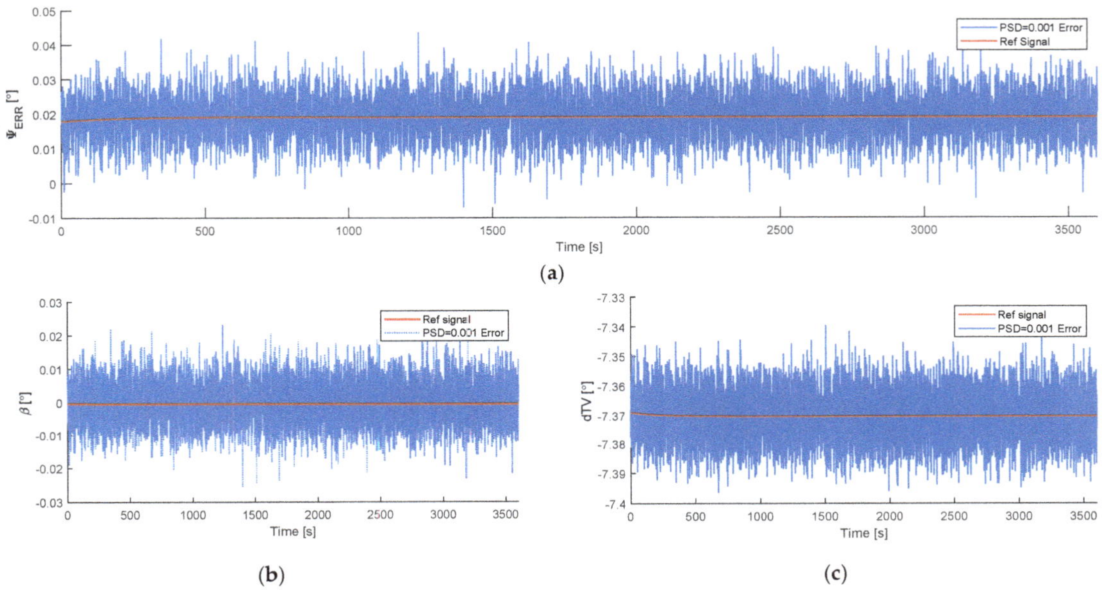

Figure 13. Influence of white noise of TSoS on a stable flight: (**a**) Comparison of heading error; (**b**) Comparison of angle of sideslip; (**c**) Comparison of rudder trimmer deflection.

The figures presented above contain data obtained during the tests. The discussion about the results and general conclusions are in the next section. Table 1 below summarizes the results shown in Figures 11–13.

Table 1. Summary of the results from Figures 11–13.

Type of Error	Value of Error	Observed Error after 1 h of Flight
Bias error	1.7435×10^{-5} [m]	56 m
Drift error	0.001 [N/s]	1.5 degrees
White noise (PSD)	0.001 [rad^2/s]	±0.02 degrees

4. Discussion and Conclusions

The results presented in this article are an answer to the question of how sensor errors affect TSS performance. During the tests, the impact of sensor errors on flight stability was assessed. The tests were limited to checking the three most common errors. Taking into account the above results, it can be concluded that IMUs with a measurement white noise of PSD equal to 1.6×10^{-9} $rad^2 \cdot s^{-2} \cdot Hz^{-1}$ do not cause malfunctioning of the system. The altitude error does not exceed a value of 1 m and heading error was less than 0.1 degree during flight. When white noise with a PSD equal to 1.6×10^{-7} [rad^2/s] was applied, the system did not work properly; the errors exceeded the values of 1 m for altitude and 0.1 degree for heading. The bias and drift errors caused malfunctioning of the system; after saturation of integrals, the heading and altitude errors exceeded the limits of the proper operation of the system. The white noise applied to torque sensor on the shaft (TSoS) caused slight disruptions in rudder trimmer deflection.

These results are the next step towards testing the TSS on real aircraft. The obtained results provided information on the system's resistance to sensor errors. They prove that white noise of IMUs, where PSD value is not greater than 1.6×10^{-9} rad^2·s^{-2}·Hz^{-1}, does not degrade the control quality of the flight control system. In the future, research may be extended to include tests of flight control system resistance to a broader range of errors.

Author Contributions: Conceptualization, M.K. and C.S.; methodology, M.W.; validation, M.K. and C.S.; formal analysis, M.W.; investigation, M.W.; writing—original draft preparation, M.W.; writing—review and editing, C.S. and M.K.; visualization, M.W. All authors have read and agreed to the published version of the manuscript.

Funding: The research described in this paper was funded by EU co-financed project number POIR.04.01.02-00-0006/17-00, titled "Innovative system of flight stabilisation with use of trimmers"—ISSLOT.

Institutional Review Board Statement: Not applicable.

Informed Consent Statement: Informed consent was obtained from all subjects involved in the study.

Conflicts of Interest: The authors declare no conflict of interest.

References

1. Krawczyk, M.; Szczepański, C.; Zajdel, A. Simulation and Testing of Flight Stabilisation System Using Trimmers. In *Automation 2021: Recent Achievements in Automation, Robotics and Measurement Techniques*; Springer: Berlin/Heidelberg, Germany, 2021; pp. 185–196. [CrossRef]
2. Szczepański, C.; Krawczyk, M.; Zajdel, A. The airplane trim system—New functionalities. *Aircr. Eng. Aerosp. Technol.* **2020**, *92*, 1401–1406. [CrossRef]
3. Krawczyk, M.; Zajdel, A. Trimming surfaces in the automatic plane control process. *J. Kones* **2017**, *24*, 223–230. [CrossRef]
4. Krawczyk, M.; Graffstein, J. A Proposition of Control Augmentation System for Dumping the Harmful Impact of Slipstream in Turboprop Airplanes. *Sci. Pap. Rzesz. Univ. Technol.-Mech.* **2013**, *288*, 287–295. [CrossRef]
5. Luu, H.; Tran, D.; Thang, N.; Duc, N.; Nguyen, T.A. Errors determination of the MEMS IMU. *VNU J. Sci.* **2007**, *XXII*, 6–12.
6. Tripathi, S.; Wagh, P.; Chaudhary, A.T. Modelling, simulation & sensitivity analysis of various types of sensor errors and its impact on Tactical Flight Vehicle navigation. In Proceedings of the International Conference on Electrical, Electronics, and Optimization Techniques (ICEEOT), Chennai, India, 3–5 March 2016; pp. 938–942. [CrossRef]
7. Ayres-Sampaio, D.; Deurloo, R.; Bos, M. Comparison Between Three IMUs for Strapdown Airborne Gravimetry. *Surv. Geophys.* **2015**, *36*, 571–586. [CrossRef]
8. Borodacz, K.; Szczepański, C.; Popowski, S. Review and selection of commercially available IMU for a short time inertial navigation. *Aircr. Eng. Aerosp. Technol.* **2021**, *24*, 223–230. [CrossRef]
9. White Noise: Simulation and Analysis Using Matlab. Available online: https://www.gaussianwaves.com/2013/11/simulation-and-analysis-of-white-noise-in-matlab/ (accessed on 29 November 2021).
10. Farrell, J.A.; Silva, F.O.; Rahman, F.; Wendel, J. IMU Error Modeling Tutorial: INS State Estimation with Real-Time Sensor Calibration. UC Riverside: Bourns College of Engineering. Available online: https://escholarship.org/uc/item/1vf7j52p (accessed on 29 November 2021).
11. Quinchia, A.G.; Falco, G.; Falletti, E.; Dovis, F.; Ferrer, C. A Comparison between Different Error Modeling of MEMS Applied to GPS/INS Integrated Systems. *Sensors* **2013**, *13*, 9549–9588. [CrossRef] [PubMed]
12. KVH. Available online: https://www.kvh.com/admin/products/gyros-imus-inss/imus/p-1775-imu/commercial-p-1775-imu (accessed on 29 November 2021).
13. Nirmal, K.; Sreejith, A.; Mathew, J.; Sarpotdar, M.; Suresh, A.; Prakash, A.; Safonova, M.; Murthy, J. Noise modeling and analysis of an IMU-based attitude sensor: Improvement of performance by filtering and sensor fusion. In *Advances in Optical and Mechanical Technologies for Telescopes and Instrumentation II*; SPIE: Edinburgh, UK, 2016; Volume 9912. [CrossRef]
14. Wen, Z.; Yang, G.; Cai, Q. An Improved Calibration Method for the IMU Biases Utilizing KF-Based AdaGrad Algorithm. *Sensors* **2021**, *21*, 5055. [CrossRef] [PubMed]

aerospace

MDPI

Article

Automated Piping in an Airbus A320 Landing Gear Bay Using Graph-Based Design Languages

Moritz Neumaier [1,*], Stefan Kranemann [2], Bernd Kazmeier [2] and Stephan Rudolph [1]

1 Institute of Aircraft Design, University of Stuttgart, Pfaffenwaldring 31, 70569 Stuttgart, Germany; rudolph@ifb.uni-stuttgart.de
2 Airbus Operations GmbH, 21129 Hamburg, Germany; stefan.kranemann@airbus.com (S.K.); bernd.kazmeier@airbus.com (B.K.)
* Correspondence: neumaier@ifb.uni-stuttgart.de

Abstract: System design in an aircraft is still a costly, manual and iterative approach. One major cost driver of changes in system installation are design efforts for creating new pipes in an earlier stage and the costs accumulated during the in service life. To reduce these costs and the time to market, an automation approach with an integrated design optimization encoded in graph-based design languages and executable in a design compiler is proposed. To generate the pipe work automatically, a set of input data (e.g., start- and end-points of a pipe with tangents and fixing positions) is given by the user. It also contains, among others, the weightings for the optimization criteria (e.g., length of the pipe resp. the weight vs. the number of bends) to influence the evaluation of the generated pipes and thereby the final solution. As an initial step in the automatic pipe generation process, a route through the installation space is searched. Subsequently, the installation space is simplified and a respective minimal distance to each obstacle which a pipe should satisfy is added. Then for each pipe an initial solution is estimated and each pipe is optimized by a simulated annealing algorithm. At last, all given requirements are automatically verified. A carried out investigation indicates a polynomial runtime behaviour of the algorithm. The capabilities of the newly developed automated piping are demonstrated on the pipe work in an Airbus A320 landing gear bay.

Keywords: automated piping; design languages; model-based systems engineering

Citation: Neumaier, M.; Kranemann, S.; Kazmeier, B.; Rudolph, S. Automated Piping in an Airbus A320 Landing Gear Bay Using Graph-Based Design Languages. *Aerospace* **2022**, *9*, 140. https://doi.org/10.3390/aerospace9030140

Academic Editors: Spiros Pantelakis, Andreas Strohmayer and Liberata Guadagno

Received: 14 January 2022
Accepted: 3 March 2022
Published: 5 March 2022

Publisher's Note: MDPI stays neutral with regard to jurisdictional claims in published maps and institutional affiliations.

1. Introduction

With the increasing complexity of aircraft systems, the design cost to modify or introduce pipes in an existing aircraft increases. The reason for this is that aircraft systems are mostly routed in the same areas, that segregation and clearances are at their limits, that the access and installation sequences need to be considered from the beginning and that the modification of one system should not lead to a cascading change in other systems.

Aircraft systems and pipe work are impacted by several design evolutions during the life-cycle of an airplane. From the first design to the subsequent iterative incremental modifications during its service life the routing of aircraft systems has to be created, modified and adapted several times in an aircraft life to meet the latest aviation requirements and manufacturing technologies or to incorporate design improvements or corrections.

Furthermore, design trades for the overall optimization of a future aircraft must consider different architectures and locations of new equipment and pipe routing. Current pipe work is a manual process of routing the pipe through the installation area, while taking into account all clearances, manufacturing and installation requirements. All these attempts increase the non-recurring cost of every change, thus reduces the profit for the aircraft manufacturer. Aircraft manufacturer are therefore constantly looking for optimization and automation of the pipe design process. For this reason, this paper explores the application of graph-based design languages for design automation in the specific case of piping design automation.

However, the task of designing the necessary pipework (i.e., the so-called "piping") of hydraulic system components inside an aircraft is not an isolated design effort with constant interfaces or constant pre- and post-conditions, but is embedded in a complex systems engineering and design process instead with constantly changing boundary conditions and goals. It is therefore worthwhile to shortly analyze the overall design process in order to understand both the position of piping in the current aircraft design hierarchy and sequence, as well as the mutual implications of physical system architecture design on piping and of piping on physical systems architecting vice versa.

1.1. Graph-Based Design Languages

Graph-based design languages are inspired by natural, human languages, in which the vocabulary and rules form a grammar. The term "design language"means that every sentence built with the available vocabulary and allowed by the grammar is a valid expression of a design. The term "graph-based"means that each node in a graph is used to represent a requirement, a product function, a solution principle, a component or an assembly, or any other arbitrary engineering concept one may encounter in or along the product life-cycle. The graphical representation of design languages is based on the Unified Modeling Language (UML), an international standardized data format in order to avoid vendor lock-in and capable of self-defined extensions.

A design language encodes the complete knowledge of the design product and its associated design processes in form of a graph-based, machine-readable and machine-executable form and is translated by a design compiler (the design compiler used to compile graph-based design languages in this paper is the Design Compiler 43® [1]). The key components of a graph-based design language are the vocabulary, the rules and the production system. In case graph-based design languages are implemented using the UML, the vocabulary is represented in form of a UML class diagram, the rules are represented as Model-to-Model (M2M) or Model-to-Text (M2T) transformations. The production system is represented in form of an UML activity diagram. The result of the compilation of a graph-based design language in a design compiler is the design graph, an UML instance diagram. This is later mapped into different domain-specific languages (DSLs) by dedicated design compiler software plug-ins for further processing, typically for visualisation purposes such a CAD-model in an external CAD-kernel or for simulation purposes such as a finite element model (FEM) simulation in an external FEM solver.

According to the Engineering as a Service (EaaS) paradigm, an engineering design activity such as the automated piping in arbitrary complex 3D geometries which will be described in this work, is encoded as an executable engineering service in a design compiler serving as an integration platform. The design compiler platform offers inbuilt digital continuity and consistency for all engineering data due to the generation of all engineering models from a single source. As known from any other formal programming language compilation run in a compiler for that language, a single design language (or several design languages combined together) can therefore be considered as one single source of truth which is compiled into different domain-specific models (CAD, FEM, ...) which are therefore automatically consistent to each other. The 3D piping algorithms are built into a design language as an executable engineering service for later reuse in a multitude of design contexts (aerospace, automotive, marine, etc.).

For this reason, the main focus of this paper is on 3D piping in arbitrary complex geometries only and the design language approach sets for the purpose of this paper just the framework of building software hulls (i.e., interfaces) around the algorithms to make them later reusable in automated process chains. The details of the 3D piping algorithms and an in-depth analysis of their performance will therefore be described in great detail later in this work. It is the algorithm which matters in theory and not the implementation, despite the fact that a thoughtful implementation will make the difference in both runtime and memory requirements in practice.

For any further explanations and properties of graph-based design languages and the design compiler approach, the reader is referred to the already quite extensive body of literature about graph-based design languages [2–5]. More specifically, any reader interested in a comparison of graph-based design languages with other related graph transformation approaches is referred to a most recent survey paper which reviews the commonalities and differences of graph transformation approaches [6] and which represents a follow-up of a previous survey paper [7]. Similarly, a further performance comparison of the design language approach with other model-based transformation and mapping concepts [8–10] is intentionally not dealt with in this paper.

1.2. The V-Model of Model-Based Systems Engineering

The V-model in Figure 1 illustrates the clearly structured approach of model-based systems engineering (MBSE) in the development of complex systems such as aircrafts: Starting from requirements (due to the central role of requirements for product shape, behavior and usage definition, one therefore often speaks of requirements engineering and, when in the verification analysis, only that has been designed and built which is specified in the requirements, it is also referred to as requirements-driven engineering), these requirements are decomposed and mapped in several decomposition steps into a functional architecture, then mapped onto a logical architecture and finally mapped into a physical architecture. On the left side of the V-model it is checked by verification measures whether the decomposition, i.e., the mapping of the requirements from the respective level to the respective subordinate level, is correct. Accordingly on the right-hand side of the V-Model, validation measures are used during integration, i.e., during the process of system assembly it is checked whether the components, subsystems and the overall assembled system meet the requirements which is checked by tests associated with each requirement.

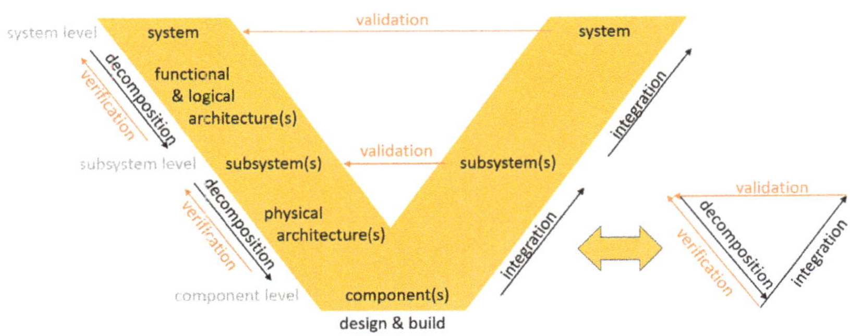

Figure 1. MBSE V-Model (**left**) including validation and verification pattern (**right**).

Piping furthermore depends heavily on the previous placement (the so-called "packing") of the hydraulic system components inside the aircraft, since any change of a component's position or orientation affects the start- and endpoints the pipe connection and their spacial orientation. Both problems of packing and piping [11,12] are known to fall mathematically speaking into the class of so-called "NP-hard problems", for which no exact algorithmic solution in polynomial time-complexity is known [13]. This means that the development and advantageous use of problem-specific search heuristics or problem-independent meta-heuristics to circumvent and overcome this existing hard theoretical bottleneck is therefore mandatory. One often chosen way of mastering this theoretical obstacle is to use an algorithm which only approximates the solution and which is therefore capable to sacrifice optimality for speed (algorithms such as the simulated annealing algorithm [14] in Section 4.2 are known to return solutions up to only several percent away from the true optimal solution while consuming only a fraction of the original polynomial runtime which is, seen the circumstances, considered as an acceptable runtime behavior).

In a complete design scenario, the design process would start on the top left of the V-model in Figure 1 with the requirements for the hydraulic systems, goes on with a functional decomposition yielding the functional architecture, and then move on to the development of a schematic of the hydraulic system as the logical architecture. Then the hydraulic system components would need to be physically allocated, which results in the so-called *package*. In the current paper, both the functional and the logical architecture as well as the package are already predetermined, since all these data are adopted from the shown industrial use-case of a A320 series solution. The series solution of the pipe work in an Airbus A320 mounting rack can be seen in Figure 2 as an example of the typical complexity in an aircraft pipe network installation. Therefore, the task of finding automatically (i.e., through a "1-click solution" by the press of a button) a set of feasible piping connections in an arbitrary complex 3D geometry remains and is described in the following sections of the paper.

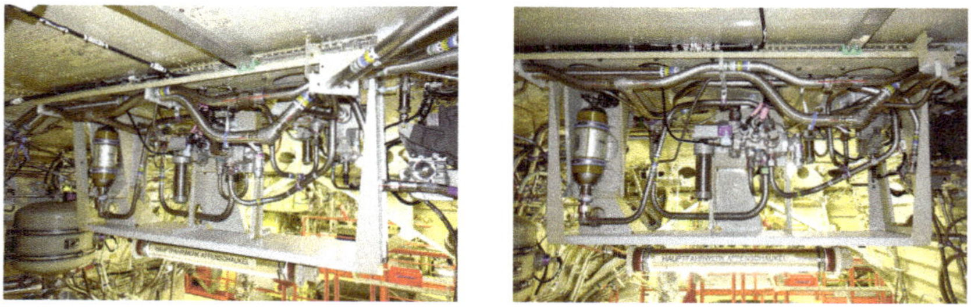

Figure 2. Airbus A320 series solution mounting rack (©2022 AIRBUS Operations GmbH).

The future vision of an automated MBSE includes therefore a machine-executable V-Model [5] which stands for the automated execution of a workflow [15] of fully interconnected and interoperable *engineering as a service* (EaaS) software modules such as automated packing and piping. The piping automation is thus just a small portion of a larger research effort [16] to achieve an intelligent design automation [3] using design languages translated by a design compiler into executable models [17].

Since the design decision making on the packing, the logical and the functional architecture are located upstream in the design process, these decisions form the future outer optimization loops where the functional architecture sets the boundary for the logical architecture and the logical architecture sets the boundary conditions for the packing. As the most inner nucleus of these nested optimization loops, the automated 3D piping described in this paper generates feasible connections and refers in this respect to the most inner validation and verification patterns in the V-model on the right hand side in Figure 1. As said in the beginning of this paragraph, that if the scope of the work would cover also the more upstream design decisions of packing, logical and functional architecture definition, the consequences of these decisions could be investigated by an exploration of the design by looping through the alternatives.

1.3. Manufacturing Technology Influence

The pipes generated by the proposed algorithm should be able to be manufactured with the rotary draw bending process for bending machines described in VDI 3430 [18]. In this process, only bends with a fixed bending radius can be bend. The bending angle can vary within a certain range determined by the bending machine. Between two bends, a straight pipe with a minimum length has to be placed. This leads to an alternating order of straights and bends. As a result of the construction of the bending machine, a minimal distance between the bends must be established if the bends should be bend independently. Furthermore, a limited, machine depending number of bending claws can be used to manufacture two bends with a smaller straight distance in between.

In Figure 3, the process is illustrated. The pipe is drawn along the fixed dies (shown in a dense hatching) and is bend around the rotating die (shown in a dotted hatching). As a result of this the bending radius is constant for a specific pipe as mentioned before because every bend is bend around the same die. If the distance between two bends is too short to retain the pipe by the rotating dies the pattern of the pipe needs to be milled into a specific bending claw.

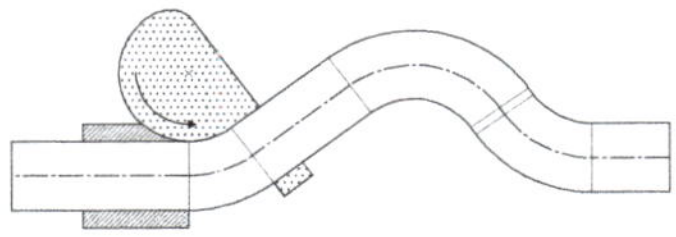

Figure 3. Rotary draw bending process.

1.4. Design Process Influence

Both the start- and the end-points with their respective directions (i.e., the tangents), the pipe diameter and the properties of the intended bending machine must be known in advance. In addition, minimal distances of the pipes to certain obstacles and to each other that must be adhered to, can be set. Furthermore, a user is able to affect the piping algorithm by modifying the weighting of the different evaluation factors which will be described in more detail in Section 4.1.

2. Related Work and Research Gap

In this section, different approaches by other researchers are discussed. The discussion is arranged from the approaches close to the manual design, like a knowledge based expert system, to multi-objective approaches with three-dimensional obstacles. Subsequently the drawbacks of the discussed approaches are worked out and the research gap, which this paper focuses on, is described.

2.1. Related Work

In the past, quite a few different approaches to solve routing problems have been investigated and published. These approaches differ in the multitude of the recognized objectives, the complexity of the obstacles which can be taken into account and the level of detail mastered in the routing process.

Simplified approaches. The approach which is most related to the manual design process is a knowledge based expert system. In a system like this empirical knowledge and documented design knowledge is used. The system proposed by Kang et al., 1999 [19] is an example. A system without the need of empirical knowledge is proposed by Niu et al., 2016 [20]. The installation space is discretized with a regular grid. The obstacles are considered by cell blocking and the pipe path is searched by a NSGA-II algorithm on the grid. As a result of this the pipe can only be bend with 90° bends. Another approach, which was proposed by Qu et al., 2018 [21], is based on an octree grid combined with an max-min ant system optimization algorithm. The obstacles are approximated by divided, deactivated grid cells and the pipe path is searched on the grid. Moreover a graph based approach is proposed by Kim et al., 2021 [22]. The graph is created with its nodes and edges and the shortest path on the graph is searched. The, manually simplified, obstacles are taken into account by deactivating the intersecting graph edges. This leads to short runtimes (in the order of minutes) but no other objectives than the length is taken into account. In addition, the diameter of the pipes is neglected.

Broader approaches. Thantulage 2009 [23] proposes in his dissertation a method based on regular and random grids for pipes with different bending angles. The paths are optimized by an pareto-based general-purpose ant colony algorithm (PSACO) developed in the thesis. Thantulage facilitates the collision detection library RAPID which can process obstacles of any shape. Liu and Jiao 2018 [24] extend an existing pipe routing algorithm based on NSGA-II with a vibration analysis. This algorithm is able to generate pipes with non-discrete bend angles. Furthermore the algorithm does not search on a grid, so the pipe path is not limited to a predefined grid. A grid based method for the generation of different pipes, which could intersect each other, is proposed by Wang et al., 2006 [25]. The method is based on an unmentioned genetic algorithm. Another suggested method is to use a system which is intentionally designed for the motion planing of a robot. In his dissertation Lee 2014 [26] uses the Motion Planning Kit developed by the Stanford AI Laboratory. In a post processing step the computed trajectory is smoothed to get a pipe path.

Sophisticated multi-objective approaches. An approach for arbitrary pipes is given by Szykman and Cagan 1996 [27]. They suggest an approach for non-orthogonal routes which are optimized by a simulated annealing algorithm. A trade-off between the length of the pipe and the number of bends is shown. The path is not limited to a grid, but the obstacles can only have the shape of a cuboid or a cylinder. Another limitation is that the segments of the pipes are assumed as one dimensional and the bends are not rounded. A combined optimization of pipes and the corresponding clamp positions is described by Liu et al., 2021 [28]. The optimization is done by an Improved Multiobjective Ant Lion Optimizer with the objectives pipe length and flow resistance, supplementary six constraints such as a maximum bend angle and a minimum length between two bends have to be considered. The algorithm takes convex hulls of obstacles into account but the design space is limited to the mathematically described surface of an aero-engine. A clash avoidance of the pipes between each other is not described. Some authors of Liu et al., 2021 [28] are also authors of Yuan et al., 2021 [29], which propose a method divided into three steps: pipe grouping and sequencing, multiple pipe layout optimization and pipe parallelization optimization. The routing problem is reduced to a 2D task on a cylindrical shaped aero-engine surface. The obstacles on the surface are modeled by a grid in which the weights of the grid cells determine if the cell is forbidden for a pipe or not. These reductions lead to small runtimes in the order of minutes. Additionally to the parallelism of the pipes the objectives pipe length, pressure loss, space occupancy and obstacle avoidance are considered.

2.2. Drawbacks and Research Gap

The shown related works have in common, that the solutions have only been used for problems with a relatively small number of pipes and/or without arbitrary complex obstacle geometries (in 3D) and/or without taking the manufacturing constraints related to the bending machine into account.

Drawbacks of grid-based approaches. Pure grid- or cell-based algorithms have an inherent resolution limitation as a result of finite computational resources, on the other hand a high resolution is necessary to be able to route through, compared to the overall size, small holes. Moreover arbitrary bend angle, which are necessary to get solutions in arbitrary complex installation spaces, also need a high resolution of the grid or cells because only discrete angles can be produced if the pipe has to follow grid points. Algorithms where the bend positions are not limited to an grid offer the possibility to overcome these difficulties.

Research gap. For an industrial application, like the pipe work in an Airbus A320 main landing gear bay, dozens of pipes, which can interact with each other and need to be able to be manufactured on a bending machine, located in an arbitrary complex installation space need to be mastered by an algorithm. Additionally the runtime and computational demand should not be too extensive otherwise it is not practical to use an algorithm in an industrial environment. This means that the runtimes should be within hours or days so that different variants can be generated and evaluated by an engineer within a week. Such requirements and the consequential challenges are not limited to the aerospace industry, Kim et al., 2021 [22] describe these also for the automotive industry using the example of the Hyundai Motor Company. In summary, it can be said that this paper focuses on the research gap to propose a multi-objective pipe routing algorithm which can solve a 3D routing problem consisting the interaction of pipes with different pipe types (diameter, minimal distances, etc.) in arbitrary complex 3D installation spaces in a runtime short enough to be used in an industrial environment.

3. Pipe Routing Process: Overview and Preprocessing

In Figure 4 an overview of the overall process is shown. The presented geometry is subsequently used to demonstrate the process, which will be described in the following sections. At first a preparation of the inputs is done (Figure 4a), after that paths through the installation space are found as a first approximation (Figure 4b) and after that the optimization of the pipes is done (Figure 4c). After the description of the piping process the automatically generated solution is compared to the manually designed solution (Figure 4d).

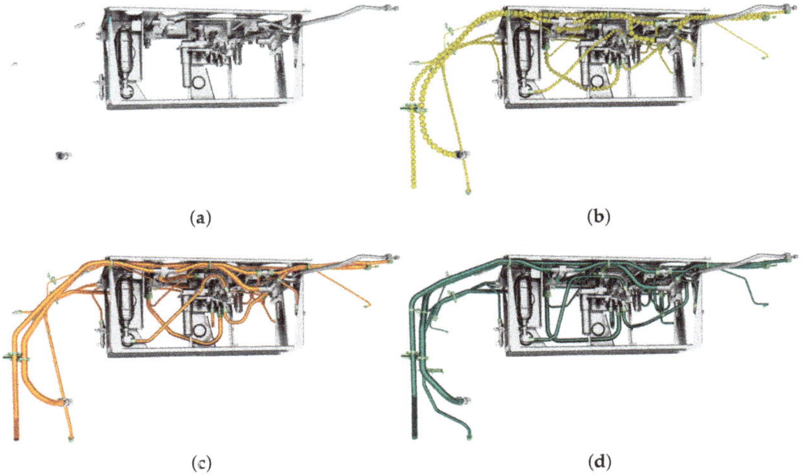

Figure 4. Exemplary installation space: Airbus A320 mounting rack. (**a**) Empty installation space, (**b**) Paths through the installation space, (**c**) Automatically generated pipes, (**d**) Manually designed pipes.

The root program of the design language is shown in Figure 5. It consists of several subordinated design languages in which the pipes are generated and optimized based on the paths through the installation space. This illustrates the fact that many steps are necessary to create automated generated pipes.

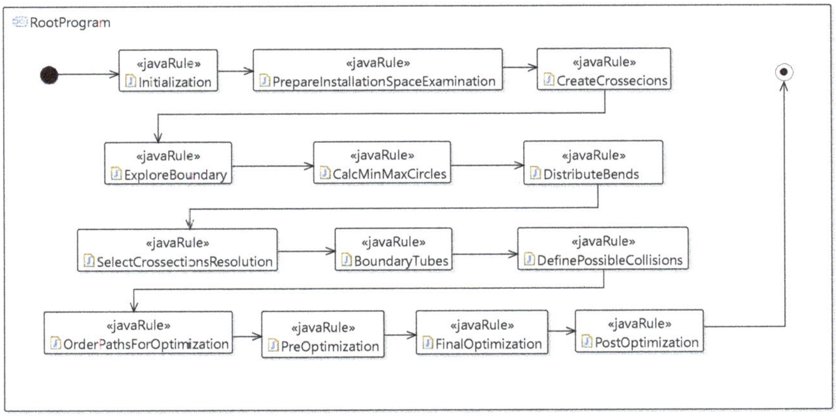

Figure 5. Root program.

As it is described in Section 1.1, this design language is compiled and executed by a design compiler. A part of the corresponding class diagram is shown later. The rules, which are shown in the root program, define how the pipe design process is carried out. An example, how a rule leads to modifications in the design graph, is given in Section 3.3 together with an exemplary design graph.

3.1. Paths through the Installation Space as a First Approximation

A routing from the desired beginning to the desired end of the pipe is required. To accomplish this task an existing plugin for wire harness design [30] is used, which is part of the industrial Design Compiler 43® software suite [1]. This plugin uses an A* algorithm and a collision detection to find the shortest route through an obstacle landscape under consideration of a wire diameter and a minimum bending radius. The result of the wire harness routing is shown in Figure 4b. As this route does not consist of discrete straights and bends it can be seen as a first, but a this stage not usable, estimation of the future correct pipe path.

3.2. Boundary Exploration for Simplified Obstacle Geometry

As the dependency between the complexity of the installation space geometry and the runtime of the algorithm described in this paper should be small, a method called 'ray tests' to reduce the obstacle geometry to the relevant information was developed. Along the path, in equidistant distances, planes orthogonal to the path are created through the installation space. On every plane rays are shot from the point where the path cuts the plane in a regular star shape. The points, where the rays collide with the obstacle geometry, are stored. If an additional (potentially individualized) margin d_m shall be applied, this margin is imposed on the geometry in advance to the ray tests.

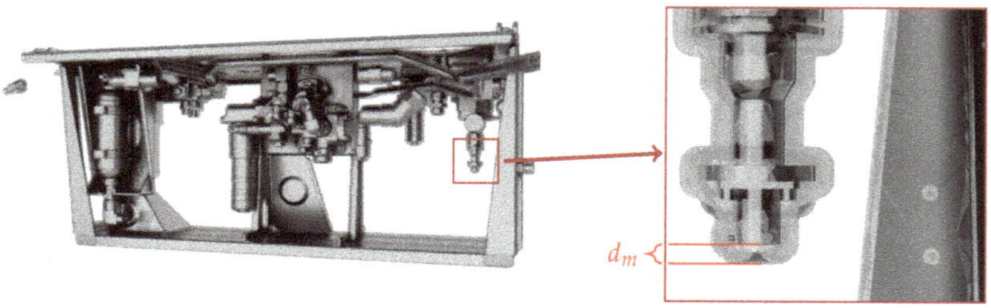

Figure 6. Airbus A320 mounting rack geometry with margin in all spatial directions.

The mounting rack with an applied margin is shown in Figure 6. This ensures that the rays hit the correct, expanded geometry. The points are connected to a polygon. An illustration of this is shown in Figure 7. The polygons are connected one after another to form a hose. In a subsequent step the number of polygons is reduced if no additional information can be gathered of these, e.g., in straight parts.

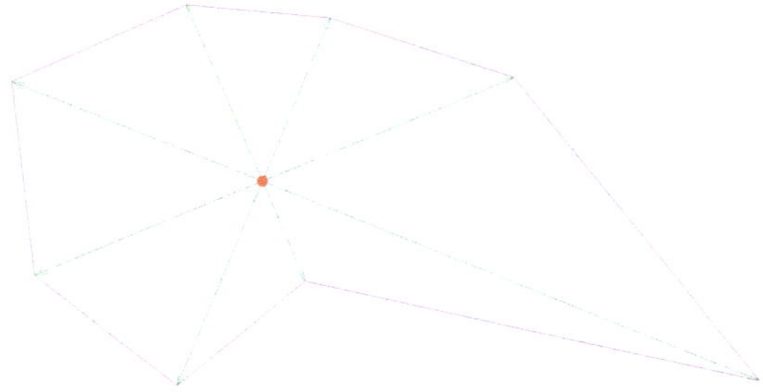

Figure 7. Boundary exploration in one crossection.

3.3. Definition of a Pipe Path

As a result of the used manufacturing technology described in Section 1.3, the pipe path is defined as a rounded polyline with fixed arc radii. The position of the points of the polyline define the pipe path. The first and the last point of the polyline will be considered as the start- and the end-point of the pipe, respectively. If a pipe is fixed by predefined fixings, the pipe is divided into independent paths between the fixings and the paths are combined together into one single pipe definition after the optimization runs have terminated. An exemplary polyline is shown in Figure 8.

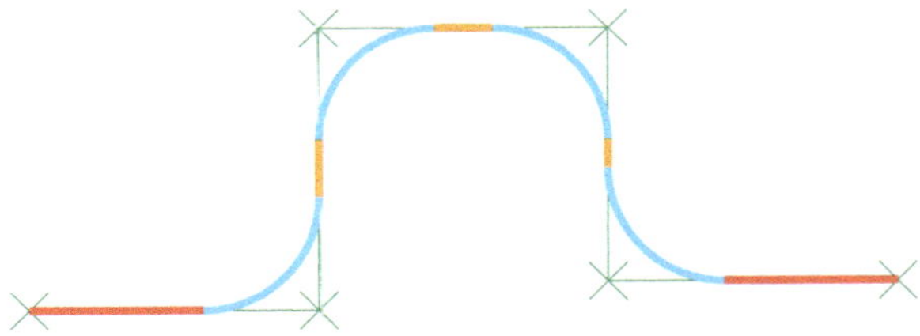

Figure 8. Exemplary rounded polyline.

The data structure used in the design language is motivated by the structure of the pipe. In the class diagram shown in Figure 9 the *PipePart* is an abstract class. This class represents the basic properties which are necessary to define a bend or a straight. As there is a *next* association from the *PipePart* to itself the *PipeParts* can be placed after one another. The *Bend* and the *Straight* inherit from the *PipePart* and add the respective properties which are necessary for the complete definition. The *PipePath* represents a possible solution for one specific pipe segment. As the pipe segments always begin and end with a straight only a *firstStraight* and a *lastStraight* association to the *Straight* is necessary to assign a chain of *PipeParts* to the correct *PipePath*.

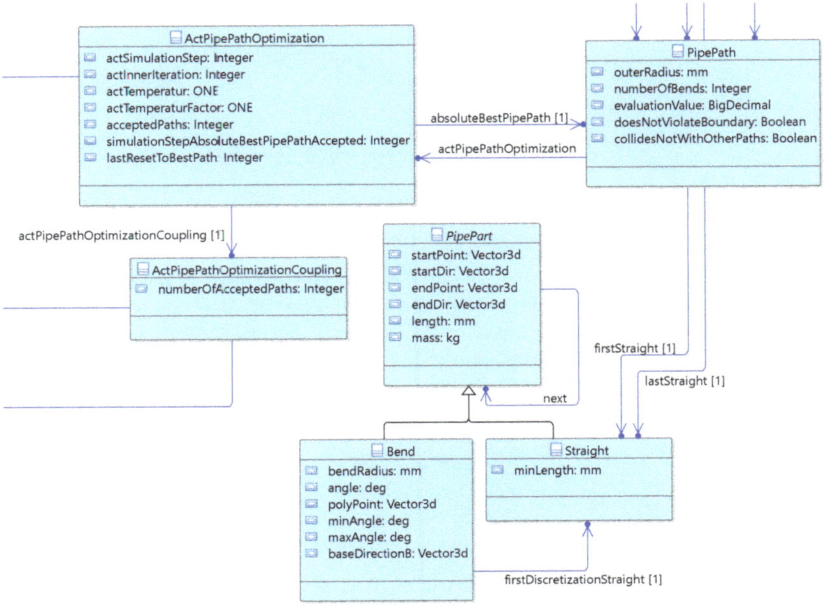

Figure 9. Class diagram (snippet).

In the following Listing 1, the instantiation of a *PipePath* (except the first *Straight*) is shown. The created instances are automatically added to the design graph by the design compiler. Therefore it is iterated over the points which define the polyline. In the iteration process *Bends* and *Straights* are created, linked to each other and the necessary information, for example based on the bending machine in use for the specific pipe, is put in the instances. As it can be seen in the class diagram in Figure 9, *Bends* and *Straights* are *PipeParts* and every *PipePart* has a next link to itself, hence the *Bends* and *Straights* can be linked to each other. The correct start and end positions together with other geometrical values about the pipe are calculated based on this information.

Listing 1. Instantiation of a *PipePath*.

```
PipePart pipePart = firstStraight;
for (int polyPIdx = 1; polyPIdx < polyPointsWithStartEnd.size() - 1;
 polyPIdx++) {
        Bend bend = Bend.create();
        bend.setBendingRadius(bendingMachine.getBendingRadius());
        bend.setMinAngle(bendingMachine.getMinAngle());
        bend.setMaxAngle(bendingMachine.getMaxAngle());
        bend.setPolyPoint(polyPointsWithStartEnd.get(polyPIdx));
        pipePart.setNext(bend);
        pipePart = bend;

        Straight straight = Straight.create();
        straight.setMinLength(bendingMachine.getMinStraightLength());
        pipePart.setNext(straight);
        pipePart = straight;
}
```

A design graph of a *PipePath* is shown in Figure 10. The instances of the *Straights* (red) and *Bends* (blue) are connected in an alternating order with *next* links. This instances and links are created by the code shown in Listing 1. The instance of the *PipePath* (orange) is via the *firstStraight* and the *lastStraight* links connected to the *PathParts*.

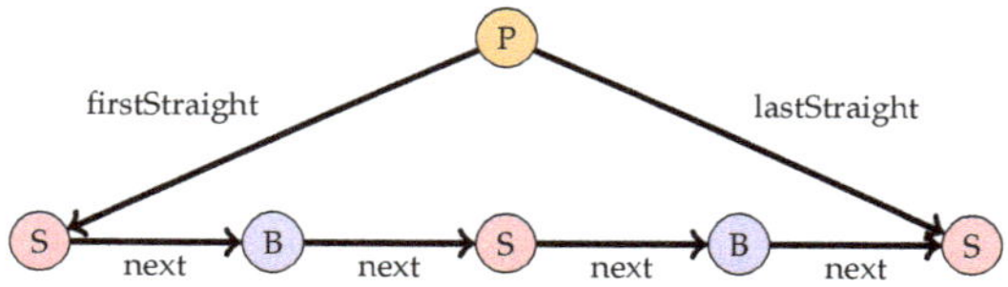

Figure 10. Exemplary graph of a *PipePath* consisting of *Straights* and *Bends*.

3.4. Estimation of Polyline Points

For an initial solution the positions of the points of the polyline must be estimated. All points, except the first and the last two, can be positioned in the three space dimensions. The algorithm must ensure that the distance between two following points is big enough that the bending radius can be applied and that the minimal distance between two bends required by the manufacturing technology can be fulfilled. Similarly, the first and last point are fixed and the next respectively the previous point must be on a straight line with the desired start and end tangents.

The number of polyline points is related to the number of bends. The user decides how many bends should be used at least and at most. For every number of bends in this range a pipe path must be initialized. The positions of the points for a given number of bends is estimated under the use of a path through the original installation space and the simplified installation space.

3.5. Selection of Depending Paths

The distance between the simplified obstacle geometries for each pipe (see Section 3.2) is calculated approximately. If the distance is smaller than the allowed distance between two pipes it is possible that the pipes could intersect or be too close to each other. Therefore these pipes cannot be generated independently. As a result of this, for all depending pipe segments all combinations of estimated paths are computed. This leads to a list which pipe combinations can be optimized independently. As long as no specific generation sequence is known to yield good results this open question is answered by the comparison of the results of multiple execution sequence candidates.

4. Pipe Routing Process: Pipe Generation and Optimization

The estimated pipe paths (see Section 3.4) fulfill the requirements imposed by the manufacturing constraints described in Section 1.3. There exist multiple criteria, which makes a trade-off necessary. These criteria need to be weighted by a user, depending on the problem specific needs, to get one specific value which describes the overall quality of the pipe path. The pipe paths are optimized based on an evaluation function by modifying the points of the polylines which describe the pipes.

4.1. Evaluation Function and Evaluation Criteria

To get one specific value $v_n \geq 0$ in every iteration step n describing the quality of the pipe path, an evaluation function is needed. A smaller value v_n stands for a better pipe path. The specific value v_n is computed by

$$v_n = \sum_i w_i (v_{n,i} + 1)^{k_i} \tag{1}$$

For each evaluation criteria a value $v_{n,i} \geq 0$ is calculated. As different criteria shall be combined, $v_{n,i}$ needs to be dimensionless. Referring to the method of least squares the value is raised to a specific power $k_i \geq 1$. Using $k_i > 1$ in combination with a $0 < v_{n,i} < 1$ would lead to a better evaluation than with $k_i = 1$. In order to prevent this, the term is modified, which leads to $(v_{n,i} + 1)^{k_i} \geq 1$. The result of this term can be weighted by a factor w_i which is chosen by a user. The evaluation can be based on the following automatically determined values:

- Length of pipe
- Aperture of bends
- Number of bends
- Boundary violation
- Distance to center line
- Distance to the path through the installation space
- Minimal distance between two bends

The calculation of the distance to the center line is based on the boundary exploration (see Section 3.2). In every plane the center of a circle is calculated which has the largest radius that fits inside the boundary polygon. The circle center points are calculated with the algorithm proposed by Garcia-Castellanos and Lombardo (2007) [31], using the euclidean distance instead of the proposed great-circle distance. The distance between the central line of the pipe and the circle center gets related to the pipe radius. For the calculation of the distance to the path through the installation space, as it is described in Section 3.1, is used.

The selection of the weight factors w_i depends on the given problem and the trade-off between the different criteria for the final solution. In a dense installation space it is often more promising to give the algorithm more freedom in using a higher number of bends and in extending the pipe length to find a good overall solution than to penalize this and to increase the complexity and the runtime to a great amount.

For this multi-criteria optimization problem, a single objective approach was chosen. The main reason for this is that to evaluate the distance between two pipes the pathway of both pipes must be known. This means that the evaluation of one pipe is depending on all pipes, with which pipes this pipe could be too close to each other. Due to this the pipes are optimized one after another so the already optimized pipes can be freezed and used for the following distance tests. If all pipes could be changed in the same optimization step simultaneously the search space would be too extensive for an engineering application. Therefore the best pipe must be selected, which can be done by a single objective approach. Generally speaking, a problem, which should be optimized in parallel, like the design of pipes in the same installation space, can be alternatively transformed into a sequential problem. In a sequential approach the boundary conditions can be added incremental (e.g., other pipes as obstacles). This reduces the degrees of freedom and therefore the runtime. This approach is well known as "staggered algorithms" for example in the simulation of the fluid-structure interaction. A utilization in the field of aeroelastics can be found in Farhat et al., 2000 [32].

4.2. Simulated Annealing Optimization

The optimization algorithm must be capable to leave a local optimum to find the global optimum. Otherwise, for example two, pipes could not be moved through each other to find a better overall solution even if a clash free solution is already known. The simulated annealing algorithm proposed by Kirkpatrick et al., 1983 [14] is able to leave local optima, therefore this algorithm was selected. For the optimization the pipes need to be modified in every step and subsequently a decision must be taken if the new solution should be accepted.

4.2.1. Generating a Modified Solution

A pipe is modified by altering the position of the points of the polyline. Additional to a random modification, more sophisticated modification schemes can be used to reach a better solution faster. A bend can, for example, be moved along its beginning and end directions, can be moved on its symmetry plane or rotated around an axis through its start and endpoint. These modification methods are illustrated in Figure 11. The coordinates of the points p can be $p_x, p_y, p_z \in \mathbb{R}$.

Figure 11. Modification strategies.

4.2.2. Acceptance of a Solution

The decision, if a solution is accepted, is based on the evaluation value in the actual simulation step v_n and on the evaluation value in the previous simulation step v_{n-1}. A solution will be accepted if $v_n < v_{n-1}$ or $v_n > v_{n-1}$ with a possibility of

$$p = e^{-\frac{v_n - v_{n-1}}{v_{n-1} \cdot t_n}}. \tag{2}$$

As a result of this, poor solutions can also be temporarily accepted to leave a local optimum. This is a key feature of the simulated annealing algorithm. It is possible to prohibit the acceptance of a solution if a solution is known which does not collide with the obstacles and/or with the other pipes. The temperature t_n in the current optimization step of the simulated annealing algorithm, referring toWeyland 2008 [33], is calculated based on an initial temperature t_0 and a factor k powered by the optimization step using

$$t_n = t_0 \cdot k^n \geq 0. \tag{3}$$

The factor k is calculated based on the desired start temperature t_0, the desired end temperature t_{end} and the user given number of optimization steps n_{steps}.

$$k = \left(\frac{t_{end}}{t_0}\right)^{\frac{1}{h \cdot n_{steps}}} \leq 1 \tag{4}$$

While $t_n > 0$, an acceptance of a poor result is temporarily possible. In the progress of the optimization p should be decreased. This ensures that to the end of the optimization the algorithm behaves increasingly like a hill climbing algorithm to make it certain that the final solution is the best possible solution in the specific area of the solution space.

4.3. Selection of Number of Bends and Optimization

The optimal number of bends for a pipe depends on the installation space as well as from the position of the start- and the end-point with their respective tangents, the bending radius, the user defined weighting of the evaluation criteria and other paths which could be intersected. Because of the combinatorial multitude of pipe sequences (in theory the piping problem is NP-complete [34]) a reliable estimation method is not known at the moment. Hence all pipe paths generated with the method described in Section 3.4 are optimized in a first step. The results of this first optimization are ranked and the best paths are selected.

The number of bends is determined accordingly. These paths get optimized eventually and usually with more optimization steps than the optimization in the first step was performed with. However, these optimizations can be done in parallel. As described in Section 4.2 the paths are optimized for one pipe connection after another. The order of this optimizations can be determined by the expected necessary volume of the pipe. This volume, as it is not known in advance, is calculated based on the outer diameter of the pipe and the length of the path trough the installation space as described in Section 3.1.

The current goal criteria including overall pipe length is by no means the only ones to be considered in piping design: further potential criteria like accessability, assemblability and exchangeability require to solve complex spatial planning problems, and pressure drop or self-induced vibration generation are physical phenomena with add further complexity to the design. They are omitted here and are subject of further research.

5. Validation of the Pipes and Final Result

In order to validate whether all pipes meet the said requirements (manufacturing, clearance, collision avoidance), a renewed boundary exploration using a 3D mesh to mesh collision algorithm is performed. The requirements, which are imposed by the manufacturing constraints, are repeatedly checked in the optimization process. It can thus be assumed that these are fulfilled. Afterwards the results of the validation checks are stored so that the user can assess them. The pipes are exported as STEP files together with the length, the number of bends and the estimated mass of the pipe. Additionally, a table with the relevant information to define the pipe is stored for every pipe. In Table 1, an exemplary bend table is shown, where the IDs mark the start (*S.*), end (*E.*) and the bend points (*B.*), which are the points defining the polyline.

Table 1. Exemplary automatically generated bend table.

ID	x	y	z	Distance	Straight Length	Bend Angle	Bend Radius
S.0	0.0	0.0	0.0	-	-	-	-
B.1	99.746	0.0	0.0	99.746	58.633	49.103°	90.0
B.2	292.924	146.471	168.194	295.06	238.27	19.763°	90.0
B.3	653.16	764.793	865.078	998.869	971.048	15.37°	90.0
B.4	1332.909	1500.0	1500.0	1185.628	1126.615	55.018°	90.0
E.5	1500.0	1500.0	1500.0	167.091	120.223	-	-

In Figure 12, an optimization result of 22 pipes with a combined length of approx. 20.7 m and 150 bends is shown in orange color. The automatic pipe generation took 4.5 h on an Intel i5-4440 and 32 GB RAM. The manually designed original solution, shown in green color for comparison, has a length of approx. 22.6 m with 107 bends for the identical pipe connections. Interesting to note is that each automatically generated pipe turns out to be shorter than its manually designed counterpart.

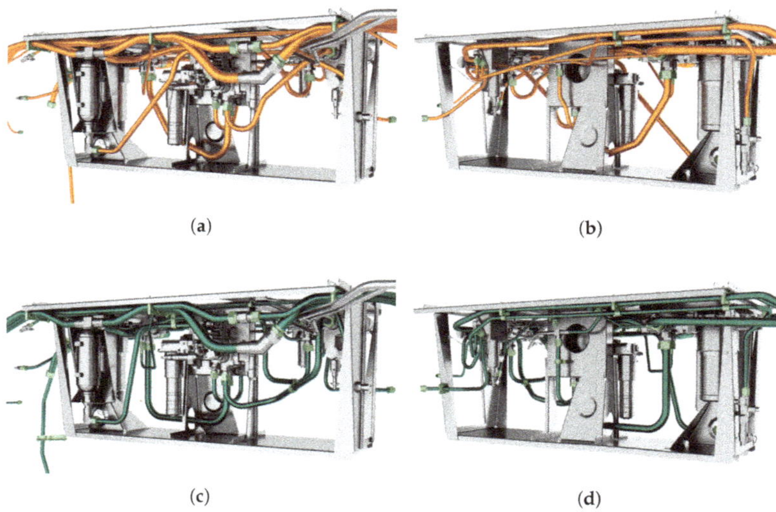

Figure 12. Automatically generated (orange) and manually designed pipes (green) in an Airbus A320 mounting rack. (**a**) Automatically generated pipes (seen from the left side), (**b**) Automatically generated pipes (seen from the right side), (**c**) Manually designed pipes (seen from the left side), (**d**) Manually designed pipes (seen from the right side).

6. Runtime Behaviour

A multitude of settings in a so-called pipe grid were used in the following to empirically investigate the runtime characteristics of the piping algorithm. According to the schematic setting shown in Figure 13, the corresponding beginnings and endings (i.e., the desired connections $A-A$, $B-B$, ...) of the pipes are placed in point-symmetric positions on two quadratic grids with a grid size of 1500 mm.

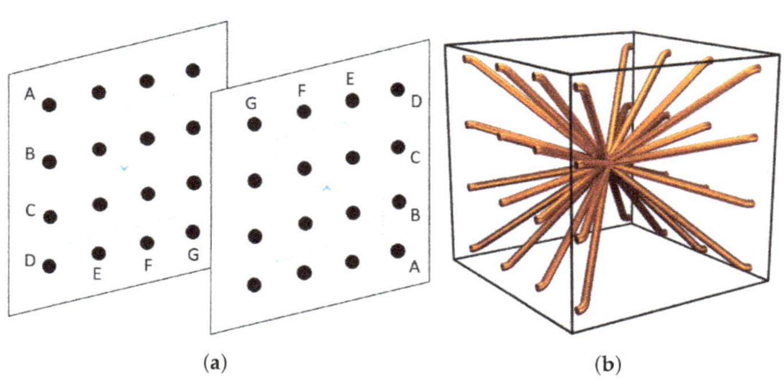

Figure 13. Connection schema $A-A, B-B, \ldots$ on opposite grids with point-symmetric positions. (**a**) Connection schema with point-symmetric positions to enforce collisions. (**b**) 4×4 pipes with enforced collisions.

Due to these boundary conditions, all the direct connections between the beginning and the ending of every pipe would theoretically intersect each other in one single point if they were drawn as mathematically straight lines. In consequence, an intended maximum of collisions will occur, thus offering an easily scalable experimental setting to study the runtime behavior of the piping algorithm.

In the following algorithmic investigations the pipes have a diameter of 60 mm, a bending radius of 90 mm and must have a distance between each other of 1 mm. For the investigation the grid size of the quadratic grid was set constant. Additionally an ambient cuboid with an edge length of 1700 mm was used as obstacle. The algorithm was run on a desktop computer with an AMD Ryzen 7 3700x processor and 64 GB RAM. These investigations of the algorithm runtimes is not only used for a scaling investigation as a function of problem size but will also be of interest in future developments of the algorithms based on additional user heuristics to improve the collision handling.

6.1. Runtime Behaviour without Obstacle Interaction

In Figure 14, the results for different number of pipes are shown. Of the ambient cuboid only the edges are shown in the figure. All the pipes of the piping algorithmic piping solution shown in Figure 14 are collision free.

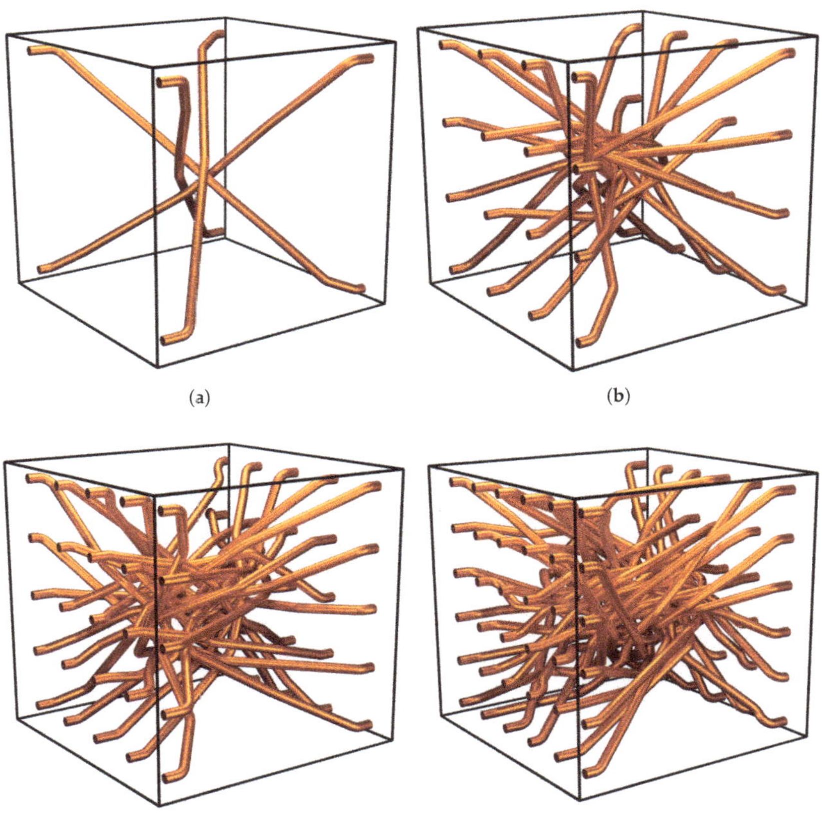

(a) (b) (c) (d)

Figure 14. Algorithmic pipe solutions in a box. (**a**) 2 × 2 pipes, (**b**) 4 × 4 pipes, (**c**) 5 × 5 pipes, (**d**) 6 × 6 pipes.

The measured runtime behaviour of the piping algorithm is shown in Figure 15. The y axes in Figure 15 are graded in logarithmic, as a result of this an exponential runtime behaviour would lead to a straight line. As it can be clearly seen in the diagram the graph of the runtime has a gradient lower than a straight line would have, therefore it can be said that the runtime is below exponential.

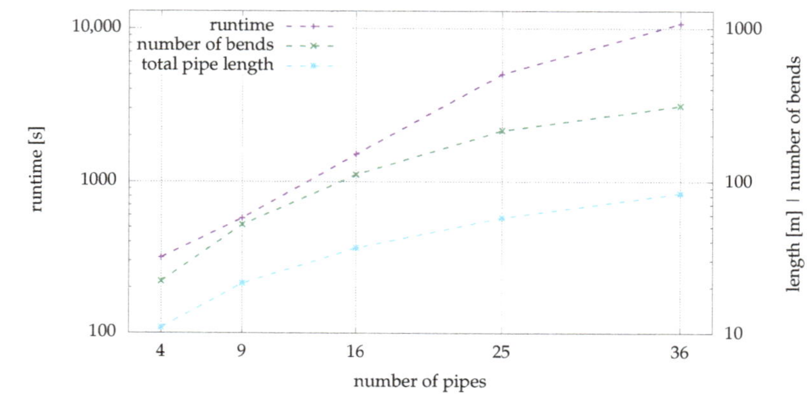

Figure 15. Relationship between the number of pipes, the runtime, the number of bends and the total pipe length.

6.2. Runtime Behaviour with Obstacle Interaction

Additionally to the ambient cuboid, a wall with hole was added. The hole has a dimension of 1000 mm × 700 mm and the wall has a thickness of 50 mm. As a result of this only approximately 24% of the crossection area is available for the pipes. The measured runtime behaviour of the piping algorithm as a function of the number of pipes routed is shown in Figure 16. Due to the shrinkage of the installation space by the obstacle, the algorithm was not able to create valid pipes for a 6 × 6 grid using the same settings like for the other pipe generation. A solution is found only through so-called "virtual fixings"to organize the pipes and thus using the remaining area more efficiently. This is explained in Section 9. For these and the previous measurements in Section 6.1 it should be taken into account that the results have been found by simple computation of the difference between the pre and post program execution time stamps.

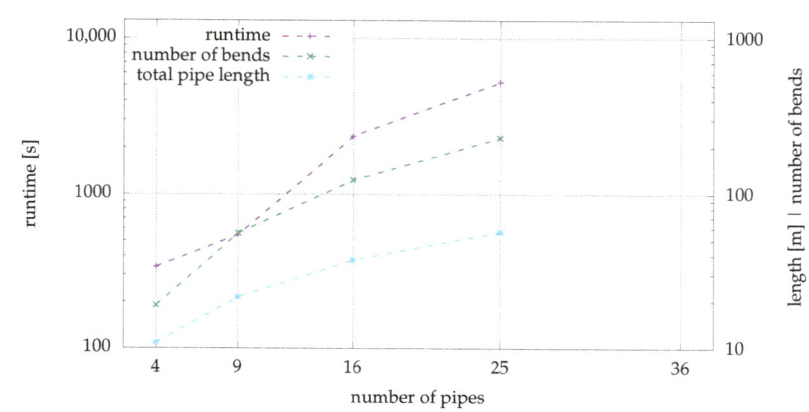

Figure 16. Relationship between the number of pipes, the runtime, the number of bends and the total pipe length with obstacle.

Similar to the diagram Figure 15 the y axes in Figure 16 are graded in logarithmic scale. As it can be seen in the diagram the graph of the runtime has a gradient lower than a straight line would have, therefore it can be said that the runtime is below exponential.

In Figure 17 the results for different number of pipes are shown. All the pipes of the algorithmic piping solution shown in Figure 17 are collision free.

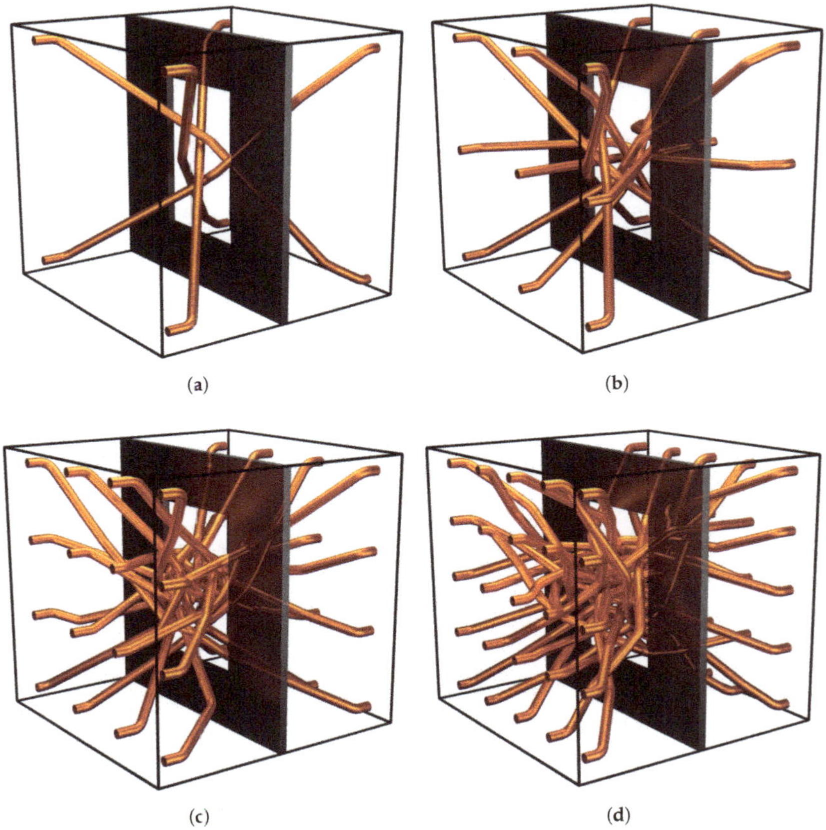

Figure 17. Algorithmic pipe solutions in a box with additional obstacle. (**a**) 2 × 2 pipes, (**b**) 3 × 3 pipes, (**c**) 4 × 4 pipes, (**d**) 5 × 5 pipes.

6.3. Comparison of the Piping Results with and without Obstacles

For the investigation of the behaviour of the piping algorithm regarding the influence of an obstacle the pipe routing with the same weightings of the objective criteria, the same number of simulation steps and the same resolution for the simplified obstacle geometry on the same computer was carried out. The results were already shown in the Sections 6.1 and 6.2. In Figure 18, the runtimes, the number of bends and the total pipe length are compared to each other. It can be seen that the runtimes for the executions with obstacle are higher and also the number of bends increases. This can be explained by the fact that a denser saturation of the available installation space leads to more interactions between the pipes and is therefore more challenging to find a good solution as well as more computational expensive as more distance calculations between the pipes need to be done. A reliable statement about the influence on the total pipe length can not be given in this context as the differences are not judged to be significant.

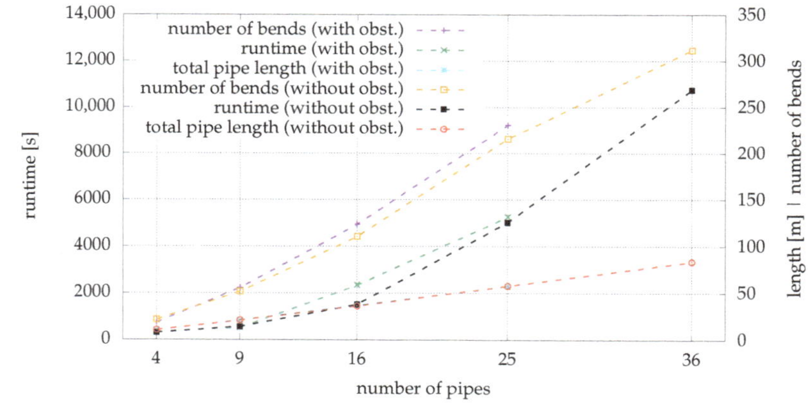

Figure 18. Comparison of the piping results without and with obstacle.

7. Automated Piping in the Airbus A320 Main Landing Gear Bay

The capabilities of the proposed automation approach are shown by generating pipes in an Airbus A320 main landing gear bay. The involute of the main landing gear and other forbidden stay out zones for the clearance of the maintenance holes have been omitted. A total number of 80 pipes with a combined length of approximately 98 m and 614 bends are shown in Figure 19.

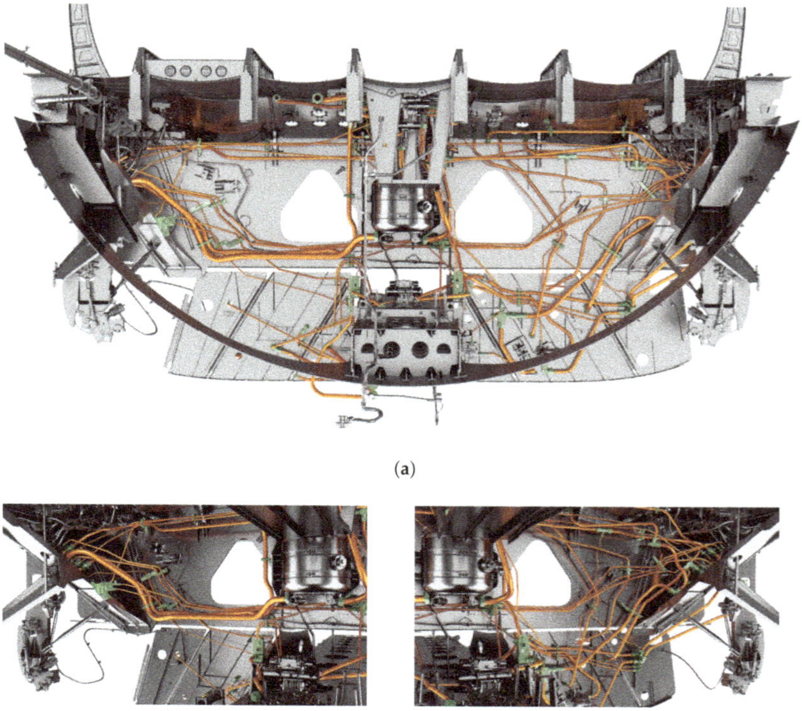

(**a**)

(**b**) (**c**)

Figure 19. Automated pipe routing in an Airbus A320 landing gear bay. (**a**) Overview in flight direction, (**b**) Detail left side, (**c**) Detail right side.

The generation of these pipes took about one day on an AMD Ryzen 7 3700x processor and 64GB RAM. The solution shown is meant to be a first version of the final pipe work, which can be used as a starting point for the user to make educated decisions to reach the final solution (e.g., adding fixings).

The industrial use cases in Figures 12 and 19 illustrate the achievement potential of the presented algorithmic piping automation approach. It could be demonstrated that dozens of pipes which interact with each other and are designed to be manufactured by a bending machine, can be generated automatically in an industrial installation space. Due to the observed runtimes below one day, the presented automated workflow can be integrated in an overall manual industrial design process. It can also be the starting point for future extensions to come closer to the future goal of an holistic model-based aircraft design process relying on a graph-based design languages methodology.

8. Conclusions

A method for an automatic pipe routing which automatically generates pipe designs with an integrated design optimization encoded in graph-based design languages and executable in a design compiler has been presented. The generated pipes can be directly manufactured in a rotary draw bending process.

This manufacturing process requires a pipe consisting of straights and bends in alternating order with specific angles and minimal lengths. Only the start and end positions with the respective direction, the pipe diameter and the properties of the intended bending machine need to be set. In addition, minimal distances of the pipes to obstacles and to each other can be predetermined.

An existing wire harness routing plugin of the Design Compiler 43®[1] was used to find a route through the installation space as a first approximation. Based on this route, ray tests are performed to obtain a simplified obstacle geometry. Given this, pipes with different number of bends are estimated. These pipes are optimized with a simulated annealing algorithm, including a just in time assessment of the constraints. The best ones are selected to be shown to the user. For the optimization an evaluation function is proposed, which takes into account multiple different criteria like the length of the pipe and the number of bends. If it occurs that several pipes are too close to each other, these pipes may be grouped and optimized together.

The evaluation criteria are weighted by the user, keeping the user in charge of the overall design decisions. Subsequently to the design optimization, the generated pipes can be automatically reviewed based on the manufacturing requirements and constraints. A runtime investigation with and without obstacles in the sightline of the start and end connection points was carried out. This investigation indicated a below exponential runtime behaviour of the algorithm. The capabilities of this approach are demonstrated and compared to the series pipe work in an Airbus A320 main landing gear bay.

9. Prospects

The automated piping algorithm shown forms a machine-executable, repeatable and reliable basis for a future overall optimization of a highly complex installation space as shown in Figure 19 compared to the manually created A320 series solution shown in Figure 20. In this condition of the algorithm different criteria, as described in Section 4.1, can be taken into account. Future implementations could also include an additional optimization focusing on assembly, integration and tests. The automation approach has the potential to reduce the pipe design effort by one to several orders of magnitude.

Furthermore, the automated piping allows to permanently shift the focus from a detailed, microscopic view on an optimal individual pipe routing back and forth towards a more system level, macroscopic view on the main routes in system installation. Furthermore, any former manually very tedious investigation of alternative equipment changes can now be automatically executed and systematically analysed in a very reasonable amount of time as the runtime for one variant like the one shown in this paper is below one day.

(a) (b)

Figure 20. Piping series solution in an Airbus A320 main landing gear bay (©2022 Airbus). (**a**) left side main landing gear bay, (**b**) right side main landing gear bay.

The obstacle geometries shown in Figure 20 are not yet semantically annotated. In this context, semantic annotation can be understood as a means to address each geometric entity (e.g., a point, a line, a triangle, etc.) with an label. This stems from the fact that the obstacle geometry is part of a manually created aircraft design of years ago. If however the aircraft structure had been designed in a design language, the whole design sequence of the aircraft geometry would be "known" to the piping algorithm which could take this information into account. In the test case described in Section 6.2 the hole could be an indication to the piping algorithm to align the pipes in the extrusion direction of the hole. This could be done in a regular pattern by virtual fixings.

Figure 21a shows an obstacle geometry (i.e., a plate with a hole) created by a rule in a design language and is thus explicitly part of the semantic model. To help "organizing"the pipes crossing the hole, the Figure 21b shows the result of a fixing rule by means of a manually predefined set of virtual fixings (i.e., the 36 green circles). Using an annotated geometry, this fixing definition could have been also generated automatically. Such additional information may (in comparison to Figure 17d) "help" the cross section of the hole to be used more efficiently since the pipes are now better aligned. As a result, more pipes can be routed through the same hole or the hole can be made smaller.

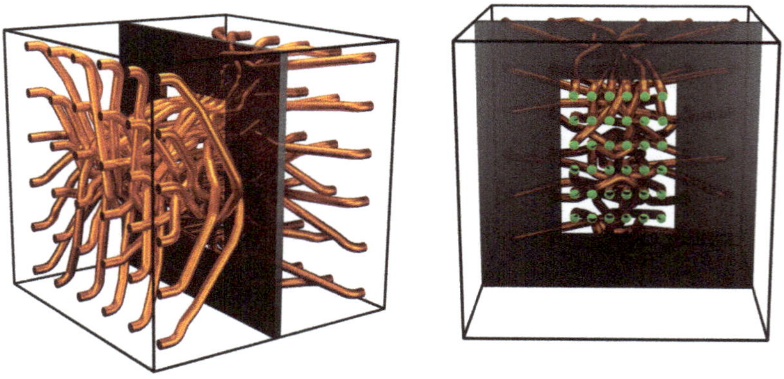

(a) (b)

Figure 21. Test case with manual defined virtual fixings. (**a**) 6 × 6 pipes, (**b**) half of 6 × 6 pipes hidden.

This semantic annotation of geometry will open in the future new ways of reasoning on geometry, since the geometry is not only accessible as geometrical entities in the form of points, lines, triangles and every other construct built upon from that. The geometric collision detection check is an important example for such a basic reasoning level on geometry and allows to deduce already important information (i.e., the information that the geometrical setting is either collision-free or exhibits a collision of x mm) from that. The aforementioned semantic annotation will allow to reason on geometry on a higher reasoning level, e.g., knowing that *"pipes can only pass through holes and never through walls or any other part of a solid structure"* will allow to automate even more complex planning tasks of pipes routes using higher levels of abstraction. Such developments will be part of future planned improvements of the automated piping algorithm.

Author Contributions: Conceptualization, M.N., S.K., B.K. and S.R.; Data curation, S.K.; Methodology, M.N. and S.R.; Project administration, B.K. and S.R.; Software, M.N.; Supervision, B.K. and S.R.; Visualization, M.N.; Writing—original draft, M.N., S.K. and S.R.; Writing—review & editing, M.N., S.K., B.K. and S.R. All authors have read and agreed to the published version of the manuscript.

Funding: The work shown here as a part of the H2020 CS2-Project PHAROS has received funding from the Clean Sky 2 Joint Undertaking (JU) under grant agreement number 865044 (see https://cordis.europa.eu/project/id/865044 (accessed on 13 January 2022) for details). The JU receives support from the European Union's Horizon 2020 research and innovation programme and the Clean Sky 2 JU members other than the Union. The content of this paper reflects only the author's view and the JU is not responsible for any use that may be made of the information it contains.

Clean Sky 2

Acknowledgments: The authors would like to thank the Airbus Operations GmbH, Hamburg, for providing the Figures 2 and 20.

Conflicts of Interest: The authors declare no conflict of interest. The funders had no role in the design of the study; in the collection, analyses, or interpretation of data; in the writing of the manuscript, or in the decision to publish the results.

References

1. IILS mbH 2021 Design Compiler 43®. Ingenieurgesellschaft für intelligente Lösungen und Systeme mbH. Available online: https://www.iils.de (accessed on 14 January 2022).
2. Groß, J.; Rudolph, S. Modeling graph-based satellite design languages. *Aerosp. Sci. Technol.* **2016**, *49*, 63–72. [CrossRef]
3. Riestenpatt, G.; Richter, M.; Rudolph, S. A scientific discourse on creativity and innovation in the formal context of graph-based design languages. In Proceedings of the 13th Anniversary Heron Island Conference Workshop on Computational and Cognitive Models of Creative Design (HI'19), Heron Island, QLD, Australia, 15–18 December 2019.
4. Tonhäuser, C.; Rudolph, S. Individual Coffee Maker Design Using Graph-Based Design Languages. In Proceedings of the 7th Conference on Design Cognition and Computing (DCC 2016), Evanston, IL, USA, 27–29 June 2016.
5. Walter, B.; Kaiser, D.; Rudolph, S. From Manual to Machine-executable Model-based Systems Engineering via Graph-based Design Languages. In Proceedings of the 7th International Conference on Model-Driven Engineering and Software Development (MODELSWARD 2019), Prague, Czech Republic, 20–22 February 2019; pp. 203–210.
6. Voss, C.; Petzold, F.; Rudolph, S. Graph Transformation in Engineering Design: An Overview of the Last Decade. *AIEDAM* 2022, *submitted*.
7. Chakrabarti, A.; Shea, K.; Stone, R.; Cagan, J.; Campbell, M.; Hernandez, N.; Wood, K. Computer-based design synthesis research: An overview. *J. Comput. Inf. Sci. Eng.* **2011**, *11*, 21003. [CrossRef]
8. Ozkaya, M.; Akdur, D. What do practitioners expect from the meta-modeling tools? A survey. *J. Comput. Lang.* **2021**, *63*, 101030. [CrossRef]
9. Sebastián, G.; Gallud, J.A.; Tesoriero, R. Code generation using model driven architecture: A systematic mapping study. *J. Comput. Lang.* **2020**, *56*, 100935. [CrossRef]
10. Nieke, M.; Hoff, A.; Seidl, C.; Schaefer, I. Augmenting metamodels with seamless support for planning, tracking, and slicing model evolution timelines. *J. Comput. Lang.* **2021**, *63*, 101031. [CrossRef]
11. Cagan, J.; Shimada, K.; Yin, S. A survey of computational approaches to three-dimensional layout problems. *Comput.-Aided Des.* **2002**, *34*, 597–611. [CrossRef]
12. De Bont, F.; Aarts; E.; Meehan, P.; O'Brian, C. Placement of shapeable blocks. *Philips J. Res.* **1988**, *43*, 1–22.

13. Goldreich, O. *P, NP and NP-Completeness: The Basics of Computational Complexity*; Cambridge University Press: Cambridge, UK, 2010.
14. Kirkpatrick, S.; Gelatt, C.; Vecchi, M. Optimization by Simulated Annealing. *Science* **1983**, *220*, 671–680. [CrossRef] [PubMed]
15. Rudolph, S. Know-How Reuse in the Conceptual Design Phase of Complex Engineering Products—Or: 'Are you still constructing manually or do you already generate automatically'? In Proceedings of the Conference on Integrated Design and Manufacture in Mechanical Engineering 2006 (IDMME 2006), Grenoble, France, 17–19 May 2006.
16. Dinkelacker, J.; Kaiser, D.; Panzeri, M.; Parmentier, P.; Neumaier, M.; Tonhäuser, C.; Rudolph, S. System Integration based on packaging, piping and harness routing automation using graph-based design languages. In Proceedings of the Deutscher Luft-und Raumfahrtkongress 2021, Bremen, Germany, 31 August–2 September 2021.
17. Rudolph, S. A Semantic Validation Scheme for Graph-Based Rudolph, S. A Semantic Validation Scheme for Graph-Based Engineering Design Grammars. In Proceedings of the Design Computing and Cognition (DCC'06), Dordrecht, The Netherlands, 10–12 July 2006; Gero, J., Ed.; Springer: Berlin/Heidelberg, Germany, 2006; pp. 541–560.
18. *VDI 3430:2014-06*; Rotary Draw Bending of Profiles; Beuth Verlag: Berlin, Germany, 2014.
19. Kang, S.; Myung, S.; Han, S. A Design expert system for auto-routing of ship pipes. *J. Ship Prod.* **1999**, *15*, 1–9. [CrossRef]
20. Niu, W.; Sui, H.; Niu, Y.;Cai, K.; Gao, W. Ship Pipe Routing Design Using NSGA-II and Coevolutionary Algorithm. *Math. Probl. Eng.* **2016**, *2016*, 7912863. [CrossRef]
21. Qu, Y.; Jiang, D.; Zhang, X. A New Pipe Routing Approach for Aero-Engines by Octree Modeling and Modified Max-Min Ant System Optimization Algorithm, *J. Mech.* **2018**, *34*, 11–19. [CrossRef]
22. Kim, S.; Choi, T.; Kim, S.; Kwon, T.; Lee, T. Sequential graph-based routing algorithm for electrical harnesses, tubes, and hoses in a commercial vehicle, *J. Intell. Manuf.* **2021**, *32*, 912–933. [CrossRef]
23. Thantulage, G. Ant Colony Optimization Based Simulation of 3D Automatic Hose/Pipe Routing. Ph.D. Thesis, School of Engineering and Design Brunel University, Uxbridge, UK, 2009.
24. Liu, Q.; Jiao, G. A Pipe Routing Method Considering Vibration for Aero-Engine Using Kriging Model and NSGA-II. *IEEE Access* **2018**, *6*, 6286–6292. [CrossRef]
25. Wang, H.; Zhao, C.; Yan, W.; Feng, X. Three-Dimensional Multi-Pipe Route Optimization Based on Genetic Algorithms. In *International Federation for Information Processing (IFIP), Knowledge Enterprise: Intelligent Strategies in Product Design, Manufacturing, and Management*; Wang, K., Kovacs, G., Wozny, M., Fang, M., Eds.; Springer: Boston, MA, USA, 2006; Volume 207, pp. 177–183.
26. Lee, K. Optimal System Design with Geometric Considerations. Ph.D. Thesis, University of Michigan, Ann Arbor, MI, USA, 2014.
27. Szykman, S.; Cagan, J. Synthesis of Optimal Nonorthogonal Routes. *ASME. J. Mech. Des.* **1996**, *118*, 419–424. [CrossRef]
28. Liu, Q.; Tang, Z.; Liu, H.; Yu, J.; Ma, H.; Yang, Y. Integrated Optimization of Pipe Routing and Clamp Layout for Aeroengine Using Improved MOALO. *Int. J. Aerosp. Eng.* **2021**, *2021*, 6681322. [CrossRef]
29. Yuan, H.; Yu, J.; Jia, D.; Liu, Q.; Ma, H. Group-based multiple pipe routing method for aero-engine focusing on parallel layout. *Front. Mech. Eng.* **2021**, 1–16. [CrossRef]
30. Eheim, M.; Kaiser, D.; Weil, R. On Automation Along the Automotive Wire Harness Value Chain. In *Advances in Automotive Production Technology—Theory and Application, Proceedings of the Stuttgart Conference on Automotive Production (SCAP2020), Stuttgart, Germany, 9–10 November 2020*; Weißgraeber, P., Heieck, F., Ackermann, C., Eds.; Springer: Berlin/Heidelberg, Germany, 2021; pp. 178–186.
31. Garcia-Castellanos, D.; Lombardo, U. Poles of Inaccessibility: A Calculation Algorithm for the Remotest Places on Earth. *Scott. Geogr. J.* **2007**, *123*, 227–233. [CrossRef]
32. Farhat, C.; Lesoinne, M. Two efficient staggered algorithms for the serial and parallel solution of three-dimensional nonlinear transient aeroelastic problems. *Comput. Methods Appl. Mech. Eng.* **2000**, *182*, 499–515. [CrossRef]
33. Weyland, D. Simulated Annealing, its Parameter Settings and the Longest Common Subsequence Problem. In Proceedings of the 10th Annual Conference on Genetic and Evolutionary Computation 2008, Atlanta, GA, USA, 12–16 July 2008; pp. 803–810.
34. Rudolph, S. 2019 Physical Architecture Optimization System (PHAROS). H2020 CS2 Grant Proposal, Part B.I (Confidential, Unpublished). Available online: https://cordis.europa.eu/project/id/865044 (accessed on 14 January 2022).

aerospace

MDPI

Article

Enhancement Opportunities for Conceptual Design in Aerospace Based on the Advanced Morphological Approach

Vladislav T. Todorov [1,*], Dmitry Rakov [2] and Andreas Bardenhagen [1]

[1] Aircraft Design and Aerostructures, Institute of Aeronautics and Astronautics, Technische Universität Berlin, 10587 Berlin, Germany; andreas.bardenhagen@tu-berlin.de

[2] Blagonravov Mechanical Engineering Research Institute (IMASH), Russian Academy of Sciences, 101990 Moscow, Russia; rdl@mail.ru

* Correspondence: vladislav.t.todorov@tu-berlin.de

Abstract: The current challenges facing the aerospace domain require unconventional solutions, which could be sought in new configurations of future aircraft and spacecraft. The choice of optimal concepts requires the consideration of a significant amount of competing engineering solutions and takes place under conditions of uncertainty. Such a problem can be addressed by enhancing existing methods for analysis and synthesis solutions, such as the Advanced Morphological Approach (AMA). It uses morphological analysis to provide a more exhaustive overview of possible problem solutions, relies on expert evaluations of alternative technological options and applies clustering to the solution space. Although an intuitive method for structured concept generation, the AMA exposes the need for more robust problem structuring, improved objectivity of options evaluation and accounting for uncertainties. The current article suggests ways to overcome these challenges and their possible integration in the process. In particular, the integration of fuzzy sets is proposed to model uncertainties during the evaluation of technological options by the experts. The Fuzzy Analytical Hierarchy Process is adapted for integration into the AMA and for the conceptual design of aerospace vehicles.

Keywords: conceptual design in aerospace; advanced morphological approach; fuzzy sets

Citation: Todorov, V.T.; Rakov, D.; Bardenhagen, A. Enhancement Opportunities for Conceptual Design in Aerospace Based on the Advanced Morphological Approach. *Aerospace* **2022**, *9*, 78. https://doi.org/10.3390/aerospace9020078

Academic Editors: Spiros Pantelakis, Andreas Strohmayer and Liberata Guadagno

Received: 15 December 2021
Accepted: 27 January 2022
Published: 1 February 2022

1. Introduction

The most important stage in the design of innovative aircraft systems is the concept design phase [1]. Mistakes at this stage might lead to significant increases in cost and lead time and could have a fatal impact on the project. The configuration of the aircraft obtained in the conceptual design stage is numerically simulated and modified in the preliminary design. In most cases, preliminary design approaches are based on development trends derived from databases and statistics [2]. However, this is not the case when aiming to develop innovative aircraft configurations, since breakthrough technologies lack experimental and statistical data. The non-metric character of design parameters, as well as the conflicting criteria, represent further challenges of conceptual design. The target function of a potential optimization problem could hardly be solved by common theoretical methods since it is discontinuous; it cannot always be defined; it exists in the operator notation; it is not based on analytical expressions; it is not differentiable, not unimodal, not separable and not additive [3]. Additionally, the circumstances in the aviation domain put forward the necessity for a technological breakthrough. The sector is forced to address the demands for higher efficiency and drastic emission reduction, while no short-term technological answer is yet in sight [4,5].

Along with heuristic methods, morphological analysis (MA) is used during the conceptual design phase [2,6]. It allows the generation of a significantly wider space of possible problem solutions than conventional methods such as brainstorming and helps to avoid the aforementioned conceptual design challenges. This is achieved by decomposing the problem into (sub-)functional and characteristic components (denoted as "attributes"). Each attribute is assigned to a set of appropriate technological options (denoted in the following as "options"). The attributes and their corresponding options are organized in a morphological matrix (MM). An example MM excerpt for the conceptual design of an unmanned aerial system is shown in Figure 1. The entity of all possible attribute–option combinations defines the solution space. The significant amount of synthesized configurations in such a way increases the possibility to find the optimal solution even for a non-conventional or demanding problem statement. In such situations, the optimal solutions can be overseen by conventional idea generation methods.

	Option 1	Option 2	Option 3	Option 4
Lift generation	Aerodynamic	Thrust	Aerostatic	
Internal Energy supply	none	Chemical, reversible (e.g. battery)	Chemical irreversible (e.g. fuel tank)	Mechanic (e.g. fly-wheel)
Power generation	Electric	Internal combustion	Gas turbine	Rocket engine
Fuselage	None	Single fuselage	Twin boom	
Altitude control	Aerodynamic/ elevators	Thrust adjustments	Aerostatic	
Wing morphing	No	Yes		

Figure 1. An excerpt from the morphological matrix for the structural synthesis of an unmanned aerial system. Data from ref. [7].

MA has been applied to the conceptual design of aerospace vehicles within the Advanced Morphological Approach (AMA), developed by Rakov and Bardenhagen [7,8]. The method extends the classical MA by integrating qualitative expert evaluations of the options and by offering an intuitive solution space exploration. The AMA steps can be summarized as follows [7–9]: (1) problem statement definition; (2) definition of the MM and a set of evaluation criteria; (3) evaluation of the technological options by domain experts according to each criterion; (4) inclusion of reference solutions based on existing aerospace vehicle configurations; (5) definition of impossible combinations of options for different attributes; (6) generation of the solution space; (7) clustering of the generated solutions based on the given expert evaluations; (8) visualization and reporting.

A simplified diagram of the methodology is presented in Figure 2. From the generated vast solution space in the MM, the method picks a limited amount of superior solutions based on their criteria evaluations [7,10]. Hence, the AMA represents a structured way to intuitively find optimal solutions to a given problem and avoid the aforementioned challenges of conceptual design. The application of AMA in the aerospace domain has been demonstrated on the conceptual design of unmanned aerial vehicles [7] and orbital reentry vehicles [11] (Figure 3).

Along with the mentioned benefits, the current AMA comes with a set of challenges and corresponding improvement possibilities, which have been presented by Todorov et al. in [12]. These are namely (a) uncertainty handling, (b) improved problem structuring and (c) multidisciplinary expert judgment elicitation.

The issue of dominating uncertainties (a) within AMA is faced during step 3 of the process: the evaluation of options by domain experts (further referred to as "experts", "decision-makers" or DMs). The method requires the DMs to evaluate each option according to a certain criterion by assigning an integer value from the qualitative scale between 1 (worst) and 9 (best). However, it is not always clear whether the quality of option A is, for example, "very good" or just "good". Such hesitation could result in a deviation of several units on the scale from the answer that is "ultimately correct" and thus reduces the reliability of the final results. This uncertainty could further increase due to the fact that the evaluation objects are innovative technologies lacking statistical and experimental data. Although the DMs are domain experts in the best-case scenario, they could naturally also lack precise knowledge on an untested or non-introduced technology.

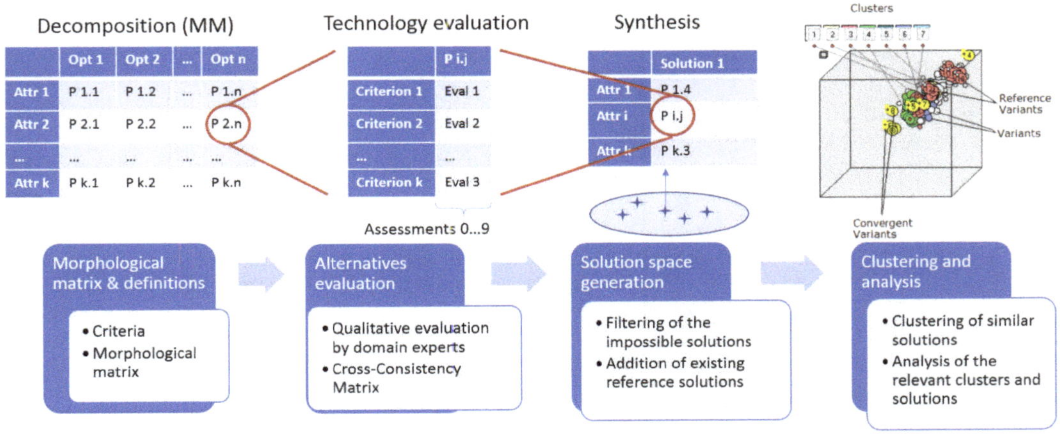

Figure 2. An overview of the AMA method. Reprinted from ref. [12].

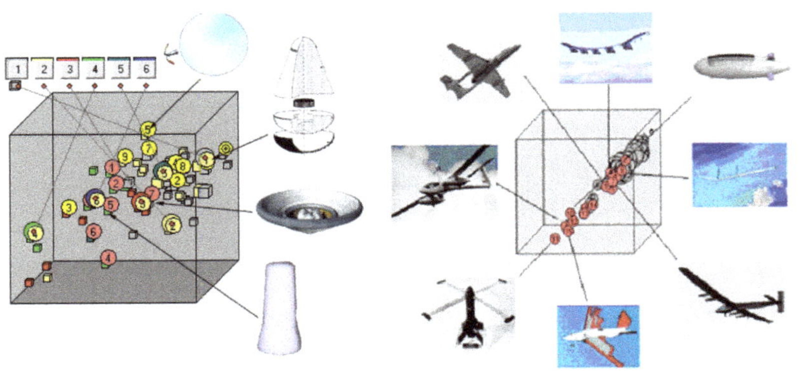

Figure 3. Reference solutions in the morphological solution space for the conceptual design tasks for an orbital reentry vehicle (**left**) and an unmanned aerial vehicle (**right**). Reprinted from refs. [7,11].

Further inaccuracy in the AMA outcome could result from the lack of consideration of technological options' interactions. Along with their separate evaluations mentioned above, certain technological combinations may additionally increase the overall value of particular configurations, which should also be reflected in the results. This aspect is acknowledged and is outlined in the future work recommendations for the improvement of the AMA.

Problem structuring (b) becomes an issue when the product to be designed represents a complex engineering solution such as an aerospace vehicle. This case requires a more thorough structural decomposition and the appropriate organization and weighting of components and criteria. This is necessary to properly reflect the complex functional division of sub-systems. The selection of appropriate problem structuring also requires the definition of evaluation aspects such as which evaluation scale to use or whether the evaluations should be absolute or pairwise comparative. Furthermore, the issue is also linked to the choice of an uncertainty modeling method.

The current article addresses the issues and the improvement possibilities for addressing uncertainties (a) and defining a robust problem structuring (b) for the AMA. Section 2 will first handle the classification of uncertainties and outline which uncertainties one faces during the option evaluations within the AMA. The choice of ordinary or Type 1 fuzzy sets as an approach for addressing the uncertainties among other possibilities is explained. After introducing their general definition, different types of fuzzy sets are briefly summarized. A way to integrate fuzzy numbers in the AMA is then suggested. Section 3 will address the issue of problem structuring. First, a classification of Multiple-Criteria Decision-Making methods is presented and the advantages of using Fuzzy Multiple-Attribute Decision-Making methods for the enhancement of the AMA are justified. Ultimately, an approach is suggested to incorporate the Fuzzy Analytic Hierarchy Process for the conceptual design of aerospace vehicles with the AMA.

2. Addressing Uncertainties

The conceptual design of a new aircraft generation is characterized by a significant amount of unknown information on innovative technologies. At the same time, the decisions made during this design phase also influence the development of the next design stages [7]. For aerospace vehicles, incorrect conceptual decisions can result in project delays and financial drawbacks. For this reason, situations associated with a significant lack of data require the anticipation of possible deviations and thus suitable models to describe uncertainties. This is the case with evaluations of options by expert groups in the AMA.

2.1. Uncertainty Classification

The most used taxonomy to classify uncertainties comes from the field of risk assessment and considers two main types: aleatory and epistemic uncertainty [13,14]. Aleatory (also stochastic) uncertainty is defined as the inherent variation of a system or the environment. Although well recognizable, it cannot be reduced or avoided. At the same time, epistemic uncertainty (also cognitive or subjective) results from the inaccurate modeling of a system or a process due to a lack of knowledge. Increasing the available data or knowledge on the matter could thus help to reduce epistemic uncertainty.

Worth noting is also the taxonomy suggested by Thunnissen [14,15], which introduces further types of uncertainty besides the ones already presented. These are the ambiguity in terms of imprecision or vagueness, as well as the uncertainty caused by the interaction of multiple disciplines or events (interaction uncertainty).

When focusing on the design of complex systems such as aerospace vehicles, further uncertainty classes have been suggested that relate rather to more precise modeling of system parameters [14]. These are the uncertainties coming from inaccurate design problem formulations, such as due to the operational environment, measurement, as well as modeling and numerical errors.

However, the conceptual design phase implies the definition of the initial vehicle structure. This task depends vastly on efficient interdisciplinary communication. The "correct" decision-making at this early stage requires proper qualitative modeling of technologies with scarce available knowledge rather than the reduction of measurement or numerical errors. For this reason, one can outline epistemic uncertainty, ambiguity and disciplinary interaction as the primary sources of deviations from the optimal conceptual design of new aerospace vehicle generations. Since the latter two categories involve a lack of knowledge and sub-optimal information exchange, these will be also considered as epistemic uncertainties in the current article.

This classification helps to identify the types of uncertainties that need to be addressed or at least acknowledged within the AMA method. Potential sources of uncertainties can be found in the evaluation of technological options. Particularly, subjective results could be obtained in the following cases:

1. When an expert estimates the quality of an option according to a certain criterion. Subjective assessments can be due to the following drawbacks:

- Epistemic uncertainty—the lack of knowledge on untested or non-introduced technologies;
- Cognitive bias—the unconscious heuristics people use to solve uncertain tasks, which may lead to systematic errors [16].

2. During group discussions of a certain technology by experts from different expertise domains (disciplinary interaction).
3. As a result of aggregation of multiple expert opinions—be it behavioral (group discussions) or mathematical.

The current article suggests that epistemic uncertainty during option evaluation could be accounted for by introducing an appropriate modeling approach, which will be discussed in the following. Handling cognitive bias, disciplinary interaction and assessment aggregation from multiple experts could be considered as research topics linked to the development, structuring and conduction of the expert workshops and are not discussed in the present work.

2.2. Uncertainty Modeling Approaches

The main requirement for the modeling of epistemic uncertainty is the ability to quantify and map the uncertain areas during decision-making onto an intuitive and robust process. It is thus necessary to find a proper mathematical and computational representation of this uncertainty type.

Zang et al. [17] summarize three ways to represent uncertainties in a computer code used for modeling and simulation. These are the interval bound, the membership function and the probability density function (PDF), shown in Figure 4. The methods incorporate different theories to describe uncertainties. Based on the classical probability theory, the PDF offers uncertainty modeling with the highest detail among the three mentioned options. However, it hides certain drawbacks when dealing with problem statements lacking statistical data [18]. Firstly, the choice of a suitable probability distribution requires prior experimental data. Secondly, PDFs aim to "represent one's degree of belief that a specific realization of a parametric value, physical process, or event can occur" [18] (p. 2). This type of uncertainty modeling has already been applied in the aerospace domain. The works of Monroe [19] and Unal et al. [20] describe an expert judgment elicitation methodology for the conceptual design of innovative launch vehicles. They use qualitative expert input in order to quantify the uncertainties through probability distributions. The parameters that were estimated by the experts referred to deterministic characteristics such as system weight and cost. Since the technologies used for the design in those works were supposed to be "extrapolated" from existing systems of that time [20], one could assume that the aerospace experts involved had some initial experience and statistical data on the technologies to lean on.

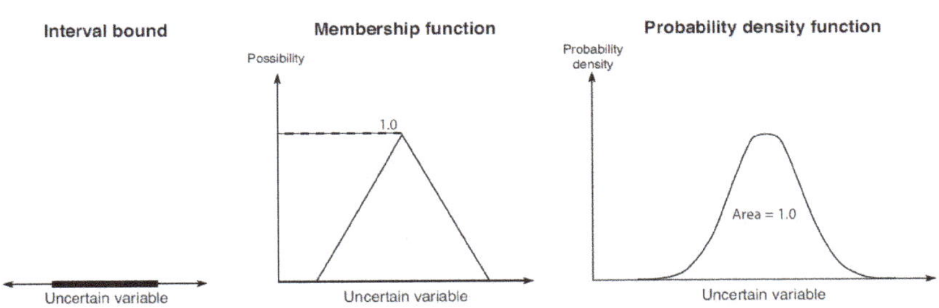

Figure 4. Ways to describe uncertainty. Reprinted from ref. [17].

However, the AMA strives not only to include technological options derived from ones already in operation, but also to consider technologies lacking any prior experimental data. Therefore, the uncertainties to be modeled are not referred to the concrete estimation of aircraft characteristics such as weight or certain flight qualities but rather a higher-level qualitative comparison of more abstract combinations of technological options.

At the same time, the interval bound defines solely the upper and lower limits of the possible value range and offers a very simplified approximation [17].

The membership function is a mathematical tool used in the fuzzy logic and fuzzy set theory to describe fuzzy sets, introduced by Lotfi Zadeh in 1965 [21], and is "particularly well-suited for handling incomplete information, the unsharpness of classes of objects or situations, or the gradualness of preference profiles" [22] (p. 4). This approach allows us to mathematically model events or statements, which can be semantically assigned to multiple parameter values simultaneously [21,22]. Such a philosophy could be used to represent subjective, ambiguous and vague statements made by experts for decision-making tasks during early conceptual design, which corresponds to the evaluation of options within the AMA.

2.3. Fuzzy Sets

An ordinary (so-called Type 1) fuzzy set $\tilde{A}$ is formally defined by a given set X and a membership function $\mu(x)$, which maps every $x \in X$ to a real value from the interval [0, 1] (Equations (1) and (2)) [22,23]. The function reflects the "grade of membership" of each x in $\tilde{A}$.

$$\tilde{A} = \{(x, \mu_{\tilde{A}}(x)) | x \in X\} \quad (1)$$

$$\mu_{\tilde{A}}(x) : X \to [0, 1] \quad (2)$$

Such a definition allows to model uncertainties by assigning a suitable membership function to an unsharp linguistic statement. This is achieved by considering the "level of truth" or the "grade of membership" of the statement for each parameter value x. For example, one could use fuzzy sets to represent the vague definition "wise" by subjectively assigning it to a person's age, as depicted in Figure 5. Since there exists no definite age threshold between defining a person as "wise" and "not wise", the fictional DM has decided to place ages between 30 and 75 in an uncertain area, where wisdom is accumulated nonlinearly. People aged 75 and older are assumed by this DM to fully correspond to the definition of "wise" with full membership of $\mu(75 \leq x) = 1$.

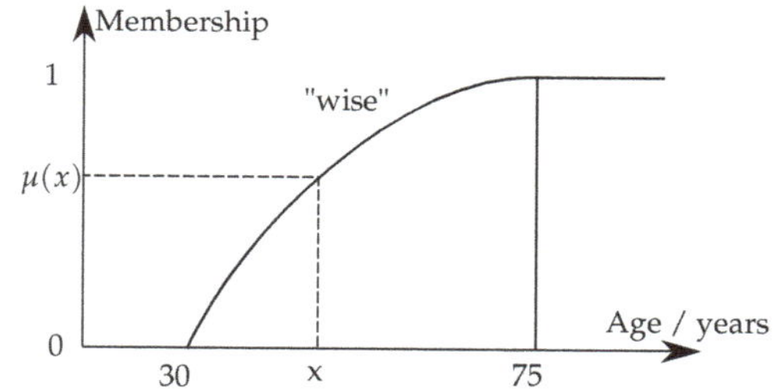

Figure 5. A fuzzy representation of the vague adjective "wise" referring to a person's age.

The last several decades have seen the active development of fuzzy sets through extensions and generalizations of the presented ordinary case. As a result, multiple new types of fuzzy sets have been introduced and have found applications in numerous domains [23]. A brief overview of the most used types is shown in Table 1.

Table 1. A summary of some of the most used fuzzy set types.

Name	Definition	Description
Type 2 fuzzy sets [21,23]	$\tilde{A} = \{((x,u), \mu_{\tilde{A}}(x,u)) \mid 0 \le \mu_{\tilde{A}}(x,u) \le 1\}$	The membership function is fuzzy itself
Intuitionistic fuzzy sets [24]	$\tilde{A} = \{\langle x, \mu_{\tilde{A}}(x), \nu_{\tilde{A}}(x) \rangle; x \in X\}$ $\mu_{\tilde{A}} : X \to [0,1]; \nu_{\tilde{A}} : X \to [0,1]$ $0 \le \mu_{\tilde{A}} + \nu_{\tilde{A}} \le 1$ $\pi_{\tilde{A}}(x) = 1 - \mu_{\tilde{A}}(x) + \nu_{\tilde{A}}(x)$	Introduction of the non-membership $\nu_{\tilde{A}}(x)$ and the hesitation margin $\pi_{\tilde{A}}$ [24,25]
Hesitant fuzzy sets [26]	$\tilde{A} = \{\langle x, h_{\tilde{A}}(x) \rangle \mid x \in X\}$ $h_{\tilde{A}} : X \to J \subset [0,1]$	The function $h_{\tilde{A}}(x)$ returns a subset of $[0,1]$

The common aspect of the given fuzzy set types is that they all introduce an additional "grade of freedom" or dimension of fuzziness to the already existing membership. One could interpret it as the uncertainty in the position of the membership value for each x. The corresponding authors justify the necessity for the additional uncertainty level with the possible hesitation of the DM when choosing the membership value $\mu_{\tilde{A}}(x)$ for each particular element x. Kahraman et al. [27] outline multiple applications of the intuitionistic and hesitant fuzzy sets within Fuzzy Multiple-Criteria Decision-Making methods.

The mentioned classification of fuzzy sets combines sophisticated ways to model multi-layer uncertainty, which vary in their definitions and depth of uncertainty representation. Therefore, it is necessary to pick the most suitable type for the particular purpose in the first place. Secondly, one should carefully evaluate whether complex and overly abstract uncertainty modeling would impact:

1. the comprehensibility of the evaluation process/the questionnaires for the experts, which could reduce the results' reliability
2. the transparency of the scientific method, thus making difficult the recreation of the process and the results.

2.4. Integration of Fuzzy Sets into the Advanced Morphological Approach

When incorporating uncertainties into the AMA for the design of aerospace vehicles, one should consider the specificity of the experts usually involved in this type of decision-making task. In particular, this tends to be a multidisciplinary group of DMs from different domains who do not necessarily have the deep statistical background encompassing the special characteristics of complex fuzzy sets. Hence, one should find a balance between the optimal capturing of uncertainty and not over-complicating the evaluation task, which could lead to even greater unexplained uncertainty in the results.

For this reason, the authors suggest to start initial studies in such a context by using simpler fuzzy sets (e.g., the ordinary one) and, if necessary, gradually increasing their complexity while studying the experts' perception.

The implementation of fuzzy sets is suitable for the mathematical and computational representation of vague expert statements or evaluations during aircraft conceptual design. At this point, an approach is suggested for the initial integration of ordinary Type 1 fuzzy sets within the AMA. It implies that the expert evaluations of a given technology could be described as membership functions. For this purpose, trapezoidal fuzzy numbers are used, which were defined in [28,29] and are graphically depicted in Figure 6. Consider the definition of a trapezoidal fuzzy number by its four distinctive points: $\tilde{A} = (k,l,m,n)$. These allow the DMs to divide their evaluation into two domains:

1. An interval of full certainty $[l,m]$ with membership $\mu_{\tilde{A}}(x) = 1$. The DM gives the highest confidence that this qualitative interval reflects the priority of the option according to a certain criterion as the best.
2. Uncertain intervals $[k,l)$ and $(m,n]$ with membership $0 < \mu_{\tilde{A}}(x) < 1$. The DM expresses hesitation about these quality values of the option.

Such a representation of the evaluations will be integrated into a suitable Multiple-Attribute Decision-Making method in the next section.

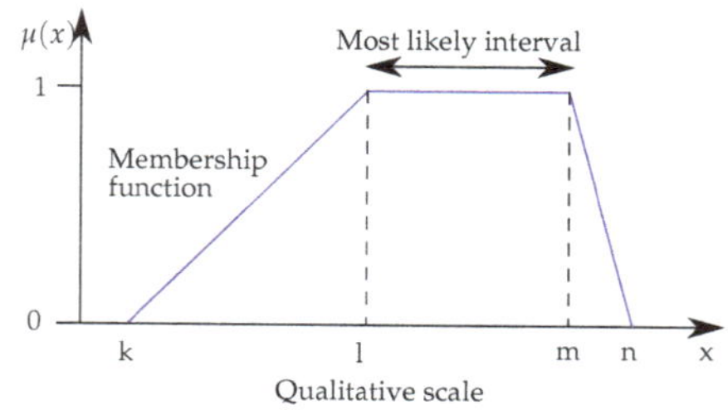

Figure 6. Own depiction of a trapezoidal fuzzy number.

3. Structuring the Evaluation Process through Hierarchies

The structural synthesis of promising engineering solutions through the AMA implies problem decomposition and the evaluation of technological options. Aerospace vehicles represent complex systems that can be defined by numerous criteria and subsystem levels. In this case, the appropriate robust structuring and organization of the system decomposition and the option evaluations represent a challenging task. It requires the ranking of discrete technological options of the attributes in order to achieve a higher-level ranking of the entire set of alternative problem solutions (in this case, aircraft configurations). The methods addressing such challenges are classified as Multiple-Attribute Decision-Making (MADM) algorithms. Following a brief overview of MADM methods, the Analytic Hierarchy Process and its potential contribution to the enhancement of the AMA process will be introduced. Furthermore, possibilities to use fuzzy logic within these methods will be investigated.

3.1. *Overview of Multi-Attribute Decision-Making Algorithms*

Within the domain of operations research (also called decision science), Multi-Criteria Decision-Making (MCDM) is a field that aims to offer mathematical tools for complex decision-making tasks [30]. Two main groups of MCDM methods are distinguished: Multiple-Objective Decision-Making (MODM) and Multiple-Attribute Decision-Making (MADM). While the first strives to solve problems containing continuous variables, MADM addresses tasks where the alternative scenarios represent a discrete decision space [30]. At the same time, these categories can be divided into methods using deterministic data, denoted as crisp methods, and such that take into account uncertainties of human evaluations, or fuzzy ones. Accordingly, the latter are marked as Fuzzy MODM (FMODM) and Fuzzy MADM (FMADM) methods. The nature of the AMA implies the definition of a discrete decision space in the form of a morphological matrix. Furthermore, one of the aims of the AMA enhancement is to account for uncertainties as well. These aspects require the application of a Fuzzy MADM approach. For a more detailed overview of crisp MCDM methods, the reader is encouraged to refer to the corresponding summary under the edition of Figueira et al. [31].

The definition of a typical MCDM method consists of three main elements [30]:

- Alternative actions/options—these correspond to the technological options for each attribute in the AMA.
- A set of criteria—for the AMA, these are the criteria against which the technological options are evaluated.
- A problematic type—one distinguishes among the following problematic types: description, choice, sorting and ranking. The expected outcome of the AMA option evaluation step is to obtain a reliable comparison of the technological options. Hence, this case corresponds to the ranking problematic.

An extensive overview of existing FMADM methods is given by Kahraman et al. [27] as well as by Chen et al. [28]. In the majority of cases, FMADM methods have been derived from crisp MADM algorithms. Some of the most prominent MADM methods for both crisp and fuzzy evaluations are classified as outranking (ELECTRE, PROMETHEE), distance-based (VIKOR, TOPSIS) and pairwise comparisons-based (Analytic Hierarchy Process, MACBETH) [27].

3.2. Analytic Hierarchy Process

The AMA enhancement seeks an intuitive way to structure a complex multi-layer engineering problem statement by considering multiple potentially contradicting criteria. In this context, the Analytical Hierarchy Process (AHP) is used to handle this challenge. The method and its potential application within the AMA will be presented in the following.

The AHP has been developed by Saaty [32] for the purpose of qualitative decision-making based on relative pairwise comparisons of alternative options/scenarios [28,32]. One of its main features is the organization of the problem structure in a hierarchical manner. An example of a hierarchy to select an airline to fly with based on the criteria ticket price, airline timetable and offered comfort is shown in Figure 7.

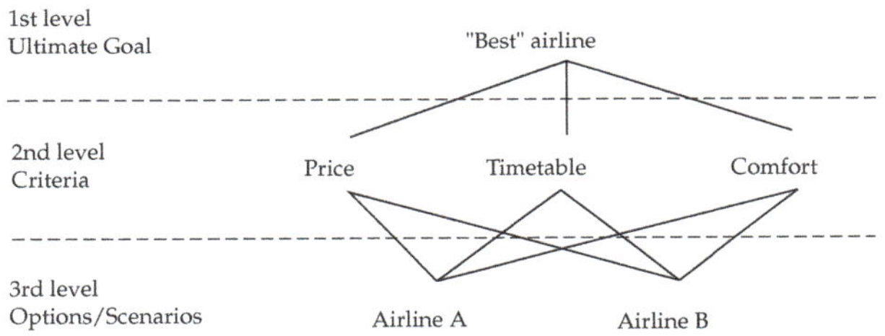

Figure 7. An exemplary hierarchy structure for the selection of an airline to fly with.

Each hierarchy level is considered as a criterion or purpose, according to which the elements of the lower level are evaluated [32]. The evaluations of alternative options from each single level are defined in a relative pairwise manner instead of as absolute values. Hence, each evaluation represents a linguistic statement on the quality of option A referred to option B against a certain criterion, e.g., option A can be much better/slightly better/equal to/slightly worse/much worse than option B. For this purpose, a scale between 1 and 9 with the following legend is used: (1) equality; (3) weak superiority; (5) strong superiority; (7) significant superiority; (9) absolute superiority. The even numbers in between stand for intermediate values. Inferiority is represented by the corresponding reciprocal values. Accordingly, the quality of option B relative to option A is then defined as the reciprocal value of the evaluation from the previous example. Such evaluations for all items of the same level according to a certain criterion are summarized in a reciprocal matrix, represented in Equation (3) [32]. Its elements are the pairwise comparisons $a_{ij} = \frac{\omega_i}{\omega_j}$, which are obtained from the expert evaluations.

$$D = \begin{bmatrix} \omega_1/\omega_1 & \omega_1/\omega_2 & \dots & \omega_1/\omega_n \\ \omega_2/\omega_1 & \omega_2/\omega_2 & \dots & \omega_2/\omega_n \\ \dots & \dots & \dots & \dots \\ \omega_n/\omega_1 & \omega_n/\omega_2 & \dots & \omega_n/\omega_n \end{bmatrix} \quad (3)$$

The next step implies using these relative evaluations a_{ij} in order to find the absolute weights ω_i of the items for the current hierarchy level [32]. In the initial crisp AHP, Saaty showed that these absolute weights correspond, with some approximations, to the dimension values of the main eigenvector of the mentioned reciprocal matrix. It can be obtained by using the so-called "eigenvector method", described by Saaty.

After establishing the absolute weights of the elements from each single hierarchy level separately, one can proceed with finding the importance of any single element of one level with respect to any element from a higher one. Considering a three-level hierarchy as in Figure 7, let ω_{ij} be the weight of the i-th element from level 3 with respect to criterion j from level 2 and b_j be the weight of the j-th element from level 2 with respect to the ultimate goal from level 1. Then, the importance U_i of the i-th element from level 3 with respect to the ultimate goal can be derived as the weighted sum from Equation (4) [28,32]:

$$U_i = \sum_{j=1}^{n} b_j \omega_{ij} \tag{4}$$

These values can be ultimately used to rank the elements from any level according to an element from any higher one.

4. Integration of a Fuzzy Analytical Hierarchy Process into the Advanced Morphological Approach

It is worth noting that Saaty himself stated that the crisp AHP already integrated a fuzzy way of thinking in some form—namely by giving the decision-makers (DMs) the opportunity to give relative option evaluations on a qualitative scale [29]. However, DMs could find themselves in situations where they cannot clearly assign a concrete statement to the technology in question [29], e.g., is technology A definitely "much better" or just "slightly better" compared to technology B? This hesitation represents the epistemic uncertainty within the expert evaluations. Instead of being classified as unnecessary and/or incomplete information, the vagueness of the expert opinions should be considered as an integral part of the detailed decision-making process due to its potential impact on the final result. Although the certainty level of such information could be considered sub-optimal, its proper aggregation for multiple experts from diverse domains has the potential to reduce epistemic uncertainty. This could therefore increase the reliability of forecasts on promising technologies without prior experimental data. Such advantages could be achieved by allowing the experts to define their evaluations as fuzzy sets.

4.1. Definition of the Main Method Parameters

To this day, multiple implementations of Fuzzy AHP (FAHP) have been introduced in the literature, such as the approach of Laarhoven and Pedrycz, as well as the one of Buckley [28,33]. Since the latter overcomes some drawbacks of the former one [28], is more intuitive and allows the use of trapezoidal fuzzy numbers, Buckley's method is used to integrate fuzzy sets into AMA at this stage.

Before introducing its actual methodology within the AMA, a short overview of the parameters will be given; these need to be set when defining an FAHP method for a particular task. According to Liu et al. [29], these are:

1. Choice of appropriate fuzzy set types as a "representation of the relative importance for pairwise comparison" [29] (p. 2);
2. Techniques for fuzzy set aggregation when multiple experts are involved in the evaluation;
3. "Defuzzification of a fuzzy set to a crisp value for final comparison" [29] (p. 2);
4. Techniques for the consistency check of the fuzzy evaluation matrix.

As already stated in Section 2.4, ordinary fuzzy sets of Type 1 in the form of trapezoidal fuzzy numbers have been selected for the initial consideration of uncertainties. Since the current paper aims to outline the global methodology of the FAHP integration into the AMA, the rest of the parameters will be left to be set in a further, more detailed definition of the method.

4.2. Representation of the Aerospace Vehicle Conceptual Design Problem as a Hierarchy

The FAHP is initialized by defining the given multi-criteria problem statement as a hierarchical tree. By following the logic outlined in Section 3.2, the attributes and the options of a given MM are used to form a comprehensible hierarchy. In this context, the current work suggests a hierarchy structure for the conceptual design of aerospace vehicles by using the attributes and options from the MM and the evaluation criteria. An example hierarchy for a sample aircraft design task is presented in Figure 8.

Since the technological options are the smallest system elements in the current representation, these are positioned at the lowest hierarchy level. The quality of each is evaluated with respect to all criteria, defined in level 3 above. These might be qualities such as Operating Empty Weight (OEW), the aerodynamic qualities of the aircraft (L/D or "L over D ratio"), Direct Operating Cost (DOC), different types of emissions, etc. Next, one obtains the importance of each criterion for the corresponding system attribute (level 2). Finally, the experts will give their priorities for the separate attributes regarding the global aircraft concept set as the ultimate hierarchy goal.

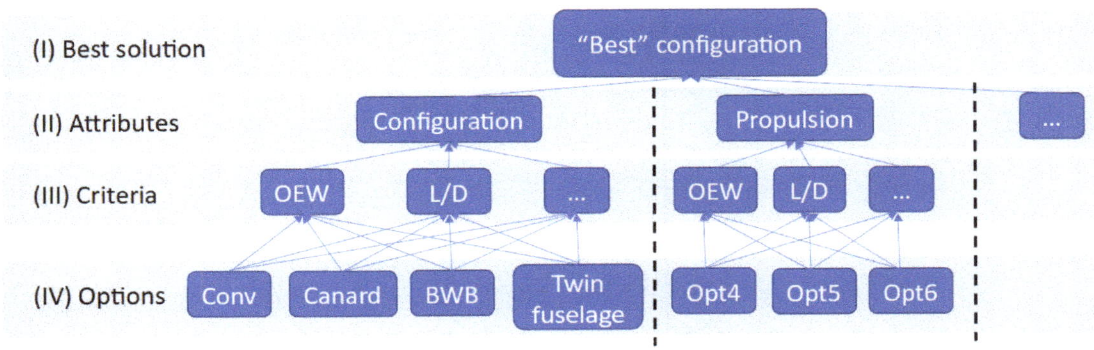

Figure 8. Suggested hierarchy structure for the evaluation process of aircraft system elements and their technological options. Abbreviations: OEW—Operating Empty Weight; L/D—"L over D" or the aerodynamic qualities; Conv—conventional configuration; BWB—Blended Wing Body; Opt—potential additional options.

One should highlight that the different attributes and their corresponding children elements are represented as separate branches in the hierarchy, which do not interconnect. This consideration was necessary due to the fact that the AHP implies pairwise comparison evaluations of elements from the same level. In this context, one cannot compare technological options for different attributes (e.g., electric propulsion and the conventional aircraft configuration) or their assigned criteria (e.g., the weight of the configuration and the weight of the propulsion subsystem).

It is important to underline that the advantage of the AHP is that the hierarchy structure is not fixed by the methodology and can be adapted according to the needs of the given problem statement. Hence, the presented hierarchy definition does not pretend to be optimal for every problematic. However, it is suggested as a structured approach to extend the MA and aid in the option evaluation, particularly in the domain of aerospace vehicle conceptual design.

4.3. Evaluation of Hierarchy Elements with Fuzzy Sets

The most commonly used fuzzy scales for pairwise importance comparisons are the nine- and five-level ones [29] (p. 16). In some studies, researchers fix the fuzzy numbers denoting the varying quality. In such cases, an exemplary nine-level scale would contain triangular fuzzy numbers with pre-defined peaks of full membership for the importance of one alternative over the other such as: equal-(1,1,2); weak-(2,3,4), strong-(4,5,6); significant-(6,7,8); absolute-(8,9,9) [29]. Here, the (a, b, c) format denotes the significant points of a triangular fuzzy number, where b is the peak with full membership of 1 and a and c are the limits of the number with membership 0 and are connected linearly to b.

However, the aim of the AMA enhancement is to give the experts more freedom in the representation of their opinions, rather than just providing them with a fixed set of pre-defined fuzzy numbers. Therefore, a nine-degree scale is suggested, where each degree is assigned a linguistic meaning. The experts are then allowed to express their preferences by freely defining and positioning a trapezoidal fuzzy number per evaluation. Figure 9 shows a single exemplary expert evaluation stating that technology A is "approximately from strongly to significantly superior" to technology B.

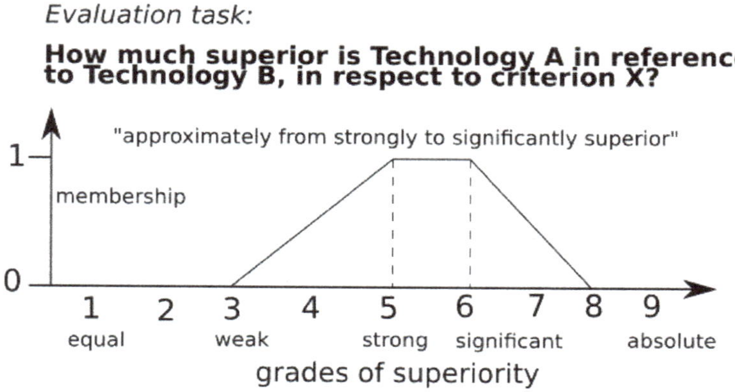

Figure 9. Suggested approach to evaluate the pairwise comparison of technological option A to option B with respect to a certain criterion.

The presented scale so far describes only the superiority degree of one option over another. As mentioned in Section 3.2, the classical crisp AHP defines option inferiority as the corresponding reciprocal crisp number. For the FAHP, Buckley [28,33] represents the evaluations of inferior options by mirroring the base values of the trapezoidal fuzzy number and taking their reciprocal values [28] (p. 359). For example, if the superiority of technology A over technology B is defined by the trapezoidal fuzzy number $\tilde{a}_{AB} = (3, 5, 6, 8)$, as shown in Figure 9, then the inferiority of technology B to technology A would be $\tilde{a}_{BA} = (1/8, 1/6, 1/5, 1/3)$. One could also use the same nine-degree scale for the expert evaluation task. However, the expert would need to specify whether the current evaluation is about an inferior option, so that the reciprocal fuzzy number is taken afterwards.

Based on the defined hierarchy structure, the experts give their pairwise relative evaluations for each element of the corresponding hierarchy level. Consider the example hierarchy presented in Figure 8. The evaluations made by a single expert of the Blended Wing Body (BWB) configuration in comparison to the other configurations (same hierarchy level) and with respect to the aircraft OEW could be as shown in Figure 10.

These evaluations are obtained for each element of each corresponding level in order to be placed in the comparison matrices. Equation (5) shows the structure of a reciprocal comparison matrix, where a_{ijk} is the fuzzy trapezoidal comparison of option i with option j made by the k-th expert (for the total of n options and p elements) [28].

$$D = \begin{bmatrix} (1,1,1,1) & a_{121}; a_{122}; \ldots; a_{12p} & \cdots & a_{1n1}; a_{1n2}; \ldots; a_{1np} \\ \cdots & \cdots & \cdots & \cdots \\ \cdots & a_{ij1}; a_{ij2}; \ldots; a_{ijp} & \cdots & \cdots \\ \cdots & \cdots & \cdots & \cdots \\ a_{n11}; a_{n12}; \ldots; a_{n1p} & a_{n21}; a_{n22}; \ldots; a_{n2p} & \cdots & (1,1,1,1) \end{bmatrix} \quad (5)$$

Subsequently, the evaluations of all experts should be aggregated for each option. As previously mentioned, the choice of the most appropriate aggregation method is left as a topic for future investigations.

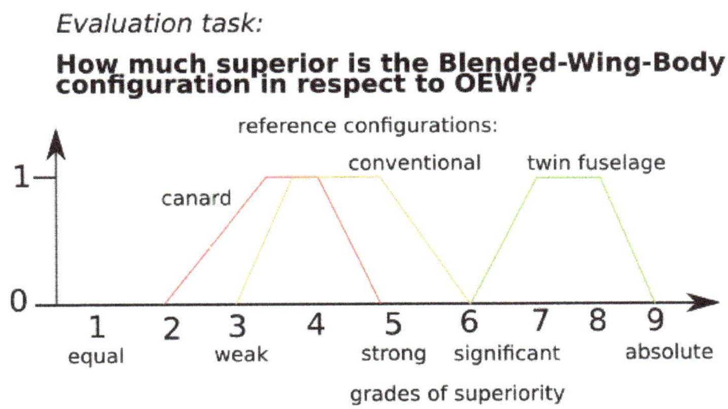

Figure 10. Example of pairwise comparisons made by a single expert of the BWB configuration referred to the configurations twin fuselage, canard and the conventional one, with respect to the OEW criterion.

4.4. Buckley's Approach for the Implementation of Fuzzy Analytic Hierarchy Process

Once all aggregated evaluations have been acquired and organized into the corresponding comparison matrices, Buckley's FAHP approach can be applied to obtain the global weights of the options [28] (p. 356). This is outlined in the following steps [28]:

(1) Calculation of the fuzzy weights ω_i for the elements of the same hierarchy level.

For this purpose, the geometric mean z_i for each row (corresponding to option i) of the reciprocal comparison matrix is estimated first (Equation (6)). The fuzzy weight ω_i is then calculated by referring the corresponding geometric mean of the row z_i to the sum of the geometric means in all rows, as shown in Equation (7). The signs $\oplus$ and $\odot$ stand for fuzzy addition and fuzzy multiplication, as defined in [28] (pp. 353–356), respectively.

$$z_i = (a_{i1} \odot \ldots \odot a_{in})^{1/n} \tag{6}$$

$$\omega_i = z_i \odot (z_1 \oplus \ldots \oplus z_n)^{-1} \tag{7}$$

(2) Estimation of the global importance of the options over all hierarchy levels. This is achieved by following the logic of the original crisp AHP. The importance of each i-th option U_i is derived as the sum of the option weights $\omega_{i,j}$ each multiplied by the weight b_j of the corresponding element j from the higher level of the hierarchy.

$$U_i = \sum_{j=1}^{n} b_j \omega_{ij} \tag{8}$$

5. Discussion

Along with the detailed descriptions of the presented methods in the sections above, it is necessary to underline the novelties suggested in the current article for the field of conceptual aircraft design. Furthermore, the findings will be positioned within the context of the existing literature on the subject.

The choice of the most suitable configuration within the AMA depends on the evaluations of technological options by a group of dedicated experts. Thus far, this process has been defined solely through the structure of the morphological matrix, namely the system attributes and their corresponding technological options, as well as by the evaluation criteria. The evaluations' reliability can be increased not only by introducing uncertainty modeling, but also by representing the complex problem definitions and the evaluation process in a structured way. For this purpose, the Fuzzy Analytic Hierarchy Process has been adapted for its use in the AMA. This was possible through the suggested novel definition of the aerospace vehicle design problem as an intuitive hierarchy structure, a concrete format for potential evaluation questions as well as an appropriate fuzzy evaluation scale.

The improved organization of criteria intuitively follows the levels of importance of attributes and subsystems. This logical yet flexible structure not only allows us to conduct the evaluation process in a formalized order; it also gives the experts the opportunity to discuss whole groups of alternatives from the same hierarchy level simultaneously. This makes it possible to widen the discussion and take into account aspects that could be left out when evaluating each option separately.

It is also important to position the findings among existing sources on the use of fuzzy sets and MADM or (F)AHP methods in the aviation domain. Thus far, these have been applied to abstract and non-deterministic problems mostly in the field of aeronautical operations or to answer organizational questions. The methods have been used, e.g., for the assessment of technical factors in aviation safety [34], the evaluation and ranking of green airlines [35], the assessment of factors to reduce fuel consumption [36] and the evaluation of aircraft efficiency [37]. Additionally, AHP has been applied to develop advanced brainstorming methods for the design of novel aircraft configurations, e.g., cryogenic planes in [38].

However, there is still a gap in the systematic idea generation and analysis of alternative aerospace vehicle concepts, particularly when using innovative technologies with dominating uncertainties. Therefore, the current article addresses these challenges by extending the AMA.

At this point, it is necessary to outline recommendations in order to continue the purposeful AMA enhancement in the specified direction. Firstly, the presented uncertainty modeling through fuzzy trapezoidal numbers and the given hierarchy structures should be integrated into a full-scale structured expert judgment elicitation framework. This will define in full detail the planning, conduction and evaluation of the expert workshops by using the presented novel approach. Secondly, a study should be conducted for the purpose of choosing the proper aggregation method for fuzzy evaluations by different experts. In this context, one could also consider the professional background of each expert, which could justify a certain weighting scheme for the marks given by experts from different domains. Furthermore, a strategy should be found to consider possible interactions of technological options contributing to the overall solution evaluation. Finally, in the event that the necessity for the deeper modeling of uncertainties arises, one could study the use of other fuzzy number types instead of the ordinary Type 1 ones.

6. Conclusions

There are currently multiple limitations and challenges during the conceptual design phase of innovative aerospace vehicles, particularly in the context of idea generation and concept evaluation. In this context, the Advanced Morphological Approach still exhibits the methodological gaps of dominating uncertainties and the lack of robust structuring for complex engineering solutions. The current article suggests concrete approaches to model uncertainties via fuzzy sets, structure the conceptual design problematic and formalize the evaluation of technological alternatives. The proposed methods have the potential to offer a new level of reliability for the decisions on innovative technologies made during the conceptual design phase in aerospace.

Existing taxonomies of uncertainties have been studied in order to identify the uncertainty types present in the AMA process. During the evaluations of technological options, one faces unmeasurable epistemic uncertainties, which are mostly due to the lack of statistical data and expert knowledge on untested innovative technologies. After outlining multiple approaches for uncertainty modeling, this work justifies the use of fuzzy sets for the representation of epistemic uncertainty within aerospace conceptual design. In particular, a suggestion has been put forward to represent unsharp evaluation statements made by experts through Type 1 trapezoidal fuzzy numbers.

Furthermore, the Fuzzy Analytical Hierarchy process has been chosen and integrated into the AMA as a method to structure the evaluation of technological alternatives. In this context, the conceptual design problematic of aerospace vehicles has been defined as a hierarchical structure, which intuitively reflects the levels of evaluation criteria. In order to justify the practical applicability of the presented approach, the current work suggests a concrete format for the evaluation questions, as well as an appropriate fuzzy scale to capture the experts' uncertainty.

The presented enhancements of the AMA method have the potential to improve the reliability of aerospace concept evaluations, especially when innovative and untested technologies are involved. Therefore, it is a step towards giving the designer more independence from already proven concepts and increasing confidence in the consideration of unconventional aerospace configurations.

The suggested methodology is defined as one of the integral stages necessary for the development of a much wider design method. The complex and abstract nature of the whole approach requires the thorough definition of further AMA improvements, which were outlined in the Section 5.

Not only does the proposed approach represent one of the multiple AMA improvement stages, but it could also be used as a "stand-alone" methodological component for similar research purposes in the aerospace domain and beyond. The authors' vision is the practical application of the improved AMA in the aerospace industry as well as in other fields with demand for non-deterministic conceptual product design. In order to ensure its smooth and intuitive application in the design process, it is planned to implement the complex logic and operations of the novel method into a software tool with formalized workshop instructions. In addition to the advantages of the AMA itself, the workshops using such a design framework would allow the experts to intuitively evaluate the technology alternatives by automatically accounting for uncertainties and reducing cognitive biases. Finally, the ultimate goal is to serve as a reliable decision-making aid when choosing the optimal concept.

Author Contributions: Conceptualization, V.T.T., D.R. and A.B.; Funding acquisition, A.B.; Investigation, V.T.T. and D.R.; Methodology, V.T.T. and D.R.; Project administration, A.B.; Resources, D.R. and A.B.; Supervision, D.R. and A.B.; Visualization, V.T.T. and D.R.; Writing—original draft, V.T.T.; Writing—review and editing, D.R. and A.B. All authors have read and agreed to the published version of the manuscript.

Funding: This research was funded by Deutsche Forschungsgemeinschaft (DFG, German Research Foundation), project number 443831887.

Conflicts of Interest: The authors declare no conflict of interest.

Abbreviations

The following abbreviations are used in this manuscript:

MA	Morphological Analysis
AMA	Advanced Morphological Analysis
MM	Morphological Matrix
Attr	Attribute
Opt	Option
DM	Decision-Maker
PDF	Probability Density Function
MCDM	Multi-Criteria Decision-Making
MODM	Multiple-Objective Decision-Making
MADM	Multiple-Attribute Decision-Making

FMADM	Fuzzy Multiple-Attribute Decision-Making
AHP	Analytical Hierarchy Process
FAHP	Fuzzy Analytical Hierarchy Process
OEW	Operating Empty Weight
L/D	"L over D ratio" or aerodynamic performance of an aircraft
DOC	Direct Operating Costs
Conv	Conventional
BWB	Blended Wing Body

References

1. Fielding, J.P. *Introduction to Aircraft Design*, 2nd ed.; Cambridge University Press: Cambridge, UK, 2017. [CrossRef]
2. Rand, O.; Khromov, V. Mission Oriented Multi-Prop UAV Analysis Using Statistical Design Trends. *Aerospace* **2021**, *8*, 321. [CrossRef]
3. Mishin, V.P.; Osin, M.I. *Introduction to Aircrafts Machine Design*; Mashinostroenie: Moscow, Russia, 1978.
4. European Commission; Directorate General for Research and Innovation; Directorate General for Mobility and Transport. *Flightpath 2050: Europe's Vision for Aviation: Maintaining Global Leadership and Serving Society's Needs*; Publications Office of the European Union: Luxembourg, 2011.
5. The Shift Project; SUPAERO DECARBO. Pouvoir voler en 2050: Quelle Aviation dans un Monde Contraint? (Flying in 2050: What Future for Aviation in a Constrained World?); Technical Report. Available online: https://theshiftproject.org/article/quelle-aviation-dans-un-monde-contraint-nouveau-rapport-du-shift/ (access on 3 March 2021).
6. Sadraey, M.H. *Aircraft Design: A Systems Engineering Approach*; Aerospace Series; Wiley: Chichestet, UK, 2013.
7. Bardenhagen, A.; Rakov, D. Analysis and Synthesis of Aircraft Configurations during Conceptual Design Using an Advanced Morphological Approach. Deutsche Gesellschaft für Luft- und Raumfahrt—Lilienthal-Oberth e.V. Available online: https://publikationen.dglr.de/?tx_dglrpublications_pi1%5bdocument_id%5d=490063 (accessed on 6 November 2019). [CrossRef]
8. Bardenhagen, A.; Rakov, D. Advanced Morphological Approach in Aerospace Design During Conceptual Stage. *Facta Univ. Ser. Mech. Eng.* **2019**, *17*, 321–332. [CrossRef]
9. Klimenko, B.; Rakov, D. Analysis and Synthesis of Innovative Engineering Solutions and Technologies Based on Advanced Morphological Approach. In *Advances in Artificial Systems for Medicine and Education*; Advances in Intelligent Systems and Computing; Hu, Z., Petoukhov, S., He, M., Eds.; Springer International Publishing: Cham, Switzerland, 2018; Volume 658, pp. 274–283. [CrossRef]
10. Rakov, D. Morphological synthesis method of search for promising technical system. *IEEE Aerosp. Electron. Syst. Mag.* **1996**, *11*, 3–8. [CrossRef]
11. Rakov, D. Superlight Reentry Vehicles. *Space Technol.* **2004**, *24*, 237–243.
12. Todorov, V.T.; Rakov, D.; Bardenhagen, A. Creation of Innovative Concepts in Aerospace Based on the Morphological Approach. In Proceedings of the 11th EASN Virtual International Conference on Innovation in Aviation & Space to the Satisfaction of the European Citizens, Salerno, Italy, 1–3 September 2021.
13. Roelofs, M.; Vos, R. Uncertainty-Based Design Optimization and Technology Evaluation: A Review. In Proceedings of the 2018 AIAA Aerospace Sciences Meeting, Kissimmee, FL, USA, 8–12 January 2018; American Institute of Aeronautics and Astronautics: Kissimmee, FL, USA, 2018. [CrossRef]
14. Yao, W.; Chen, X.; Luo, W.; van Tooren, M.; Guo, J. Review of uncertainty-based multidisciplinary design optimization methods for aerospace vehicles. *Prog. Aerosp. Sci.* **2011**, *47*, 450–479. [CrossRef]
15. Thunnissen, D.P. Propagating and Mitigating Uncertainty in the Design of Complex Multidisciplinary Systems. Ph.D. Thesis, California Institute of Technology: Pasadena, CA, USA, 2005. [CrossRef]
16. Tversky, A.; Kahneman, D. Judgment under Uncertainty: Heuristics and Biases. *Science* **1974**, *185*, 1124–1131. [CrossRef] [PubMed]
17. Zang, T.A.; Hemsch, M.J.; Hilburger, M.W.; Kenny, S.P.; Luckring, J.M.; Maghami, P.; Padula, S.L.; Stroud, W.J. *Needs and Opportunities for Uncertainty-Based Multidisciplinary Design Methods for Aerospace Vehicles*; Technical Memorandum (TM) NASA/TM-2002-211462; Issue: NASA/TM-2002-211462; Langley Research Center, National Aeronautics and Space Administration: Hampton, VA, USA, 2002.
18. Oberkampf, W.; Helton, J.; Sentz, K. Mathematical representation of uncertainty. In Proceedings of the 19th AIAA Applied Aerodynamics Conference, Anaheim, CA, USA, 11–14 June 2001; American Institute of Aeronautics and Astronautics: Anaheim, CA, USA, 2001. [CrossRef]
19. Monroe, R.W. A Synthesized Methodology for Eliciting Expert Judgment for Addressing Uncertainty in Decision Analysis. Ph.D. Thesis, Old Dominion University Libraries, Norfolk, VA, USA, 1997. [CrossRef]
20. Unal, R.; Keating, C.; Conway, B.; Chytka, T. *Development of an Expert Judgement Elicitation Methodology Using Calibration and Aggregation for Risk Analysis in Conceptual Vehicle Design*; Technical Report 20040016143; Issue: 20040016143; NASA Langley Research Center, Engineering Management Department Old Dominion University: Norfolk, VA, USA, 2004.
21. Zadeh, L. The concept of a linguistic variable and its application to approximate reasoning—I. *Inf. Sci.* **1975**, *8*, 199–249. [CrossRef]
22. Dubois, D.; Prade, H. *Fundamentals of Fuzzy Sets*; Springer: Boston, MA, USA, 2000.

23. Kahraman, C.; Öztaysi, B.; Onar, S.C. A Comprehensive Literature Review of 50 Years of Fuzzy Set Theory. *Int. J. Comput. Intell. Syst.* **2016**, *9*, 3. [CrossRef]
24. Atanassov, K.T. Intuitionistic fuzzy sets. *Fuzzy Sets Syst.* **1986**, *20*, 87–96. [CrossRef]
25. Szmidt, E.; Kacprzyk, J. A consensus-reaching process under intuitionistic fuzzy preference relations. *Int. J. Intell. Syst.* **2003**, *18*, 837–852. [CrossRef]
26. Torra, V. Hesitant fuzzy sets. *Int. J. Intell. Syst.* **2010**, *25*, 529–539. [CrossRef]
27. Kahraman, C.; Onar, S.C.; Oztaysi, B. Fuzzy Multicriteria Decision-Making: A Literature Review. *Int. J. Comput. Intell. Syst.* **2015**, *8*, 637–666. [CrossRef]
28. Chen, S.J.; Hwang, C.L.; Hwang, F.P. *Fuzzy Multiple Attribute Decision Making: Methods And Applications*; Number 375 in Lecture Notes in Economics and Mathematical Systems; Springer: Berlin, Germany; New York, NY, USA, 1992.
29. Liu, Y.; Eckert, C.M.; Earl, C. A review of fuzzy AHP methods for decision-making with subjective judgements. *Expert Syst. Appl.* **2020**, *161*, 113738. [CrossRef]
30. Mota, P.; Campos, A.R.; Neves-Silva, R. First Look at MCDM: Choosing a Decision Method. *Adv. Smart Syst. Res.* **2013**, *3*, 25–30.
31. Figueira, J.; Greco, S.; Ehrgott, M. (Eds.) *Multiple Criteria Decision Analysis: State of the Art Surveys*; Springer: New York, NY, USA, 2005.
32. Saaty, T. *Decision-Making. Analytic Hierarchy Process*; Radio i Svyaz: Moscow, Russia, 1993.
33. Buckley, J. Fuzzy hierarchical analysis. *Fuzzy Sets Syst.* **1985**, *17*, 233–247. [CrossRef]
34. Chen, C.J.; Yang, S.M.; Chang, S.C. A model integrating fuzzy AHP with QFD for assessing technical factors in aviation safety. *Int. J. Mach. Learn. Cybern.* **2014**, *5*, 761–774. [CrossRef]
35. Alkhatib, S.F.; Migdadi, Y.K.A.A. A novel technique for evaluating and ranking green airlines: Benchmarking-base comparison. *Manag. Environ. Qual. Int. J.* **2020**, *32*, 210–226. [CrossRef]
36. Singh, J.; Sharma, S.K.; Srivastava, R. AHP-Entropy based priority assessment of factors to reduce aviation fuel consumption. *Int. J. Syst. Assur. Eng. Manag.* **2019**, *10*, 212–227. [CrossRef]
37. Grzesik, N. Fuzzy sets in aircraft system efficiency evaluation. *Aircr. Eng. Aerosp. Technol.* **2016**, *88*, 707–716. [CrossRef]
38. Ucler, C. Brainstorming the cryoplane layout by using the iterative AHP-QFD-AHP approach. *Aviation* **2018**, *21*, 55–63. [CrossRef]

aerospace

MDPI

Article

Design of an ATC Tool for Conflict Detection Based on Machine Learning Techniques

Javier Alberto Pérez-Castán *, Luis Pérez-Sanz, Lidia Serrano-Mira , Francisco Javier Saéz-Hernando, Irene Rodríguez Gauxachs and Víctor Fernando Gómez-Comendador

ETSI Aeronáutica y del Espacio, Universidad Politécnica de Madrid, Plaza del Cardenal Cisneros, 28008 Madrid, Spain; l.perez@upm.es (L.P.-S.); lidia.serrano@upm.es (L.S.-M.); javier.saez.hernando@alumnos.upm.es (F.J.S.-H.); irene.rodriguez.gauxachs@alumnos.upm.es (I.R.G.); fernando.gcomendador@upm.es (V.F.G.-C.)

* Correspondence: javier.perez.castan@upm.es

Abstract: Given the ongoing interest in the application of Machine Learning (ML) techniques, the development of new Air Traffic Control (ATC) tools is paramount for the improvement of the management of the air transport system. This article develops an ATC tool based on ML techniques for conflict detection. The methodology develops a data-driven approach that predicts separation infringements between aircraft within airspace. The methodology exploits two different ML algorithms: classification and regression. Classification algorithms denote aircraft pairs as a Situation of Interest (SI), i.e., when two aircraft are predicted to cross with a separation lower than 10 Nautical Miles (NM) and 1000 feet. Regression algorithms predict the minimum separation expected between an aircraft pair. This data-driven approach extracts ADS-B trajectories from the OpenSky Network. In addition, the historical ADS-B trajectories work as 4D trajectory predictions to be used as inputs for the database. Conflict and SI are simulated by performing temporary modifications to ensure that the aircraft pierces into the airspace in the same time period. The methodology is applied to Switzerland's airspace. The results show that the ML algorithms could perform conflict prediction with high-accuracy metrics: 99% for SI classification and 1.5 NM for RMSE.

Keywords: air traffic; conflict detection; machine learning; ATC tool

Citation: Pérez-Castán, J.A.; Pérez-Sanz, L.; Serrano-Mira, L.; Saéz-Hernando, F.J.; Rodríguez Gauxachs, I.; Gómez-Comendador, V.F. Design of an ATC Tool for Conflict Detection Based on Machine Learning Techniques. *Aerospace* **2022**, *9*, 67. https://doi.org/10.3390/aerospace9020067

Academic Editors: Spiros Pantelakis, Andreas Strohmayer and Liberata Guadagno

Received: 26 November 2021
Accepted: 20 January 2022
Published: 26 January 2022

1. Introduction

In the same way that air traffic is constantly increasing, technological development is needed to deal with this air traffic demand. Single European Sky ATM Research (SESAR) is the European solution to tackle the evolution of the Air Traffic Management (ATM) system. One of the goals of SESAR is to research the feasibility of using new technologies in the ATM system [1] and, in particular, the Air Traffic Control (ATC) system. Automation is one of the pillars to ensure that traffic increases are managed safely [2]. This paper deals with this issue by developing a new data-driven approach for conflict detection between aircraft.

An accident is the worst event that can occur in ATM [3]. An accident means that the whole system and all safety barriers have failed. A conflict is a precursor of an accident. The ICAO defines a conflict as any situation involving aircraft and hazards in which the applicable separation minima may be compromised [4]. Then, it is crucial to avoid separation infringements that could lead, eventually, to a collision. Since the 1960s, several authors have developed different methods to study conflict and collision risk. Conflict and collision risk are metrics that analyse the level of safety for the ATM system. Reich [5,6] pioneered the development of collision risk modelling (CRM) for parallel routes in oceanic airspace. His work was paramount because it settled the pillars of collision risk based on random flight errors (positioning and velocity). Other authors have employed more or less complex techniques to evaluate collision and conflict risk [7–11]. The authors recommend the work of Netjasov and Janic [12] for a deeper review of CRM.

However, conflict detection differs from conflict risk because it is based on a tactical level (separation provision) instead of a strategic level (airspace design). Conflict detection tools aim to help Air Traffic Controllers (ATCos) to identify conflicts in the airspace. Besides this, conflict detection is typically studied at three different levels: long, mid-, and short term. This work analyses the mid- and short-term horizon (2–10 min) for ATC tools. ATC tools automatize conflict detection, helping ATCos in their labour to avoid separation infringements and reducing their workload. There are many different ATC tool models which were developed for conflict detection. Krozel et al. [13] stated that they can be divided into four areas depending on being static, being dynamic, uncertainty, and probability. Paielli et al. [14] developed a new approach based on tactical pairwise trajectory analysis. Other authors found their studies on complex probabilistic models to understand trajectory uncertainty [15–18].

Here, the applicability of data-driven approaches is analysed for conflict detection. Machine Learning (ML) is one of the branches that takes advantage of historical data. The European Aviation AI high-level group defines ML as "the ability of algorithms to learn from the input and output data that characterise them" [19]. The goal of ML algorithms is to learn patterns from historical situations (databases) that underline those situations. The output of the algorithm is to apply the learned rules to predict new situations. Typically, there are three types of ML algorithms depending on the input/output data required: supervised, non-supervised, and reinforcement learning. Supervised learning demands the output associated with the input data, unsupervised learning acquires patterns from the input data without knowing the output data, and reinforcement learning makes the model learn a sequence of decisions based on rewards and penalties. In addition, ML algorithms can be divided in turn into classification and regression problems. Classification problems separate samples into different classes (binary or multiclassification). Regression problems provide numerical predictions. The authors recommend the works [20,21] to understand these topics.

Aviation is a field with a huge amount of data that is increasingly available for research purposes. ML techniques take advantage of the data, and have recently been applied to different topics: trajectory prediction [22–24], airspace performance metrics [25,26] and atmospheric models [27,28]. One of the issues of employing a data-driven approach is conflict generation. Up until now, conflict simulation has used different methods. Top-down models generate customised conflicts in advance, based on predefined situations [29]. Other approaches perform simulations adding uncertainty to different operational variables (wind, weight, velocity, etc.) [29–32]. The goal is to obtain an extensive database with enough separation infringements, which is not trivial.

Therefore, this work aims to develop the pillars of a further ATC tool for conflict detection based on ML techniques. One of the strong points is the addition of Four-Dimension Trajectory (4DT) predictions based on historical ADS-B trajectories. This allows us to improve the ability to predict conflicts based on 4DT predictions. After the introduction, the framework for the ATC tool based on ML techniques is presented, and the data-driven approach is explained for the extraction of trajectories to constitute the database. Section 3 details the selected ML techniques and the process followed in order to apply them. Section 4 shows the results obtained for the classification and regression techniques. Conclusions are presented in Section 5.

2. Framework for ATC Tools Based on ML Techniques

A conflict detection ATC tool aims to provide information to the ATCo about aircraft pairs which are expected to infringe the separation minima. The ATC tool developed in this work is based on a data-driven approach using ML techniques. Here, it is essential to note that the ATC tool provides predictions focusing on conflict detection but not trajectory prediction. This means that the ML model does not perform the trajectory prediction, nor does it afterwards analyse if a conflict could occur; it performs the conflict prediction based on what happened in previous situations. This novel approach employs ML algorithms to learn the underlying factors that lead to separation infringement by including historical 4D trajectories as predictions. The primary characteristics and hypothesis are the following:

1. The ATC tool makes predictions based on the evolution of the aircraft throughout the airspace. The predictions depend on the time (t) and the operational features of both aircraft. The ATC tool updates the prediction every c minutes.
2. The ATC tool obtains the aircraft within the sector and the aircraft in proximity of the airspace that will penetrate the airspace in the following m minutes. c and m values should be defined by Air Navigation Service Providers (ANSPs) in advance.
3. The ATC tool receives a 4DT prediction for each aircraft based on historical data and the operational features (state vector) based on ADS-B data. The ML model uses both inputs.
4. The ATC tool was developed for the tactical ATCo, which continuously monitors the state of the aircraft throughout the airspace.
5. The ATC tool does not provide information on conflict resolution.

Figure 1 represents the operational concept of the ATC tool. Both aircraft i and j are flying in the airspace. The ATC tool performs conflict prediction between them, along with other aircraft flying in the airspace and the aircraft that will pierce in m minutes. The ATC tool demands two types of information: the state vector at the moment of the prediction and 4DT predictions. The state vector of each aircraft is accessible by different data sources based on ADS-B. However, the main problem is obtaining 4DT predictions of the aircraft. Typically, 4DT predictions are calculated by the ground server or provided by the aircraft throughout the flight. We do not have access to this type of prediction, and it should be one of the topics to research in further work. Hence, 4DT predictions are obtained by historical 4DT trajectories that fly in the same airspace. It is assumed that aircraft have flown similar trajectories in the airspace in a recent period of time.

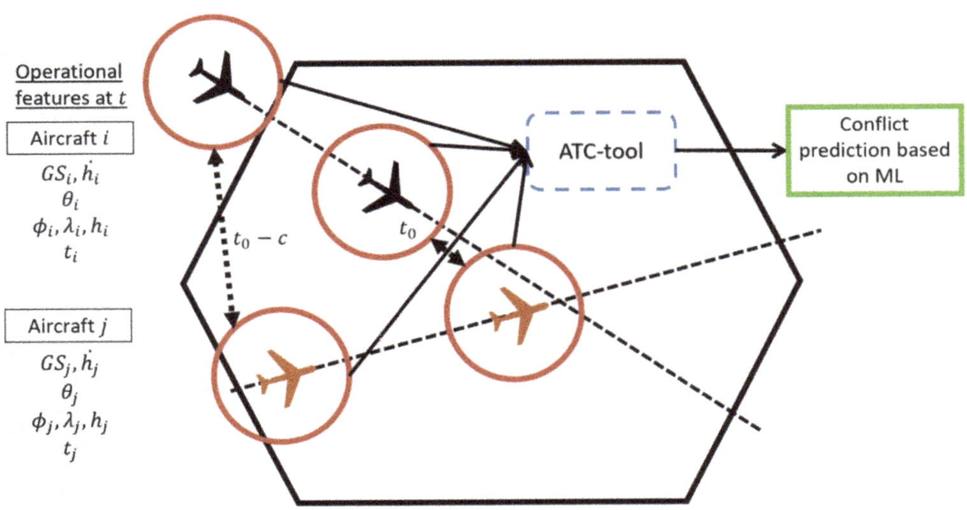

Figure 1. Operational concept of the ATC tool.

2.1. Conflict Detection Principles

Conflict detection aims to identify or detect aircraft pairs which are expected to infringe the separation minima in the short term (2–5 min). The current separation minima in the en-route airspace are 5 Nautical Miles (NM) horizontally (S_{min}) and 1000 feet (ft) vertically (H_{min}). The concept of the short term is ambiguous because it depends on the time frame considered. The strategical time frame (from several hours to one year in advance) is different from the tactical time frame (during the aircraft operation). This work focuses on a tactical time frame because it tries to develop a tool to help ATCos detect conflicts in the airspace.

One of the primary issues that affects ATC tools for conflict detection is the definition of the limits to inform us about whether there is a conflict or not. The separation monitoring of an aircraft pair throughout the airspace is not accurate because it presents several uncertainties based on navigation and surveillance. Two types of situations can arise due to the poor performance of conflict detection.

- Missed Alerts: The ATC tool does not inform us about the separation infringement between an aircraft pair because the system erroneously calculates that no separation infringement will occur, when in fact it will.
- False Alerts: The ATC tool informs us about the separation infringement between an aircraft pair because the system erroneously estimates a separation infringement when in fact it is not one.

Typically, ATC tools expand the limits of separation infringement to avoid critical missed alerts that identify aircraft pairs that cross with a separation larger than the current separation minima. To this end, this work uses the concept of a Situation of Interest (SI). One SI is a situation in which an aircraft pair is expected to intersect with a horizontal separation smaller than a predefined distance. Typically, this predefined separation is specified by the ANSP, and is larger than the current separation minima. In this work, an SI was defined as 10 NM.

Therefore, this work evaluates two metrics regarding the separation minima reached by a pair of aircraft.

1. Minimum Distance ($MinDis$): This is the minimum separation reached by an aircraft pair $(a_i,\ a_j)$. This variable considers the horizontal separation (s) and vertical separation (Δh), and provides information about the severity of the SI. The $MinDis$ considers two cases depending on the vertical separation—(1) the aircraft pairs cross with a vertical separation lower than the separation minima (H_{min}), and (2) the aircraft pairs intersect with a vertical separation higher than the vertical separation minima—and it is transformed considering a horizontal scale:

$$\begin{aligned} MinDis &= \min(s(a_i,\ a_j)\ if\ \Delta h < H_{min} \\ otherwise\ MinDis & \\ &= s(s(a_i,\ a_j) = \min(s(a_i,\ a_j)) \\ &+ \Delta h(s(a_i,\ a_j) = \min(s(a_i,\ a_j)))S_{min}/H_{min} \end{aligned} \tag{1}$$

2. Aircraft pairs are denoted as being of interest (SI) when they cross with a vertical separation lower than H_{min} and a horizontal separation $s < s_{SI}$:

$$\begin{aligned} if\ \Delta h(a_i,\ a_j) < H_{min}\ and\ s < s_{SI}\ for\ the\ same\ t \rightarrow \\ SI = 1,\ otherwise\ SI = 0 \end{aligned} \tag{2}$$

2.2. Data-Driven Approach

One requirement for ML techniques is to acquire a database from which the ML algorithm could learn the underlying patterns in order to perform predictions. The database must have enough historical situations based on actual aircraft pair trajectories. The selected data source affects the constitution and performance because each provides different information. Here, the data source is ADS-B trajectories from the OpenSky Network [33].

The first step is to train an ML predictor to perform conflict predictions for each pair of aircraft. To this end, a database must be used to train the ML models. This database should have enough situations to represent the different situations that can arise between pairs of aircraft. As such, the number of cases considered in the database is paramount because the larger the number of them, the better the prediction ability will be. On the contrary, increasing the number of samples means a higher computational time.

Typically, the database is split into two sets: the training set and the testing set. The ML model uses the training set to learn the underlying relations and develop a mathematical model to make predictions. The testing set is used to evaluate the performance of the trained model for new instances. The features considered in this work come from the ADS-B information:

- Position: Longitude (λ), latitude (ϕ) and altitude (h).
- Velocity: Ground speed (GS) and vertical rate ($\dot{h}$).
- Heading (θ).
- Target variables based on 4DT predictions.

Although there are other features from ADS-B, they do not provide information about the state vector of the aircraft. The labels or targets (variables to predict) are:

- Minimum distance ($MinDis$): A numerical variable of the minimum distance expected to reach between an aircraft pair.
- Situation of interest (SI): Binary variable that classifies aircraft pairs as SI or $No\ SI$.

2.3. ADS-B Traffic from the OpenSky Network

This work uses ADS-B trajectories from the OpenSky Network [33]. Raw data downloaded from the OpenSky network cannot be fed immediately to the ML algorithm. ML algorithms require previous filtering and adjustment to be used. Data preparation performs the following actions:

- Cleaning trajectories that present errors in the ADS-B data. Errors in the ADS-B data are identified by the OpenSky functions [33] and represent faulty data such as callsigns or repeated positions. Trajectories with more than 10 ADS-B errors are removed. Duplicate trajectories are also removed.
- Trajectories are gathered for each day and time period. One operational day is split into nine time periods based on the traffic distribution throughout the operational day. The first period covers from 0 to 8 a.m., and the traffic is removed because the LSAZM567 is not opened (due to the low traffic); the rest of the time periods are sets of two hours, and assume similar weather conditions in the simulations.
- The generation of aircraft pairs based on the combinatorial problem.
- Relative variables that combine operational features for each aircraft pair are calculated.
- The generation of conflicts between aircraft pairs and the calculation of conflict metrics.

Figure 2 summarises the whole process.

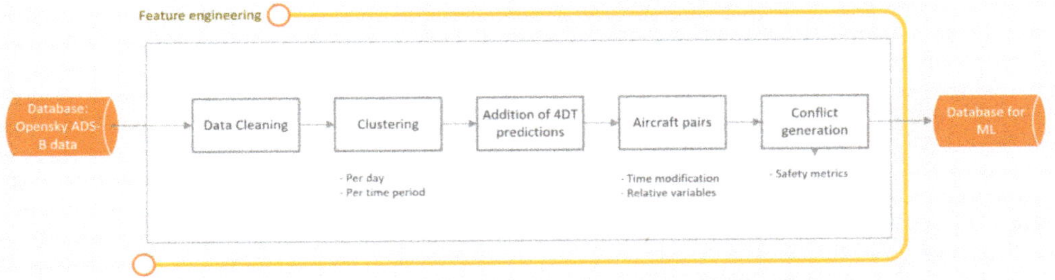

Figure 2. Flow of the feature engineering process.

Weather or operational variables of airspace status were not considered, and should be included in further work.

2.4. Historial ADS-B Database as 4DT Predictions

The operational concept considers 4DT predictions of the aircraft trajectories as inputs for the ML models. Each aircraft provides a trajectory prediction from the most basic information (the flight plan) until a 4DT prediction is calculated by the ground system. It was not possible to obtain this type of prediction because there was no access to the ground server or flight plans correlated with the ADS-B data. Therefore, the 4DT predictions were assumed to be historical trajectories stored from ADS-B data. It was assumed that the 4DT prediction can match with a similar trajectory stored in the ADS-B database. In case the system has another type of prediction, the process will be the same.

The main problem of using stored trajectories as 4DT predictions of other aircraft is ensuring the similarity between them. When one aircraft pierces into the airspace, the algorithm performs a search in the historical database in order to identify the most suitable trajectory that fulfils the following requirements:

1. The first filtering is about selecting one trajectory with the same callsign. Typically, aircraft repeat their trajectories.
2. The second filtering considers operational restrictions such as the ground speed, heading, and location of the entry point:
   - The GS difference must be lower than 10 knots.
   - The heading difference must be lower than 2°.
   - The location difference of the entry points must be lower than 5 NM.
3. In case there is more than one trajectory that fulfils the previous restrictions, it will select the trajectory with a lower heading difference.
4. If none of the trajectories with the same callsign satisfy the previous restrictions or there are no trajectories with the same callsign, the algorithm extends the search to other callsigns. The main issue is that the computational time to find a similar trajectory increases exponentially.

Figure 3 represents the selection of a 4DT prediction from the historical database that is the most similar to one random trajectory when the aircraft pierces into the airspace.

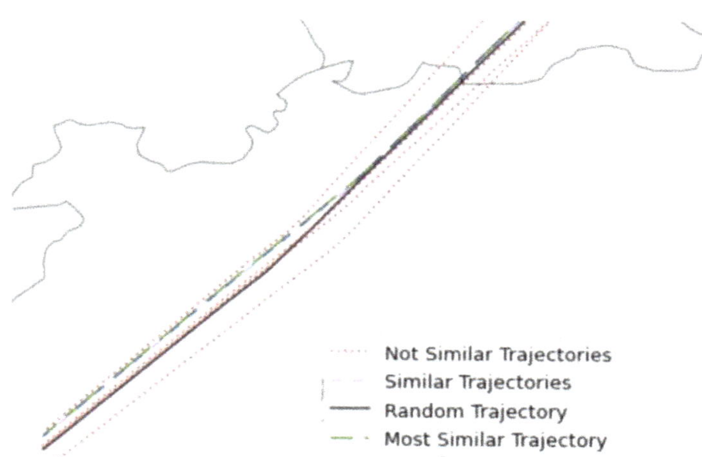

Figure 3. Representation of the selection of 4DT predictions.

Once the 4DT prediction is selected from the ADS-B database, two modifications are made to ease the interoperability. The first modification is about a resample of the 4DT prediction to reduce the number of samples that constitute the prediction. The goal of the resample is to reduce the computational workload and the size of the final database to consider due to the high number of samples. The resample is not an issue in the case where the aircraft fly straight lines, but if the aircraft is turning or performing a manoeuvre, it will not be reflected correctly.

The second modification aims to introduce uncertainty by adapting the 4DT prediction entry time to the actual trajectory. This uncertainty is a temporal deviation (τ) following a random distribution of ± 20 s. This is a value selected by the authors, and its suitability should be studied in further work. The higher this value, the larger the error between the actual trajectory and the 4DT prediction. The result is a database of 4DT trajectories composed of the actual trajectory and a similar trajectory considered as the 4DT prediction.

2.5. Generation of the Aircraft Pairs and Conflicts

The main limitation of the analysis of aircraft pairs extracted from ADS-B data is the lack of conflicts in real situations. Separation infringements safely occur in rare situations and under specific circumstances [34]. Conversely to these rare situations, ML models need a large and diverse database in order to learn the patterns that underlie separation infringement. Therefore, the first need is to generate a database with sufficient samples of conflict situations.

There are two ways to simulate conflicts. First, we can define the conflict parameters and then simulate backwards in order to obtain the conflicts which were previously defined. However, this solution does not encompass some operational parameters, such as airspace design or air traffic distribution. The second solution considers a real scenario and introduces modifications to the real trajectories in order to force separation infringements. These simulated conflicts are not real conflicts, but represent situations that could occur. These can be assumed to be pre-tactical conflicts based on real trajectories that consider many operational parameters, such as airspace design, speed uncertainty, and wind, etc. This work adopts the second approach and performs simulations that modify the entry time of the aircraft in a limited time period. In this way, an algorithm was developed to generate conflict situations by modifying their entry time. The process is as follows:

(1) First, each aircraft modifies its entry time randomly between the temporary boundaries (t_{entry}^{init}, t_{entry}^{end}). The temporary restriction is that every aircraft should be in the airspace between the limits t_{entry}^{init} and t_{entry}^{end}. The new entry times are calculated randomly for each aircraft.
(2) Secondly, the aircraft pair generation is constituted. Suppose that there is a set of n trajectories that constitutes $(n-1)n$ aircraft pairs. For each aircraft i is generated an $(n-1)$ aircraft pair to evaluate.
(3) For each aircraft pair at each time t, the database is constituted by storing the operational features of each aircraft pair.
(4) Relative variables and conflict metrics are calculated for each aircraft pair, both for the actual trajectory and the 4DT prediction. Conflicts when one aircraft pierces into the airspace are discarded because this would imply that the ATCo from the previous airspace would not have done his work correctly.

Figure 4 represents this temporary modification and the generation of pairs for aircraft i. Aircraft i pierces at time t_{a_i}, and the rest of aircraft receives aleatory new entry times between the time boundaries. The constitution of the aircraft pairs is repeated for every aircraft, avoiding duplicity.

Relative variables provide information about any situation that combines the operational features of a pair of aircraft. Relative variables are calculated at each timestamp when operational information is available. The mathematical descriptions of the relative variables are as follows:

- Horizontal separation ($s_{i,j}(t)$) is the horizontal separation between the locations of an aircraft pair:

$$s_{i,j}(t) = position_i(t) - position_j(t) \tag{3}$$

- Vertical separation ($\Delta h_{i,j}$) is the altitude variation (h_i, h_j) between an aircraft pair:

$$\Delta h_{i,j}(t) = h_i(t) - h_j(t) \tag{4}$$

- Course ($\gamma_{i,j}$) is the course that links the locations between an aircraft pair:

$$\gamma_{i,j}(t) = \gamma_i(t) - \gamma_j(t) \tag{5}$$

- Track variation ($\Delta\theta_{i,j}$) is the track variation (θ_i, θ_j) between an aircraft pair:

$$\Delta\theta_{i,j}(t) = \theta_i(t) - \theta_j(t) \tag{6}$$

- Ground Speed (GS) variation ($\Delta GS_{i,j}$): is the difference in the module of the GS (GS_i, GS_j) between an aircraft pair:

$$\Delta GS_{i,j}(t) = |GS_i(t) - GS_j(t)| = \sqrt{\left(GS_i^x - GS_j^x\right)^2 + \left(GS_i^y - GS_j^y\right)^2} \tag{7}$$

- Vertical rate variation ($\Delta \dot{h}_{i,j}$) is the variation of the vertical rate ($\dot{h}_i, \dot{h}_j$) between an aircraft pair.

$$\Delta \dot{h}_{i,j}(t) = \dot{h}_i(t) - \dot{h}_j(t) \tag{8}$$

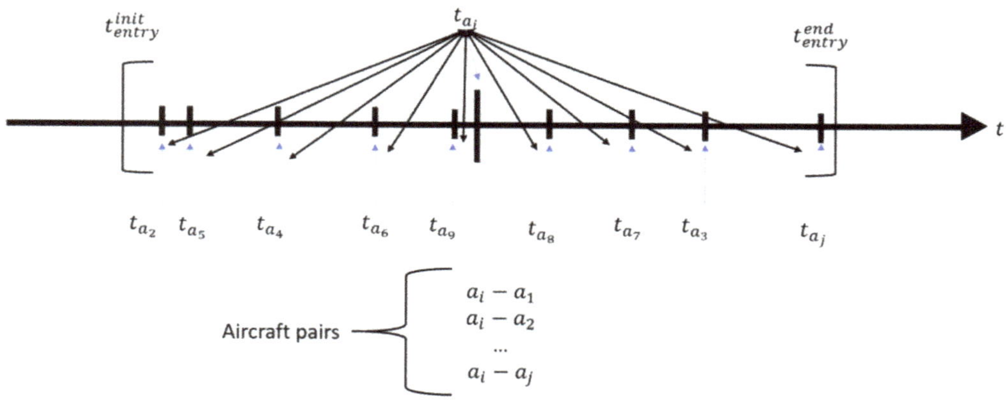

Figure 4. Generation of the aircraft pairs with temporary modifications.

The library *geographiclib* was used to calculate the relative variables based on aircraft positioning (longitude and latitude) because the library performs the calculations without transforming the aircraft positioning to Cartesian coordinates [35]. Therefore, 40 variables constitute the ML database: 12 ADS-B variables (six for each aircraft), six relative variables, two labels, and the same for the 4DT prediction.

2.6. Database for ML

The above process allows the performance of simulations to generate a database that could be used for ML models. The ML database presents situations of aircraft pairs that could constitute SI or not.

This work was developed by the AISA project, and focused on the LSAZM567 sector of the Switzerland airspace (from FL355 to FL660) [36]. LSAZM567 airspace is an en-route airspace where most of the airways are composed of straight lines. The 4DT were downloaded from the AIRAC cycle from 20 June 2019 to 04 July 2019 (15 days). Approximately 12,500 trajectories were downloaded, containing more than 7e6 samples (each sample represents an aircraft position). Table 1 shows the results for all of the simulations.

Table 1. Results of the simulations.

Variables	Number
Number of Aircraft	12,549
Number of pairs generated	1,257,234
Number of samples considered	675,3283
Samples with *SI* (%)	1,215,152 (18%)
Samples with conflict (%)	270,132 (4%)
Computational time (hours)	420

The number of aircraft varied from 80 to 120 for a one-hour period. The number of pairs generated was far from the whole combinatorial number of aircraft pairs because the simulations were performed for specific time periods. The total computational time required to perform the simulations was more than 420 h. This shows one of the issues that should be improved in future developments. Lastly, the simulations were performed in computer with Intel® Core™ i5-6600 CPU @ 3.30 GHz, RAM 8.00 GB, and 64 bits.

Figure 5 shows the histograms of the regression and classification targets.

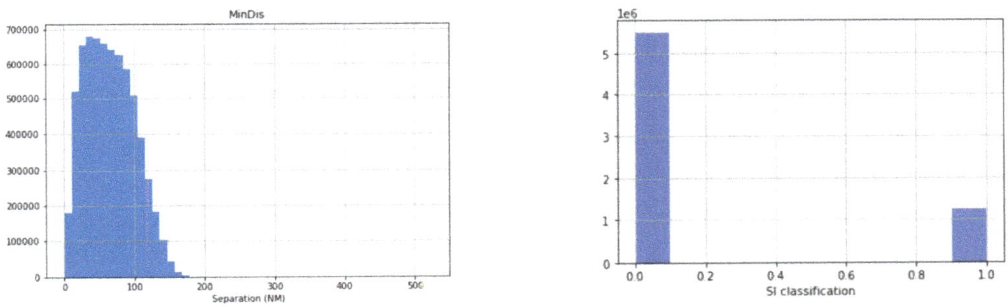

Figure 5. Histograms of the regression and classification targets.

As is shown in Figure 5, the *MinDis* range of the distribution is from zero to more than 150 NM. The most significant density is between 10 and 20 NM. The *MinDis* located at 0 NM refers to aircraft pairs that move away. Regarding the classification, there is a high imbalance between the *SI* and *No SI* samples and conflict samples. These results show the problem of generating conflicts despite using specific simulations for this purpose.

3. Machine Learning Techniques

One of the current technological developments is the implementation of data-driven techniques, and particularly ML, which is one of the most promising technologies in data prediction. As explained above, this approach tackles two types of ML problems: classification and regression. The classification problem aims to classify aircraft pairs into *SI* or *No SI* depending on the minimum separation that they are expected to reach. The regression problem seeks to predict the numerical value of the minimum separation they are expected to reach. It is important to note that both problems are solved independently. This allows one to analyse the validity of the combinations based on both predictions.

ML models are currently implemented in different programming languages. This work was developed in Python®. Scikit-learn is the library which was used to implement ML algorithms [37], and PyCaret is another library that eases this process [38].

3.1. ML Experiments

As the ML database shows, there is a high imbalance between the *SI* and *No SI* samples. This is an issue in classification problems because the ML algorithm performs best with balanced databases. In order to tackle this problem, we considered a Hybrid model in which a filtering process based on operational restrictions was introduced to reduce the strong imbalance of the dataset. Then, there are two experiments:

1. Pure model: The ML model considers the whole database without any modification.
2. Hybrid model: The ML model considers a previously filtered database. The filtering is performed on the basis of operational restrictions. The database only contains aircraft pairs that cross with a horizontal separation of less than 20 NM and a vertical separation of less than 1000 ft.

The ML process performs the following steps to obtain the best ML model:

1. The analysis of different ML algorithms: The first step is to evaluate the performance of different ML algorithms in order to identify the ML model that provides better results. In total, 15 ML algorithms were assessed for classification and regression.
2. Feature selection: The feature selection identifies the influence of the different features that build the database, and analyses how they affect the model. The feature selection is performed based on graphical analysis and Recursive Feature Elimination (RFE) with CV [39]. RFE analysis evaluates the impact of the features on the model accuracy. Finally, features without influence on the ML model are removed from the database.
3. ML optimisation: The optimisation process aims to improve the algorithm's performance based on a grid search of the hyperparameters. The hyperparameters must be set up in advance of the training model, and can be defined as the settings of an algorithm. The metric to be optimised depends on the classification and regression problem.

The result of this process is the readiness of two trained and tested ML algorithms for conflict detection.

3.2. ML Database Preparation

The ML process requires the preparation of the database prior to feeding it to the algorithms. Data preparation modifies the raw data into valuable data for ML algorithms. Here, the following preprocessing activities are applied:

- The training and testing sets: The dataset is divided into the training and testing datasets. The testing set (also known as the hold-out set) works as a proxy for new data. It is not used in model training, and can be used to evaluate the model metrics' suitability. The database is split into 70% for training and 30% for the testing set.
- Normalisation: Normalisation rescales the values of the numeric columns following a normal distribution. It does not distort the differences in the range of values.
- Shuffling: This distributes the samples randomly into the training and testing datasets.

1. Stratification: Apart from randomly distributing the samples, the stratification process spreads the samples whilst maintaining the SI statistical distribution.
2. Cross-validation: This is a process which is used to avoid the overfitting of the ML model [39]. The training set is divided into k folds: one is considered the validation set, and the rest are considered the training set. This process is repeated for every k-fold, and the metrics are calculated based on the mean and standard deviation.

Table 2 shows the number of samples for both experiments; the rates of $SI/No\ SI$ samples appear in parentheses. Each sample represents the situation of one aircraft pair at one specific time.

Table 2. Number of samples for each experiment and set.

Model	Training Set	Testing Set
Pure model (18/82)	4,254,567	1,823,387
Hybrid model (68/32)	1,140,470	488,773

3.3. ML Algorithms and Metrics

Currently, there are several state-of-the-art ML algorithms, and we evaluate some of them. The evaluation of different algorithms allows one to identify the best among them. The goal of this section is not to explain the algorithms employed, but to denote which were trained: Extra Trees, Random Forest, Extreme Gradient Boosting, Light Gradient Boosting, CatBoost, Decision Tree, K Neighbors, Gradient Boosting, Ada Boost, Naïve Bayes, SVM—Linear Kernel, Logistic Regression, Ridge Classifier, Linear Discriminant Analysis, and Quadratic Discriminant Analysis. The authors encourage readers interested in the mathematical ML definitions to see [20]. Other ML techniques such as Deep Learning could improve the results, and should be evaluated in further work.

ML metrics are different for classification and regression problems:

- Classification problems include accuracy, precision, recall and F1. The most important is the F1 metric that leverages precision and recall. Accuracy is not a good option because the database is highly imbalanced.
- Regression problems include the Mean Absolute Error (MAE), Root Mean Square Error (RMSE), R2 and Root Mean Square Logarithmic Error (RMSLE). The most important metric is the RMSE, which increases the impact of the prediction distance from the true values.

4. Results

This section provides the results of the experiments, and are split into classification and regression. In addition, the performance of both predictors is analysed for different separation infringement stretches.

4.1. ML Techniques for Classification

ML classification techniques predict whether an aircraft pair is SI or not. The first step is to analyse the performance of different ML algorithms, aiming to identify the best one. The results of the different ML algorithms are shown in 'Appendix A ML results' in order to improve the reading of the paper. Both experiments agree that the three top algorithms vary between Extra Trees, Random Forest and Decision Tree, i.e., ensemble models. The random forest provided the best results for both experiments. The models were evaluated on the basis of stratified cross-validation techniques in order to avoid overfitting.

The feature selection for both experiments concluded that the most influential features were the prediction about the minimum distance and the SI prediction. The sum of both of them reached around 70% of the influence of the variable. The variation of separation is also important (up to 10%), and the rest of the features provide a similar impact between 1 and 5%. As can be seen in Figure 6, the feature selection is similar for both experiments (Pure and Hybrid models). This similarity confirms the significance of the prediction variables in the ML algorithm. The variables labelled '_n' refer to the aircraft trajectory, and '_re' refers to the 4DT prediction.

The next step is to optimise the Random Forest algorithm. The database is imbalanced in both experiments because the distribution of SI samples is not balanced. This issue is tackled by implementing cost-sensitive techniques during the optimisation process. We studied the optimisation of recall, precision and F1. However, the most balanced results were obtained by optimising the F1 metric. Table 3 shows the results of both experiments in the training set applying a fivefold cross-validation.

Table 3. Results of the classification experiments.

Experiment	Accuracy (std)	Recall (std)	Precision (std)	F1 (std)
Pure model	0.990 (0.01)	0.987 (0.01)	0.989 (0.02)	0.988 (0.01)
Hybrid model	0.991 (0.001)	0.990 (0.001)	0.991(0.001)	0.990 (0.002)

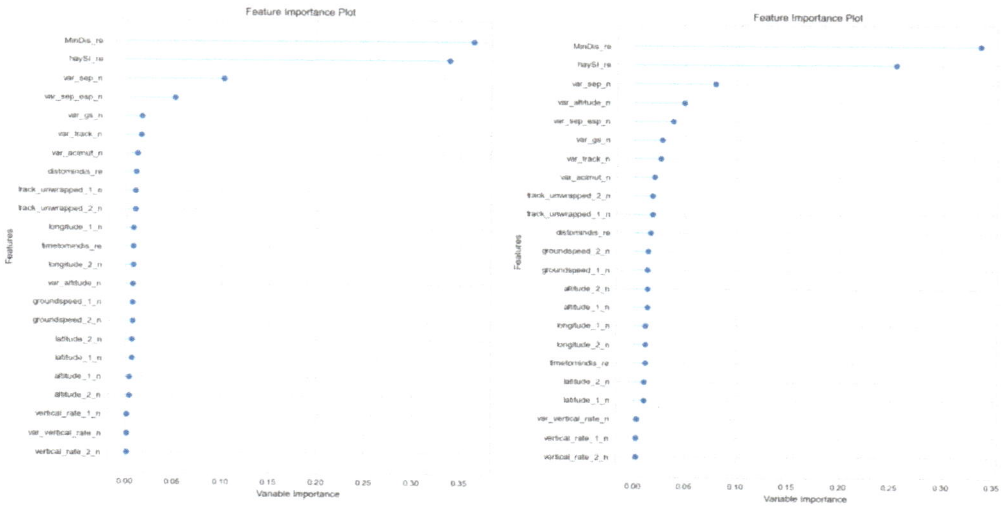

Figure 6. Feature importance of the Pure model (**left**) and Hybrid model (**right**) for classification.

The following conclusions were obtained:

- The optimisation process barely improved the initial metrics of the ensemble models. The initial values obtained for the ensemble models were very high, and the improvements obtained by optimising the hyperparameters were reduced by up to 1%. This implies that identifying the correct model is crucial because the subsequent optimisation process does not provide substantial gains.
- The metrics of both models were extremely high and similar; their rates were almost 99% for every metric. Although the Hybrid model provides better metrics than the Pure model, the difference was less than 2%. This implies that the usage of unbalanced datasets does not greatly affect classification purposes by implementing cost-sensitive techniques.

These results confirm the goal of this research on the introduction of ML techniques. Therefore, ML algorithms learn from the initial prediction and improve the quality of conflict prediction.

4.2. ML Techniques for Regression

ML regression techniques predict the numerical value of the minimum separation to be reached by an aircraft pair. The first step is to analyse the performance of different ML algorithms, aiming to identify the best one. The results of the different ML algorithms are shown in 'Appendix A ML results' in order to improve the reading of the paper. Both experiments agree that the three top algorithms vary between Extreme Gradient, Light Gradient and Gradient Boosting, i.e., ensemble models. Extreme Gradient provides the best results for both experiments. The models were evaluated based on stratified cross-validation techniques in order to avoid overfitting the results.

Both experiments concluded that the most influential feature is the prediction of the minimum distance. However, the influence varied from the Hybrid model (over 60%) to the Pure model (almost 100%). The rest of the variables provided influence values lower than 1% for the Pure model; as such, the Pure model was analysed without them. However, the Hybrid model requires more variables, such as the variation of the altitude and the horizontal separation among others. Figure 7 shows these results and does not confirm the similarity identified in the feature importance of the classification experiment. In addition, the analysis shows a performance decrease of more than 2%, which does not recommend removing the features.

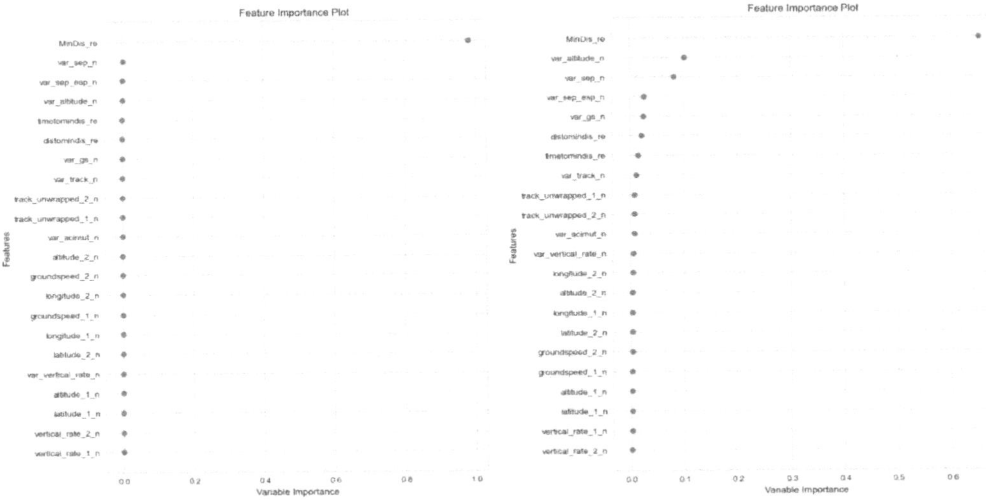

Figure 7. Feature importance of the Pure model (**left**) and the Hybrid model (**right**) for regression.

The next step is to optimise the Extreme Gradient algorithm. Only the optimisation of RMSE was studied. Table 4 shows the results of both experiments on the training set applying a fivefold cross-validation.

Table 4. Results (NM) of the regression experiments.

Experiment	RMSE	MAE	R2	RMSLE
Pure model	2.579	1.601	0.994	0.178
Hybrid model	1.473	0.992	0.934	0.235

The following conclusions were obtained:

- The optimisation process improves the initial metrics by up to 10% in some experiments. This implies that the behaviour of the Extreme Gradient model greatly depends on the optimisation process, conversely to the random forest during the classification optimisation.
- The results differed between the Pure and Hybrid models. The best results were obtained for the Hybrid model (1.5 NM). The Pure model worsens the RMSE over 1 NM, which implies an increase of over 40% compared to the Hybrid model.
- These results are different from the classification problem because cost-sensitive techniques cannot be applied to the Pure model. The higher imbalance of the Pure dataset means a clear decrease in the metrics of the regression predictors.

4.3. Analysis of ML Technique Prediction

This section analyses the performance of the predictions of the ML techniques between them in the testing set. The goal is to identify which predictions the ML models (classification and regression) tend to fail or guess right and match. Therefore, the predictions of both models will be compared considering only the Hybrid model that provided the best results. The predictions are split into four ranges in order to differentiate between SI and conflict.

Table 5 shows the allocation of the classification and regression predictions in the four ranges. The analysis concludes that the predictions fail at the 10 NM border. In case both ML predictors confirm a conflict, the conflict probability is 100%. In opposition, the probability is 100% when both predictors discard an SI. However, the samples with a predicted value of 10 to 15 NM only guess 82% right. The samples with a predicted value from 5 to 10 NM guess 97% right. It can be concluded that the model only fails in the classification when the predicted value of $MinDis$ is between 5 and 15 NM, especially if it is between 10 and 15 NM.

Table 5. Distribution of the classification and regression predictions.

Experiments	*MinDis_ML* = (0,5]	*MinDis_ML* = (5,10]	*MinDis_ML* = (10,15]	*MinDis_ML* = (15,20]
$SI\_ML = 1$	100%	97%	18%	0%
$SI\_ML = 0$	0%	3%	82%	100%

Table 6 makes the same comparison between the classification predictions and the real values of the samples.

Table 6. Distribution of the classification predictions and real values.

Experiments	*MinDis_Real* = (0,5]	*MinDis_Real* = (5,10]	*MinDis_Real* = (10,15]	*MinDis_Real* = (15,20]
$SI\_ML = 1$	100%	91%	6%	5%
$SI\_ML = 0$	0%	9%	94%	95%

Although the values are quite similar, there are some differences compared to Table 5:

- The percentages of correctly classified samples between 0 and 5 NM are the same (100%). These results confirm that training the model using SI instead of conflict provides a 100% guess of conflict.
- The model predictions fail at the 10 NM limit. The predictions from 5 to 15 NM present higher misclassification errors. However, when we use both predictions (classification and regression), the misclassification errors decrease from 5 to 10 NM by increasing the false rate from 10 to 15 NM.

The errors considered from 5 to 10 NM are samples considered from 10 to 15 NM; these errors do not impact the aircraft pairs classified as SI.

5. Conclusions

This article develops the pillars for the further development of an ATC tool based on ML techniques. This work is framed on a Technology Readiness Level 1–2 based on the European project to which it belongs. It was developed in order to analyse the viability of introducing ML techniques for conflict detection purposes by analysing their accuracy and expected errors. The introduction of ML techniques in safety-critical areas in air traffic management is very complex, and must be properly analysed in order to ensure the feasibility of these techniques. This paper develops a data-driven approach that considers the evolution of the aircraft within the airspace, and the separation evaluation along with it. The trajectories were extracted from the OpenSky Network based on ADS-B information. One of the difficulties was the generation of conflict. One limitation of using real ADS-B trajectories is that the trajectories that suffered a tactical modification by the ATC action could not be removed.

Real ADS-B trajectories were temporarily modified, considering specific aircraft sets in the same time period, to constitute the SI. In addition, historical ADS-B trajectories were used for 4DT predictions as inputs for the ML models. The introduction of 4DT predictions as an input is crucial in order to obtain such high metrics. One of the strong points of this approach is the use of real ADS-B trajectories to constitute the database. Using real trajectories means that this work could be developed for real scenarios. Moreover, the introduction of 4DT trajectories presents a twofold implication: (1) it shows the clear improvement provided by introducing this information, and (2) this methodology could be developed for other types of predictions (such as pretactical trajectories from the Network Manager or calculated by the ATC ground system).

The results confirmed that ML models can be applied to perform conflict prediction. The best algorithms for both problems are the ensemble methods. Notably, Random

Forest for classification and Extreme Gradient for regression provided the best results in the experiments. The main problem is the large dataset used to train the models, which demanded substantial computational time and computer resources. Classification techniques reach a success rate of more than 99% for SI prediction and regression techniques up to 1.5 NM in RMSE. In addition, both ML models were identified to have ensured 100% success in conflict detection (a separation infringement lower than 5 NM) combining both predictions. These results confirm the appropriateness of using 10 NM as the boundary to train the model instead of 5 NM.

Finally, this work brings to light the possibility of introducing ML techniques for conflict detection purposes, although it deals with some limitations in different areas due to its novelty. Further work should focus on: (1) the introduction of Deep Learning based on neural networks, (2) the introduction of operational and environmental variables not considered herein, and (3) the introduction of other sources for 4DT predictions as an input.

Author Contributions: Conceptualization, J.A.P.-C. and L.P.-S.; methodology, J.A.P.-C., L.P.-S. and L.S.-M.; software, F.J.S.-H. and I.R.G.; validation, F.J.S.-H. and I.R.G.; investigation, J.A.P.-C. and L.S.-M.; data curation, F.J.S.-H. and I.R.G.; writing—original draft: J.A.P.-C.; writing and review—L.P.-S., L.S.-M. and V.F.G.-C.; funding acquisition, J.A.P.-C. and V.F.G.-C. All authors have read and agreed to the published version of the manuscript.

Funding: This paper is part of a project that has received funding from the SESAR Joint Undertaking under grant agreement No 892618 under the European Union's Horizon 2020 research and innovation programme.

Institutional Review Board Statement: Not applicable.

Informed Consent Statement: Not applicable.

Data Availability Statement: Data is available upon request.

Acknowledgments: Notably, the authors would like to acknowledge Tomislav Radisic from FTTS and Lars Schmidt from TUBS.

Conflicts of Interest: The authors declare no conflict of interest.

Appendix A. ML Results

Table A1 shows the results for the different ML classification algorithms obtained for the Pure model.

Table A1. Classification metrics for the Pure model.

Model	Accuracy	AUC	Recall	Prec.	F1	Kappa
Random Forest Classifier	0.987	0.999	0.954	0.971	0.962	0.952
Extra Trees Classifier	0.986	0.999	0.955	0.972	0.961	0.951
CatBoost Classifier	0.984	0.998	0.953	0.962	0.957	0.948
Decision Tree Classifier	0.983	0.973	0.955	0.954	0.955	0.945
Extreme Gradient Boosting	0.981	0.998	0.943	0.953	0.948	0.936
Light Gradient Boosting Machine	0.978	0.997	0.934	0.943	0.939	0.925
Gradient Boosting Classifier	0.975	0.995	0.927	0.935	0.931	0.916
Ada Boost Classifier	0.973	0.993	0.921	0.928	0.925	0.908
K Neighbors Classifier	0.970	0.987	0.910	0.923	0.916	0.898
SVM—Linear Kernel	0.966	0.000	0.902	0.909	0.906	0.884
Ridge Classifier	0.966	0.000	0.902	0.909	0.906	0.884
Linear Discriminant Analysis	0.966	0.984	0.902	0.909	0.906	0.884
Logistic Regression	0.965	0.987	0.902	0.908	0.905	0.884
Naive Bayes	0.965	0.979	0.904	0.902	0.903	0.882
Quadratic Discriminant Analysis	0.960	0.980	0.906	0.877	0.891	0.866

Table A2 shows the results for the different ML classification algorithms obtained for the Hybrid model.

Table A2. Classification metrics for the Hybrid model.

Model	Accuracy	AUC	Recall	Prec.	F1	Kappa
Random Forest Classifier	0.984	0.998	0.984	0.989	0.985	0.952
Extra Trees Classifier	0.982	0.989	0.982	0.988	0.981	0.949
Decision Tree Classifier	0.972	0.967	0.979	0.979	0.979	0.935
CatBoost Classifier	0.964	0.994	0.973	0.974	0.974	0.918
Extreme Gradient Boosting	0.954	0.991	0.966	0.967	0.966	0.894
K Neighbors Classifier	0.951	0.987	0.965	0.963	0.964	0.886
Light Gradient Boosting Machine	0.943	0.987	0.958	0.959	0.959	0.870
Gradient Boosting Classifier	0.935	0.982	0.954	0.952	0.953	0.851
Ada Boost Classifier	0.926	0.976	0.941	0.950	0.946	0.831
Logistic Regression	0.894	0.949	0.914	0.930	0.922	0.759
Naive Bayes	0.894	0.943	0.905	0.937	0.921	0.759
SVM—Linear Kernel	0.894	0.000	0.902	0.941	0.921	0.762
Ridge Classifier	0.894	0.000	0.902	0.941	0.921	0.762
Linear Discriminant Analysis	0.894	0.942	0.902	0.941	0.921	0.762
Quadratic Discriminant Analysis	0.893	0.929	0.905	0.936	0.920	0.757

Table A3 shows the results for the different ML regression algorithms obtained for the Pure model.

Table A3. Regression metrics for the Pure model.

Model	MAE	RMSE	R2	RMSLE
Extreme Gradient Boosting	1.832	2.954	0.991	0.199
Light Gradient Boosting Machine	1.991	3.173	0.990	0.209
Gradient Boosting Regressor	2.179	3.464	0.988	0.224
Linear Regression	2.313	3.763	0.986	0.254
Ridge Regression	2.313	3.763	0.986	0.254
Bayesian Ridge	2.313	3.763	0.986	0.254
Least Angle Regression	2.315	3.765	0.986	0.255
Orthogonal Matching Pursuit	2.293	3.777	0.986	0.256
Huber Regressor	2.263	3.781	0.986	0.251
Decision Tree Regressor	2.024	3.825	0.986	0.236
Lasso Regression	2.506	3.921	0.985	0.269
K Neighbors Regressor	3.975	5.720	0.968	0.368
Passive Aggressive Regressor	4.145	5.549	0.969	0.382
AdaBoost Regressor	7.510	9.111	0.919	0.637
Elastic Net	8.130	10.242	0.897	0.605
Lasso Least Angle Regression	26.181	31.984	-0.000	1.082

Table A4 shows the results for the different ML regression algorithms obtained for the Hybrid model.

Table A4. Regression metrics for the Hybrid model.

Model	MAE	RMSE	R2	RMSLE
Extreme Gradient Boosting	1.281	1.881	0.892	0.294
Light Gradient Boosting Machine	1.371	2.014	0.876	0.312
Gradient Boosting Regressor	1.548	2.216	0.850	0.336
Decision Tree Regressor	1.174	2.231	0.848	0.336
K Neighbors Regressor	1.883	2.560	0.801	0.395
AdaBoost Regressor	2.299	2.835	0.755	0.419
Linear Regression	2.489	3.393	0.650	0.467
Least Angle Regression	2.490	3.393	0.650	0.467
Ridge Regression	2.490	3.393	0.650	0.467
Bayesian Ridge	2.490	3.393	0.650	0.467
Orthogonal Matching Pursuit	2.493	3.410	0.646	0.466
Huber Regressor	2.265	3.665	0.591	0.425
Lasso Regression	2.892	3.668	0.590	0.525
Elastic Net	3.218	3.904	0.536	0.576
Passive Aggressive Regressor	3.618	4.782	0.302	0.629
Lasso Least Angle Regression	4.799	5.732	-0.000	0.770

References

1. SESAR Joint Undertaking SESAR 2020 Concept of Operations. 2017. Available online: https://ec.europa.eu/research/participants/documents/downloadPublic?documentIds=080166e5b6d2b912&appId=PPGMS (accessed on 25 November 2021).
2. SESAR Joint Undertaking. *European Atm Master Plan—Digitalising Europe's Aviation Infraestructure*; SESAR Joint Undertaking: Brussels, Belgium, 2019.
3. ICAO. *Air Traffic Management—Doc 4444*; ICAO: Montreal, QC, Canada, 2012.
4. ICAO. *Doc 9689-AN/953—Manual on Airspace Planning Methodology for the Determination of Separation Minima*; ICAO: Montreal, QC, Canada, 1998; pp. 1–205.
5. Reich, P.G. Analysis of long-Range air traffic systems—Separation Standards I. *J. Inst. Navig.* **1966**, *50*, 436–477. [CrossRef]
6. Reich, P.G. Analysis of long-range air Separation Standards—II. *J. Navig.* **1966**, *19*, 169–186. [CrossRef]
7. Kim, S.H. Conflict Risk Assessment of Structured and Unstructured Traffic of Small Unmanned Aircraft Systems. In Proceedings of the 2018 Aviation Technology, Integration, and Operations Conference, Atlanta, GA, USA, 25–29 June 2018; Volume 15, pp. 25–29.
8. Pérez-Castán, J.A.; Gómez Comendador, F.; Rodríguez-Sanz, Á.; Arnaldo Valdés, R.M. Conflict-risk assessment model for continuous climb operations. *Aerosp. Sci. Technol.* **2019**, *84*, 812–820. [CrossRef]
9. Pérez-Castán, J.A.; Gómez Comendador, F.; Rodríguez-Sanz, A.; Armas Cabrera, I.; Torrecilla, J. RPAS conflict-risk assessment in non-segregated airspace. *Saf. Sci.* **2019**, *111*, 7–16. [CrossRef]
10. Siddiqee, W. A mathematical model for predicting the number of potential conflict situations at intersecting air routes. *Transp. Sci.* **1973**, *7*, 571–577. [CrossRef]
11. Sunil, E.; Ellerbroek, J.; Hoekstra, J.M.; Maas, J. Three-dimensional conflict count models for unstructured and layered airspace designs. *Transp. Res. Part C Emerg. Technol.* **2018**, *95*, 295–319. [CrossRef]
12. Netjasov, F.; Janic, M. A review of research on risk and safety modelling in civil aviation. *J. Air Transp. Manag.* **2008**, *14*, 213–220. [CrossRef]
13. Krozel, J.; Peters, M.E.; Hunter, G. *Conflict Detection and Resolution for Future Air Transportation Management—TR97138-01*; NASA Ames Research Center: Moffett Field, CA, USA, 1997.
14. Paielli, R.A.; Erzberger, H. Trajectory Specification for Terminal Air Traffic: Pairwise Conflict Detection and Resolution. In Proceedings of the 17th AIAA Aviation Technology, Integration, and Operations Conference, Denver, CO, USA, 5–9 June 2017.
15. Kuchar, J.K.; Yang, L.C. Conflict detection and resolution, air traffic control, alerting systems, warning systems. *IEEE Trans. Intell. Transp. Syst.* **2000**, *1*, 179–189. [CrossRef]
16. Erzberger, H.; Lauderdale, T.A.; Chu, Y.-C. Automated conflict resolution, arrival management, and weather avoidance for air traffic management. *Proc. Inst. Mech. Eng. Part G J. Aerosp. Eng.* **2012**, *226*, 930–949. [CrossRef]
17. Pérez-Castán, J.A.; Comendador, F.G.; Rodríguez-Sanz, Á.; Barragán, R.; Arnaldo-Valdés, R.M. Design of a conflict-detection air traffic control tool for the implementation of continuous climb operations: A case study at Palma TMA. *Proc. Inst. Mech. Eng. Part G J. Aerosp. Eng.* **2019**, *233*, 4839–4852. [CrossRef]
18. Tang, H.; Robinson, J.E.; Denery, D.G. Tactical Conflict Detection in Terminal Airspace. *J. Guid. Control Dyn.* **2011**, *34*, 403–413. [CrossRef]
19. EAAI-HLG. *The FLY AI Report Demystifying and Accelerating AI in Aviation/ATM*; EAAI-HLG: Bournemouth, UK, 2020.
20. Geron, A. *Hands-On Machine Learning with Scikit-Learn and TensorFlow*; O'Reilly Media, Inc.: Newton, MA, USA, 2017; ISBN 9781491962299.
21. Mitchell, T.M. *Machine Learning*; McGraw-Hill Science/Engineering/Math: New York, NY, USA, 1997; ISBN 0-07-042807-07.
22. Alligier, R.; Gianazza, D.; Durand, N. Machine Learning and Mass Estimation Methods for Ground-Based Aircraft Climb Prediction. *IEEE Trans. Intell. Transp. Syst.* **2015**, *16*, 3138–3149. [CrossRef]
23. Alligier, R.; Gianazza, D. Learning aircraft operational factors to improve aircraft climb prediction: A large scale multi-airport study. *Transp. Res. Part C Emerg. Technol.* **2018**, *96*, 72–95. [CrossRef]
24. Fernández, E.C.; Cordero, J.M.; Vouros, G.; Pelekis, N.; Kravaris, T.; Georgiou, H.; Fuchs, G.; Andrienko, N.; Andrienko, G.; Casado, E.; et al. DART: A machine-learning approach to trajectory prediction and demand-capacity balancing. *SESAR Innov. Days* **2017**. Available online: https://www.sesarju.eu/sites/default/files/documents/sid/2017/SIDs_2017_paper_65.pdf (accessed on 25 November 2021).
25. Sridhar, B. Applications of machine learning techniques to aviation operations: Promises and challenges. In Proceedings of the 2020 International Conference on Artificial Intelligence and Data Analytics for Air Transportation (AIDA-AT), Singapore, 3–4 February 2020; pp. 1–12.
26. Galar, M.; Fernandez, A.; Barrenechea, E.; Bustince, H.; Herrera, F. A review on ensembles for the class imbalance problem: Bagging-, boosting-, and hybrid-based approaches. *IEEE Trans. Syst. Man Cybern. Part C Appl. Rev.* **2012**, *42*, 463–484. [CrossRef]
27. Cho, D.; Yoo, C.; Im, J.; Cha, D.H. Comparative Assessment of Various Machine Learning-Based Bias Correction Methods for Numerical Weather Prediction Model Forecasts of Extreme Air Temperatures in Urban Areas. *Earth Sci.* **2020**, *7*, 1–18. [CrossRef]
28. Haupt, S.E.; Cowie, J.; Linden, S.; McCandless, T.; Kosovic, B.; Alessandrini, S. Machine learning for applied weather prediction. In Proceedings of the IEEE 14th International Conference on e-Science (e-Science), Amsterdam, The Netherlands, 29 October–1 November 2018; pp. 276–277.

29. Chaimatanan, S.; Delahaye, D.; Mongeau, M. Aircraft 4D trajectories planning under uncertainties. In Proceedings of the IEEE Symposium Series on Computational Intelligence, Cape Town, South Africa, 8–10 December 2015; pp. 51–58.
30. Persiani, C.A.; Bagassi, S. Route planner for unmanned aerial system insertion in civil non-segregated airspace. *Proc. Inst. Mech. Eng. Part G J. Aerosp. Eng.* **2013**, *227*, 687–702. [CrossRef]
31. Pérez-Castán, J.A.; Rodríguez-Sanz, Á.; Pérez Sanz, L.; Arnaldo Valdés, R.M.; Gomez Comendador, V.F.; Greatti, C.; Serrano-Mira, L. Probabilistic Strategic Conflict Management for 4D Trajectories in Free Route Airspace. *Entropy* **2020**, *22*, 159. [CrossRef] [PubMed]
32. Alam, S.; Shafi, K.; Abbass, H.A.; Barlow, M. An ensemble approach for conflict detection in Free Flight by data mining. *Transp. Res. Part C Emerg. Technol.* **2009**, *17*, 298–317. [CrossRef]
33. The OpenSky Network. The OpenSky Network API. Available online: https://opensky-network.org/apidoc/index.html (accessed on 7 October 2020).
34. Boeing Commercial Airplanes Statistical Summary of Commercial Jet Airplane Accidents World Wide Operations 1959–2018. 2019. Available online: http://www.boeing.com/commercial/safety/investigate.html (accessed on 25 November 2021).
35. Karney, C.F.F. Geographiclib. Available online: https://geographiclib.sourceforge.io/html/python (accessed on 10 September 2020).
36. AISA Concept of Operations for AI Situational Awareness. 2021. Available online: https://aisa-project.eu/downloads/AISA_D2.1_CONOPS.pdf (accessed on 25 November 2021).
37. Buitinck, L.; Louppe, G.; Blondel, M.; Pedregosa, F.; Mueller, A.; Grisel, O.; Niculae, V.; Prettenhofer, P.; Gramfort, A.; Grobler, J.; et al. API design for machine learning software: Experiences from the scikit-learn project. *arXiv* **2013**, arXiv:1309.0238.
38. Ali, M. PyCaret. Available online: https://pycaret.org (accessed on 30 November 2020).
39. Varoquaux, G.; Buitinck, L.; Louppe, G.; Grisel, O.; Pedregosa, F.; Mueller, A. Scikit-learn. *GetMobile Mob. Comput. Commun.* **2015**, *19*, 29–33. [CrossRef]

aerospace

MDPI

Review

Structural Batteries for Aeronautic Applications—State of the Art, Research Gaps and Technology Development Needs

Helmut Kühnelt [1,*], Alexander Beutl [1], Francesco Mastropierro [1], Frederic Laurin [2], Sebastian Willrodt [3], Alexander Bismarck [4], Michele Guida [5] and Fulvio Romano [6]

1 AIT Austrian Institute of Technology GmbH, Giefinggasse 2, 1210 Vienna, Austria; alexander.beutl@ait.ac.at (A.B.); francesco.mastropierro@ait.ac.at (F.M.)
2 ONERA, The French Aerospace Lab, Centre de Palaiseau, Chemin de la Huniere, F-91761 Palaiseau, France; frederic.laurin@onera.fr
3 Customcells Holding GmbH, Fraunhoferstraße 1b, 25524 Itzehoe, Germany; sebastian.willrodt@customcells.de
4 Polymer & Composite Engineering Group, Institute of Materials Chemistry & Research, University Vienna, Währinger Straße 42, 1090 Vienna, Austria; alexander.bismarck@univie.ac.at
5 Dipartimento di Ingegneria Industriale, Università degli Studi di Napoli Frederico II, Via Claudio, 21, 80125 Napoli, Italy; michele.guida@unina.it
6 CIRA—Italian Aerospace Research Centre, Via Maiorise, 81043 Capua, CE, Italy; f.romano@cira.it
* Correspondence: helmut.kuehnelt@ait.ac.at

Abstract: Radical innovations for all aircraft systems and subsystems are needed for realizing future carbon-neutral aircraft, with hybrid-electric aircraft due to be delivered after 2035, initially in the regional aircraft segment of the industry. Electrical energy storage is one key element here, demanding safe, energy-dense, lightweight technologies. Combining load-bearing with energy storage capabilities to create multifunctional structural batteries is a promising way to minimize the detrimental impact of battery weight on the aircraft. However, despite the various concepts developed in recent years, their viability has been demonstrated mostly at the material or coupon level, leaving many open questions concerning their applicability to structural elements of a relevant size for implementation into the airframe. This review aims at providing an overview of recent approaches for structural batteries, assessing their multifunctional performance, and identifying gaps in technology development toward their introduction for commercial aeronautic applications. The main areas where substantial progress needs to be achieved are materials, for better energy storage capabilities; structural integration and aircraft design, for optimizing the mechanical-electrical performance and lifetime; aeronautically compatible manufacturing techniques; and the testing and monitoring of multifunctional structures. Finally, structural batteries will introduce novel aspects to the certification framework.

Keywords: multifunctional materials and structures; structural batteries; aerospace engineering; electric aircraft; composites

Citation: Kühnelt, H.; Beutl, A.; Mastropierro, F.; Laurin, F.; Willrodt, S.; Bismarck, A.; Guida, M.; Romano, F. Structural Batteries for Aeronautic Applications—State of the Art, Research Gaps and Technology Development Needs. *Aerospace* **2022**, 9, 7. https://doi.org/10.3390/aerospace9010007

Academic Editors: Spiros Pantelakis, Andreas Strohmayer and Liberata Guadagno

Received: 26 November 2021
Accepted: 18 December 2021
Published: 23 December 2021

1. Introduction

The necessity to unite environmental sustainability and economic growth is a major challenge for the aviation industry [1]. These needs resulted in policy roadmaps in Europe and the US [2–4], and led to consistent long-term research efforts. In 2019, the European Green Deal [5] set the ambitious goal of achieving climate neutrality in all sectors of the EU economy, including air transport, by 2050, raising the bar significantly above the previous environmental targets.

One of the main strategies to limit in-flight emissions of greenhouse gases (GHG) and pollutants is the increased use of electrical energy onboard aircraft for both non-propulsive (e.g., secondary systems) and propulsive purposes, leading to the concepts of "more electric aircraft" (MEA), "hybrid electric aircraft" (HEA) and "all-electric aircraft" (AEA).

In addition, the electrification of aircraft may result in lower operative and maintenance costs, and may open new markets that are not served with conventional aircraft, such as urban air mobility with electric vertical take-off and landing vehicles (e-VTOL) [6]. The use of electrical energy is at the same time a key enabler, i.e., benefiting from the highly energy-efficient electric drive train, enabling distributed electric propulsion concepts, and also a limiting factor toward the large-scale introduction of electrified aircraft, due to the limited energy and power density of current battery technologies, thus introducing a substantial weight penalty.

An alternative approach to storing electrical energy in a conventional battery system installed in the aircraft is to combine energy storage and load-bearing capabilities in multifunctional structures, or structural batteries (SB), which have come to the forefront of research since the late 1990s.

After this introduction, Section 2 discusses the latest research on structural batteries, with the different approaches undertaken and considerations according to the level of integration, as well as the manufacturing of structural battery cells; Section 3 reviews their integration into composite structures, including structural materials, manufacturing processes, integration strategies and the testing of SB; Section 4 reviews the certification framework that is likely to apply to SB; Sections 5 and 6 look at first studies on structural batteries integration at aircraft level and collect expectations and first recommendations made by aeronautic industry experts. Finally, Section 7 evaluates the state of current approaches versus 2030+ aeronautic targets, while Section 8 aims to give recommendations for developing further structural batteries for aeronautic applications.

2. Research into the State of the Art of Structural Battery Approaches

2.1. Classification and Performance Metrics

Structural batteries or multifunctional electrochemical energy storage systems combine energy storage capabilities, like conventional lithium-ion (Li-ion) batteries, with the mechanical properties of composite materials like carbon-fiber-reinforced polymers (CFRPs). They can be classified into different categories depending on their degree of structural integration. This review adopts a convenient 5-class system, introduced in [7] and shown in Figure 1, starting from a side-by-side combination of a structural element and a conventional battery (zero degrees of integration) and increasing the degree of integration from integrating battery cells into structural components (type I), thin-film approaches (type II), up to a fully integrated system, for which the structural element (type III) or material (type IV) also functions as an energy storage unit.

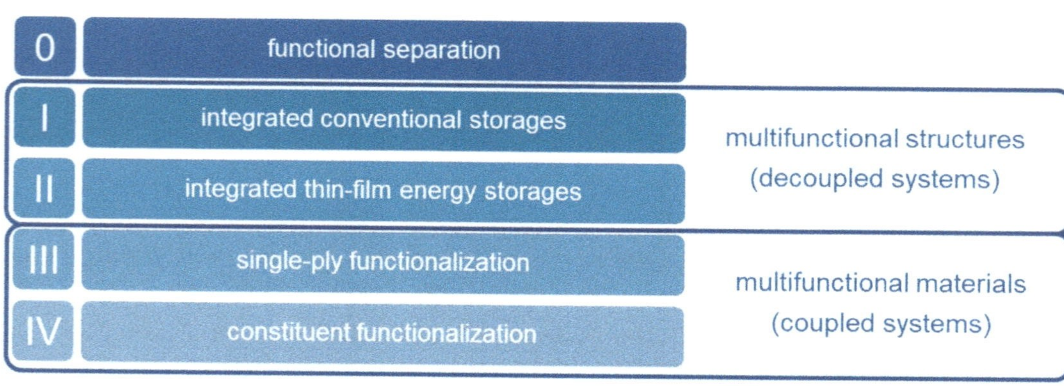

Figure 1. Classification system of structural batteries, adopted from [7].

Structural batteries can be also grouped into two additional categories, namely, multifunctional structures (or decoupled systems) and multifunctional materials (or coupled systems). In the former, different materials fulfill only one function (either energy storage or load-bearing); however, the overall composite is multifunctional. In the latter, all materials adopt multiple functions (i.e., energy storage and load-bearing). Thus, multifunctional structures are synonymous with type-I and type-II structural batteries, whereas multifunctional materials include type-III and type-IV structural batteries [8].

A simple approach to evaluate the performance of structural batteries has been proposed by Synder et al. [9,10]:

$$\eta_{mf} = \eta_e + \eta_s,\ \eta_e = \frac{E_{mf}}{E_e},\ \eta_s = \frac{G_{mf}}{G_s}, \tag{1}$$

where η_{mf} is the overall multifunctional efficiency of the structural battery, consisting of the electrochemical efficiency η_e and the structural efficiency η_s. The electrochemical efficiency (η_e) represents the ratio of the electrochemical performance of the multifunctional material/structure (E_{mf}) and a non-structural reference system (E_e), whereas η_s represents the ratio of the specific moduli of the multifunctional material/structure (G_{mf}) and a purely structural reference system (G_s), or, in the case of multiple relevant performance indicators, the minimum of their ratios. The structural battery yields an improvement against the non-integrated reference and, thus, a relative weight reduction, only for $\eta_{mf} > 1$.

The approach represented by Equation (1) is simple but quickly provides an initial generalized basis for comparing at a materials level the different approaches, as presented in Section 7. Equation (1) can be generalized into a more advanced, but application-specific metric, defined by a value function:

$$V = \sum\nolimits_n^i \alpha_i P_i, \tag{2}$$

with weightings α_i and performance indicators P_i, one that is minimized by balancing the conflicting objectives with well-tuned weightings. For SB integration into complex structures and components, the application-specific design degrees of freedom from material to element levels, and trade-offs would need to be considered.

The evaluation of the electrochemical performance of structural batteries usually uses straightforward parameters as performance indicators, such as specific energy or gravimetric energy density (in Wh/kg), volumetric energy density (in Wh/L), or specific power (in W/kg). In some cases, the exact material for which data are reported is not explicitly stated (all, only active, etc.), making comparative analyses of data difficult. However, reasonable assumptions can usually be made from the complementary data provided. On the other hand, the evaluation of the mechanical capabilities of structural batteries takes more effort because applied testing methods are usually closely related to the envisioned application. As such, a large variety of different mechanical properties are reported, among them: elastic modulus [11–16], tensile strength [11,14], bending modulus [16], stiffness [17,18], performance under compressive stress [19], tensile stress [20], compression modulus [21], and shear modulus [22]. This considerably complicates any comparison between different studies. However, the elastic modulus is recently becoming an accepted representative of the overall structural performance of the structural battery [11,23]. For aeronautic applications, additional mechanical properties should be considered, e.g., the onset of damage or the strengths for different loadings, such as bending or compression. Common performance indicators for the evaluation of structural batteries are compiled in Figure 2.

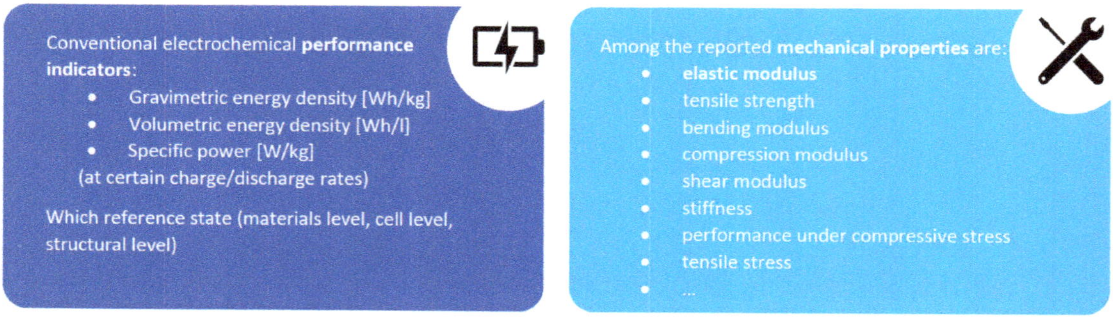

Figure 2. Common parameters for evaluation of structural batteries.

2.2. *Summary of the State of Art for Structural Batteries*

This section summarizes structural battery approaches reported in the literature from 2002 to 2021, according to the classification system introduced in Section 2.1. Beforehand, recent general reviews are summarized.

Asp et al. [24] review the most recent works on SB as multifunctional materials (types III and IV), without reference to a particular application. Data and research experiences on carbon fibers as negative electrodes and composite polymer electrolytes, tested in half-cells, are gathered. The authors conclude that the technology presents no clear showstoppers, due to recent advancements in structural electrolytes. Nonetheless, incomplete data on complete cells are available, due to the unavailability of structural positive electrodes. This impacts the information available for simulations, manufacturing scalability, lifecycle and recycling. Danzi et al. [25] published a comprehensive review on structural batteries that covers materials for anodes, cathodes, manufacturing techniques and tests, again not focusing on specific applications. On top of presenting existing research on electrode cells and embedded Li-ion batteries, Kalnaus et al. [26] describe approaches toward designs focused on safety (features to improve the resistance of SB to damage or to prevent thermal runaway). The work also lists the more significant research gaps: the creation of structural electrolytes with high ionic conductivity; the improvement of mechanical performance after cycling; the execution of extensive testing programs for various load conditions; the research on the system-level requirements of SB (thermal management, electrical connections).

2.2.1. Functional Separation (Type 0)

Structural batteries need to compete against a classical, functionally separate system of commercial batteries and conventional load-bearing elements. Currently, consistent research efforts are dedicated to increasing the energy density of current Li-ion batteries, which lies in the region of 260 Wh/kg and 700 Wh/L at cell level (e.g., Panasonic NCR18650B cylindrical Li-ion cells) [27].

CFRP composites are usually considered as most suitable for load-bearing components. They provide an elastic modulus of around 130–340 GPa for unidirectional (UD) fibers and around 70–90 GPa for fabrics, a compression strength of around 800–1600 MPa (UD) and 700–800 MPa (fabrics), a shear modulus of 4–5.5 GPa and in-plane shear strengths of 80–95 MPa (average composite properties with 60% fiber volume, data compiled from [28,29]).

2.2.2. Integrated Conventional Storage (Type I)

The integration of commercial Li-ion batteries into a dedicated structural element, as shown in Figure 3, presents the lowest degree of multifunctionality. It was the first [30] and remains the most widely investigated approach. Considered applications are unmanned vehicles [30–32], the automotive industry [21], aeronautic industry [33], and maritime sector [18], as well as for space technology [17,34,35], especially microsatellite applications [36,37].

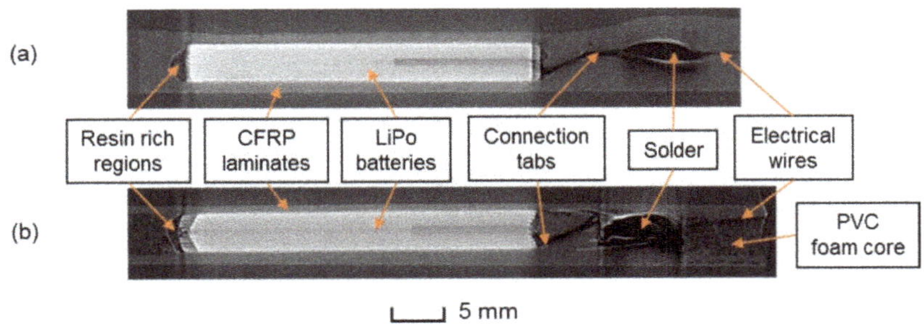

Figure 3. Examples of integrated conventional storage: cross-sectional X-ray CT images of an embedded LiPo pouch cell in (**a**) CFRP laminate and (**b**) sandwich composite [38].

The low multifunctionality of this type of structural battery limits the savings in mass, often restricted to the weight of the casing/packaging, which is often replaced by a mechanically reinforcing material like CFRP to increase the degree of multifunctionality [7]. Attempts to further increase the structural capabilities of the composite include modifications of the batteries themselves by introducing, e.g., rivets [39–43] or perforations filled with epoxy [44]. On the other hand, feasible processing and the availability of cheap components (i.e., battery cells) render this approach the most cost-effective, although Roberts et al. [17] stated that the beneficial effects of structural batteries might be offset by the additional cost of manual or semi-automatic manufacturing of battery designs tailored for integration into structural parts. This issue has already been mentioned in the early works of Thomas et al. [31], who identified the procurement of custom-made battery cells as the bottleneck of this approach.

The mechanical and electrochemical performance of type-I structural batteries depends on how many battery cells are incorporated per unit of area. An observed general trend is that the more batteries are incorporated, the higher the achieved energy density [45], but the lower the mechanical properties [30,38,43]. This is expected, as some of the mechanically reinforcing material with a high tensile modulus (around 50 GPa) is removed to fit the battery cells with a low modulus (around 150 MPa, i.e., ~330 times lower) in the structural composite [38]. Therefore, the main challenge remains to provide a proper balance between the reduction of mechanical properties and the achievable energy density. A common reference value for the electrochemical performance of type-I structural batteries was established by Thomas et al. [18], who set a goal of 50 Wh/l for marine systems applications. To reach this target, a reduction of the specific mechanical properties of around 40% for laminate composites and 25–30% for sandwich composites can be expected [38,46]. In addition, it was shown that different mechanical properties, like tension, bending, and compression, are differently affected by the incorporation of battery cells into a structural sandwich element [47] and that the impact is largest on the compressive properties.

With type-I approaches, gravimetry energy densities of between 20 and 139 Wh/kg (and volumetric energy density of between 20 and 276 Wh/L) could be achieved, the maximum being realized by integrating custom-made planar multi-electrode pair cells, with conventional-production NMC and graphite active materials and electrodes with organic liquid electrolyte, that were reinforced with rivets as a sandwich structure into CFRP face sheets [43].

The weight savings gained by this approach, compared to functional separation, will become less relevant as battery processing technologies improves. The relative amounts of passive components (like the casings) in conventional batteries are more and more reduced by employing advanced battery-processing methods (e.g., thicker electrode coatings) and optimizations of the cell design (e.g., the recent development of 21,700 cylindrical cells compared to 18,650 cells [48]).

2.2.3. Integrated Thin-Film Energy Storage (Type-II)

Integrating thin-film batteries rather than conventional cells into structural elements has the advantage of minimizing the impact of the battery on the mechanical properties of the composite. Examples of type-II concepts are shown in Figure 4. Unfortunately, the obtainable energy density is quite limited, with reported values in the range of 10–1000 μAh/cm^2. Average discharge voltages of around 3.5 V and energy densities of only 35–3500 μWh/cm^2 can be expected. In addition, thin-film batteries are significantly more expensive than conventional cells [18], increasing the total cost of the structural composite. Planar [49] or coaxial [50] configurations have been performed and tested, with applications mainly focused on spacecraft appliances.

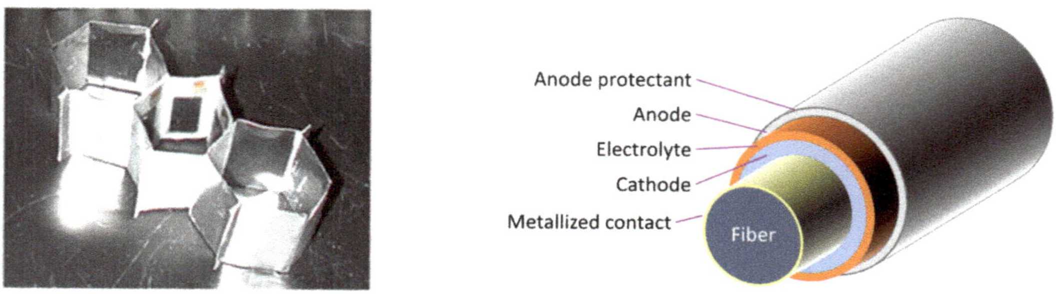

Figure 4. Examples of type-II structural batteries: planar [49] (**left**) and coaxial configurations, after [50] (**right**).

This renders the class of "integrated thin-film energy storages", i.e., type II, not fit for large-scale aeronautic application and is, therefore, not considered further in this review. However, a concise summary of type-II approaches is given in [18].

2.2.4. Single-Ply Functionalization (Type III)

Type-III structural batteries represent the first step toward true multifunctionality. The basic idea behind this approach is to substitute as many elements as possible of a conventional battery cell, especially the passive ones (i.e., the casing, current collectors, separator, and electrolyte), with load-bearing elements, as shown in Figure 5.

Promising substitutes can be found for most of the components, except for the cathode material. Therefore, the cathode is usually prepared from either conventional laminate electrodes [11] or functionalizing carbon fibers with active materials like $LiFePO_4$ [51,52].

Carbon fibers can be used for the anode, as they show similar performances as conventional graphitic materials [53]. However, volumetric changes, leading to elongations of up to 1% and 10% in both the longitudinal and transverse directions are observed [54]. Carbon fiber weaves and cloths have also been employed for anodes [52,55], showing, in general, lower performance than spread single fibers [52,56]. To mitigate these issues, coated or functionalized fibers can be employed [54].

The polyolefine separator usually encountered in conventional Li-ion batteries is replaced with a glass-fiber separator, either as a filter-like material or a weave [11,20]. Due to the increase in thickness (by a factor of 3 to 10) and in density (by a factor of around 2.8) of the separator and of the increased amount of stored electrolyte, the gravimetric energy density of the cell decreases up to a factor of 4 (estimated with data from [57]).

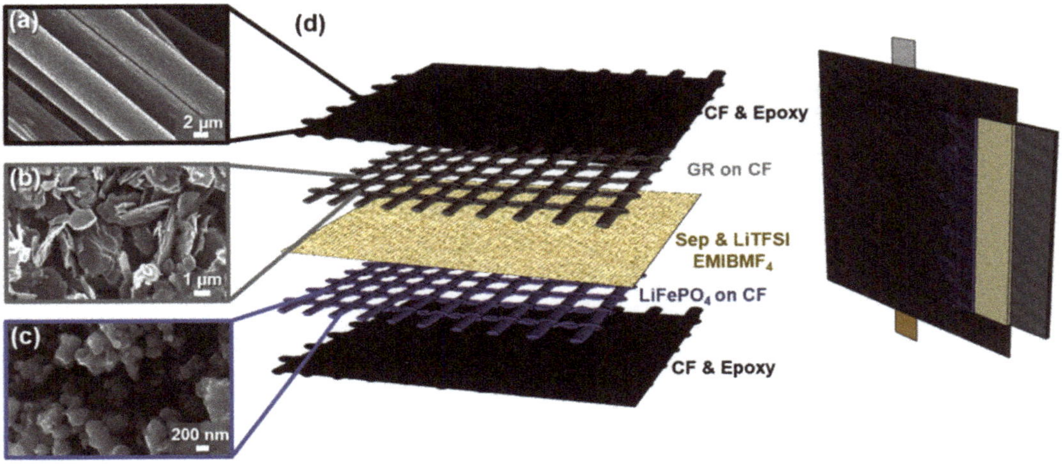

Figure 5. Example of a type-III structural battery [20].

The electrolyte is one of the most crucial components for type-III structural batteries. Conventional liquid carbonate electrolytes are still employed due to the lack of alternative highly conducting electrolytes with good mechanical and structural properties. Reports on polymer and co-polymer electrolytes showed that a mono-phasic electrolyte cannot satisfy both high load-bearing and high ion conductivity requirements. Therefore, biphasic or bicontinuous structural electrolytes are used. These comprise a load-bearing component, usually an epoxy-based polymer/resin matrix, and an ionically conductive component, which is either a conventional liquid electrolyte or an ionic liquid electrolyte [58–61].

One of the advantages of type-III structural batteries, apart from higher mass savings due to their higher degree of multifunctionality, is the feasible processing of the different components. Conventional methods like tape casting [15] or vacuum-assisted resin transfer molding [62] can usually be applied for the preparation of the battery cells. However, substantial research efforts are still needed to develop the different multifunctional components and full structural battery assemblies.

Working type-III structural batteries have shown specific energies of between 12 and 58 Wh/kg and elastic or tensile modulus in the GPa range (see the references within this section).

2.2.5. Constituent Multifunctionalization (Type IV)

For type-IV structural batteries, as depicted in Figure 6, different setups were proposed; these can be classified into coaxial [63–65] and layered approaches [22,66,67].

The coaxial approach envisions a fiber-shaped battery, which is composed of a carbon-fiber core onto which an electrolyte layer is deposited; afterward, the fibers are immersed into a matrix acting as the cathode material [63–65].

In the layered approach, carbon fibers act as the anode material as well, with coated carbon fibers acting as the cathode material. They are aligned to form electrode bands, separated by a thin layer of electrolyte [22,66,67].

Research has primarily focused on the development of the different components. No functional full cell of a type-IV structural battery offering significant energy storage and load-bearing capabilities has yet been demonstrated. However, some recent publications showed proof-of-concept-type cells: a coaxial, type-IV micro-battery was developed by Thakur et al. [13], who used a continuous carbon fiber coextrusion method to produce battery filaments. The use of a liquid electrolyte, as well as a great deal of insulating photopolymer (which increases the passive weight), resulted in negligible mechanical as well as electrochemical performance.

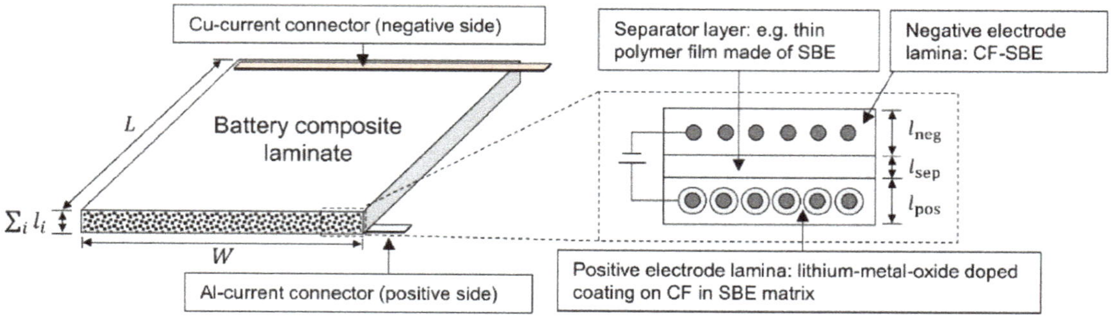

Figure 6. Example of a type-IV structural battery concept [66].

Asp et al. [11] recently assembled a structural battery using the layered approach, with carbon fiber acting as an anode and either a Whatman glass microfiber separator or two 0°/90° woven GF fabrics as the separator. They obtained meaningful energy density and structural properties, i.e., 11.6 or 23.6 Wh/kg @ 0.05 C and an elastic modulus of 18 or 25 GPa, although the use of a conventional laminate $LiFePO_4$ cathode would qualify this approach as closer to a type-III battery.

Performance evaluations and possible shortcomings of type-IV cells have been studied with analytical and numerical models: an inverse relationship between the shear modulus and the achievable energy density has been predicted [22]. Energy densities of around 146 Wh/kg @ C/10 are expected, assuming NMC622 and CF as active materials and a separator thickness of 25 μm, but reducing the elastic properties of the composite by approx. 50% in a longitudinal direction and by 80–90% in a transverse direction [63,64].

The potential of type-IV batteries for use in different vehicles, including aircraft, has also been estimated: for an interior-foam-core sandwich panel, commonly found in commercial aircraft, a potential mass saving of 4% was stated when functionalizing the CFRP face sheets with a full-layer, single electrode-pair structural battery [68].

2.2.6. Structural Batteries Using Different Cell Chemistries (Other Than Li-Ion)

Cell chemistries other than lithium-ion (such as $LiFePO_4$ or NMC) have been employed to realize structural batteries. Among them are Ni-Fe [14] or Zn-MnO_2 [69], Zn-air and Ni-Fe aqueous batteries, which are safer than lithium-ion ones [23,70].

2.3. *Considerations for the Manufacturing of Structural Battery Cells*

Type-I and type-III/IV structural battery approaches have very different requirements concerning their production. This section links the established Li-ion battery cell production with the needs of structural batteries. Type-I concepts share similarities to established battery cell production, i.e., by manufacturing cell stacks of electrodes, separators and electrolytes and later integrating them into structural elements. The stacking process is already industrially established in the field of pouch cell design. Two major differences to the industrially established Li-ion batteries will influence the production process of structural batteries: the use of a structural, (quasi-)solid-state electrolyte and the housing/casing of the cell stacks.

For conventional pouch cells, the process chain of the current manufacturing process, as shown in Figure 7, is generally divided into three sections: electrode manufacturing, cell assembly and finishing.

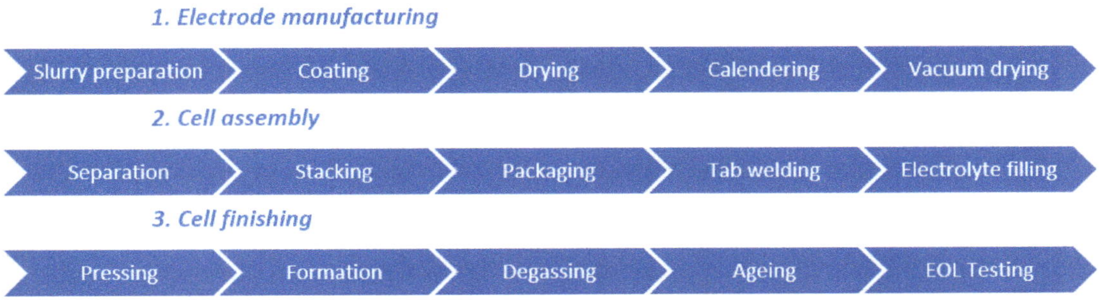

Figure 7. Process chain of conventional pouch cell production.

For type-I concepts, the electrode production would not require significant changes from the current industrial process, as wet-chemically-manufactured electrodes could be produced via an already established process. Adaptations would possibly be needed with respect to the specific cell chemistry.

Most processes could also be adapted for the cell assembly, with some changes in the details. The challenges that appear in the subsequent drying, winding and separation processes would need to be investigated in more detail. The use of a structural, quasi-solid-state electrolyte would eliminate the electrolyte filling process. The electrodes would then best be coated before electrode separation as this step could be efficiently carried out on the roll. As conventional pouches are mechanically sensitive, structural casing alternatives that can be integrated into the structural element seamlessly need to be considered. Laminates have been proposed as a casing around the stack, e.g., by Thomas et al. [18], using carbon glass composite stabilized with styrene foam and epoxy glue directly applied to the cells to improve adhesion, or by Ladpli et al. [43], who suggested using carbon fiber composites with reinforcement pins for lamination. In the latter approach, the cell stack has recesses to accommodate the pins. For industrial applications, the question remains whether these cut-outs should be created before or after stacking.

As the process steps for cell finishing are mainly determined by the cell chemistry, no substantial differences are to be expected. The stability of the housing may be of concern for transport from battery production to structural integration (which most likely will be performed on different sites). If safety issues would arise here, cell finishing could be carried out after transport.

For type-III/IV structural battery concepts, using (coated) carbon fibers as electrodes and structural electrolytes, scalability of their production would be the main question. Electrophoretic deposition and emulsion polymerization are already industrially established processes. However, their interplay and scale-up would need to be evaluated in the context of structural batteries.

The process chains for both type-I and type-III/IV concepts appear to be scalable at first glance. Already established cell production processes can be used for type-I concepts, with adaptations as demanded by the chosen materials and designs. Processes that are already used industrially elsewhere are proposed for type-III/IV concepts but these still need to be developed in detail.

3. Structural Integration

In this section, the state of the art of structural batteries is assessed from the perspective of integrating electrical energy storage within aeronautical composite structures, considering: (i) the choice of the elementary constituents, (ii) the associated manufacturing processes, (iii) the integration strategies and the considered structures, and finally, (iv) the different kinds of mechanical tests applied to structural battery composites.

3.1. Choice of Elementary Constituents and Architecture

The choice of the elementary constituents for structural battery composites is critical for the battery's mechanical properties but must be also compatible with the manufacturing process of the battery materials. Here, composite materials constituted with fibers embedded in a polymer matrix are considered.

Carbon fibers are mostly used as a way to obtain high mechanical properties for the composite components. The carbon fibers considered in the literature are T300, UMS50, T800H, IMS65, UMS45, as summarized in [71]. For type-III and type-IV structural battery concepts employing individual carbon fibers as an anode, the IMS65 fibers (strength 6 GPa, modulus 290 GPa) are widely used [67,72] as well as the rather older-generation T300 fibers (strength 3.5 GPa, modulus 230 GPa) [43,73], because they both present an interesting trade-off between rigidities/strengths in the fiber direction and specific capacity, as reported in Figure 8. Additionally, they could be subjected to different treatments to improve their mechanical properties, especially the adhesion between the matrix and fibers.

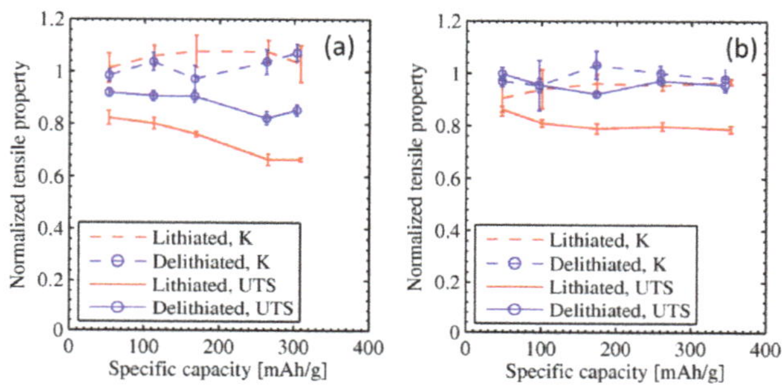

Figure 8. Mechanical properties and specific capacity of (**a**) IMS65, (**b**) T800H carbon fibers [74].

Type-III concepts are mostly constituted of 2D woven composites [43,74,75], as reported in Figure 9, mainly with twill or satin architectures that are commonly found for automotive applications and that allow the easy manufacturing of complex shapes. Multi-layered composites with 0° and 45° 2D-woven plies have been considered in order to obtain interesting properties not only for tension/compression in the warp and weft directions but also for in-plane shear loading (due to ±45° plies).

Moreover, laminated unidirectional plies can be used [7,67,76] in order to obtain higher compressive strength than 2D woven composites, which is a critical point for aeronautical applications. In methods proposed by Asp et al. [76], and by Adam et al. [7], simple cross-ply laminates [0°/90°], closely related to 2D woven composite architecture, have been considered, as reported in Figure 9b.

The matrix associated with the fibers not only needs to have a high adhesion with carbon fibers, its curing cycle must also be compatible with the battery materials' maximum allowable temperature, e.g., PVDF (Polyvinylidene fluoride, used as a binder in the battery electrodes) has a melting point of 177 °C. Therefore, thermoset epoxy matrices with curing temperatures below 170 °C are most commonly considered, such as bisphenol-A [43], an MTS57 [73] matrix, or epoxy resin based on Araldite LY556/hardener XB3473 [77]. Epoxy matrices are also relevant for use with the resin transfer-molding (RTM) process (involving the injection of a liquid matrix) and, thus, are well adapted to 2D-woven composite structures with complex shapes.

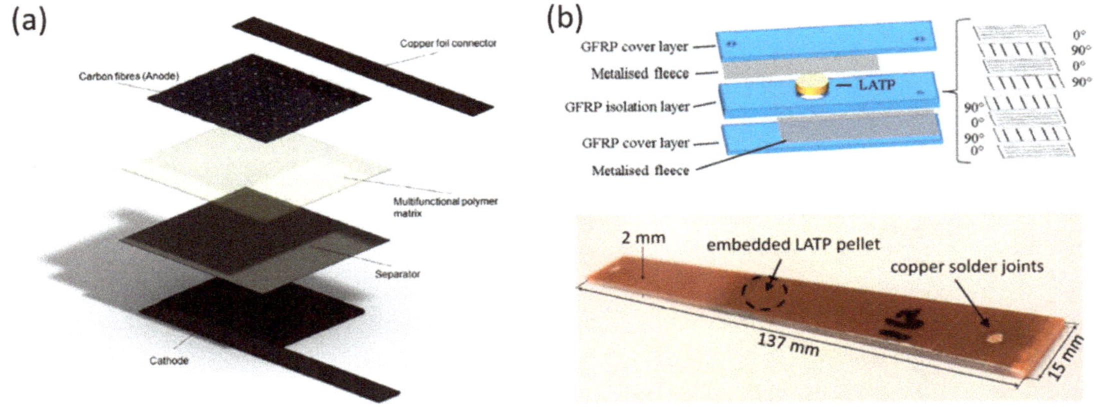

Figure 9. Representation of 2D-woven composite material: 2 × 2 twill architecture for the anode and cathode [71] (**a**), and laminate unidirectional glass fiber plies with an embedded LATP (solid-state electrolyte) pellet, adapted from [7] (**b**).

High-performance thermoplastic matrices cannot be used, as their processing temperatures of 330 °C and above (e.g., for PAEK (polyaryletherketone), PEKK (polyetherketoneketone) or PEEK (polyetheretherketone)) are way above the maximum temperature tolerated by the battery materials. Thus, no study using these materials is found in the literature.

The material chosen for the interface of the battery cell (e.g., copper or aluminum foil) with the carbon fiber is also important, i.e., carbon-coated copper foils are used to avoid galvanic corrosion [76,78].

3.2. Manufacturing Process

The manufacturing process is generally performed using a vacuum bag for pre-compaction, and the curing is performed at low temperature and low pressure. In the study by Ladpli et al. [43], the composite plate (2D-woven plies) is cured at ambient temperature for 24 h, and post-cured at 90 °C for a few hours after that. For Carlson [75], the maximal temperature applied during the curing cycle is 120 °C or 150 °C, respectively, and the composite part is embedded in a vacuum bag, as reported in Figure 10a. Curing at a low temperature requires increasing the curing duration or applying a post-curing process to obtain the same polymerization state.

It is important to note that, typically, a low level of pressure has been applied (about 1 bar) during the curing cycle. However, curing composite parts at low pressure may result in the formation of voids, as reported in Figure 10b, that generate transverse cracks or delamination, decreasing the rigidity and strength of the composite part. This is clearly an issue for aeronautical applications.

Composite plates for aeronautical applications are usually manufactured with an autoclave, where the pressure (up to 7 bars) and the temperature (<180 °C) can be controlled. The application of a higher level of pressure could be an issue for any liquid phase of the structural battery (e.g., liquid electrolyte), which may be pressed out into the composite part. However, the autoclave curing of structural batteries has not been reported so far.

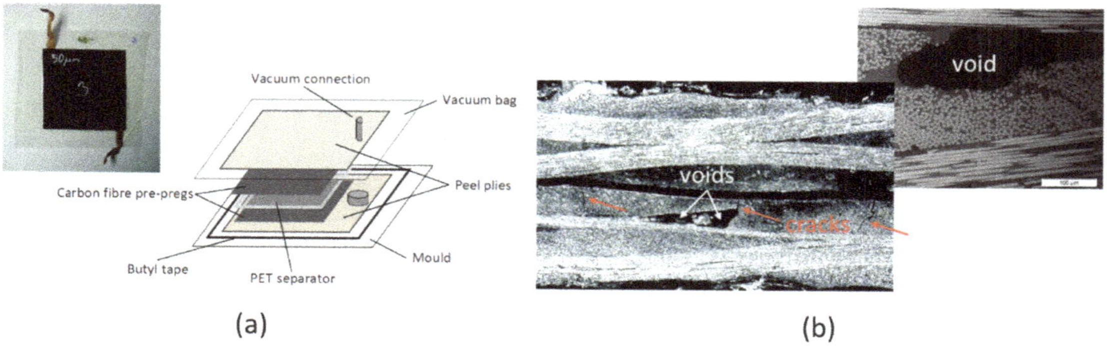

Figure 10. Manufacturing process with a vacuum bag and low-temperature curing (**a**), voids observed after the manufacturing process and possible interaction with cracks during the loading (**b**), adapted from [75].

3.3. Integration Strategies and Considered Structures

This section reviews proposed methodologies for integrating battery cells within composite structures, aiming at a minimum degradation of the structural mechanical properties (stiffness and strength). Most of the studies [8,43,68] integrated the battery cells into sandwich-like material constituted of two composite skins and a core (the actual battery cell), as reported in Figures 3 and 11a. With this integration strategy, only plain plates with large dimensions have been manufactured and characterized to determine how the introduction of a battery cell influences the mechanical performance of a composite structure.

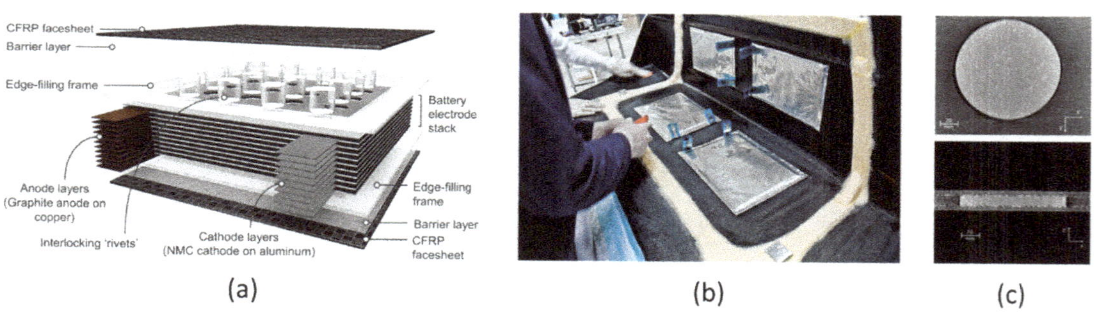

Figure 11. Sandwich-like composite material with a core battery cell [43] (**a**), rectangular-shaped supercapacitor cells introduced into a large L-angle composite specimen [79] (**b**), and a supercapacitor LATP pellet of cylindrical shape introduced into a flat composite panel (adapted from [7]) (**c**).

For large structures, the battery cells can be considered as inserts within composite parts [43,78] with a rectangular shape (with smooth corners to limit stress concentrations in [78] or with sharp ones [43]), similar to the S80 Bootlid demonstrator, as reported in Figure 11b, conceived for the automotive industry, but that includes large and thin ($200 \times 300 \times 2$ mm^3) supercapacitors as electrical energy storage. Multi-layer battery cells of the type shown in Figure 11a, with dimensions of $200 \times 50 \times 25$ mm^3, have been introduced into a very thick structural I-beam, typically used for civil engineering applications [43].

One alternative, proposed by Adam et al. [7], consists of introducing a cylindrical insert into a composite plate with a thickness that equals the thickness of a block of unidirectional plies, in order to reduce stress concentration around the insert and limit damage, as reported in Figure 11c. The pellet is about 10 mm thick with a diameter of around 100 mm, consists of a solid (LATP) electrolyte, and works as a supercapacitor, not as a battery. Only plain plates with a single pellet located at the center of the specimen have been realized.

As per the authors' knowledge, no published study has considered the optimization of the shape, size, or position of an SB cell within a composite plate. More importantly, there are no studies available on how to introduce and optimally distribute SB cells within (aeronautic) composite structures, aiming at limiting the degradation of mechanical properties due to SB introduction.

3.4. Mechanical Testing

The mechanical tests applied to the composite parts, including battery cells, have been designed to measure the decrease in mechanical properties and the electrical performance of the battery cell. Performing quasi-static tests is preferred for their simplicity and for the limited quantity of composite material required.

In order to determine the decrease in the out-of-plane shear strength due to the introduction of a battery cell, three-point bending tests have been widely performed [43,68,75,80]. An interlaminar shear strength (ILSS) test, based on European standards [81], with a ratio length between the support span over the thickness of the plate that is equal to 5 is widely used, as reported in Figure 12a. The typical dimensions of the tested specimens are about 300 mm × 25 mm. For this test, the bending rigidity and the failure load have been carefully studied. The introduction of pins into the battery cell (see Figure 11a) has been proposed by Ladpli et al. [43] to increase the shear rigidity and strength of the tested composite plates.

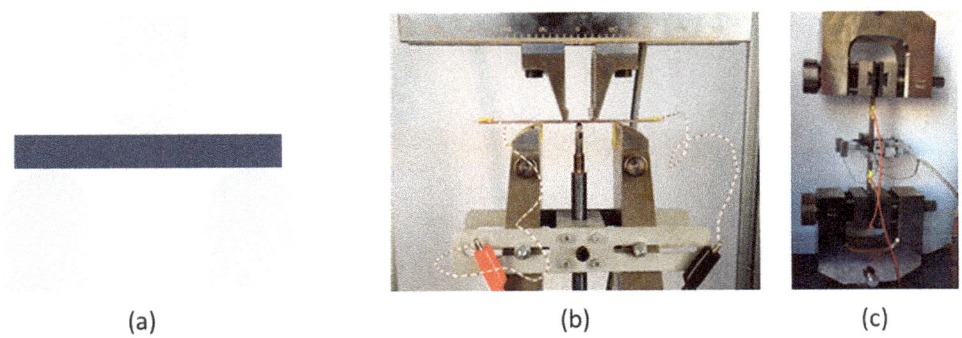

Figure 12. Mechanical tests mostly applied for structural batteries: Short span three-point bending test (**a**), four-point bending test on plain plates (**b**) and tensile test on plain coupons (**c**) [7].

The influence of the introduction of a battery cell on the bending rigidity and the failure load is studied via four-point bending tests [7], as reported in Figure 12b. It ensures a pure in-plane stress in the central part of the tested sample where the battery cell has been introduced. The typical dimensions of the specimen for such a test are 137 mm × 15 mm.

Tensile tests on plain coupons [7] with typical dimensions of 200 mm × 20 mm, consistent with the industrial standards [82], have been performed as illustrated (Figure 12c). The analysis of such a test is rather simple, and macroscopic behavior and strength can be determined easily.

Only the study of Adam et al. [7] has performed post-mortem analyses in order to establish the damage scenario due to the introduction of an insert, i.e., the battery cell, into the structure and to determine its influence on the failure mode.

The electrical performances of the battery are measured after the application of mechanical loading, to quantify their degradation compared to the no-load case. There is no clear link established between the damage evolution in the composite part around the battery cell and the evolution of its electrical performances. Moreover, only a few studies, such as [46,74], measure the electrical performance of structural batteries while the loading applied to the composite structures is maintained.

Finally, few fatigue tests [43,83] have been performed in order to determine the influence of the integration of battery cells on the fatigue limit and on the fatigue lifetime. The fatigue tests are rather similar to those performed for quasi-static loadings and mostly consist of three- or four-point bending tests. Such tests are also well adapted to measure the evolution of electrical performances as a function of the number of applied cycles.

4. Airworthiness

4.1. Certification Framework of Composite Aeronautical Structures

Structural batteries, due to their own multifunctional nature, are expected to comply with standards for both structural components and energy storage at the same time.

The certification requirements of aeronautical structures are substantially the same whatever the material used for their construction. However, composite products present specific issues unencountered in metallic elements (aluminum, titanium, and steel alloys). Due to the wide variability in the physical properties of their constituents, composite parts require accurate monitoring during the procurement of initial materials, storage, and manufacturing, as they may influence the final mechanical behavior and the relative failure modes.

Regulations issued by the FAA in the US are usually considered the benchmark for all other entities. The current reference document issued by the FAA is the Advisory Circular (AC) AC 20-107B [84], which integrates the requirements outlined in 14 CFR, parts 21/23/25/27/29 [85]. Its European equivalent is EASA's certification memorandum, CM-S-010 [86]. The AC refers particularly to critical structures (essential in maintaining overall flight safety) and specifies that such structures, if made of composites, must not subject aircraft operators to higher risks than those they accept by relying on metallic materials. It is the designer's responsibility to ensure these levels of safety. The mechanical properties of the material under examination are determined by conducting targeted experimental tests, carried out in climatic and environmental conditions as close as possible to the operational scenarios. A "building-block" approach, derived from industrial practice, is generally employed to reduce the testing costs and avoid early failures: datasets from a generic specimen are used for elementary components (e.g., a single-stringer), while complete tests tend to be performed on complex structures (e.g., a wing box). The tests typically required for the certification of a composite structure are the following:

- Static test, in which the structure is subjected to 150% of the design limit load (DLL), i.e., up to the ultimate load; these tests evaluate the resistance and long-term dimensional stability, especially at high temperatures, by subjecting the material samples to lower percentages of the static breaking stress σ_u (typically, $10 \div 50\%$ of σ_u).
- Fatigue tests on primary structures—the cyclical load stresses measure the resistance to degradation and breakage due to loads varying over time. The frequency of application is generally low (between 5 Hz and 10 Hz) to avoid excessive heating of the specimens. The cyclic loads can be of a constant amplitude or follow a load pattern representative of the real operating conditions of a particular structure.
- Damage tolerance compliance and impact resistance of primary structures. These tests quantify the resistance capacity of the material to impact and penetration. The specimens are impacted by means of indenters and impactors of various sizes.

The high dispersion that affects the experimental datasets of composites requires the execution of tests of the same type on distinct batches of material. A guideline for standardized testing methodologies and data reporting is provided in MIL-HDBK-17-1 [87], which recommends testing multiple specimens from different lots for each property and for each environmental condition of interest. The mechanical tests must be carried out at different ambient temperatures and different levels of absorbed humidity. The effects of a wide variety of environmental conditions, such as immersion in solvents and prolonged exposure to ultraviolet radiation, must also be investigated.

4.2. Certification Framework and Perspectives of Li-Ion Batteries for Aviation

Consistent and standardized test methods are necessary to facilitate the certification of new aircraft designs that incorporate permanently installed, rechargeable Li-ion batteries. The FAA's circular AC 20-184 [88] provides manufacturers and installers with indications of acceptable means of compliance to meet the installation, operation, maintenance, and airworthiness requirements for the installation of lithium batteries and battery systems on-board aircraft. The AC integrates the standard RTCA DO-311A [89] and RTCA DO-347 [90]. The former, incorporated into the FAA's Technical Standard Order TSO-C179b [91], contains the procedure for abuse tests for a single cell of a permanently installed, rechargeable lithium-ion battery until thermal runaway. The installed battery must demonstrate the mitigation of all hazardous propagation effects to other cells and the release of electrolytes, fire, or explosive debris outside the battery case. The tests replicate the battery installation on the aircraft and are conducted under conditions that are considered to produce the most severe outcome. Note that the required capacity of aircraft propulsive batteries is far beyond what the AC20-184 defines as a large battery.

The FAA is reviewing the methods of compliance used to certify permanently installed rechargeable lithium-ion batteries, to determine design/installation and certification details. RTCA DO-311A and RTCA DO-347 are used as temporary means of compliance. The following aspects require particular consideration:

- Health-monitoring for the saving storage and operation; avoiding the overcharging of individual cells may lead to thermal runaway and undercharging, whereas overvoltage and undervoltage can hamper battery life.
- Flammability: aircraft batteries must meet the applicable requirements for insulation materials, flammable fluids, and electrical systems, in compliance with 14 CFR part 25 Appendix F (equivalent to EU CS-25 Appendix F). The testing methodologies are outlined in the AC 25.856.1 [92] and in the *FAA Aircraft Materials Fire Test Handbook* [93].

Since current regulations historically focus on non-propulsive batteries, special conditions are currently issued for new aircraft designs, including rechargeable Li-ion or other Li-based propulsive batteries. In Europe, the Special Condition SC-LSA-F2840-01 [94] states the general requirements for propulsion batteries in light sports aircraft. In addition to the points above, it explicitly mentions that: battery packs must have an automatic system to monitor the SoC and manage the power absorption from the engine, if the specifications include this operational mode (e.g., during windmilling); vapor and fluid leakages, hazardous electromagnetic fields and electric shocks must be avoided; limits for the energy available should be indicated, as well as instructions for reservicing if these are violated in case of an emergency. If batteries are considered part of the propulsion system, they must also comply with the provisions of the Special Condition SC E-19 [95], although the definition itself of a propulsive battery is not always clear. In particular, the control systems must avoid sudden power oscillations and should isolate the battery if this is hazardous.

A currently debated topic is whether DO-311A can be applied to the testing of small modular systems, the combination of which enables larger systems [96], a point that is of specific interest for structural battery modules interconnecting many SB cells.

Despite the existing and advancing certification framework, structural batteries present important novel aspects that have as yet not been addressed. From the point of view of energy storage, the existence of a regulation gap for battery systems for high energy/power is acknowledged [96] because of (a) their peculiar function and their critical importance for continued flight; and (b) the amount of energy stored, which is significantly higher than in current products. It is reasonable to assume that general Li-ion battery certification standards will apply to SB as well, in addition to the specific structural requirements of the particular component utilizing SB.

5. Preliminary Studies on the Aeronautic Applications of Structural Batteries

Several works assess the impact of structural batteries in an aircraft context. Yang [97] provides a brief overview of the implementation techniques and materials used in structural batteries, collecting illustrative examples for automotive, aerospace, and marine applications, highlighting where weight advantages can be obtained.

Adam et al. [7] estimate the potential of integrating the energy capacity of a conventional battery with a weight of 10–40% of the MTOW in the aircraft structure, resulting in a range extension of between 11% and 66%, assuming ideal and full substitution.

Scholz et al. [8] assess the feasibility of structural battery integration for two small all-electric aircraft (two-seaters with 600 and 750 kg MTOW). Replacing the entire propulsive battery (with a capacity of around 20 kWh), a minimum energy density of around one-third to one-half with respect to the conventional battery pack could be accepted, depending on the degree of structural degradation, i.e., at minimum, 52 Wh/kg for the structural battery vs. 176 Wh/kg for the conventional battery pack.

Riboldi et al. [98] present an initial methodology for the preliminary (structural) design of an aircraft equipped with structural batteries. The use of SB panels is envisaged for the fuselage and parts of the wings and stabilizers, where compressive loads are not excessive. A case study based on a hybrid electric aircraft for general aviation (CS-23) is performed. Assuming an energy density of the structural battery of 125 Wh/kg, i.e., 50% of the conventional battery pack, and equipping about 46 wt % of the structure with it so as to store about 54% of the total electric energy in the structure (keeping all other parts of the propulsion system unchanged), the weight of the aircraft structure plus the battery is projected to be reduced by about 20% compared to a conventional CFRP structure, and about 29% compared to an Al-alloy structure (and using a conventional battery pack).

Nguyen et al. [99] assess the possibility of using structural batteries as a replacement for the floor panels, to power in-flight entertainment systems on an A220-type aircraft. It is found that the minimum requirements are a specific energy of 144 Wh/kg, a specific power of 290 W/kg, an in-plane elastic modulus of 28 GPa, and a compressive strength of 219 MPa.

Karadotcheva et al. [100] assess the potential of structural power composites for propulsive applications in more-electric, hybrid-electric and all-electric short-and-medium-range (A320-type) aircraft configurations, focusing on the required energy and power densities of structural batteries, as well as achievable fuels and GHG reduction. For a 1500 km mission with MEA, integrating SB into 50% of the airframe with a minimum energy and power density of 90 Wh/kg and 55 W/kg resulted in a 5.6% fuel efficiency improvement. For the HEA, SB energy and power density values would need to exceed 200 Wh/kg and 120 W/kg (in a 100% SB airframe and a main, conventional battery pack with 400 to 600 Wh/kg), and for AEA, these values increase to 400 Wh/kg SB and 700 to 800 Wh/kg battery pack. (HEA and AEA were assessed, assuming optimized strut-braced and box wing configurations.)

These values provide an indication of the acceptable energy density ratio between a main (conventional) battery pack and structural battery of around 3:1 to 2:1, i.e., a maximum of 33% to 50% degradation of the electrical performance would, however, be acceptable, assuming low mechanical degradation (20%) of the SB and up to 100% of utilization in the airframe.

6. View of Aeronautic Industry Experts on Structural Batteries

The authors of this review met with the external expert advisory group of the EU CleanSky project SOLIFLY (GA 101007577) in the first half-year of 2021, concluding with a workshop on 7 June 2021. This meeting included experts from two European aircraft manufacturers (Piaggio Aerospace and Pipistrel, both developing hybrid-electric CS-23 class commuter aircraft concepts) and a supplier (FACC, who manufacture composite parts for large commercial aircraft) in order to collect and evaluate the expectations and first recommendations of the aeronautics industry regarding structural batteries. The outcome of this discussion is summarized in this section.

6.1. Aircraft Application and Integration Cases for Structural Batteries

The application cases of interest from structural batteries range from low-power, non-propulsive applications, such as decentralized energy storage for cabin infotainment or electric hat-rack systems, over secondary systems, such as auxiliary power units or energy supply for e-taxiing (to enable zero-emission, low-noise ground operation), to primary propulsive applications, i.e., structural batteries contributing to the aircraft's electric propulsion. As mentioned in the previous chapter, the progressive introduction of structural batteries, first in non-safety-critical, and later, in safety-critical applications could also be a pathway for certificating this novel technology.

Considering the integration of structural batteries into specific aircraft parts, the wing would be of particular interest due to its surface area and volume. However, it needs to be evaluated to what extent their performance might be limited under compression loads. This would restrict their adoption in components like the upper panels and stringers, in contrast to the lower panels that are usually not subject to compression during the flight.

The fuselage is not expected to cause particular issues regarding the stress limits due to the limited aerodynamic loading.

Incorporating structural batteries in the control surfaces might be demanding and not efficient as these movable parts need to be connected to the onboard electrical system and are often of small size (e.g., for small aircraft).

In general, it remains an open question how structural batteries would be maintained and replaced at the end of their lifetime. Manageable concepts will be needed that might impact the application areas of structural batteries, e.g., by limiting them to the most accessible parts of the aircraft.

6.2. Sizing, Design with Structural Batteries, and Multi-Functional Performance

The integration of structural batteries in new aircraft would imply novel sizing procedures, particularly to account for their multifunctional properties. From a structural modeling point of view, structural batteries could be considered as composites, and stress limit/failure criteria for composite materials could also be safely applied in the case of structural batteries in a preliminary design.

Structural battery components made of composite materials such as CFRP could be reinforced with a core having specific characteristics (Nomex honeycomb, etc.), to increase the bending stiffness of the structural battery panels and to improve the buckling characteristics. Furthermore, sandwich composites would be particularly attractive as they could integrate electric energy storage in both the core and the face skins.

For an initial assessment of the energy storage capacity that is achievable with structural batteries, in addition to the main electrochemical key performance indicators such as gravimetric and volumetric energy and power density, establishing the areal energy density (in Wh/m^2) would be useful. However, such a parameter, when taking into account the thickness portion of the multifunctional material or structure that is usable for electric energy storage, would be linked to a specific design of the structural battery element that is most likely highly application-dependent and, thus, hardly generalizable.

The use of a simple preliminary indicator for structural performance, such as the commonly used elastic modulus, is most likely not sufficient for complex aeronautic applications. Other mechanical properties, e.g., bending and compression, as well as the onset of damage strength should be considered. Strategies to avoid or at least mitigate stress concentration around the battery cells during the integration phase are crucial as eventual crack propagation can lead to substantial degradation of the mechanical properties, up to premature structural failure.

A full picture needs to be established of how mechanical stresses impact the electrical performance and cycle life of structural batteries, and vice versa, how the electrochemistry (e.g., due to the volume change of the electrodes during charge and discharge) impacts structural performance. The availability of a global parameter or method to estimate overall performance, as well as an indicator for the multifunctional state-of-health of structural batteries would be very desirable and beneficial.

6.3. Manufacturing Recommendations

The scalability of the design and manufacturing process, from the laboratory structural battery cell coupons up to the aeronautic component modules consisting of multiple interconnected cells, is one further point to investigate and improve. Therefore, parameters related to the manufacturing processes and the integration strategies of the several SB constituents are also of mandatory importance. On this topic, some preliminary recommendations should be considered, among others:

- For type-III and type-IV concepts utilizing carbon fibers as electrodes, good adhesion with the polymer matrix needs to be guaranteed.
- The fibers' architecture should be designed in order to improve their tensile/compression strength and the in-plane shear-loading resistance.
- A polymer matrix with a particular curing cycle and temperature should be selected that is compatible with the battery active materials, avoiding their thermal decomposition or deterioration. Application of pressure during the curing process, generally required in the aeronautic industries to prevent void/porosity formation, needs to be carefully evaluated with respect to the potential discharge of any liquid phase (e.g., ionic liquids) embedded in the structural battery cell, which could escape into the CFRP. On the other hand, curing at low pressure could introduce voids and porosity in the composite parts that will decrease the mechanical properties.
- Pre-curing of the battery cells before manufacturing and curing the composite parts could be relevant.

The aforementioned industrial experts have pointed out the increasing importance of recyclability and the weldability of composite materials for aeronautic use. From this viewpoint, the use of thermoplastic polymers would be beneficial. However, as already pointed out in Section 3.1, no reference with these materials is available in the literature.

6.4. Safety Issues

The main, general concerns related to the safety and storing/handling conditions of batteries that should also be considered in the context of structural batteries are thermal runway, the leakage of electrolyte materials, fire safety and susceptibility to debris impact. Structural batteries, as with any Li-ion rechargeable battery, should be stored and handled according to the manufacturer's instructions (that need to be established), especially in terms of the state of charge, temperature and humidity exposure, avoidance of overcharging of individual cells, and flammability requirements. In any case, the use of non-flammable constituents (e.g., ionic liquids instead of conventional organic electrolyte, which is flammable) should be considered to increase battery safety.

7. Evaluation and Target Setting for Aeronautic Applications

In order to assess the status and compare the performance of state-of-the-art (SotA) structural battery approaches in a simple way, two key performance indicators, the gravimetric energy density and the elastic modulus, are plotted against each other in Figure 13, using data reported by [11,23]. Note that this SotA data is far from being complete, as several references do not report the elastic modulus but instead give other mechanical quantities.

As a SotA base-line, functionally separated systems are considered, using for their electrical energy storage current commercial, high energy lithium-iron-phosphate (LFP) and lithium-nickel-manganese-cobalt-oxide (NMC) battery cells with a gravimetric energy density of around 160 Wh/kg [101], up to 210 Wh/kg [102] for LFP and 260 Wh/kg [27] for NMC, and, for the structural part, CF fabric and unidirectional (UD) composite materials (60% fiber with resin) with an elastic modulus of between 70 GPa (fabrics) and 150 Gpa (UD) [28,29]. These provide the boundaries for the SotA area (the gray area in Figure 13) within which structural batteries would have a multifunctional efficiency $\eta_{mf} = 1$, depending on the respective constituents used.

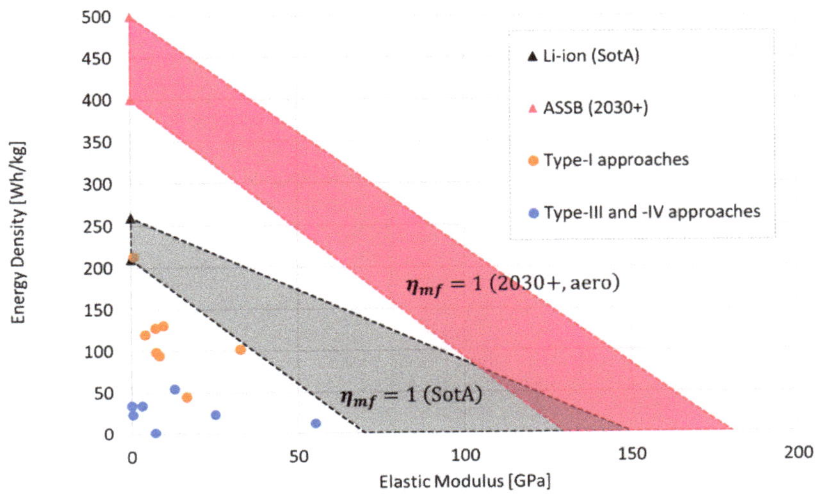

Figure 13. Specific energy vs. elastic modulus plots of structural batteries, as reported in the literature (complied from data reported in [11,23]) and structural battery targets: SotA (grey area) and for aeronautic applications in 2030+ (pink area).

Although higher mass savings are predicted for high degrees of structural integration, i.e., for type-III and type-IV batteries, available results show that structural batteries with a low degree of multifunctionality, i.e., type I, exhibit better multifunctional performances. Several type-I concepts approach the lower boundary of the target area, showing, in general, higher energy density but having a lower elastic modulus compared to type-III and type-IV concepts that have higher elastic moduli but that have shown rather limited energy density, between 1/10 to 1/3 of the density achievable in conventional battery cells. In fact, current type-I concepts often incorporate liquid electrolyte battery cells in a pouch format, with low load-bearing capabilities, while type-III and type-IV concepts are far from being optimized in their content and the weight of passive components.

Furthermore, Hopkins et al. [23] highlight that decoupled (type I), sandwich-type approaches can achieve higher flexural rigidity values than those for comparable coupled (types III and IV) versions as, in the former, the structural material is concentrated in the face skins.

Projecting the requirements for aeronautic SB application forward to 2030+, the baseline for electrical energy storage will shift, with upcoming battery technologies having substantially higher energy density, such as solid-state batteries with an expected gravimetric energy density of between 400 and 500 Wh/kg at the cell level. Such energy density is required as a minimum for propelling future hybrid-electric regional aircraft (CS-25, 40–50 Pax) and fully electric commuter aircraft (CS-23, 19 Pax), as forecasted per [103], while fully electrical regional aircraft would require energy densities of beyond 500 Wh/kg [104]. For aeronautic structures, composite materials from carbon UD fibers (not woven fabrics) should be considered, with elastic moduli of between about 130 and 180 GPa [29]. This shifts the multifunctional efficiency target $\eta_{mf}^{2030+} = 1$ farther, as outlined by the pink area in Figure 13. Note that the increase of rigidity in aeronautic structures will not be achieved by further strengthening the composite material (for cost reasons) but instead by optimizing the ply-stacking sequence.

8. Final Recommendations and Conclusions

In order to develop further structural batteries intended for aeronautic application, research efforts have to be undertaken in the fields of materials, integration, testing and monitoring.

On the materials side, the SB cell net energy density needs to be increased by (a) utilizing energy-dense electrochemistry, taking up ongoing developments in battery materials research, specifically in the field of all-solid-state batteries, and transferring them to structural batteries, as well as by (b) optimizing the SB cell with respect to the active material, loading and reducing the content and weight of passive components, such as the current collectors and separator. The intrinsic structural capability of the SB cell needs to be improved by superseding the conventional liquid electrolyte found in many type-I concepts with a performant structural or solid-state electrolyte, with high adhesion between electrolytes and electrodes.

SB integration concepts specifically intended for aeronautic applications need to be developed, considering UD fibers instead of the commonly used 2D-woven fabrics. Aeronautic ply stacking sequences need to be considered and concepts established that optimize the position, shape, and distribution of SB cells within the structure while minimizing the onset of damage. Reliable electrical wiring and connections will need to be integrated into the structure. Thermal management will depend on the final electrochemistry used in the SB. A comprehensive understanding of the influence of SB integration on mechanical properties needs to be developed not only at a material level but at the level of aeronautic structures. SB integration needs to advance from the current materials level to a structural level, i.e., stepping up in the structural testing and certification pyramid [84] from coupons over elements to (sub-)components.

This need calls for advanced characterization and testing methods beyond basic mechanical and electrical tests, to assess the multifunctional performance of aeronautic structural battery composite parts in complex-load cases. These could include multi-instrumented test setups with (high-speed/real-time) digital image correlation, X-ray tomography, acoustic emission tests, and infrared thermography to monitor the damage evolution, especially around the SB cells' locations.

Furthermore, the state-of-health of the resulting multifunctional structures during manufacturing, testing and operation will need advanced sensing arrays and monitoring with embedded sensors, such as fiber Bragg gratings or piezoelectric sensors, or using novel miniaturized battery-cell-based sensing devices.

Author Contributions: State of the art review, A.B. (Alexander Beutl), F.L., S.W. and M.G.; interaction with industrial experts: F.R.; writing—draft preparation, A.B. (Alexander Beutl), F.M., H.K., F.L., S.W., F.R. and M.G.; writing—review and editing, H.K. and A.B. (Alexander Bismarck); project coordination and funding acquisition, H.K. All authors have read and agreed to the published version of the manuscript.

Funding: This review was prepared within the project SOLIFLY that received funding from the *Clean Sky* 2 Joint Undertaking (JU) under grant agreement No. 101007577.

Institutional Review Board Statement: Not applicable.

Informed Consent Statement: Not applicable.

Acknowledgments: The authors thank Umberto Mercurio and Guido Saccone at CIRA for their support in interacting with the industrial experts and collecting their inputs.

Conflicts of Interest: The authors declare no conflict of interest.

References

1. National Academy of Sciences, Transportation Research Board. *Transportation Research Circular E-C271: Critical Issues in Aviation and the Environment 2021*; National Academy of Sciences, Transportation Research Board: Washington, DC, USA, 2021.
2. European Commission. *Flightpath 2050—Europe's Vision for Aviation: Maintaining Global Leadership and Serving Society's Needs*; Publications Office of the European Union: Luxembourg, 2012; ISBN 978-92-79-26229-6.

3. European Commission. Clean Sky Benefits. Available online: https://www.cleansky.eu/benefits (accessed on 15 September 2021).
4. NASA. *NASA Aeronautics—Strategic Implementation Plan—2019 Update*; NASA: Washington, DC, USA, 2019.
5. European Commission. Communication from the Commission—The European Green Deal. Available online: https://eur-lex.europa.eu/legal-content/EN/TXT/?uri=COM%3A2019%3A640%3AFIN (accessed on 15 September 2021).
6. UBER Elevate. Fast-Forwarding to a Future of On-Demand Urban Air Transportation. Available online: https://evtol.news/__media/PDFs/UberElevateWhitePaperOct2016.pdf (accessed on 25 September 2021).
7. Adam, T.J.; Liao, G.; Petersen, J.; Geier, S.; Finke, B.; Wierach, P.; Kwade, A.; Wiedemann, M. Multifunctional Composites for Future Energy Storage in Aerospace Structures. *Energies* **2018**, *11*, 335. [CrossRef]
8. Scholz, A.E.; Hermanutz, A.; Hornung, M. Feasibility Analysis and Comparative Assessment of Structural Power Technology in All-Electric Composite Aircraft. In Proceedings of the Deutscher Luft- und Raumfahrtkongress 2018, Friedrichshafen, Germany, 4–6 September 2018. [CrossRef]
9. Snyder, J.F.; O'Brien, D.J.; Wetzel, E.D. Structural Batteries, Capacitors and Supercapacitors. In *Handbook of Solid State Batteries*, 2nd ed.; Materials and Energy; World Scientific: Singapore, 2015; Volume 6, pp. 657–699, ISBN 978-981-4651-89-9.
10. Snyder, J.; Gienger, E.; Wetzel, E. Performance Metrics for Structural Composites with Electrochemical Multifunctionality. *J. Compos. Mater.* **2015**, *49*, 1835–1848. [CrossRef]
11. Asp, L.E.; Bouton, K.; Carlstedt, D.; Duan, S.; Harnden, R.; Johannisson, W.; Johansen, M.; Johansson, M.K.G.; Lindbergh, G.; Liu, F.; et al. A Structural Battery and Its Multifunctional Performance. *Adv. Energy Sustain. Res.* **2021**, *2*, 2000093. [CrossRef]
12. Moyer, K.; Boucherbil, N.A.; Zohair, M.; Eaves-Rathert, J.; Pint, C.L. Polymer Reinforced Carbon Fiber Interfaces for High Energy Density Structural Lithium-Ion Batteries. *Sustain. Energy Fuels* **2020**, *4*, 2661–2668. [CrossRef]
13. Thakur, A.; Dong, X. Printing with 3D Continuous Carbon Fiber Multifunctional Composites via UV-Assisted Coextrusion Deposition. *Manuf. Lett.* **2020**, *24*, 1–5. [CrossRef]
14. Meng, C.; Muralidharan, N.; Teblum, E.; Moyer, K.E.; Nessim, G.D.; Pint, C.L. Multifunctional Structural Ultrabattery Composite. *Nano Lett.* **2018**, *18*, 7761–7768. [CrossRef] [PubMed]
15. Liu, P.; Sherman, E.; Jacobsen, A. Design and Fabrication of Multifunctional Structural Batteries. *J. Power Sources* **2009**, *189*, 646–650. [CrossRef]
16. Zhao, Y.; Zhao, D.; Zhang, T.; Li, H.; Zhang, B.; Zhenchong, Z. Preparation and Multifunctional Performance of Carbon Fiber-Reinforced Plastic Composites for Laminated Structural Batteries. *Polym. Compos.* **2020**, *41*, 3023–3033. [CrossRef]
17. Roberts, S.C.; Aglietti, G.S. Structural Performance of a Multifunctional Spacecraft Structure Based on Plastic Lithium-Ion Batteries. *Acta Astronaut.* **2010**, *67*, 424–439. [CrossRef]
18. Thomas, J.; Qidwai, S.; Pogue, W.; Pham, G. Multifunctional Structure-Battery Composites for Marine Systems. *J. Compos. Mater.* **2013**, *47*, 5–26. [CrossRef]
19. Huang, W.; Wang, P.; Liao, X.; Chen, Y.; Borovilas, J.; Jin, T.; Li, A.; Cheng, Q.; Zhang, Y.; Zhai, H.; et al. Mechanically-Robust Structural Lithium-Sulfur Battery with High Energy Density. *Energy Storage Mater.* **2020**, *33*, 416–422. [CrossRef]
20. Moyer, K.; Meng, C.; Marshall, B.; Assal, O.; Eaves, J.; Perez, D.; Karkkainen, R.; Roberson, L.; Pint, C.L. Carbon Fiber Reinforced Structural Lithium-Ion Battery Composite: Multifunctional Power Integration for CubeSats. *Energy Storage Mater.* **2020**, *24*, 676–681. [CrossRef]
21. Attar, P.; Galos, J.; Best, A.S.; Mouritz, A.P. Compression Properties of Multifunctional Composite Structures with Embedded Lithium-Ion Polymer Batteries. *Compos. Struct.* **2020**, *237*, 111937. [CrossRef]
22. Carlstedt, D.; Johannisson, W.; Zenkert, D.; Linde, P.; Asp, L.E. Conceptual Design Framework for Laminated Structural Battery Composites. In Proceedings of the ECCM 2018—18th European Conference on Composite Materials, Athens, Greece, 24 June 2018; pp. 1–8.
23. Hopkins, B.J.; Long, J.W.; Rolison, D.R.; Parker, J.F. High-Performance Structural Batteries. *Joule* **2020**, *4*, 2240–2243. [CrossRef]
24. Asp, L.E.; Johansson, M.; Lindbergh, G.; Xu, J.; Zenkert, D. Structural Battery Composites: A Review. *Funct. Compos. Struct.* **2019**, *1*, 042001. [CrossRef]
25. Danzi, F.; Salgado, R.M.; Oliveira, J.E.; Arteiro, A.; Camanho, P.P.; Braga, M.H. Structural Batteries: A Review. *Molecules* **2021**, *26*, 2203. [CrossRef] [PubMed]
26. Kalnaus, S.; Asp, L.E.; Li, J.; Veith, G.M.; Nanda, J.; Daniel, C.; Chen, X.C.; Westover, A.; Dudney, N.J. Multifunctional Approaches for Safe Structural Batteries. *J. Energy Storage* **2021**, *40*, 102747. [CrossRef]
27. Armand, M.; Axmann, P.; Bresser, D.; Copley, M.; Edström, K.; Ekberg, C.; Guyomard, D.; Lestriez, B.; Novák, P.; Petranikova, M.; et al. Lithium-Ion Batteries—Current State of the Art and Anticipated Developments. *J. Power Sources* **2020**, *479*, 228708. [CrossRef]
28. Hexcel Corporation. HexPly® Prepreg Technology. Available online: https://www.hexcel.com/user_area/content_media/raw/Prepreg_Technology.pdf (accessed on 15 September 2021).
29. Toray Carbon Fibers Europe. Carbon Fibers and Composite Materials Datasheets. Available online: https://toray-cfe.com/en/e-documents/ (accessed on 15 September 2011).
30. Thomas, J.P.; Keennon, M.T.; DuPasquier, A.; Qidwai, M.A.; Matic, P. Multifunctional Structure-Battery Materials for Enhanced Performance in Small Unmanned Air Vehicles. In Proceedings of the ASME 2003 International Mechanical Engineering Congress and Exposition, Washington, DC, USA, 15–21 November 2003; pp. 289–292.

31. Thomas, J.P.; Qidwai, M.A. The Design and Application of Multifunctional Structure-Battery Materials Systems. *JOM* **2005**, *57*, 18–24. [CrossRef]
32. Thomas, J.P.; Qidwai, M.A. Mechanical Design and Performance of Composite Multifunctional Materials. *Acta Mater.* **2004**, *52*, 2155–2164. [CrossRef]
33. Schlichting, A.D.; Eisenbeiser, K. *Multifunctional Power Systems for Improved Size, Weight, and Power (SWaP) in Portable Electronic Systems*; Technical Report MTR1-5-00-2-9; MITRE: McLean, VA, USA, 2015.
34. Roberts, S.C.; Aglietti, G.S. Satellite Multi-Functional Power Structure: Feasibility and Mass Savings. *Proc. Inst. Mech. Eng. Part G J. Aerosp. Eng.* **2008**, *222*, 41–51. [CrossRef]
35. Wang, Y.; Peng, C.; Zhang, W. Thermal Analysis of Multifunctional Structural Battery for Satellite Applications. *Appl. Therm. Eng.* **2015**, *78*, 209–216. [CrossRef]
36. Grzesik, B.; Liao, G.; Vogt, D.; Froböse, L.; Kwade, A.; Linke, S.; Stoll, E. Integration of Energy Storage Functionalities into Fiber Reinforced Spacecraft Structures. *Acta Astronaut.* **2020**, *166*, 172–179. [CrossRef]
37. Capovilla, G.; Cestino, E.; Reyneri, L.M.; Romeo, G. Modular Multifunctional Composite Structure for CubeSat Applications: Preliminary Design and Structural Analysis. *Aerospace* **2020**, *7*, 17. [CrossRef]
38. Pattarakunnan, K.; Galos, J.; Das, R.; Mouritz, A.P. Tensile Properties of Multifunctional Composites Embedded with Lithium-Ion Polymer Batteries. *Compos. Part A Appl. Sci. Manuf.* **2020**, *136*, 105966. [CrossRef]
39. Ladpli, R.; Nardari, R.; Wang, Y.; Hernandez-Gallegos, P.A.; Rewari, R.; Kuo, H.T.; Kopsaftopoulos, F.; Kepler, K.D.; Lopez, H.A.; Chang, F.-K. Multifunctional Energy Storage Composites for SHM Distributed Sensor Networks. In Proceedings of the Tenth International Workshop on Structural Health Monitoring 2015, Stanford, CA, USA, 1–3 September 2015; p. 938.
40. Ladpli, P.; Nardari, R.; Kopsaftopoulos, F.; Wang, Y.; Chang, F.-K. Design of Multifunctional Structural Batteries with Health Monitoring Capabilities. In Proceedings of the 8th European Workshop on Structural Health Monitoring (EWSHM 2016), Bilbao, Spain, 5–8 July 2016; p. 416.
41. Ladpli, P.; Nardari, R.; Rewari, R.; Liu, H.; Slater, M.; Kepler, K.; Wang, Y.; Kopsaftopoulos, F.; Chang, F.-K. Multifunctional Energy Storage Composites: Design, Fabrication, and Experimental Characterization. In Proceedings of the ASME 2016 10th International Conference on Energy Sustainability collocated with the ASME 2016 Power Conference and the ASME 2016 14th International Conference on Fuel Cell Science, Engineering and Technology, Charlotte, CA, USA, 26–30 June 2016.
42. Ladpli, P.; Nardari, R.; Liu, H.; Slater, M.; Kepler, K.; Wang, Y.; Kopsaftopoulos, F.; Chang, F.-K. Multitfunctional Energy Storage Composites—Electrochemical and Mechanical Cycling Characterization. In Proceedings of the Battery Congress 2016, Troy, MI, USA, 16–18 May 2016; pp. 1–12.
43. Ladpli, P.; Nardari, R.; Kopsaftopoulos, F.; Chang, F.-K. Multifunctional Energy Storage Composite Structures with Embedded Lithium-Ion Batteries. *J. Power Sources* **2019**, *414*, 517–529. [CrossRef]
44. Mullenax, J.; Browning, P.; Huebsch, W.; Gautam, M.; Sabolsky, E.M. Composite Multifunctional Lithium-Ion Batteries. *ECS Trans.* **2012**, *41*, 175. [CrossRef]
45. Galos, J.; Khatibi, A.A.; Mouritz, A.P. Vibration and Acoustic Properties of Composites with Embedded Lithium-Ion Polymer Batteries. *Compos. Struct.* **2019**, *220*, 677–686. [CrossRef]
46. Galos, J.; Best, A.S.; Mouritz, A.P. Multifunctional Sandwich Composites Containing Embedded Lithium-Ion Polymer Batteries under Bending Loads. *Mater. Des.* **2020**, *185*, 108228. [CrossRef]
47. Shalouf, S.M.; Zhang, J.; Wang, C.H. Effects of Mechanical Deformation on Electric Performance of Rechargeable Batteries Embedded in Load Carrying Composite Structures. *Plast. Rubber Compos.* **2014**, *43*, 98–104. [CrossRef]
48. Waldmann, T.; Scurtu, R.-G.; Richter, K.; Wohlfahrt-Mehrens, M. 18650 vs. 21700 Li-Ion Cells—A Direct Comparison of Electrochemical, Thermal, and Geometrical Properties. *J. Power Sources* **2020**, *472*, 228614. [CrossRef]
49. Marcelli, D.; Summers, J.; Neudecker, B. LiBaCore II: Power Storage in Primary Structure. In Proceedings of the 43rd AIAA/ASME/ASCE/AHS/ASC Structures, Structural Dynamics, and Materials Conference, Denver, CO, USA, 22–25 April 2002; American Institute of Aeronautics and Astronautics: Denver, CO, USA, 2002.
50. Neudecker, B.J.; Benson, M.H.; Emerson, B.K. *Power Fibers: Thin-Film Batteries on Fiber Substrates*; ITN Energy Systems Inc.: Littleton, CO, USA, 2003.
51. Hagberg, J.; Maples, H.A.; Alvim, K.S.P.; Xu, J.; Johannisson, W.; Bismarck, A.; Zenkert, D.; Lindbergh, G. Lithium Iron Phosphate Coated Carbon Fiber Electrodes for Structural Lithium Ion Batteries. *Compos. Sci. Technol.* **2018**, *162*, 235–243. [CrossRef]
52. Park, H.-W.; Jang, M.-S.; Choi, J.-S.; Pyo, J.; Kim, C.-G. Characteristics of Woven Carbon Fabric Current Collector Electrodes for Structural Battery. *Compos. Struct.* **2021**, *256*, 112999. [CrossRef]
53. Hagberg, J.; Leijonmarck, S.; Lindbergh, G. High Precision Coulometry of Commercial PAN-Based Carbon Fibers as Electrodes in Structural Batteries. *J. Electrochem. Soc.* **2016**, *163*, A1790. [CrossRef]
54. Jacques, E.; Hellqvist Kjell, M.; Zenkert, D.; Lindbergh, G.; Behm, M. Expansion of Carbon Fibres Induced by Lithium Intercalation for Structural Electrode Applications. *Carbon* **2013**, *59*, 246–254. [CrossRef]
55. Park, M.Y.; Kim, J.-H.; Kim, D.K.; Kim, C.G. Perspective on Carbon Fiber Woven Fabric Electrodes for Structural Batteries. *Fibers Polym.* **2018**, *19*, 599–606. [CrossRef]
56. Kjell, M.H.; Jacques, E.; Zenkert, D.; Behm, M.; Lindbergh, G. PAN-Based Carbon Fiber Negative Electrodes for Structural Lithium-Ion Batteries. *J. Electrochem. Soc.* **2011**, *158*, A1455. [CrossRef]

57. Zhu, P.; Gastol, D.; Marshall, J.; Sommerville, R.; Goodship, V.; Kendrick, E. A Review of Current Collectors for Lithium-Ion Batteries. *J. Power Sources* **2021**, *485*, 229321. [CrossRef]
58. Snyder, J.F.; Carter, R.H.; Wong, E.L.; Nguyen, P.A.; Xu, K.; Ngo, E.H.; Wetzel, E.D. Multifunctional Structural Composite Batteries. In Proceedings of the Society for the Advancement of Material and Process Engineering (SAMPE) 2006 Fall Technical Conference, Orlando, FL, USA, 27–30 November 2006; Army Research Laboratory: Aberdeen Proving Ground, MD, USA, 2006.
59. Snyder, J.F.; Carter, R.H.; Wetzel, E.D. Electrochemical and Mechanical Behavior in Mechanically Robust Solid Polymer Electrolytes for Use in Multifunctional Structural Batteries. *Chem. Mater.* **2007**, *19*, 3793–3801. [CrossRef]
60. Snyder, J.F.; Baechle, D.M.; Wetzel, E.D.; Xu, K. Multifunctional Structural Composite Batteries for U.S. Army Applications. In Proceedings of the Army Science Conference (26th), Orlando, FL, USA, 1–4 December 2008; Army Research Laboratory: Aberdeen Proving Ground, MD, USA, 2008.
61. Snyder, J.F.; Wetzel, E.D.; Watson, C.M. Improving Multifunctional Behavior in Structural Electrolytes through Copolymerization of Structure- and Conductivity-Promoting Monomers. *Polymer* **2009**, *50*, 4906–4916. [CrossRef]
62. Javaid, A.; Ali, M.Z. Multifunctional Structural Lithium Ion Batteries for Electrical Energy Storage Applications. *Mater. Res. Express* **2018**, *5*, 055701. [CrossRef]
63. Carlstedt, D.; Marklund, E.; Asp, L.E. Effects of State of Charge on Elastic Properties of 3D Structural Battery Composites. *Compos. Sci. Technol.* **2019**, *169*, 26–33. [CrossRef]
64. Carlstedt, D.; Asp, L.E. Thermal and Diffusion Induced Stresses in a Structural Battery under Galvanostatic Cycling. *Compos. Sci. Technol.* **2019**, *179*, 69–78. [CrossRef]
65. Xu, J.; Lindbergh, G.; Varna, J. Multiphysics Modeling of Mechanical and Electrochemical Phenomena in Structural Composites for Energy Storage: Single Carbon Fiber Micro-Battery. *J. Reinf. Plast. Compos.* **2018**, *37*, 701–715. [CrossRef]
66. Carlstedt, D.; Asp, L.E. Performance Analysis Framework for Structural Battery Composites in Electric Vehicles. *Compos. Part B Eng.* **2020**, *186*, 107822. [CrossRef]
67. Asp, L.E.; Leijonmarck, S.; Carlson, T.; Lindbergh, G. Realisation of Structural Battery Composite Materials. In Proceedings of the 20th International Conference on Composite Materials, Copenhagen, Sweden, 19–24 July 2015.
68. Johannisson, W.; Zenkert, D.; Lindbergh, G. Model of a Structural Battery and Its Potential for System Level Mass Savings. *Multifunct. Mater.* **2019**, *2*, 035002. [CrossRef]
69. Chen, J.; Zhou, Y.; Islam, M.S.; Cheng, X.; Brown, S.A.; Han, Z.; Rider, A.N.; Wang, C.H. Carbon Fiber Reinforced Zn–MnO2 Structural Composite Batteries. *Compos. Sci. Technol.* **2021**, *209*, 108787. [CrossRef]
70. Wang, M.; Emre, A.; Tung, S.; Gerber, A.; Wang, D.; Huang, Y.; Cecen, V.; Kotov, N.A. Biomimetic Solid-State Zn2+ Electrolyte for Corrugated Structural Batteries. *ACS Nano* **2019**, *13*, 1107–1115. [CrossRef]
71. Asp, L.E.; Greenhalgh, E.S. Structural Power Composites. *Compos. Sci. Technol.* **2014**, *101*, 41–61. [CrossRef]
72. Asp, L.E. Multifunctional Composite Materials for Energy Storage in Structural Load Paths. *Plast. Rubber Compos.* **2013**, *42*, 144–149. [CrossRef]
73. Greenhalgh, E.S.; Ankersen, J.; Asp, L.E.; Bismarck, A.; Fontana, Q.P.V.; Houlle, M.; Kalinka, G.; Kucernak, A.; Mistry, M.; Nguyen, S.; et al. Mechanical, Electrical and Microstructural Characterisation of Multifunctional Structural Power Composites. *J. Compos. Mater.* **2015**, *49*, 1823–1834. [CrossRef]
74. Jacques, E.; Kjell, M.H.; Zenkert, D.; Lindbergh, G. The Effect of Lithium-Intercalation on the Mechanical Properties of Carbon Fibres. *Carbon* **2014**, *68*, 725–733. [CrossRef]
75. Carlson, T. Multifunctional Composite Materials Design, Manufacture and Experimental Characterisation. Ph.D. Thesis, Luleå University of Technology, Luleå, Sweden, 2013.
76. Asp, L.E. Structural Battery Materials. In Proceedings of the ECCM15—15th European Conference on Composite Materials, Venice, Italy, 24–28 June 2012; p. 1061.
77. Jiang, Q.; Zhang, H.; Rusakov, D.; Yousefi, N.; Bismarck, A. Additive Manufactured Carbon Nanotube/Epoxy Nanocomposites for Heavy-Duty Applications. *ACS Appl. Polym. Mater.* **2021**, *3*, 93–97. [CrossRef] [PubMed]
78. Asp, L.E.; Greenhalgh, E.S. Chapter 20: Multifunctional Structural Battery and Supercapacitor Composites. In *Multifunctionality of Polymer Composites*; William Andrew Publishing: Norwich, NY, USA, 2015.
79. Volvo Car Group. Volvo Car Group Makes Conventional Batteries a Thing of the Past. Available online: https://www.media.volvocars.com/global/en-gb/media/pressreleases/134235/volvo-car-group-makes-conventional-batteries-a-thing-of-the-past (accessed on 15 September 2021).
80. Johannisson, W.; Ihrner, N.; Zenkert, D.; Johansson, M.; Carlstedt, D.; Asp, L.E.; Sieland, F. Multifunctional Performance of a Carbon Fiber UD Lamina Electrode for Structural Batteries. *Compos. Sci. Technol.* **2018**, *168*, 81–87. [CrossRef]
81. ASTM. *Standard Test Method for Short-Beam Strength of Polymer Matrix Composite Materials and Their Laminates*; ASTM International: West Conshohocken, PA, USA, 2016.
82. AITM. *Airbus Test Method Standard: Determination of Plain, Open Hole and Filled Hole Tensile Strength*; Norm AITM 1-0008 Issue 3; AITM: Blagnac, France, 2004; pp. 1–16.
83. Zhang, Y.; Ma, J.; Singh, A.K.; Cao, L.; Seo, J.; Rahn, C.D.; Bakis, C.E.; Hickner, M.A. Multifunctional Structural Lithium-Ion Battery for Electric Vehicles. *J. Intell. Mater. Syst. Struct.* **2017**, *28*, 1603–1613. [CrossRef]
84. Federal Aviation Administration. *AC 20-107B—Composite Aircraft Structure*; Federal Aviation Administration: Washington, DC, USA, 2009.

85. Federal Aviation Administration. Standard Airworthiness Certification—Regulations. Available online: https://www.faa.gov/aircraft/air_cert/airworthiness_certification/std_awcert/std_awcert_regs/regs/ (accessed on 15 September 2021).
86. European Union Aviation Safety Agency. *EASA CM-S-010 Composite Materials—The Safe Design and Use of Monocoque Sandwich Structures in Principal Structural Element Application*; European Union Aviation Safety Agency: Cologne, Germany, 2018.
87. U.S. Department of Defense. *Military Handbook—MIL-HDBK-17-1F: Composite Materials Handbook, Volume 1—Polymer Matrix Composites Guidelines for Characterization of Structural Materials*; U.S. Department of Defense: Washington, DC, USA, 2002.
88. Federal Aviation Administration. *AC 20-184—Guidance on Testing and Installation of Rechargeable Lithium Battery and Battery Systems on Aircraft*; Federal Aviation Administration: Washington, DC, USA, 2015.
89. RTCA. *DO-311A—Minimum Operational Performance Standards for Rechargeable Lithium Batteries and Battery Systems*; RTCA: Washington, DC, USA, 2017.
90. RTCA. *DO-347—Certification Test Guidance for Small and Medium Sized Rechargeable Lithium Batteries and Battery Systems*; RTCA: Washington, DC, USA, 2013.
91. Federal Aviation Administration. *TSO-C179b—Rechargeable Lithium Batteries and Battery Systems*; Federal Aviation Administration: Washington, DC, USA, 2018.
92. Federal Aviation Administration. *AC 25.856-1—Thermal/Acoustic Insulation Flame Propagation Test Method Details—Document Information*; Federal Aviation Administration: Washington, DC, USA, 2005.
93. Federal Aviation Administration. *Aircraft Materials Fire Test Handbook, Revision 3*; Federal Aviation Administration: Washington, DC, USA, 2019.
94. European Union Aviation Safety Agency. *Special Condition LSA Propulsion Lithium Batteries SC-LSA-Fx2480*; European Union Aviation Safety Agency: Cologne, Germany, 2017.
95. European Union Aviation Safety Agency. *Special Condition SC E-19—Electric/Hybrid Propulsion System—Issue 01*; European Union Aviation Safety Agency: Cologne, Germany, 2021.
96. Rajamani, R.; Mracek Dietrich, A. *Unsettled Issues Regarding the Certification of Electric Aircraft*; SAE International: Warrendale, PA, USA, 2021; pp. 1–40.
97. Yang, H. A Review of Structural Batteries Implementations and Applications. In Proceedings of the 2020 IEEE Transportation Electrification Conference & Expo (ITEC), Chicago, IL, USA, 23–25 June 2020; pp. 223–228.
98. Riboldi, C.E.D.; Trainelli, L.; Biondani, F. Structural Batteries in Aviation: A Preliminary Sizing Methodology. *J. Aerosp. Eng.* **2020**, *33*, 04020031. [CrossRef]
99. Nguyen, S.N.; Millereux, A.; Pouyat, A.; Greenhalgh, E.S.; Shaffer, M.S.P.; Kucernak, A.R.J.; Linde, P. Conceptual Multifunctional Design, Feasibility and Requirements for Structural Power in Aircraft Cabins. *J. Aircr.* **2021**, *58*, 677–687. [CrossRef]
100. Karadotcheva, E.; Nguyen, S.N.; Greenhalgh, E.S.; Shaffer, M.S.P.; Kucernak, A.R.J.; Linde, P. Structural Power Performance Targets for Future Electric Aircraft. *Energies* **2021**, *14*, 6006. [CrossRef]
101. Lithium System AG. *Datasheet LFP EP-LFP-277Ah*; Lithium System AG: Effretikon, Switzerland, 2021.
102. Gotion High-Tech. Gotion High-Tech Presents New LFP Pouch Cell with 210 Wh/Kg. Available online: https://www.gotion.com.cn/welcome/newsdetails/486.html (accessed on 15 September 2021).
103. European Commission—Batteries Europe ETIP. *WG5: Application and Integration: Mobile—Roadmap, Air Transport Chapter*; European Commission: Brussels, Belgium, *in publication*.
104. Bills, A.; Sripad, S.; Fredericks, W.L.; Singh, M.; Viswanathan, V. Performance Metrics Required of Next-Generation Batteries to Electrify Commercial Aircraft. *ACS Energy Lett.* **2020**, *5*, 663–668. [CrossRef]

MDPI
St. Alban-Anlage 66
4052 Basel
Switzerland
Tel. +41 61 683 77 34
Fax +41 61 302 89 18
www.mdpi.com

Aerospace Editorial Office
E-mail: aerospace@mdpi.com
www.mdpi.com/journal/aerospace

www.ingramcontent.com/pod-product-compliance
Lightning Source LLC
Chambersburg PA
CBHW061547220926
51850CB00005B/269
9783036563466